Universitext

Andreas E. Kyprianou

Introductory Lectures on Fluctuations of Lévy Processes with Applications

With 22 Figures

Springer

Andreas E. Kyprianou
Department of Mathematical Sciences
University of Bath
Bath BA2 7AY
UK
e-mail: a.kyprianou@bath.ac.uk

The background text on the front cover is written in old (pre-1941) Mongolian scripture. It is a translation of the words 'stochastic processes with stationary and independent increments' and should be read from the top left hand corner of the back cover to the bottom right hand corner of the front cover.

Mathematics Subject Classification (2000): 60G50, 60G51, 60G52

Library of Congress Control Number: 2006924567

ISBN-10 3-540-31342-7 Springer Berlin Heidelberg New York
ISBN-13 978-3-540-31342-7 Springer Berlin Heidelberg New York

Springer is a part of Springer Science+Business Media
springer.com

Printed in Germany

Cover design: Erich Kirchner, Heidelberg
Typesetting by the author and SPi using a Springer LaTeX macro package

Printed on acid-free paper SPIN 11338987 41/3100/SPi - 5 4 3 2 1 0

[illegible]
[illegible]
[illegible]
[illegible]
[illegible]
[illegible]
[illegible]
[illegible]
[illegible]

Preface

In 2003 I began teaching a course entitled *Lévy processes* on the Amsterdam-Utrecht masters programme in stochastics and financial mathematics. Quite naturally, I wanted to expose my students to my own interests in Lévy processes; that is, the role that certain subtle behaviour concerning their fluctuations play in explaining different types of phenomena appearing in a number of classical models of applied probability. Indeed, recent developments in the theory of Lévy processes, in particular concerning path fluctuation, have offered the clarity required to revisit classical applied probability models and improve on well established and fundamental results.

Whilst teaching the course I wrote some lecture notes which have now matured into this text. Given the audience of students, who were either engaged in their 'afstudeerfase'[1] or just starting a Ph.D., these lecture notes were originally written with the restriction that the mathematics used would not surpass the level that they should in principle have reached. Roughly speaking that means the following: experience to the level of third year or fourth year university courses delivered by a mathematics department on

- foundational real and complex analysis,
- basic facts about L^p spaces,
- measure theory, integration theory and measure theoretic probability theory,
- elements of the classical theory of Markov processes, stopping times and the Strong Markov Property.
- Poisson processes and renewal processes,
- Brownian motion as a Markov process and elementary martingale theory in continuous time.

For the most part this affected the way in which the material was handled compared to the classical texts and research papers from which almost all of the results and arguments in this text originate. A good example of this is

[1]The afstudeerfase is equivalent to at least a European masters-level programme.

the conscious exclusion of calculations involving the master formula for the Poisson point process of excursions of a Lévy process from its maximum.

There are approximately 80 exercises and likewise these are pitched at a level appropriate to the aforementioned audience. Indeed several of the exercises have been included in response to some of the questions that have been asked by students themselves concerning curiosities of the arguments given in class. Arguably the exercises are at times quite long. Such exercises reflect some of the other ways in which I have used preliminary versions of this text. A small number of students in Utrecht also used the text as an individual reading/self-study programme contributing to their 'kleinescripite' (extended mathematical essay) or 'onderzoekopdracht' (research option); in addition, some exercises were used as (take-home) examination questions. The exercises in the first chapter in particular are designed to show the reader that the basics of the material presented thereafter is already accessible assuming basic knowledge of Poisson processes and Brownian motion.

There can be no doubt, particularly to the more experienced reader, that the current text has been heavily influenced by the outstanding books of Bertoin (1996) and Sato (1999), and especially the former which also takes a predominantly pathwise approach to its content. It should be reiterated however that, unlike the latter two books, this text is *not* intended as a research monograph nor as a reference manual for the researcher.

Writing of this text began whilst I was employed at Utrecht University, The Netherlands. In early 2005 I moved to a new position at Heriot Watt University in Edinburgh, Scotland, and in the final stages of completion of the book to The University of Bath. Over a period of several months my presence in Utrecht was phased out and my presence in Edinburgh was phased in. Along the way I passed through the Technical University of Munich and The University of Manchester. I should like to thank these four institutes and my hosts for giving me the facilities necessary to write this text (mostly time and a warm, dry, quiet room with an ethernet connection). I would especially like to thank my colleagues at Utrecht for giving me the opportunity and environment in which to develop this course, Ron Doney during his two-month absence for lending me the key to his office and book collection whilst mine was in storage and Andrew Cairns for arranging to push my teaching duties into 2006 allowing me the focus to finalise this text.

Let me now thank the many, including several of the students who took the course, who have made a number of remarks, corrections and suggestions (minor and major) which have helped to shape this text. In alphabetical order these are: Larbi Alili, David Applebaum, Johnathan Bagley, Erik Baurdoux, M.S. Bratiychuk, Catriona Byrne, Zhen-Qing Chen, Gunther Cornelissen, Irmingard Erder, Abdelghafour Es-Saghouani, Serguei Foss, Uwe Franz, Shota Gugushvili, Thorsten Kleinow, Paweł Kliber, Claudia Klüppelberg, V.S. Korolyuk, Ronnie Loeffen, Alexander Novikov, Zbigniew Palmowski, Goran Peskir, Kees van Schaik, Sonja Scheer, Wim Schoutens, Budhi Arta Surya, Enno Veerman, Maaike Verloop, Zoran Vondraček. In particular I would also like to thank, Peter Andrew, Jean Bertoin, Ron Doney, Niel Farricker, Alexander Gnedin, Amaury Lambert, Antonis Papapantoleon and Martijn Pistorius rooted out many errors from extensive sections of the text and provided valuable criticism. Antonis Papapantoleon very kindly produced some simulations of the paths of Lévy processes which have been included in Chap. 1. I am most grateful to Takis Konstantopoulos who read through earlier drafts of the entire text in considerable detail, taking the time to discuss with me at length many of the issues that arose. The front cover was produced in consultation with Hurlee Gonchigdanzan and Jargalmaa Magsarjav. All further comments, corrections and suggestions on the current text are welcome.

Finally, the deepest gratitude of all goes to Jagaa, Sophia and Sanaa for whom the special inscription is written.

Edinburgh
June 2006

Andreas E. Kyprianou

Contents

1

Lévy Processes and Applications

In this chapter we define a Lévy process and attempt to give some indication of how rich a class of processes they form. To illustrate the variety of processes captured within the definition of a Lévy process, we explore briefly the relationship of Lévy processes with infinitely divisible distributions. We also discuss some classical applied probability models, which are built on the strength of well-understood path properties of elementary Lévy processes. We hint at how generalisations of these models may be approached using more sophisticated Lévy processes. At a number of points later on in this text we handle these generalisations in more detail. The models we have chosen to present are suitable for the course of this text as a way of exemplifying fluctuation theory but are by no means the only applications.

1.1 Lévy Processes and Infinite Divisibility

Let us begin by recalling the definition of two familiar processes, a Brownian motion and a Poisson process.

A real-valued process $B = \{B_t : t \geq 0\}$ defined on a probability space $(\Omega, \mathcal{F}, \mathbb{P})$ is said to be a Brownian motion if the following hold:

(i) The paths of B are $\mathbb{P}$-almost surely continuous.
(ii) $\mathbb{P}(B_0 = 0) = 1$.
(iii) For $0 \leq s \leq t$, $B_t - B_s$ is equal in distribution to B_{t-s}.
(iv) For $0 \leq s \leq t$, $B_t - B_s$ is independent of $\{B_u : u \leq s\}$.
(v) For each $t > 0$, B_t is equal in distribution to a normal random variable with variance t.

A process valued on the non-negative integers $N = \{N_t : t \geq 0\}$, defined on a probability space $(\Omega, \mathcal{F}, \mathbb{P})$, is said to be a Poisson process with intensity $\lambda > 0$ if the following hold:

(i) The paths of N are $\mathbb{P}$-almost surely right continuous with left limits.
(ii) $\mathbb{P}(N_0 = 0) = 1$.

(iii) For $0 \leq s \leq t$, $N_t - N_s$ is equal in distribution to N_{t-s}.
(iv) For $0 \leq s \leq t$, $N_t - N_s$ is independent of $\{N_u : u \leq s\}$.
(v) For each $t > 0$, N_t is equal in distribution to a Poisson random variable with parameter λt.

On first encounter, these processes would seem to be considerably different from one another. Firstly, Brownian motion has continuous paths whereas a Poisson process does not. Secondly, a Poisson process is a non-decreasing process and thus has paths of bounded variation over finite time horizons, whereas a Brownian motion does not have monotone paths and in fact its paths are of unbounded variation over finite time horizons.

However, when we line up their definitions next to one another, we see that they have a lot in common. Both processes have right continuous paths with left limits, are initiated from the origin and both have stationary and independent increments; that is properties (i), (ii), (iii) and (iv). We may use these common properties to define a general class of stochastic processes, which are called Lévy processes.

Definition 1.1 (Lévy Process). *A process $X = \{X_t : t \geq 0\}$ defined on a probability space $(\Omega, \mathcal{F}, \mathbb{P})$ is said to be a Lévy process if it possesses the following properties:*

(i) The paths of X are $\mathbb{P}$-almost surely right continuous with left limits.
(ii) $\mathbb{P}(X_0 = 0) = 1$.
(iii) For $0 \leq s \leq t$, $X_t - X_s$ is equal in distribution to X_{t-s}.
(iv) For $0 \leq s \leq t$, $X_t - X_s$ is independent of $\{X_u : u \leq s\}$.

Unless otherwise stated, from now on, when talking of a Lévy process, we shall always use the measure $\mathbb{P}$ (with associated expectation operator $\mathbb{E}$) to be implicitly understood as its law.

The term "Lévy process" honours the work of the French mathematician Paul Lévy who, although not alone in his contribution, played an instrumental role in bringing together an understanding and characterisation of processes with stationary independent increments. In earlier literature, Lévy processes can be found under a number of different names. In the 1940s, Lévy himself referred to them as a sub-class of *processus additif* (additive processes), that is processes with independent increments. For the most part however, research literature through the 1960s and 1970s refers to Lévy processes simply as *processes with stationary independent increments.* One sees a change in language through the 1980s and by the 1990s the use of the term "Lévy process" had become standard.

From Definition 1.1 alone it is difficult to see just how rich a class of processes the class of Lévy processes forms. De Finetti (1929) introduced the notion of an *infinitely divisible* distribution and showed that they have an intimate relationship with Lévy processes. This relationship gives a reasonably good impression of how varied the class of Lévy processes really is. To this end, let us now devote a little time to discussing infinitely divisible distributions.

Definition 1.2. *We say that a real-valued random variable Θ has an infinitely divisible distribution if for each $n = 1, 2, ...$ there exist a sequence of i.i.d. random variables $\Theta_{1,n}, ..., \Theta_{n,n}$ such that*

$$\Theta \stackrel{d}{=} \Theta_{1,n} + \cdots + \Theta_{n,n}$$

where $\stackrel{d}{=}$ is equality in distribution. Alternatively, we could have expressed this relation in terms of probability laws. That is to say, the law μ of a real-valued random variable is infinitely divisible if for each $n = 1, 2, ...$ there exists another law μ_n of a real valued random variable such that $\mu = \mu_n^{*n}$. (Here μ_n^* denotes the n-fold convolution of μ_n).

In view of the above definition, one way to establish whether a given random variable has an infinitely divisible distribution is via its characteristic exponent. Suppose that Θ has characteristic exponent $\Psi(u) := -\log \mathbb{E}(\mathrm{e}^{\mathrm{i}u\Theta})$ for all $u \in \mathbb{R}$. Then Θ has an infinitely divisible distribution if for all $n \geq 1$ there exists a characteristic exponent of a probability distribution, say Ψ_n, such that $\Psi(u) = n\Psi_n(u)$ for all $u \in \mathbb{R}$.

The full extent to which we may characterise infinitely divisible distributions is described by the characteristic exponent Ψ and an expression known as the Lévy–Khintchine formula.

Theorem 1.3 (Lévy–Khintchine formula). *A probability law μ of a real-valued random variable is infinitely divisible with characteristic exponent Ψ,*

$$\int_{\mathbb{R}} \mathrm{e}^{\mathrm{i}\theta x} \mu(\mathrm{d}x) = \mathrm{e}^{-\Psi(\theta)} \text{ for } \theta \in \mathbb{R},$$

if and only if there exists a triple (a, σ, Π), where $a \in \mathbb{R}$, $\sigma \geq 0$ and Π is a measure concentrated on $\mathbb{R}\backslash\{0\}$ satisfying $\int_{\mathbb{R}} (1 \wedge x^2)\, \Pi(\mathrm{d}x) < \infty$, such that

$$\Psi(\theta) = \mathrm{i}a\theta + \frac{1}{2}\sigma^2\theta^2 + \int_{\mathbb{R}} (1 - \mathrm{e}^{\mathrm{i}\theta x} + \mathrm{i}\theta x \mathbf{1}_{(|x|<1)})\Pi(\mathrm{d}x)$$

for every $\theta \in \mathbb{R}$.

Definition 1.4. *The measure Π is called the Lévy (characteristic) measure.*

The proof of the Lévy–Khintchine characterisation of infinitely divisible random variables is quite lengthy and we choose to exclude it in favour of moving as quickly as possible to fluctuation theory. The interested reader is referred to Lukacs (1970) or Sato (1999), to name but two of many possible references.

A special case of the Lévy–Khintchine formula was established by Kolmogorov (1932) for infinitely divisible distributions with second moments.

However it was Lévy (1934) who gave a complete characterisation of infinitely divisible distributions and in doing so he also characterised the general class of processes with stationary independent increments. Later, Khintchine (1937) and Itô (1942) gave further simplification and deeper insight to Lévy's original proof.

Let us now discuss in further detail the relationship between infinitely divisible distributions and processes with stationary independent increments.

From the definition of a Lévy process we see that for any $t > 0$, X_t is a random variable belonging to the class of infinitely divisible distributions. This follows from the fact that for any $n = 1, 2, ...,$

$$X_t = X_{t/n} + (X_{2t/n} - X_{t/n}) + \cdots + (X_t - X_{(n-1)t/n}) \tag{1.1}$$

together with the fact that X has stationary independent increments. Suppose now that we define for all $\theta \in \mathbb{R}$, $t \geq 0$,

$$\Psi_t(\theta) = -\log \mathbb{E}\left(\mathrm{e}^{\mathrm{i}\theta X_t}\right)$$

then using (1.1) twice we have for any two positive integers m, n that

$$m\Psi_1(\theta) = \Psi_m(\theta) = n\Psi_{m/n}(\theta)$$

and hence for any rational $t > 0$,

$$\Psi_t(\theta) = t\Psi_1(\theta). \tag{1.2}$$

If t is an irrational number, then we can choose a decreasing sequence of rationals $\{t_n : n \geq 1\}$ such that $t_n \downarrow t$ as n tends to infinity. Almost sure right continuity of X implies right continuity of $\exp\{-\Psi_t(\theta)\}$ (by dominated convergence) and hence (1.2) holds for all $t \geq 0$.

In conclusion, any Lévy process has the property that for all $t \geq 0$

$$\mathbb{E}\left(\mathrm{e}^{\mathrm{i}\theta X_t}\right) = \mathrm{e}^{-t\Psi(\theta)},$$

where $\Psi(\theta) := \Psi_1(\theta)$ is the characteristic exponent of X_1, which has an infinitely divisible distribution.

Definition 1.5. *In the sequel we shall also refer to $\Psi(\theta)$ as the characteristic exponent of the Lévy process.*

It is now clear that each Lévy process can be associated with an infinitely divisible distribution. What is not clear is whether given an infinitely divisible distribution, one may construct a Lévy process X, such that X_1 has that distribution. This latter issue is affirmed by the following theorem which gives the Lévy–Khintchine formula for Lévy processes.

Theorem 1.6 (Lévy–Khintchine formula for Lévy processes). *Suppose that $a \in \mathbb{R}$, $\sigma \geq 0$ and Π is a measure concentrated on $\mathbb{R}\backslash\{0\}$ such that $\int_{\mathbb{R}}(1 \wedge x^2)\Pi(\mathrm{d}x) < \infty$. From this triple define for each $\theta \in \mathbb{R}$,*

$$\Psi(\theta) = \mathrm{i}a\theta + \frac{1}{2}\sigma^2\theta^2 + \int_{\mathbb{R}} (1 - \mathrm{e}^{\mathrm{i}\theta x} + \mathrm{i}\theta x \mathbf{1}_{(|x|<1)})\Pi(\mathrm{d}x).$$

Then there exists a probability space $(\Omega, \mathcal{F}, \mathbb{P})$ on which a Lévy process is defined having characteristic exponent Ψ.

The proof of this theorem is rather complicated but very rewarding as it also reveals much more about the general structure of Lévy processes. Later, in Chap. 2, we will prove a stronger version of this theorem, which also explains the path structure of the Lévy process in terms of the triple (a, σ, Π).

1.2 Some Examples of Lévy Processes

To conclude our introduction to Lévy processes and infinite divisible distributions, let us proceed to some concrete examples. Some of these will also be of use later to verify certain results from the forthcoming fluctuation theory we will present.

1.2.1 Poisson Processes

For each $\lambda > 0$ consider a probability distribution μ_λ which is concentrated on $k = 0, 1, 2...$ such that $\mu_\lambda(\{k\}) = \mathrm{e}^{-\lambda}\lambda^k/k!$. That is to say the Poisson distribution. An easy calculation reveals that

$$\begin{aligned}\sum_{k\geq 0} \mathrm{e}^{\mathrm{i}\theta k}\mu_\lambda(\{k\}) &= \mathrm{e}^{-\lambda(1-\mathrm{e}^{\mathrm{i}\theta})} \\ &= \left[\mathrm{e}^{-\frac{\lambda}{n}(1-\mathrm{e}^{\mathrm{i}\theta})}\right]^n.\end{aligned}$$

The right-hand side is the characteristic function of the sum of n independent Poisson processes, each of which with parameter λ/n. In the Lévy–Khintchine decomposition we see that $a = \sigma = 0$ and $\Pi = \lambda\delta_1$, the Dirac measure supported on $\{1\}$.

Recall that a Poisson process, $\{N_t : n \geq 0\}$, is a Lévy process with distribution at time $t > 0$, which is Poisson with parameter λt. From the above calculations we have

$$\mathbb{E}(\mathrm{e}^{\mathrm{i}\theta N_t}) = \mathrm{e}^{-\lambda t(1-\mathrm{e}^{\mathrm{i}\theta})}$$

and hence its characteristic exponent is given by $\Psi(\theta) = \lambda(1 - \mathrm{e}^{\mathrm{i}\theta})$ for $\theta \in \mathbb{R}$.

1.2.2 Compound Poisson Processes

Suppose now that N is a Poisson random variable with parameter $\lambda > 0$ and that $\{\xi_i : i \geq 1\}$ is an i.i.d. sequence of random variables (independent of N)

with common law F having no atom at zero. By first conditioning on N, we have for $\theta \in \mathbb{R}$,

$$
\begin{aligned}
E(\mathrm{e}^{\mathrm{i}\theta \sum_{i=1}^{N} \xi_i}) &= \sum_{n \geq 0} E(\mathrm{e}^{\mathrm{i}\theta \sum_{i=1}^{n} \xi_i})\mathrm{e}^{-\lambda}\frac{\lambda^n}{n!} \\
&= \sum_{n \geq 0} \left(\int_{\mathbb{R}} \mathrm{e}^{\mathrm{i}\theta x} F(\mathrm{d}x)\right)^n \mathrm{e}^{-\lambda}\frac{\lambda^n}{n!} \\
&= \mathrm{e}^{-\lambda \int_{\mathbb{R}} (1-\mathrm{e}^{\mathrm{i}\theta x}) F(\mathrm{d}x)}. \qquad (1.3)
\end{aligned}
$$

Note we use the convention here that $\sum_1^0 = 0$. We see from (1.3) that distributions of the form $\sum_{i=1}^{N} \xi_i$ are infinitely divisible with triple $a = -\lambda \int_{0<|x|<1} xF(\mathrm{d}x)$, $\sigma = 0$ and $\Pi(\mathrm{d}x) = \lambda F(\mathrm{d}x)$. When F has an atom of unit mass at 1 then we have simply a Poisson distribution.

Suppose now that $\{N_t : t \geq 0\}$ is a Poisson process with intensity λ and consider a compound Poisson process $\{X_t : t \geq 0\}$ defined by

$$X_t = \sum_{i=0}^{N_t} \xi_i, \ t \geq 0.$$

Using the fact that N has stationary independent increments together with the mutual independence of the random variables $\{\xi_i : i \geq 1\}$, for $0 \leq s < t < \infty$, by writing

$$X_t = X_s + \sum_{i=N_s+1}^{N_t} \xi_i$$

it is clear that X_t is the sum of X_s and an independent copy of X_{t-s}. Right continuity and left limits of the process N also ensure right continuity and left limits of X. Thus compound Poisson processes are Lévy processes. From the calculations in the previous paragraph, for each $t \geq 0$ we may substitute N_t for the variable N to discover that the Lévy–Khintchine formula for a compound Poisson process takes the form $\Psi(\theta) = \lambda \int_{\mathbb{R}} (1 - \mathrm{e}^{\mathrm{i}\theta x})F(\mathrm{d}x)$. Note in particular that the Lévy measure of a compound Poisson process is always finite with total mass equal to the rate λ of the underlying process N.

Compound Poisson processes provide a direct link between Lévy processes and random walks; that is discrete time processes of the form $S = \{S_n : n \geq 0\}$ where

$$S_0 = 0 \text{ and } S_n = \sum_{i=1}^{n} \xi_i \text{ for } n \geq 1.$$

Indeed a compound Poisson process is nothing more than a random walk whose jumps have been spaced out with independent and exponentially distributed periods.

1.2.3 Linear Brownian Motion

Take the probability law

$$\mu_{s,\gamma}(\mathrm{d}x) := \frac{1}{\sqrt{2\pi s^2}}\mathrm{e}^{-(x-\gamma)^2/2s^2}\mathrm{d}x$$

supported on $\mathbb{R}$ where $\gamma \in \mathbb{R}$ and $s > 0$; the well-known Gaussian distribution with mean γ and variance s^2. It is well known that

$$\begin{aligned}\int_{\mathbb{R}} \mathrm{e}^{\mathrm{i}\theta x}\mu_{s,\gamma}(\mathrm{d}x) &= \mathrm{e}^{-\frac{1}{2}s^2\theta^2+\mathrm{i}\theta\gamma} \\ &= \left[\mathrm{e}^{-\frac{1}{2}(\frac{s}{\sqrt{n}})^2\theta^2+\mathrm{i}\theta\frac{\gamma}{n}}\right]^n\end{aligned}$$

showing again that it is an infinitely divisible distribution, this time with $a = -\gamma$, $\sigma = s$ and $\Pi = 0$.

We immediately recognise the characteristic exponent $\Psi(\theta) = s^2\theta^2/2 - \mathrm{i}\theta\gamma$ as also that of a scaled Brownian motion with linear drift,

$$X_t := sB_t + \gamma t,\ t \geq 0,$$

where $B = \{B_t : t \geq 0\}$ is a standard Brownian motion; that is to say a linear Brownian motion with parameters $\sigma = 1$ and $\gamma = 0$. It is a trivial exercise to verify that X has stationary independent increments with continuous paths as a consequence of the fact that B does.

1.2.4 Gamma Processes

For $\alpha, \beta > 0$ define the probability measure

$$\mu_{\alpha,\beta}(\mathrm{d}x) = \frac{\alpha^\beta}{\Gamma(\beta)}x^{\beta-1}\mathrm{e}^{-\alpha x}\mathrm{d}x$$

concentrated on $(0,\infty)$; the gamma-(α,β) distribution. Note that when $\beta = 1$ this is the exponential distribution. We have

$$\begin{aligned}\int_0^\infty \mathrm{e}^{\mathrm{i}\theta x}\mu_{\alpha,\beta}(\mathrm{d}x) &= \frac{1}{(1-\mathrm{i}\theta/\alpha)^\beta} \\ &= \left[\frac{1}{(1-\mathrm{i}\theta/\alpha)^{\beta/n}}\right]^n\end{aligned}$$

and infinite divisibility follows. For the Lévy–Khintchine decomposition we have $\sigma = 0$ and $\Pi(\mathrm{d}x) = \beta x^{-1}\mathrm{e}^{-\alpha x}\mathrm{d}x$, concentrated on $(0,\infty)$ and $a = -\int_0^1 x\Pi(\mathrm{d}x)$. However this is not immediately obvious. The following lemma proves to be useful in establishing the above triple (a, σ, Π). Its proof is Exercise 1.3.

Lemma 1.7 (Frullani integral). *For all $\alpha, \beta > 0$ and $z \in \mathbb{C}$ such that $\Re z \leq 0$ we have*

$$\frac{1}{(1 - z/\alpha)^\beta} = \mathrm{e}^{-\int_0^\infty (1-\mathrm{e}^{zx})\beta x^{-1}\mathrm{e}^{-\alpha x}\mathrm{d}x}.$$

To see how this lemma helps note that the Lévy–Khintchine formula for a gamma distribution takes the form

$$\Psi(\theta) = \beta \int_0^\infty (1 - \mathrm{e}^{\mathrm{i}\theta x})\frac{1}{x}\mathrm{e}^{-\alpha x}\mathrm{d}x = \beta \log(1 - \mathrm{i}\theta/\alpha)$$

for $\theta \in \mathbb{R}$. The choice of a in the Lévy–Khintchine formula is the necessary quantity to cancel the term coming from $\mathrm{i}\theta \mathbf{1}_{(|x|<1)}$ in the integral with respect to Π in the general Lévy–Khintchine formula.

According to Theorem 1.6 there exists a Lévy process whose Lévy–Khintchine formula is given by Ψ, the so-called *gamma process*.

Suppose now that $X = \{X_t : t \geq 0\}$ is a gamma process. Stationary independent increments tell us that for all $0 \leq s < t < \infty$, $X_t = X_s + \widetilde{X}_{t-s}$ where $\widetilde{X}_{t-s}$ is an independent copy of X_{t-s}. The fact that the latter is strictly positive with probability one (on account of it being gamma distributed) implies that $X_t > X_s$ almost surely. Hence a gamma process is an example of a Lévy process with almost surely non-decreasing paths (in fact its paths are strictly increasing). Another example of a Lévy process with non-decreasing paths is a compound Poisson process where the jump distribution F is concentrated on $(0, \infty)$. Note however that a gamma process is not a compound Poisson process on two counts. Firstly, its Lévy measure has infinite total mass unlike the Lévy measure of a compound Poisson process, which is necessarily finite (and equal to the arrival rate of jumps). Secondly, whilst a compound Poisson process with positive jumps does have paths, which are almost surely non-decreasing, it does not have paths that are almost surely strictly increasing.

Lévy processes whose paths are almost surely non-decreasing (or simply non-decreasing for short) are called *subordinators*. We will return to a formal definition of this subclass of processes in Chap. 2.

1.2.5 Inverse Gaussian Processes

Suppose as usual that $B = \{B_t : t \geq 0\}$ is a standard Brownian motion. Define the first passage time

$$\tau_s = \inf\{t > 0 : B_t + bt > s\}, \tag{1.4}$$

that is, the first time a Brownian motion with linear drift $b > 0$ crosses above level s. Recall that τ_s is a stopping time[1] with respect to the filtration $\{\mathcal{F}_t : t \geq 0\}$ where $\mathcal{F}_t$ is generated by $\{B_s : s \leq t\}$. Otherwise said, since Brownian motion has continuous paths, for all $t \geq 0$,

$$\{\tau_s \leq t\} = \bigcup_{u \in [0,t] \cap \mathbb{Q}} \{B_u + bu > s\}$$

and hence the latter belongs to the sigma algebra $\mathcal{F}_t$.

Recalling again that Brownian motion has continuous paths we know that $B_{\tau_s} + b\tau_s = s$ almost surely. From the Strong Markov Property,[2] it is known that $\{B_{\tau_s+t} + b(\tau_s + t) - s : t \geq 0\}$ is equal in law to B and hence for all $0 \leq s < t$,

$$\tau_t = \tau_s + \widetilde{\tau}_{t-s},$$

where $\widetilde{\tau}_{t-s}$ is an independent copy of τ_{t-s}. This shows that the process $\tau := \{\tau_t : t \geq 0\}$ has stationary independent increments. Continuity of the paths of $\{B_t + bt : t \geq 0\}$ ensures that τ has right continuous paths. Further, it is clear that τ has almost surely non-decreasing paths, which guarantees its paths have left limits as well as being yet another example of a subordinator. According to its definition as a sequence of first passage times, τ is also the almost sure right inverse of the path of the graph of $\{B_t + bt : t \geq 0\}$ in the sense of (1.4). From this τ earns its name as the inverse Gaussian process.

According to the discussion following Theorem 1.3 it is now immediate that for each fixed $s > 0$, the random variable τ_s is infinitely divisible. Its characteristic exponent takes the form

$$\Psi(\theta) = s(\sqrt{-2\mathrm{i}\theta + b^2} - b)$$

for all $\theta \in \mathbb{R}$ and corresponds to a triple $a = -2sb^{-1}\int_0^b (2\pi)^{-1/2}\mathrm{e}^{-y^2/2}\mathrm{d}y$, $\sigma = 0$ and

$$\Pi(\mathrm{d}x) = s\frac{1}{\sqrt{2\pi x^3}}\mathrm{e}^{-\frac{b^2 x}{2}}\mathrm{d}x$$

concentrated on $(0, \infty)$. The law of τ_s can also be computed explicitly as

$$\mu_s(\mathrm{d}x) = \frac{s}{\sqrt{2\pi x^3}}\mathrm{e}^{sb}\mathrm{e}^{-\frac{1}{2}(s^2x^{-1}+b^2x)}$$

for $x > 0$. The proof of these facts forms Exercise 1.6.

---

[1] We assume that the reader is familiar with the notion of a stopping time for a Markov process. By definition, the random time τ is a stopping time with respect to the filtration $\{\mathcal{G}_t : t \geq 0\}$ if for all $t \geq 0$,

$$\{\tau \leq t\} \in \mathcal{G}_t.$$

[2] The Strong Markov Property will be dealt with in more detail for a general Lévy process in Chap. 3.

1.2.6 Stable Processes

Stable processes are the class of Lévy processes whose characteristic exponents correspond to those of stable distributions. Stable distributions were introduced by Lévy (1924, 1925) as a third example of infinitely divisible distributions after Gaussian and Poisson distributions. A random variable, Y, is said to have a stable distribution if for all $n \geq 1$ it observes the distributional equality

$$Y_1 + \cdots + Y_n \overset{d}{=} a_n Y + b_n, \tag{1.5}$$

where $Y_1, \ldots, Y_n$ are independent copies of Y, $a_n > 0$ and $b_n \in \mathbb{R}$. By subtracting b_n/n from each of the terms on the left-hand side of (1.5) one sees in particular that this definition implies that any stable random variable is infinitely divisible. It turns out that necessarily $a_n = n^{1/\alpha}$ for $\alpha \in (0, 2]$; see Feller (1971), Sect. VI.1. In that case we refer to the parameter α as the *index*. A smaller class of distributions are the *strictly stable* distributions. A random variable Y is said to have a strictly stable distribution if it observes (1.5) but with $b_n = 0$. In that case, we necessarily have

$$Y_1 + \cdots + Y_n \overset{d}{=} n^{1/\alpha} Y. \tag{1.6}$$

The case $\alpha = 2$ corresponds to zero mean Gaussian random variables and is excluded in the remainder of the discussion as it has essentially been dealt with in Sect. 1.2.3.

Stable random variables observing the relation (1.5) for $\alpha \in (0, 1) \cup (1, 2)$ have characteristic exponents of the form

$$\Psi(\theta) = c|\theta|^\alpha (1 - \mathrm{i}\beta \tan \frac{\pi\alpha}{2} \operatorname{sgn} \theta) + \mathrm{i}\theta\eta, \tag{1.7}$$

where $\beta \in [-1, 1]$, $\eta \in \mathbb{R}$ and $c > 0$. Stable random variables observing the relation (1.5) for $\alpha = 1$, have characteristic exponents of the form

$$\Psi(\theta) = c|\theta|(1 + \mathrm{i}\beta \frac{2}{\pi} \operatorname{sgn} \theta \log |\theta|) + \mathrm{i}\theta\eta, \tag{1.8}$$

where $\beta \in [-1, 1]$ $\eta \in \mathbb{R}$ and $c > 0$. Here we work with the definition of the sign function $\operatorname{sgn} \theta = \mathbf{1}_{(\theta > 0)} - \mathbf{1}_{(\theta < 0)}$. To make the connection with the Lévy–Khintchine formula, one needs $\sigma = 0$ and

$$\Pi(\mathrm{d}x) = \begin{cases} c_1 x^{-1-\alpha} \mathrm{d}x & \text{for } x \in (0, \infty) \\ c_2 |x|^{-1-\alpha} \mathrm{d}x & \text{for } x \in (-\infty, 0), \end{cases} \tag{1.9}$$

where $c = c_1 + c_2$, $c_1, c_2 \geq 0$ and $\beta = (c_1 - c_2)/(c_1 + c_2)$ if $\alpha \in (0, 1) \cup (1, 2)$ and $c_1 = c_2$ if $\alpha = 1$. The choice of $a \in \mathbb{R}$ in the Lévy–Khintchine formula is then implicit. Exercise 1.4 shows how to make the connection between Π and Ψ with the right choice of a (which depends on α). Unlike the previous examples, the distributions that lie behind these characteristic exponents are heavy tailed in the sense that the tails of their distributions decay slowly enough to zero so that they only have moments strictly less than α. The value of the parameter β gives a measure of asymmetry in the Lévy measure and likewise for the distributional asymmetry (although this latter fact is not immediately obvious). The densities of stable processes are known explicitly in the form of convergent power series. See Zolotarev (1986), Sato (1999) and (Samorodnitsky and Taqqu, 1994) for further details of all the facts given in this paragraph. With the exception of the defining property (1.6) we shall generally not need detailed information on distributional properties of stable processes in order to proceed with their fluctuation theory. This explains the reluctance to give further details here.

Two examples of the aforementioned power series that tidy up to more compact expressions are centred Cauchy distributions, corresponding to $\alpha = 1$, $\beta = 0$ and $\eta = 0$, and stable-$\frac{1}{2}$ distributions, corresponding to $\alpha = 1/2$, $\beta = 1$ and $\eta = 0$. In the former case, $\Psi(\theta) = c|\theta|$ for $\theta \in \mathbb{R}$ and its law is given by

$$\frac{c}{\pi}\frac{1}{(x^2+c^2)}\mathrm{d}x \tag{1.10}$$

for $x \in \mathbb{R}$. In the latter case, $\Psi(\theta) = c|\theta|^{1/2}(1 - \mathrm{i}\,\mathrm{sgn}\,\theta)$ for $\theta \in \mathbb{R}$ and its law is given by

$$\frac{c}{\sqrt{2\pi x^3}}\mathrm{e}^{-c^2/2x}\mathrm{d}x.$$

Note then that an inverse Gaussian distribution coincides with a stable-$\frac{1}{2}$ distribution for $a = c$ and $b = 0$.

Suppose that $\mathcal{S}(c, \alpha, \beta, \eta)$ is the distribution of a stable random variable with parameters c, α, β and η. For each choice of $c > 0$, $\alpha \in (0, 2)$, $\beta \in [-1, 1]$ and $\eta \in \mathbb{R}$ Theorem 1.6 tells us that there exists a Lévy process, with characteristic exponent given by (1.7) or (1.8) according to this choice of parameters. Further, from the definition of its characteristic exponent it is clear that at each fixed time the α-stable process will have distribution $\mathcal{S}(ct, \alpha, \beta, \eta)$.

In this text, we shall henceforth make an abuse of notation and refer to an α-stable process to mean a Lévy process based on a strictly stable distribution.

Necessarily this means that the associated characteristic exponent takes the form

$$\Psi(\theta) = \begin{cases} c|\theta|^\alpha(1 - \mathrm{i}\beta\tan\frac{\pi\alpha}{2}\mathrm{sgn}\,\theta) & \text{for } \alpha \in (0,1)\cup(1,2) \\ c|\theta| + \mathrm{i}\eta. & \text{for } \alpha = 1, \end{cases}$$

where the parameter ranges for c and β are as above. The reason for the restriction to strictly stable distribution is essentially that we shall want to make use of the following fact. If $\{X_t : t \geq 0\}$ is an α-stable process, then from its characteristic exponent (or equivalently the scaling properties of strictly stable random variables) we see that for all $\lambda > 0$ $\{X_{\lambda t} : t \geq 0\}$ has the same law as $\{\lambda^{1/\alpha} X_t : t \geq 0\}$.

1.2.7 Other Examples

There are many more known examples of infinitely divisible distributions (and hence Lévy processes). Of the many known proofs of infinitely divisibility for specific distributions, most of them are non-trivial, often requiring intimate knowledge of special functions. A brief list of such distributions might include generalised inverse Gaussian (see Good (1953) and Jørgensen (1982)), truncated stable (see Tweedie (1984), Hougaard (1986), Koponen (1995), Boyarchenko and Levendorskii (2002a) and Carr et al. (2003)), generalised hyperbolic (see Halgreen (1979)), Meixner (see Schoutens and Teugels (1998)), Pareto (see Steutel (1970) and Thorin (1977a)), F-distributions (see Ismail (1979)), Gumbel (see Johnson and Kotz (1970) and Steutel (1973)), Weibull (see Johnson and Kotz (1970) and Steutel (1970)), lognormal (see Thorin (1977b)) and Student t-distribution (see Grosswald (1976) and Ismail (1977)).

Despite our being able to identify a large number of infinitely divisible distributions and hence associated Lévy processes, it is not clear at this point what the paths of Lévy processes look like. The task of giving a mathematically precise account of this lies ahead in Chap. 2. In the meantime let us make the following informal remarks concerning paths of Lévy processes.

Exercise 1.1 shows that a linear combination of a finite number of independent Lévy processes is again a Lévy process. It turns out that one may consider any Lévy process as an independent sum of a Brownian motion with drift and a countable number of independent compound Poisson processes with different jump rates, jump distributions and drifts. The superposition occurs in such a way that the resulting path remains almost surely finite at all times and, for each $\varepsilon > 0$, the process experiences at most a countably infinite number of jumps of magnitude ε or less with probability one and an almost surely finite number of jumps of magnitude greater than ε, over all fixed finite time intervals. If in the latter description there is always an almost surely finite number of jumps over each fixed time interval then it is necessary and sufficient that one has the linear independent combination of a Brownian motion with drift and a compound Poisson process. Depending on the underlying structure of the jumps and the presence of a Brownian motion in the described linear combination, a Lévy process will either have paths of bounded variation on all finite time intervals or paths of unbounded variation on all finite time intervals.

Below we include five computer simulations to give a rough sense of how the paths of Lévy processes look. Figs. 1.1 and 1.2 depict the paths of Poisson process and a compound Poisson process, respectively. Figs. 1.3 and 1.4 show the paths of a Brownian motion and the independent sum of a Brownian motion and a compound Poisson process, respectively. Finally Figs. 1.5 and 1.6 show the paths of a variance gamma process and a normal inverse Gaussian processes. Both are pure jump processes (no Brownian component as described above). Variance gamma processes are discussed in more detail later in Sect. 2.7.3 and Exercise 1.5, normal inverse Gaussian

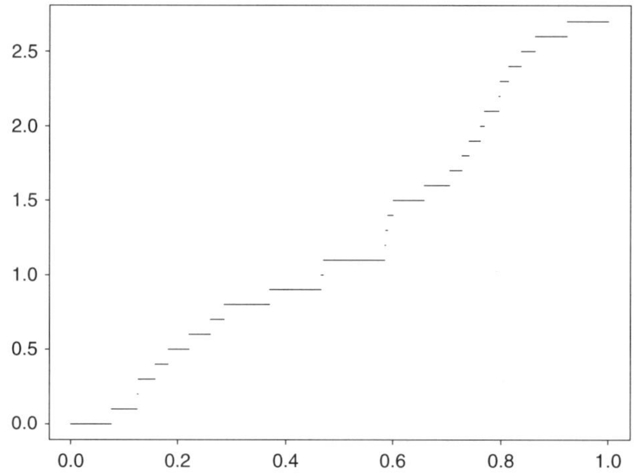

Fig. 1.1. A sample path of a Poisson process; $\Psi(\theta) = \lambda(1 - \mathrm{e}^{\mathrm{i}\theta})$ where λ is the jump rate.

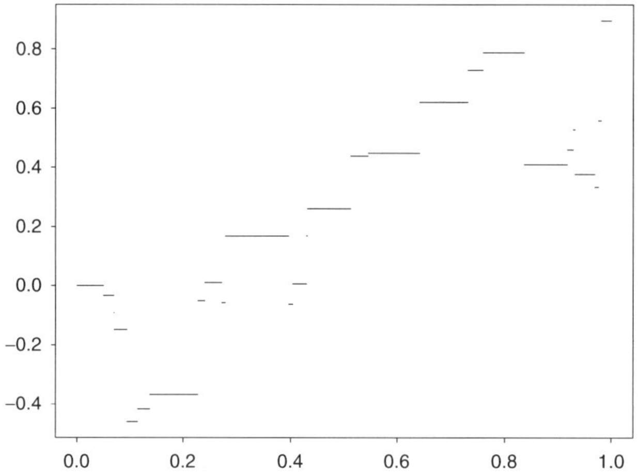

Fig. 1.2. A sample path of a compound Poisson process; $\Psi(\theta) = \lambda \int_{\mathbb{R}}(1 - \mathrm{e}^{\mathrm{i}\theta x})F(\mathrm{d}x)$ where λ is the jump rate and F is the common distribution of the jumps.

processes are Lévy processes whose jump measure is given by $\Pi(\mathrm{d}x) = (\delta\alpha/\pi|x|)\exp\{\beta x\}K_1(\alpha|x|)\mathrm{d}x$ for $x \in \mathbb{R}$ where $\alpha, \delta > 0$, $\beta \leq |\alpha|$ and $K_1(x)$ is the modified Bessel function of the third kind with index 1 (the precise definition of the latter is not worth the detail at this moment in the text). Both experience an infinite number of jumps over a finite time horizon. However, variance gamma processes have paths of bounded variation whereas normal inverse Gaussian processes have paths of unbounded variation. The reader should be warned however that computer simulations ultimately can only

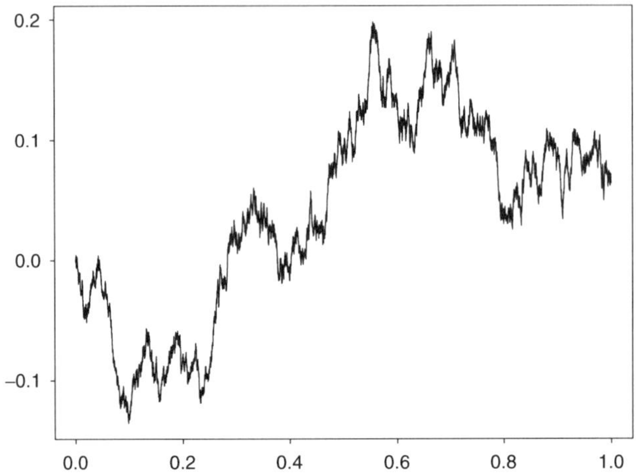

Fig. 1.3. A sample path of a Brownian motion; $\Psi(\theta) = \theta^2/2$.

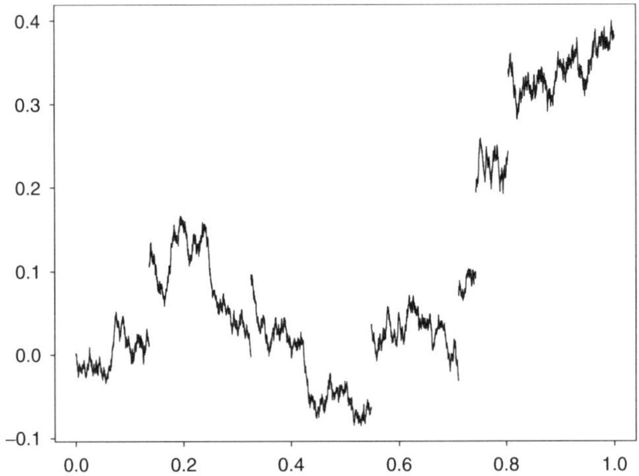

Fig. 1.4. A sample path of the independent sum of a Brownian motion and a compound Poisson process; $\Psi(\theta) = \theta^2/2 + \int_{\mathbb{R}}(1 - \mathrm{e}^{\mathrm{i}\theta x})F(\mathrm{d}x)$.

depict a finite number of jumps in any given path. All figures were very kindly produced by Antonis Papapantoleon for the purpose of this text.

1.3 Lévy Processes and Some Applied Probability Models

In this section we introduce some classical applied probability models, which are structured around basic examples of Lévy processes. This section provides a particular motivation for the study of fluctuation theory that follows in

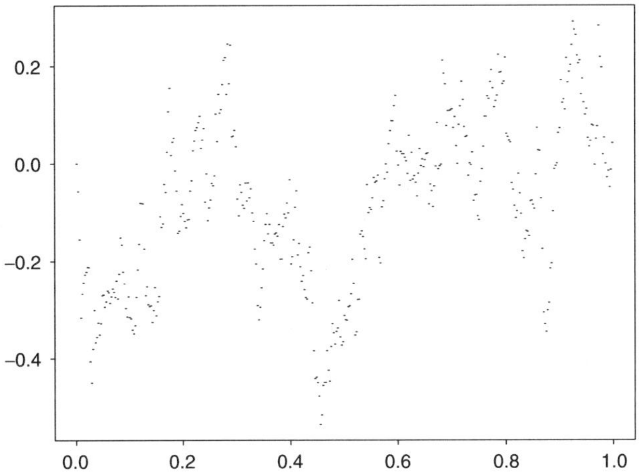

Fig. 1.5. A sample path of a variance gamma processes. The latter has characteristic exponent given by $\Psi(\theta) = \beta \log(1 - \mathrm{i}\theta c/\alpha + \beta^2\theta^2/2\alpha)$ where $c \in \mathbb{R}$ and $\beta > 0$.

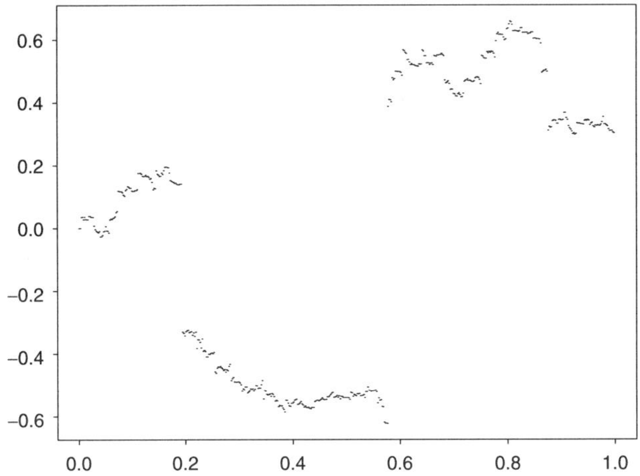

Fig. 1.6. A sample path of a normal inverse Gaussian process; $\Psi(\theta) = \delta(\sqrt{\alpha^2 - (\beta + \mathrm{i}\theta)^2} - \sqrt{\alpha^2 - \beta^2})$ where $\alpha, \delta > 0$, $|\beta| < \alpha$.

subsequent chapters. (There are of course other reasons for wanting to study fluctuation theory of Lévy processes.) With the right understanding of particular features of the models given below in terms of the path properties of the underlying Lévy processes, much richer generalisations of the aforementioned models may be studied for which familiar and new phenomena may be observed. At different points later on in this text we will return to these models and reconsider these phenomena in the light of the theory that has been presented along the way. In particular all of the results either stated or alluded to below will be proved in greater generality in later chapters.

1.3.1 Cramér–Lundberg Risk Process

Consider the following model of the revenue of an insurance company as a process in time proposed by Lundberg (1903). The insurance company collects premiums at a fixed rate $c > 0$ from its customers. At times of a Poisson process, a customer will make a claim causing the revenue to jump downwards. The size of claims is independent and identically distributed. If we call X_t the capital of the company at time t, then the latter description amounts to

$$X_t = x + ct - \sum_{i=1}^{N_t} \xi_i, \ t \geq 0,$$

where $x > 0$ is the initial capital of the company, $N = \{N_t : t \geq 0\}$ is a Poisson process with rate $\lambda > 0$, and $\{\xi_i : i \geq 1\}$ is a sequence of positive, independent and identically distributed random variables also independent of N. The process $X = \{X_t : t \geq 0\}$ is nothing more than a compound Poisson process with drift of rate c, initiated from $x > 0$.

Financial ruin in this model (or just *ruin* for short) will occur if the revenue of the insurance company drops below zero. Since this will happen with probability one if $\mathbb{P}(\liminf_{t\uparrow\infty} X_t = -\infty) = 1$, an additional assumption imposed on the model is that

$$\lim_{t\uparrow\infty} X_t = \infty.$$

A sufficient condition to guarantee the latter is that the distribution of ξ has finite mean, say $\mu > 0$, and that

$$\frac{\lambda\mu}{c} < 1,$$

the so-called *net profit condition.* To see why this presents a sufficient condition, note that the Strong Law of Large Numbers and the obvious fact that $\lim_{t\uparrow\infty} N_t = \infty$ imply that

$$\lim_{t\uparrow\infty} \frac{X_t}{t} = \lim_{t\uparrow\infty} \left(\frac{x}{t} + c - \frac{N_t}{t} \frac{\sum_{i=1}^{N_t} \xi_i}{N_t} \right) = c - \lambda\mu > 0,$$

Under the net profit condition it follows that ruin will occur only with probability less than one. Fundamental quantities of interest in this model thus become the distribution of the time to ruin and the deficit at ruin; otherwise identified as

$$\tau_0^- := \inf\{t > 0 : X_t < 0\} \text{ and } X_{\tau_0^-} \text{ on } \{\tau_0^- < \infty\}$$

when the process X drifts to infinity.

The following classic result links the probability of ruin to the conditional distribution

$$\eta(x) = \mathbb{P}(-X_{\tau_0^-} \leq x | \tau_0^- < \infty).$$

Theorem 1.8 (Pollaczek–Khintchine formula). *Suppose that $\lambda\mu/c < 1$. For all $x \geq 0$,*

$$1 - \mathbb{P}(\tau_0^- < \infty | X_0 = x) = (1-\rho) \sum_{k \geq 0} \rho^k \eta^{k*}(x), \tag{1.11}$$

where $\rho = \mathbb{P}(\tau_0^- < \infty)$.

Formula (1.11) is missing quite some details in the sense that we know nothing of the constant ρ, nor of the distribution η. It turns out that the unknowns ρ and η in the Pollaczek–Khintchine formula can be identified explicitly as the next theorem reveals.

Theorem 1.9. *In the Cramér–Lundberg model (with $\lambda\mu/c < 1$), $\rho = \lambda\mu/c$ and*

$$\eta(x) = \frac{1}{\mu} \int_0^x F(y, \infty) \mathrm{d}y, \tag{1.12}$$

where F is the distribution of ξ_1.

This result can be derived by a classical path analysis of random walks. This analysis gives some taste of the general theory of fluctuations of Lévy processes that we will spend quite some time with in this book. The proof of Theorem 1.9 can be found in Exercise 1.8.

The Pollaczek–Khintchine formula together with some additional assumptions on F gives rise to an interesting asymptotic behaviour of the probability of ruin. Specifically we have the following result.

Theorem 1.10. *Suppose that $\lambda\mu/c < 1$ and there exists a $0 < \nu < \infty$ such that $\mathbb{E}\left(\mathrm{e}^{-\nu X_1}\right) = 1$, then*

$$\mathbb{P}\left(\tau_0^- < \infty\right) \leq \mathrm{e}^{-\nu x}$$

for all $x > 0$ where $\mathbb{P}_x(\cdot)$ denotes $\mathbb{P}(\cdot | X_0 = x)$. If further, the distribution of F is non-lattice, then

$$\lim_{x \uparrow \infty} \mathrm{e}^{\nu x} \mathbb{P}_x\left(\tau_0^- < \infty\right) = \left(\frac{\lambda\nu}{c - \lambda\mu} \int_0^\infty x \mathrm{e}^{\nu x} F(x, \infty) \mathrm{d}x\right)^{-1}$$

where the right-hand side should be interpreted as zero if the integral is infinite.

In the above theorem, the parameter ν is known as the *Lundberg exponent.* See Cramér (1994a,b) for a review of the appearance of these results.

In more recent times, authors have extended the idea of modelling with compound Poisson processes with drift and moved to more general classes of Lévy processes for which the measure Π is concentrated on $(-\infty, 0)$ and hence processes for which there are no positive jumps. See for example Huzak et al. (2004a,b), Chan (2004) and Klüppelberg et al. (2004). It turns out that working with this class of Lévy processes preserves the idea that the revenue of the insurance company is the aggregate superposition of lots of independent claims sequentially through time offset against a deterministic increasing process corresponding to the accumulation of premiums, even when there are an almost surely infinite number of jumps downwards (claims) in any fixed time interval. We will provide a more detailed interpretation of the latter class in Chap. 2. In Chaps. 4 and 7, amongst other things, we will also re-examine the Pollaczek–Khintchine formula and the asymptotic probability of ruin given in Theorem 1.10 in light of these generalised risk models.

1.3.2 The $M/G/1$ queue

Let us recall the definition of the $M/G/1$ queue. Customers arrive at a service desk according to a Poisson process and join a queue. Customers have service times that are independent and identically distributed. Once served, they leave the queue.

The workload, W_t, at each time $t \geq 0$, is defined to be the time it will take a customer who joins the back of the queue at that moment to reach the service desk, that is to say the amount of processing time remaining in the queue at time t. Suppose that at an arbitrary moment, which we shall call time zero, the server is not idle and the workload is equal to $w > 0$. On the event that t is before the first time the queue becomes empty, we have that W_t is equal to

$$w + \sum_{i=1}^{N_t} \xi_i - t, \tag{1.13}$$

where, as with the Cramér–Lundberg risk process, $N = \{N_t : t \geq 0\}$ is a Poisson process with intensity $\lambda > 0$ and $\{\xi_i : i \geq 0\}$ are positive random variables that are independent and identically distributed with common distribution F and mean $\mu < \infty$. The process N represents the arrivals of new customers and $\{\xi_i : i \geq 0\}$ are understood as their respective service times that are added to the workload. The negative unit drift simply corresponds to the decrease in time as the server deals with jobs. Thanks to the lack of memory property, once the queue becomes empty, the queue remains empty for an exponentially distributed period of time with parameter λ after which a new arrival incurs a jump in W, which has distribution F. The process proceeds as the compound Poisson process described above until the queue next empties and so on.

The workload is clearly not a Lévy process as it is impossible for $W_t : t \geq 0$ to decrease in value from the state zero where as it can decrease in value from any other state $x > 0$. However, it turns out that it is quite easy to link the workload to a familiar functional of a Lévy process, which is also a Markov process. Specifically, suppose we define X_t equal to precisely the same Lévy process given in the Cramér–Lundberg risk model with $c = 1$ and $x = 0$, then

$$W_t = (w \vee \overline{X}_t) - X_t,\ t \geq 0,$$

where the process $\overline{X} := \{\overline{X}_t : t \geq 0\}$ is the running supremum of X, hence $\overline{X}_t = \sup_{u \leq t} X_u$. Whilst it is easy to show that the pair $(\overline{X}, X)$ is a Markov process, with a little extra work it can be shown that W is a Strong Markov Process (this is dealt with in more detail in Exercise 3.2). Clearly then, under $\mathbb{P}$, the process W behaves like $-X$ until the random time

$$\tau_w^+ := \inf\{t > 0 : X_t > w\}.$$

The latter is in fact a stopping time since $\{\tau_w^+ \leq t\} = \{\overline{X}_t \geq w\}$ belongs to the filtration generated by the process X. At the time τ_w^+, the process $W = \{W_t : t \geq 0\}$ first becomes zero and on account of the Strong Markov Property and the lack of memory property, it remains so for a period of time, which is exponentially distributed with parameter λ since during this period $w \vee \overline{X}_t = \overline{X}_t = X_t$. At the end of this period, X makes another negative jump distributed according to F and hence W makes a positive jump with the same distribution and so on thus matching the description in the previous paragraph; see Fig. 1.7.

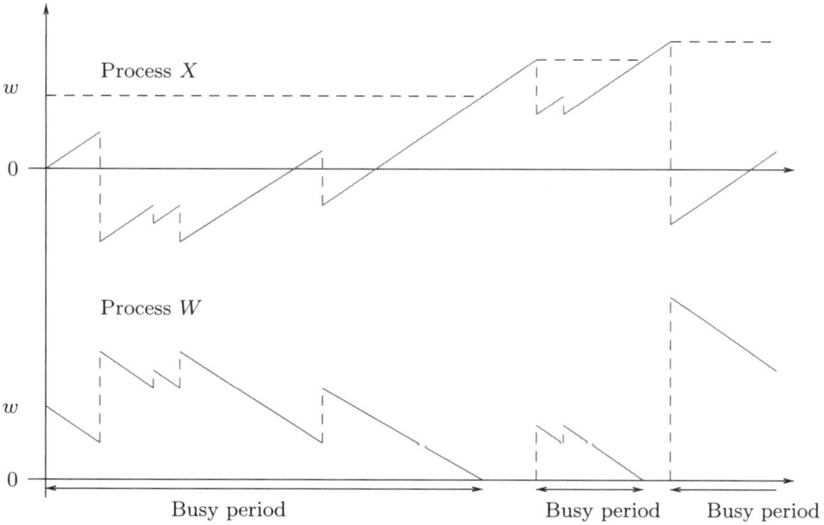

Fig. 1.7. Sample paths of X and W.

Note that this description still makes sense when $w = 0$ in which case for an initial period of time, which is exponentially distributed, W remains equal to zero until X first jumps (corresponding to the first arrival in the queue).

There are a number of fundamental points of interest concerning both local and global behavioural properties of the $M/G/1$ queue. Take for example the time it takes before the queue first empties; in other words τ_w^+. It is clear from a simple analysis of the paths of X and W that the latter is finite with probability one if the underlying process X drifts to infinity with probability one. Using similar reasoning to the previous example, with the help of the Strong Law of Large Numbers it is easy to deduce that this happens when $\lambda\mu < 1$. Another common situation of interest in this model corresponds to the case that the server is only capable of dealing with a maximum workload of z units of time. The first time the workload exceeds the buffer level z

$$\sigma_z := \inf\{t > 0 : W_t > z\}$$

therefore becomes of interest. In particular the probability of $\{\sigma_z < \tau_w^+\}$ which corresponds to the event that the workload exceeds the buffer level before the server can complete a busy period.

The following two theorems give some classical results concerning the idle time of the $M/G/1$ queue and the stationary distribution of the work load. Roughly speaking they say that when there is heavy traffic ($\lambda\mu > 1$) eventually the queue never becomes empty and the workload grows to infinity and the total time that the queue remains empty is finite with a particular distribution. Further, when there is light traffic ($\lambda\mu < 1$) the queue repeatedly becomes empty and the total idle time grows to infinity whilst the workload process converges in distribution. At the critical value $\lambda\mu = 1$ the workload grows to arbitrary large values but nonetheless the queue repeatedly becomes empty and the total idle time grows to infinity. Ultimately all these properties are a reinterpretation of the long-term behaviour of a special class of reflected Lévy processes.

Theorem 1.11. *Suppose that W is the workload of an $M/G/1$ queue with arrival rate λ and service distribution F having mean μ. Define the total idle time*

$$I = \int_0^\infty \mathbf{1}_{(W_t=0)} \mathrm{d}t.$$

(i) Suppose that $\lambda\mu > 1$. Let

$$\psi(\theta) = \theta - \lambda \int_{(0,\infty)} (1 - \mathrm{e}^{-\theta x}) F(\mathrm{d}x), \ \theta \geq 0,$$

and define θ^* to be the largest root of the equation $\psi(\theta) = 0$. Then[3]

$$\mathbb{P}\left(I \in \mathrm{d}x | W_0 = w\right) = (1 - \mathrm{e}^{-\theta^* w})\delta_0\left(\mathrm{d}x\right) + \theta^* \mathrm{e}^{-\theta^*(w+x)}\mathrm{d}x.$$

[3]Following standard notation, the measure δ_0 assigns a unit atom to the point 0.

(ii) If $\lambda\mu \leq 1$ then I is infinite with probability one.

Note that the function ψ given above is nothing more than the Laplace exponent of the underlying Lévy process

$$X_t = t - \sum_{i=1}^{N_t} \xi_i, \ t \geq 0$$

which drives the process W and fulfils the relation $\psi(\theta) = \log \mathbb{E}(\mathrm{e}^{\theta X_1})$. It is easy to check by differentiating it twice that ψ is a strictly convex function, which is zero at the origin and tends to infinity at infinity. Further $\psi'(0+) < 0$ under the assumption $\lambda\mu > 1$ and hence θ^* exists, is finite and is in fact the only solution to $\psi(\theta) = 0$ other than $\theta = 0$.

Theorem 1.12. *Let W be the same as in Theorem 1.11.*

(i) Suppose that $\lambda\mu < 1$. Then for all $w \geq 0$ the virtual waiting time has a stationary distribution,

$$\lim_{t\uparrow\infty} \mathbb{P}(W_t \leq x | W_0 = w) = (1-\rho)\sum_{k=0}^{\infty} \rho^k \eta^{*k}(x),$$

where

$$\eta(x) = \frac{1}{\mu}\int_0^x F(y,\infty)\mathrm{d}y \textit{ and } \rho = \lambda\mu.$$

(ii) If $\lambda\mu \geq 1$ then $\limsup_{t\uparrow\infty} W_t = \infty$ with probability one.

Some of the conclusions in the above two theorems can already be obtained with basic knowledge of compound Poisson processes. Theorem 1.11 is proved in Exercise 1.9 and gives some feeling of the fluctuation theory that will be touched upon later on in this text. The remarkable similarity between Theorem 1.12 part (i) and the Pollaczek–Khintchine formula is of course no coincidence. The principles that are responsible for the latter two results are embedded within the general fluctuation theory of Lévy processes. Indeed we will revisit Theorems 1.11 and 1.12 but for more general versions of the workload process of the $M/G/1$ queue known as general storage models. Such generalisations involve working with a general class of Lévy process with no positive jumps (that is $\Pi(0,\infty) = 0$) and defining as before $W_t = (w \vee \overline{X}_t) - X_t$. When there are an infinite number of jumps in each finite time interval the latter process may be thought of as modelling a processor that deals with an arbitrarily large number of small jobs and occasional large jobs. The precise interpretation of such a generalised $M/G/1$ workload process and issues concerning the distribution of the busy period, the stationary distribution of the workload, time to buffer overflow and other related quantities will be dealt with later on in Chaps. 2, 4 and 8.

1.3.3 Optimal Stopping Problems

A fundamental class of problems motivated by applications from physics, optimal control, sequential testing and economics (to name but a few) concern optimal stopping problems of the form: Find $v(x)$ and a stopping time, τ^*, belonging to a specified family of stopping times, $\mathcal{T}$, such that

$$v(x) = \sup_{\tau \in \mathcal{T}} \mathbb{E}_x(\mathrm{e}^{-q\tau} G(X_\tau)) = \mathbb{E}_x(\mathrm{e}^{-q\tau^*} G(X_{\tau^*})) \tag{1.14}$$

for all $x \in \mathcal{R} \subseteq \mathbb{R}$, where X is a $\mathcal{R}$-valued Markov process with probabilities $\{\mathbb{P}_x : x \in \mathcal{R}\}$ (with the usual understanding that $\mathbb{P}_x$ is the law of X given that $X_0 = x$), $q \geq 0$ and $G : \mathcal{R} \to [0, \infty)$ is a function suitable to the application at hand. The optimal stopping problem (1.14) is not the most general class of such problems that one may consider but will suffice for the discussion at hand.

In many cases it turns out that the optimal strategy takes the form

$$\tau^* = \inf\{t > 0 : (t, X_t) \in D\},$$

where $D \subset [0, \infty) \times \mathbb{R}$ is a domain in time–space called the *stopping region*. Further still, there are many examples within the latter class for which $D = [0, \infty) \times I$ where I is an interval or the complement of an interval. In other words the optimal strategy is the first passage time into I,

$$\tau^* = \inf\{t > 0 : X_t \in I\}. \tag{1.15}$$

A classic example of an optimal stopping problem in the form (1.14) for which the solution agrees with (1.15) is the following taken from McKean (1965),

$$v(x) = \sup_{\tau \in \mathcal{T}} \mathbb{E}_x(\mathrm{e}^{-q\tau}(K - \mathrm{e}^{X_\tau})^+), \tag{1.16}$$

where now $q > 0$, $\mathcal{T}$ is the family of stopping times with respect to the filtration $\mathcal{F}_t := \sigma(X_s : s \leq t)$ and $\{X_t : t \geq 0\}$ is a linear Brownian motion, $X_t = \sigma B_t + \gamma t$, $t \geq 0$ (see Sect. 1.2.3). Note that we use here the standard notation $y^+ = y \vee 0$. This particular example models the optimal time to sell a risky asset for a fixed value K when the asset's dynamics are those of an exponential linear Brownian motion. Optimality in this case is determined via the expected discounted gain at the selling time. On account of the underlying source of randomness being Brownian motion and the optimal strategy taking the simple form (1.15), the solution to (1.16) turns out to be explicitly computable as follows.

Theorem 1.13. *The solution* (v, τ^*) *to (1.16) is given by*

$$\tau^* = \inf\{t > 0 : X_t < x^*\},$$

where

$$e^{x^*} = K\left(\frac{\Phi(q)}{1+\Phi(q)}\right),$$

$\Phi(q) = (\sqrt{\gamma^2 + 2\sigma^2 q} + \gamma)/\sigma^2$ *and*

$$v(x) = \begin{cases} (K - e^x) & \text{if } x < x^* \\ (K - e^{x^*})e^{-\Phi(q)(x-x^*)} & \text{if } x \geq x^*. \end{cases}$$

The solution to this problem reflects the intuition that the optimal time to stop should be at a time when X is as negative as possible taking into consideration that taking too long to stop incurs an exponentially weighted penalty. Note that in $(-\infty, x^*)$ the value function $v(x)$ is equal to the *gain function* $(K - e^x)^+$ as the optimal strategy τ^* dictates that one should stop immediately. A particular curiosity of the solution to (1.16) is the fact that at x^*, the value function v joins smoothly to the gain function. In other words,

$$v'(x^*-) = -e^{x^*} = v'(x^*+).$$

A natural question in light of the above optimal stopping problem is whether one can characterise the solution to (1.16) when X is replaced by a general Lévy process. Indeed, if the same strategy of first passage below a specified level is still optimal, one is then confronted with needing information about the distribution of the overshoot of a Lévy process when first crossing below a barrier in order to compute the function v. The latter is of particular interest if one would like to address the question as to whether the phenomenon of smooth fit is still to be expected in the general Lévy process setting.

Later in Chap. 9 we give a brief introduction to some general principles appearing in the theory of optimal stopping and apply them to a handful of examples where the underlying source of randomness is provided by a Lévy process. The first of these examples being the generalisation of (1.16) as mentioned above. All of the examples presented in Chap. 9 can be solved (semi-) explicitly thanks to a degree of simplicity in the optimal strategy such as (1.15) coupled with knowledge of fluctuation theory of Lévy processes. In addition, through these examples, we will attempt to give some insight into how and when smooth pasting occurs as a consequence of a subtle type of path behaviour of the underlying Lévy process.

1.3.4 Continuous-State Branching Processes

Originating in part from the concerns of the Victorian British upper classes that aristocratic surnames were becoming extinct, the theory of branching processes now forms a cornerstone of classical applied probability. Some of the earliest work on branching processes dates back to Watson and Galton (1874). However, approximately 100 years later, it was discovered by Heyde and Seneta (1977) that the less well-exposed work of I.J. Bienaymé, dated around 1845, contained many aspects of the later work of Galton and Watson. The *Bienaymé–Galton–Watson* process, as it is now known, is a discrete time Markov chain with state space $\{0, 1, 2, ...\}$ described by the sequence $\{Z_n : n = 0, 1, 2, ...\}$ satisfying the recursion $Z_0 > 0$ and

$$Z_n = \sum_{i=1}^{Z_{n-1}} \xi_i^{(n)}$$

for $n = 1, 2, ...$ where $\{\xi^{(n)} : i = 1, 2, ...\}$ are independent and exponentially distributed on $\{0, 1, 2, ...\}$. We use the usual notation $\sum_{i=1}^{0}$ to represent the empty sum. The basic idea behind this model is that Z_n is the population count in the nth generation and from an initial population Z_0 (which may be randomly distributed) individuals reproduce asexually and independently with the same distribution of numbers of offspring. The latter reproductive properties are referred to as the *branching property.* Note that as soon as $Z_n = 0$ it follows from the given construction that $Z_{n+k} = 0$ for all $k = 1, 2, ...$ A particular consequence of the branching property is that if $Z_0 = a + b$ then Z_n is equal in distribution to $Z_n^{(1)} + Z_n^{(2)}$ where $Z_n^{(1)}$ and $Z_n^{(2)}$ are independent with the same distribution as an nth generation Bienaymé–Galton–Watson process initiated from population sizes a and b, respectively.

A mild modification of the Bienaymé–Galton–Watson process is to set it into continuous time by assigning life lengths to each individual which are independent and identically distributed with parameter $\lambda > 0$. Individuals reproduce at their moment of death in the same way as described previously for the Bienaymé-Galton-Watson process. If $Y = \{Y_t : t \geq 0\}$ is the $\{0, 1, 2, ...\}$-valued process describing the population size then it is straightforward to see that the lack of memory property of the exponential distribution implies that for all $0 \leq s \leq t$,

$$Y_t = \sum_{i=1}^{Y_s} Y_{t-s}^{(i)},$$

where given $\{Y_u : u \leq s\}$ the variables $\{Y_{t-s}^{(i)} : i = 1, ..., Y_s\}$ are independent with the same distribution as Y_{t-s} conditional on $Y_0 = 1$. In that case, we may talk of Y as a continuous-time Markov chain on $\{0, 1, 2, ...\}$, with probabilities, say, $\{P_y : y = 0, 1, 2, ...\}$ where P_y is the law of Y under the assumption that $Y_0 = y$. As before, the state 0 is absorbing in the sense that if $Y_t = 0$ then $Y_{t+u} = 0$ for all $u > 0$. The process Y is called the *continuous time Markov branching process.* The branching property for Y may now be formulated as follows.

Definition 1.14 (Branching property). *For any $t \geq 0$ and y_1, y_2 in the state space of Y, Y_t under $P_{y_1+y_2}$ is equal in law to the independent sum $Y_t^{(1)} + Y_t^{(2)}$ where the distribution of $Y_t^{(i)}$ is equal to that of Y_t under P_{y_i} for $i = 1, 2$.*

So far there appears to be little connection with Lévy processes. However a remarkable time transformation shows that the path of Y is intimately linked to the path of a compound Poisson process with jumps whose distribution is supported in $\{-1,0,1,2,...\}$, stopped at the first instant that it hits zero. To explain this in more detail let us introduce the probabilities $\{\pi_i : i = -1,0,1,2,...\}$, where $\pi_i = P(\xi = i+1)$ and ξ has the same distribution as the typical family size in the Bienaymé–Galton–Watson process. To avoid complications let us assume that $\pi_0 = 0$ so that a transition in the state of Y always occurs when an individual dies. When jumps of Y occur, they are independent and always distributed according to $\{\pi_i : i = -1,0,1,...\}$. The idea now is to adjust time accordingly with the evolution of Y in such a way that these jumps are spaced out with inter-arrival times that are independent and exponentially distributed. Crucial to the following exposition is the simple and well-known fact that the minimum of $n \in \{1,2,...\}$ independent and exponentially distributed random variables is exponentially distributed with parameter λn. Further, that if $\mathbf{e}_\alpha$ is exponentially distributed with parameter $\alpha > 0$ then for $\beta > 0$, $\beta\mathbf{e}_\alpha$ is equal in distribution to $\mathbf{e}_{\alpha/\beta}$.

Write for $t \geq 0$,

$$J_t = \int_0^t Y_u \mathrm{d}u$$

set

$$\varphi_t = \inf\{s \geq 0 : J_s > t\}$$

with the usual convention that $\inf \emptyset = \infty$ and define

$$X_t = Y_{\varphi_t} \tag{1.17}$$

with the understanding that when $\varphi_t = \infty$ we set $X_t = 0$. Now observe that when $Y_0 = y \in \{1,2,...\}$ the first jump of Y occurs at a time, say T_1 (the minimum of y independent exponential random variables, each with parameter $\lambda > 0$) which is exponentially distributed with parameter λy and the size of the jump is distributed according to $\{\pi_i : i = -1,0,1,2,...\}$. However, note that $J_{T_1} = yT_1$ is the first time that the process $X = \{X_t : t \geq 0\}$ jumps. The latter time is exponentially distributed with parameter λ. The jump at this time is independent and distributed according to $\{\pi_i : i = -1,0,1,2,...\}$.

Given the information $\mathcal{G}_1 = \sigma(Y_t : t \leq T_1)$, the lack of memory property implies that the continuation $\{Y_{T_1+t} : t \geq 0\}$ has the same law as Y under P_y with $y = Y_{T_1}$. Hence if T_2 is the time of the second jump of Y then conditional on $\mathcal{G}_1$ we have that $T_2 - T_1$ is exponentially distributed with parameter λY_{T_1} and $J_{T_2} - J_{T_1} = Y_{T_1}(T_2 - T_1)$ which is again exponentially distributed with parameter λ and further, is independent of $\mathcal{G}_1$. Note that J_{T_2} is the time of the second jump of X and the size of the second jump is again independent and distributed according to $\{\pi_i : i = -1,0,1,...\}$. Iterating in this way it becomes clear that X is nothing more than a compound Poisson process with arrival rate λ and jump distribution

$$F(\mathrm{d}x) = \sum_{i=-1}^{\infty} \pi_i \delta_i(\mathrm{d}x) \tag{1.18}$$

stopped on first hitting the origin.

A converse to this construction is also possible. Suppose now that $X = \{X_t : t \geq 0\}$ is a compound Poisson process with arrival rate $\lambda > 0$ and jump distribution $F(\mathrm{d}x) = \sum_{i=-1}^{\infty} \pi_i \delta_i(\mathrm{d}x)$. Write

$$I_t = \int_0^t X_u^{-1} \mathrm{d}u$$

and set

$$\theta_t = \inf\{s \geq 0 : I_s > t\}. \tag{1.19}$$

again with the understanding that $\inf \emptyset = \infty$. Define

$$Y_t = X_{\theta_t \wedge \tau_0^-}$$

where $\tau_0^- = \inf\{t > 0 : X_t < 0\}$. By analysing the behaviour of $Y = \{Y_t : t \geq 0\}$ at the jump times of X in a similar way to above one readily shows that the process Y is a continuous time Markov branching process. The details are left as an exercise to the reader.

The relationship between compound Poisson processes and continuous time Markov branching processes described above turns out to have a much more general setting. In the work of Lamperti (1967a, 1976b) it is shown that there exists a correspondence between a class of branching processes called continuous-state branching processes and Lévy processes with no negative jumps ($\Pi(-\infty, 0) = 0$). In brief, a continuous-state branching process is a $[0,\infty)$-valued Markov process having paths that are right continuous with left limits and probabilities $\{P_x : x > 0\}$ that satisfy the branching property in Definition 1.14. Note in particular that now the quantities y_1 and y_2 may be chosen from the non-negative real numbers. Lamperti's characterisation of continuous-state branching processes shows that they can be identified as time changed Lévy processes with no negative jumps precisely via the transformations given in (1.17) with an inverse transformation analogous to (1.19). We explore this relationship in more detail in Chap. 10 by looking at issues such as explosion, extinction and conditioning on survival.

Exercises

1.1. Using Definition 1.1, show that the sum of two (or indeed any finite number of) independent Lévy processes is again a Lévy process.

1.2. Suppose that $S = \{S_n : n \geq 0\}$ is any random walk and $\mathbf{\Gamma}_p$ is an independent random variable with a geometric distribution on $\{0, 1, 2, ...\}$ with parameter p.

(i) Show that $\mathbf{\Gamma}_p$ is infinitely divisible.
(ii) Show that $S_{\mathbf{\Gamma}_p}$ is infinitely divisible.

1.3 (Proof of Lemma 1.7). In this exercise we derive the Frullani identity.

(i) Show for any function f such that f' exists and is continuous and $f(0)$ and $f(\infty)$ are finite, that

$$\int_0^\infty \frac{f(ax) - f(bx)}{x} \mathrm{d}x = (f(0) - f(\infty)) \log \left(\frac{b}{a}\right),$$

where $b > a > 0$.

(ii) By choosing $f(x) = \mathrm{e}^{-x}$, $a = \alpha > 0$ and $b = \alpha - z$ where $z < 0$, show that

$$\frac{1}{(1 - z/\alpha)^\beta} = \mathrm{e}^{-\int_0^\infty (1-\mathrm{e}^{zx})\frac{\beta}{x}\mathrm{e}^{-\alpha x}\mathrm{d}x}$$

and hence by analytic extension show that the above identity is still valid for all $z \in \mathbb{C}$ such that $\Re z \leq 0$.

1.4. Establishing formulae (1.7) and (1.8) from the Lévy measure given in (1.9) is the result of a series of technical manipulations of special integrals. In this exercise we work through them. In the following text we will use the gamma function $\Gamma(z)$, defined by

$$\Gamma(z) = \int_0^\infty t^{z-1}\mathrm{e}^{-t}\mathrm{d}t$$

for $z > 0$. Note the gamma function can also be analytically extended so that it is also defined on $\mathbb{R}\backslash\{0, -1, -2, ...\}$ (see Lebedev (1972)). Whilst the specific definition of the gamma function for negative numbers will not play an important role in this exercise, the following two facts that can be derived from it will. For $z \in \mathbb{R}\backslash\{0, -1, -2, ...\}$ the gamma function observes the recursion $\Gamma(1 + z) = z\Gamma(z)$ and $\Gamma(1/2) = \sqrt{\pi}$.

(i) Suppose that $0 < \alpha < 1$. Prove that for $u > 0$,

$$\int_0^\infty (\mathrm{e}^{-ur} - 1)r^{-\alpha-1}\mathrm{d}r = \Gamma(-\alpha)u^\alpha$$

and show that the same equality is valid when $-u$ is replaced by any complex number $w \neq 0$ with $\Re w \leq 0$. Conclude by considering $w = i$ that

$$\int_0^\infty (1 - \mathrm{e}^{\mathrm{i}r})r^{-\alpha-1}\mathrm{d}r = -\Gamma(-\alpha)\mathrm{e}^{-\mathrm{i}\pi\alpha/2} \tag{1.20}$$

as well as the complex conjugate of both sides being equal. Deduce (1.7) by considering the integral

$$\int_0^\infty (1 - \mathrm{e}^{\mathrm{i}\xi\theta r})r^{-\alpha-1}\mathrm{d}r$$

for $\xi = \pm 1$ and $\theta \in \mathbb{R}$. Note that you will have to take $a = \eta - \int_{\mathbb{R}} x\mathbf{1}_{(|x|<1)}\Pi(\mathrm{d}x)$, which you should check is finite.

(ii) Now suppose that $\alpha = 1$. First prove that

$$\int_{|x|<1} \mathrm{e}^{\mathrm{i}\theta x}(1-|x|)\mathrm{d}x = 2\left(\frac{1-\cos\theta}{\theta^2}\right)$$

for $\theta \in \mathbb{R}$ and hence by Fourier inversion,

$$\int_0^\infty \frac{1-\cos r}{r^2}\mathrm{d}r = \frac{\pi}{2}.$$

Use this identity to show that for $z > 0$,

$$\int_0^\infty (1 - \mathrm{e}^{\mathrm{i}rz} + \mathrm{i}zr\mathbf{1}_{(r<1)})\frac{1}{r^2}\mathrm{d}r = \frac{\pi}{2}z + \mathrm{i}z\log z - \mathrm{i}kz$$

for some constant $k \in \mathbb{R}$. By considering the complex conjugate of the above integral establish the expression in (1.8). Note that you will need a different choice of a to part (i).

(iii) Now suppose that $1 < \alpha < 2$. Integrate (1.20) by parts to reach

$$\int_0^\infty (\mathrm{e}^{\mathrm{i}r} - 1 - \mathrm{i}r)r^{-\alpha-1}\mathrm{d}r = \Gamma(-\alpha)\mathrm{e}^{-\mathrm{i}\pi\alpha/2}.$$

Consider the above integral for $z = \xi\theta$, where $\xi = \pm 1$ and $\theta \in \mathbb{R}$ and deduce the identity (1.7) in a similar manner to the proof in (i) and (ii).

1.5. Prove for any $\theta \in \mathbb{R}$ that

$$\exp\{\mathrm{i}\theta X_t + t\Psi(\theta)\},\ t \geq 0$$

is a martingale where $\{X_t : t \geq 0\}$ is a Lévy process with characteristic exponent Ψ.

1.6. In this exercise we will work out in detail the features of the inverse Gaussian process discussed earlier on in this chapter. Recall that $\tau = \{\tau_s : s \geq 0\}$ is a non-decreasing Lévy process defined by $\tau_s = \inf\{t \geq 0 : B_t + bt > s\}$, $s \geq 0$, where $B = \{B_t : t \geq 0\}$ is a standard Brownian motion and $b > 0$.

(i) Argue along the lines of Exercise 1.5 to show that for each $\lambda > 0$,

$$\mathrm{e}^{\lambda B_t - \frac{1}{2}\lambda^2 t},\ t \geq 0$$

is a martingale. Use Doob's Optimal Stopping Theorem to obtain

$$\mathbb{E}(\mathrm{e}^{-(\frac{1}{2}\lambda^2 + b\lambda)\tau_s}) = \mathrm{e}^{-\lambda s}.$$

Use analytic extension to deduce further that τ_s has characteristic exponent

$$\Psi(\theta) = s(\sqrt{-2\mathrm{i}\theta + b^2} - b)$$

for all $\theta \in \mathbb{R}$.

(ii) Defining the measure $\Pi(\mathrm{d}x) = (2\pi x^3)^{-1/2}\mathrm{e}^{-xb^2/2}\mathrm{d}x$ on $x > 0$, check using (1.20) from Exercise 1.4 that

$$\int_0^\infty (1 - \mathrm{e}^{\mathrm{i}\theta x})\Pi(\mathrm{d}x) = \Psi(\theta)$$

for all $\theta \in \mathbb{R}$. Confirm that the triple (a, σ, Π) in the Lévy–Khintchine formula are thus $\sigma = 0$, Π as above and $a = -2sb^{-1}\int_0^b (2\pi)^{-1/2}\mathrm{e}^{-y^2/2}\mathrm{d}y$.

(iii) Taking

$$\mu_s(\mathrm{d}x) = \frac{s}{\sqrt{2\pi x^3}}\mathrm{e}^{sb}\mathrm{e}^{-\frac{1}{2}(s^2x^{-1}+b^2x)}\mathrm{d}x$$

on $x > 0$ show that

$$\begin{aligned}\int_0^\infty \mathrm{e}^{-\lambda x}\mu_s(\mathrm{d}x) &= \mathrm{e}^{bs - s\sqrt{b^2+2\lambda}}\int_0^\infty \frac{s}{\sqrt{2\pi x^3}}\mathrm{e}^{-\frac{1}{2}(\frac{s}{\sqrt{x}} - \sqrt{(b^2+2\lambda)x})^2}\mathrm{d}x \\ &= \mathrm{e}^{bs - s\sqrt{b^2+2\lambda}}\int_0^\infty \sqrt{\frac{2\lambda + b^2}{2\pi u}}\mathrm{e}^{-\frac{1}{2}(\frac{s}{\sqrt{u}} - \sqrt{(b^2+2\lambda)u})^2}\mathrm{d}u.\end{aligned}$$

Hence by adding the last two integrals together deduce that

$$\int_0^\infty \mathrm{e}^{-\lambda x}\mu_s(\mathrm{d}x) = \mathrm{e}^{-s(\sqrt{b^2+2\lambda}-b)}$$

confirming both that $\mu_s(\mathrm{d}x)$ is a probability distribution as well as being the probability distribution of τ_s.

1.7. Show that for a simple Brownian motion $B = \{B_t : t > 0\}$ the first passage process $\tau = \{\tau_s : s > 0\}$ (where $\tau_s = \inf\{t \geq 0 : B_t \geq s\}$) is a stable process with parameters $\alpha = 1/2$ and $\beta = 1$.

1.8 (Proof of Theorem 1.9). As we shall see in this exercise, the proof of Theorem 1.9 follows from the proof of a more general result given by the conclusion of parts (i)–(v) below for random walks.

(i) Suppose that $S = \{S_n : n \geq 0\}$ is a random walk with $S_0 = 0$ and jump distribution μ. By considering the variables $S_k^* := S_n - S_{n-k}$ for $k = 0, 1, ..., n$ and noting that the joint distributions of $(S_0, ..., S_n)$ and $(S_0^*, ..., S_n^*)$ are identical, show that for all $y > 0$ and $n \geq 1$,

$$\begin{aligned}&P(S_n \in \mathrm{d}y \text{ and } S_n > S_j \text{ for } j = 0, ..., n-1) \\ &= P(S_n \in \mathrm{d}y \text{ and } S_j > 0 \text{ for } j = 1, ..., n).\end{aligned}$$

[Hint: it may be helpful to draw a diagram of the path of the first n steps of S and to rotate it by 180°.]

(ii) Define

$$T_0^- = \inf\{n > 0 : S_n \leq 0\} \text{ and } T_0^+ = \inf\{n > 0 : S_n > 0\}.$$

By summing both sides of the equality

$$P(S_1 > 0, ..., S_n > 0, S_{n+1} \in \mathrm{d}x)$$
$$= \int_{(0,\infty)} P(S_1 > 0, ..., S_n > 0, S_n \in \mathrm{d}y)\mu(\mathrm{d}x - y)$$

over n show that for $x \leq 0$,

$$P(S_{T_0^-} \in \mathrm{d}x) = \int_{[0,\infty)} V(\mathrm{d}y)\mu(\mathrm{d}x - y),$$

where for $y \geq 0$,

$$V(\mathrm{d}y) = \delta_0(\mathrm{d}y) + \sum_{n\geq 1} P(H_n \in \mathrm{d}y)$$

and $H = \{H_n : n \geq 0\}$ is a random walk with $H_0 = 0$ and step distribution given by $P(S_{T_0^+} \in \mathrm{d}z)$ for $z \geq 0$.

(iii) Embedded in the Cramér–Lundberg model is a random walk S whose increments are equal in distribution to the distribution of $c\mathbf{e}_\lambda - \xi_1$, where $\mathbf{e}_\lambda$ is an independent exponential random variable with mean $1/\lambda$. Noting (using obvious notation) that $c\mathbf{e}_\lambda$ has the same distribution as $\mathbf{e}_\beta$ where $\beta = \lambda/c$ show that the step distribution of this random walk satisfies

$$\mu(z,\infty) = \left(\int_0^\infty \mathrm{e}^{-\beta u} F(\mathrm{d}u)\right) \mathrm{e}^{-\beta z} \text{ for } z \geq 0$$

and

$$\mu(-\infty, -z) = E(\overline{F}(\mathbf{e}_\beta + z)) \text{ for } z > 0,$$

where $\overline{F}$ is the tail of the distribution function F of ξ_1 and E is expectation with respect to the random variable $\mathbf{e}_\beta$.

(iv) Since upward jumps are exponentially distributed in this random walk, use the lack of memory property to reason that

$$V(\mathrm{d}y) = \delta_0(\mathrm{d}y) + \beta \mathrm{d}y.$$

Hence deduce from part (iii) that

$$P(-S_{T_0^-} > z) = E\left(\overline{F}(\mathbf{e}_\beta) + \int_x^\infty \beta \overline{F}(\mathbf{e}_\beta + z)\mathrm{d}z\right)$$

and so by writing out the latter with the density of the exponential distribution, show that the conclusions of Theorem 1.9 hold.

1.9 (Proof of Theorem 1.11). Suppose that X is a compound Poisson process of the form

$$X_t = t - \sum_{i=1}^{N_t} \xi_i,$$

where the process $N = \{N_t : t \geq 0\}$ is a Poisson process with rate $\lambda > 0$ and $\{\xi_i : i \geq 1\}$ positive, independent and identically distributed with common distribution F having mean μ.

(i) Show by analytic extension from the Lévy–Khintchine formula or otherwise that $\mathbb{E}(\mathrm{e}^{\theta X_t}) = \mathrm{e}^{\psi(\theta)t}$ for all $\theta \geq 0$, where

$$\psi(\theta) = \theta - \lambda \int_{(0,\infty)} (1 - \mathrm{e}^{-\theta x})F(\mathrm{d}x).$$

Show that ψ is strictly convex, is equal to zero at the origin and tends to infinity at infinity. Further, show that $\psi(\theta) = 0$ has one additional root in $[0, \infty)$ other than $\theta = 0$ if and only if $\psi'(0) < 0$.

(ii) Show that $\{\mathrm{e}^{\theta X_t - \psi(\theta)t} : t \geq 0\}$ is a martingale and hence so is $\{\mathrm{e}^{\theta^* X_{t\wedge\tau_x^+}} : t \geq 0\}$ where $\tau_x^+ = \inf\{t > 0 : X_t > x\}$, $x > 0$ and θ^* is the largest root described in the previous part of the question. Show further that

$$\mathbb{P}(\overline{X}_\infty > x) = \mathrm{e}^{-\theta^* x}$$

for all $x > 0$.

(iii) Show that for all $t \geq 0$,

$$\int_0^t \mathbf{1}_{(W_s=0)}\mathrm{d}s = (\overline{X}_t - w) \vee 0,$$

where $W_t = (w \vee \overline{X}_t) - X_t$.

(iv) Deduce that $I := \int_0^\infty \mathbf{1}_{(W_s=0)}\mathrm{d}s = \infty$ if $\lambda\mu \leq 1$.

(v) Assume that $\lambda\mu > 1$. Show that

$$\mathbb{P}(I \in \mathrm{d}x; \tau_w^+ = \infty | W_0 = w) = (1 - \mathrm{e}^{-\theta^* w})\delta_0(\mathrm{d}x).$$

Next use the lack of memory property to deduce that

$$\mathbb{P}(I \in \mathrm{d}x; \tau_w^+ < \infty | W_0 = w) = \theta^* \mathrm{e}^{-\theta^*(w+x)}\mathrm{d}x.$$

1.10. Here we solve a slightly simpler optimal stopping problem than (1.16). Suppose, as in the aforementioned problem, that X is a linear Brownian motion with scaling parameter $\sigma > 0$ and drift $\gamma \in \mathbb{R}$. Fix $K > 0$ and let

$$v(x) = \sup_{a\in\mathbb{R}} \mathbb{E}_x(\mathrm{e}^{-q\tau_a^-}(K - \mathrm{e}^{X_{\tau_a^-}})^+), \tag{1.21}$$

where

$$\tau_a^- = \inf\{t > 0 : X_t < a\}.$$

(i) Following similar arguments to those in Exercise 1.5 shows that $\{\exp\{\theta X_t - \psi(\theta)t\} : t \geq 0\}$ is a martingale, where $\psi(\theta) = \sigma^2\theta^2/2 + \gamma\theta$.
(ii) By considering the martingale in part (i) at the stopping time $t \wedge \tau_x^+$ and then letting $t \uparrow \infty$, deduce that

$$\mathbb{E}(\mathrm{e}^{-q\tau_x^+}) = \mathrm{e}^{-x(\sqrt{\gamma^2+2\sigma^2 q}-\gamma)/\sigma^2}$$

and hence deduce that for $a \geq 0$,

$$\mathbb{E}(\mathrm{e}^{-q\tau_{-a}^-}) = \mathrm{e}^{-a(\sqrt{\gamma^2+2\sigma^2 q}+\gamma)/\sigma^2}.$$

(iii) Let $v(x,a) = \mathbb{E}_x(\mathrm{e}^{-q\tau_{-a}^-}(K - \exp\{X_{\tau_{-a}^-}\}))$. For each fixed x differentiate $v(x,a)$ in the variable a and show that the solution to 1.21 is the same as the solution given in Theorem 1.13.

1.11. In this exercise, we characterise the Laplace exponent of the continuous time Markov branching process Y described in Sect. 1.3.4.

(i) Show that for $\phi > 0$ and $t \geq 0$ there exists some function $u_t(\phi) > 0$ satisfying

$$\mathbb{E}_y(\mathrm{e}^{-\phi Y_t}) = \mathrm{e}^{-y u_t(\phi)},$$

where $y \in \{1, 2, ...\}$.
(ii) Show that for $s, t \geq 0$,

$$u_{t+s}(\phi) = u_s(u_t(\phi)).$$

(iii) Appealing to the infinitesimal behaviour of the Markov chain Y show that

$$\frac{\partial u_t(\phi)}{\partial t} = \psi(u_t(\phi))$$

and $u_0(\phi) = \phi$ where

$$\psi(q) = \lambda \int_{[-1,\infty)} (1 - \mathrm{e}^{-qx}) F(\mathrm{d}x)$$

and F is given in (1.18).

2

The Lévy–Itô Decomposition and Path Structure

The main aim of this chapter is to establish a rigorous understanding of the structure of the paths of Lévy processes. The way we shall do this is to prove the assertion in Theorem 1.6 that given any characteristic exponent, Ψ, belonging to an infinitely divisible distribution, there exists a Lévy process with the same characteristic exponent. This will be done by establishing the so-called Lévy–Itô decomposition which describes the structure of a general Lévy process in terms of three independent auxiliary Lévy processes, each with different types of path behaviour. In doing so it will be necessary to digress temporarily into the theory of Poisson random measures and associated square integrable martingales. Understanding the Lévy–Itô decomposition will allow us to distinguish a number of important general subclasses of Lévy processes according to their path type. The chapter is concluded with a discussion of the interpretation of the Lévy–Itô decomposition in the context of some of the applied probability models mentioned in Chap. 1.

2.1 The Lévy–Itô Decomposition

According to Theorem 1.3, any characteristic exponent Ψ belonging to an infinitely divisible distribution can be written, after some simple reorganisation, in the form

$$\begin{aligned}\Psi(\theta) &= \left\{ \mathrm{i}a\theta + \frac{1}{2}\sigma^2\theta^2 \right\} \\ &\quad + \left\{ \Pi(\mathbb{R}\backslash(-1,1)) \int_{|x|\geq 1} (1 - \mathrm{e}^{\mathrm{i}\theta x}) \frac{\Pi(\mathrm{d}x)}{\Pi(\mathbb{R}\backslash(-1,1))} \right\} \\ &\quad + \left\{ \int_{0<|x|<1} (1 - \mathrm{e}^{\mathrm{i}\theta x} + \mathrm{i}\theta x)\Pi(\mathrm{d}x) \right\} \qquad (2.1)\end{aligned}$$

for all $\theta \in \mathbb{R}$ where $a \in \mathbb{R}$, $\sigma \geq 0$ and Π is a measure on $\mathbb{R}\backslash\{0\}$ satisfying $\int_{\mathbb{R}}(1 \wedge x^2)\Pi(\mathrm{d}x) < \infty$. Note that the latter condition on Π implies that

$\Pi(A) < \infty$ for all Borel A such that 0 is in the interior of A^c and particular that $\Pi(\mathbb{R}\backslash(-1,1)) \in [0,\infty)$. In the case that $\Pi(\mathbb{R}\backslash(-1,1)) = 0$ one should think of the second bracket in (2.1) as absent. Call the three brackets in (2.1) $\Psi^{(1)}$, $\Psi^{(2)}$ and $\Psi^{(3)}$. The essence of the Lévy–Itô decomposition revolves around showing that $\Psi^{(1)}$, $\Psi^{(2)}$ and $\Psi^{(3)}$ all correspond to the characteristic exponents of three different types of Lévy processes. Therefore Ψ may be considered as the characteristic exponent of the independent sum of these three Lévy processes which is again a Lévy process (cf. Exercise 1.1). Indeed, as we have already seen in Chap. 1, $\Psi^{(1)}$ and $\Psi^{(2)}$ correspond, respectively, to a linear Brownian motion with drift, $X^{(1)} = \{X_t^{(1)} : t \geq 0\}$ where

$$X_t^{(1)} = \sigma B_t - at, \ t \geq 0 \tag{2.2}$$

and a compound Poisson process, say $X^{(2)} = \{X_t^{(2)} : t \geq 0\}$, where,

$$X_t^{(2)} = \sum_{i=1}^{N_t} \xi_i, \ t \geq 0, \tag{2.3}$$

$\{N_t : t \geq 0\}$ is a Poisson process with rate $\Pi(\mathbb{R}\backslash(-1,1))$ and $\{\xi_i : i \geq 1\}$ are independent and identically distributed with distribution $\Pi(\mathrm{d}x)/\Pi(\mathbb{R}\backslash(-1,1))$ concentrated on $\{x : |x| \geq 1\}$ (unless $\Pi(\mathbb{R}\backslash(-1,1)) = 0$ in which case $X^{(2)}$ is the process which is identically zero).

The proof of existence of a Lévy process with characteristic exponent given by (2.1) thus boils down to showing the existence of a Lévy process, $X^{(3)}$, whose characteristic exponent is given by $\Psi^{(3)}$. Noting that

$$\begin{aligned}
&\int_{0<|x|<1} (1 - \mathrm{e}^{\mathrm{i}\theta x} + \mathrm{i}\theta x)\Pi(\mathrm{d}x) \\
&= \sum_{n\geq 0} \left\{ \lambda_n \int_{2^{-(n+1)}\leq|x|<2^{-n}} (1 - \mathrm{e}^{\mathrm{i}\theta x})F_n(\mathrm{d}x) \right. \\
&\qquad\qquad \left. + \mathrm{i}\theta\lambda_n \left(\int_{2^{-(n+1)}\leq|x|<2^{-n}} xF_n(\mathrm{d}x) \right) \right\},
\end{aligned} \tag{2.4}$$

where $\lambda_n = \Pi(\{x : 2^{-(n+1)} \leq |x| < 2^{-n}\})$ and $F_n(\mathrm{d}x) = \Pi(\mathrm{d}x)/\lambda_n$ (again with the understanding that the nth integral is absent if $\lambda_n = 0$). It would appear from (2.4) that the process $X^{(3)}$ consists of the superposition of (at most) a countable number of independent compound Poisson processes with different arrival rates and additional linear drift. To understand the mathematical sense of this superposition we shall need to establish some facts concerning Poisson random measures and related martingales. Hence Sects. 2.2 and 2.3 are dedicated to the study of the latter processes. The precise construction of $X^{(3)}$ is given in Sect. 2.5.

The identification of a Lévy process, X as the independent sum of processes $X^{(1)}$, $X^{(2)}$ and $X^{(3)}$ is attributed to Lévy (1954) and Itô (1942) (see also Itô (2004)) and is thus known as the *Lévy–Itô decomposition.* Formally speaking and in a little more detail we quote the Lévy–Itô decomposition in the form of a theorem.

Theorem 2.1. *(Lévy–Itô decomposition) Given any $a \in \mathbb{R}$, $\sigma \geq 0$ and measure Π concentrated on $\mathbb{R}\backslash\{0\}$ satisfying*

$$\int_{\mathbb{R}} (1 \wedge x^2)\Pi(\mathrm{d}x) < \infty,$$

there exists a probability space on which three independent Lévy processes exist, $X^{(1)}$, $X^{(2)}$ and $X^{(3)}$ where $X^{(1)}$ is a linear Brownian motion with drift given by (2.2), $X^{(2)}$ is a compound Poisson process given by (2.3) and $X^{(3)}$ is a square integrable martingale with an almost surely countable number of jumps on each finite time interval which are of magnitude less than unity and with characteristic exponent given by $\Psi^{(3)}$. By taking $X = X^{(1)} + X^{(2)} + X^{(3)}$ we see that the conclusion of Theorem 1.6 holds, that there exists a probability space on which a Lévy process is defined with characteristic exponent

$$\Psi(\theta) = a\mathrm{i}\theta + \frac{1}{2}\sigma^2\theta^2 + \int_{\mathbb{R}} (1 - \mathrm{e}^{\mathrm{i}\theta x} + \mathrm{i}\theta x \mathbf{1}_{(|x|<1)})\Pi(\mathrm{d}x) \tag{2.5}$$

for $\theta \in \mathbb{R}$.

2.2 Poisson Random Measures

Poisson random measures turn out to be the right mathematical mechanism to describe the jump structure embedded in any Lévy process. Before engaging in an abstract study of Poisson random measures however, we give a rough idea of how they are related to the jump structure of Lévy processes by considering the less complicated case of a compound Poisson process.

Suppose then that $X = \{X_t : t \geq 0\}$ is a compound Poisson process with a drift taking the form

$$X_t = dt + \sum_{i=1}^{N_t} \xi_i, \quad t \geq 0,$$

where $d \in \mathbb{R}$ and as usual $\{\xi_i : i \geq 1\}$ are independent and identically distributed random variable with common distribution function F. Further, let $\{T_i : i \geq 0\}$ be the times of arrival of the Poisson process $N = \{N_t : t \geq 0\}$ with rate $\lambda > 0$. See Fig. 2.1.

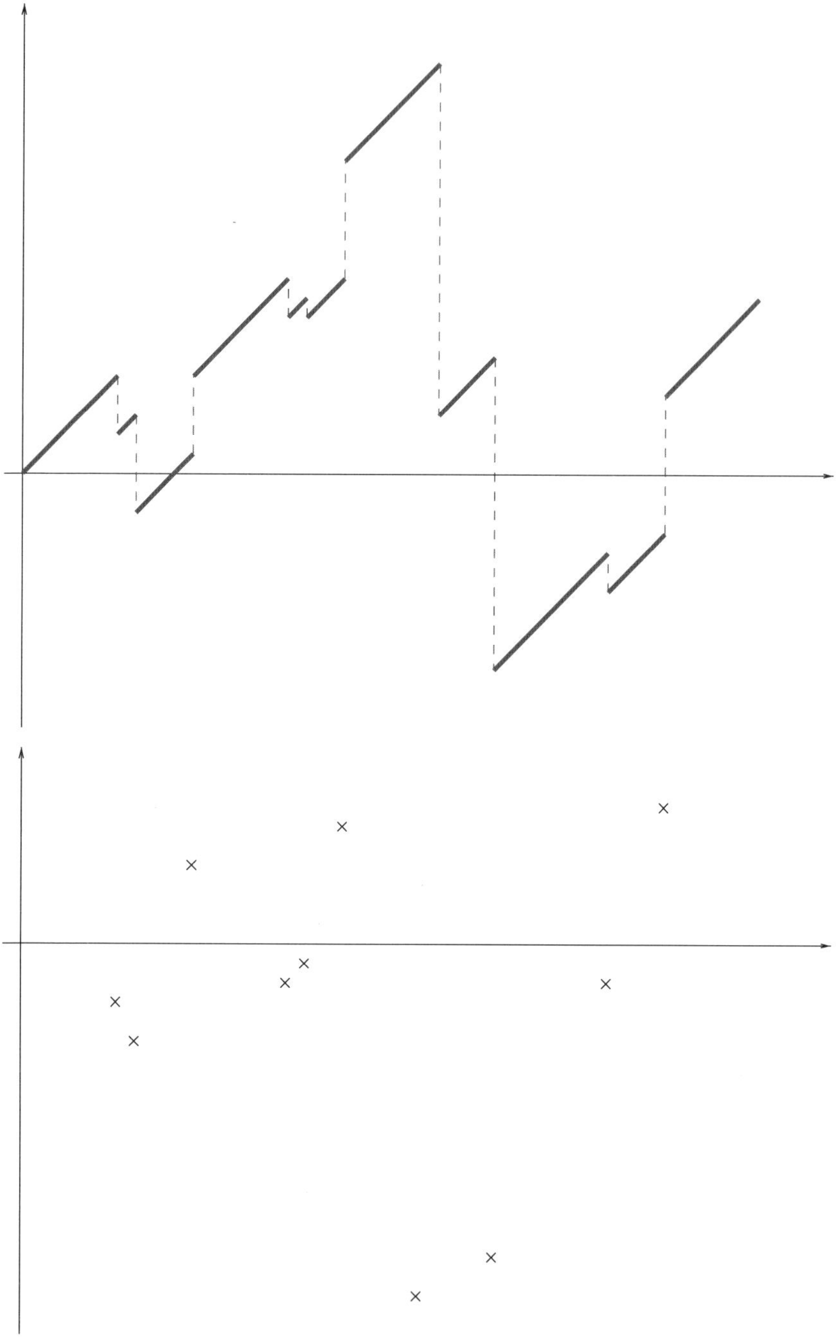

Fig. 2.1. The initial period of a sample path of a compound Poisson process with drift $\{X_t : t \geq 0\}$ and the field of points it generates.

Suppose now that we pick any set in $A \in \mathcal{B}[0,\infty) \times \mathcal{B}(\mathbb{R}\backslash\{0\})$. Define

$$N(A) = \#\{i \geq 0 : (T_i, \xi_i) \in A\} = \sum_{i=1}^{\infty} \mathbf{1}_{((T_i,\xi_i)\in A)}. \tag{2.6}$$

Clearly since X experiences an almost surely finite number of jumps over a finite period of time it follows that $N(A) < \infty$ almost surely where for $t \geq 0$, $A \subseteq \mathcal{B}[0,t) \times \mathcal{B}(\mathbb{R}\backslash\{0\})$.

Lemma 2.2. *Choose $k \geq 1$. If $A_1, ..., A_k$ are disjoint sets in $\mathcal{B}[0,\infty) \times \mathcal{B}(\mathbb{R}\backslash\{0\})$ then $N(A_1), ..., N(A_k)$ are mutually independent and Poisson distributed with parameters $\lambda_i := \lambda \int_{A_i} \mathrm{d}t \times F(\mathrm{d}x)$, respectively. Further, for $\mathbb{P}$-almost every realisation of X, $N : \mathcal{B}[0,\infty) \times \mathcal{B}(\mathbb{R}\backslash\{0\}) \to \{0,1,2,...\} \cup \{\infty\}$ is a measure.*[1]

Proof. First recall a classic result concerning the Poisson process $\{N_t : t \geq 0\}$. That is, the law of $\{T_1, ..., T_n\}$ conditional on the event $\{N_t = n\}$ is the same as the law of an ordered independent sample of size n from the uniform distribution on $[0,t]$. (Exercise 2.2 has the full details). This latter fact together with the fact that the variables $\{\xi_i : i = 1, ..., k\}$ are independent and identically distributed with common law F implies that conditional on $\{N_t = n\}$, the joint law of the pairs $\{(T_i, \xi_i) : i = 1, ..., k\}$ is that of n independent bivariate random variables with common distribution $t^{-1}\mathrm{d}s \times F(\mathrm{d}x)$ on $[0,t] \times \mathbb{R}$ ordered in time. In particular, for any $A \in \mathcal{B}[0,t] \times \mathcal{B}(\mathbb{R}\backslash\{0\})$, the random variable $N(A)$ conditional on the event $\{N_t = n\}$ is a Binomial random variable with probability of success given by $\int_A t^{-1}\mathrm{d}s \times F(\mathrm{d}x)$. A generalisation of this latter statement for the k-tuple $(N(A_1), ..., N(A_k))$, where $A_1, ..., A_k$ are mutually disjoint and chosen from $\mathcal{B}[0,t] \times \mathcal{B}(\mathbb{R})$, is the following. Suppose that $A_0 = \{[0,t] \times \mathbb{R}\}\backslash\{A_1 \cup ... \cup A_k\}$, $\sum_{i=1}^k n_i \leq n$, $n_0 = n - \sum_{i=1}^k n_i$ and $\lambda_0 = \int_{A_0} \lambda \mathrm{d}s \times F(\mathrm{d}x) = \lambda t - \lambda_1 - ... - \lambda_k$ then $(N(A_1), ..., N(A_k))$ has the following multinomial law

$$\begin{aligned}&\mathbb{P}(N(A_1) = n_1, ..., N(A_k) = n_k | N_t = n) \\ &= \frac{n!}{n_0! n_1! ... n_k!} \prod_{i=0}^{k} \left(\frac{\lambda_i}{\lambda t}\right)^{n_i}.\end{aligned}$$

[1] Specifically, $\mathbb{P}$-almost surely, $N(\emptyset) = 0$ and for disjoint $A_1, A_2, ...$ in $\mathcal{B}[0,\infty) \times \mathcal{B}(\mathbb{R}\backslash\{0\})$ we have

$$N\left(\bigcup_{i\geq 1} A_i\right) = \sum_{i\geq 1} N(A_i).$$

Integrating out the conditioning on N_t it follows that

$$\begin{aligned}
&\mathbb{P}(N(A_1) = n_1, ..., N(A_k) = n_k) \\
&= \sum_{n \geq \Sigma_{i=1}^k n_i} \mathrm{e}^{-\lambda t} \frac{(\lambda t)^n}{n!} \frac{n!}{n_0! n_1! ... n_k!} \prod_{i=0}^{k} \left(\frac{\lambda_i}{\lambda t} \right)^{n_i} \\
&= \sum_{n \geq \Sigma_{i=1}^k n_i} \mathrm{e}^{-\lambda_0} \frac{(\lambda_0)^{(n - \Sigma_{i=1}^k n_i)}}{(n - \Sigma_{i=1}^k n_i)!} \left(\prod_{i=1}^{k} \mathrm{e}^{-\lambda_i} \frac{(\lambda_i)^{n_i}}{n_i!} \right) \\
&= \prod_{i=1}^{k} \mathrm{e}^{-\lambda_i} \frac{(\lambda_i)^{n_i}}{n_i!}
\end{aligned}$$

showing that $N(A_1), ..., N(A_k)$ are independent and Poisson distributed as claimed.

To complete the proof for arbitrary disjoint $A_1, ..., A_k$, in other words to remove the restriction on the time horizon, suppose without loss of generality that A_1 belongs to $\mathcal{B}[0, \infty) \times \mathcal{B}(\mathbb{R}\backslash\{0\})$. Write A_1 as a countable union of disjoint sets and recall that the sum of an independent sequence of Poisson random variables is Poisson distributed with the sum of their rates. If we agree that a Poisson random variable with infinite rate is infinite with probability one (see Exercise 2.1), then the proof is complete.

Finally the fact that N is a measure $\mathbb{P}$-almost surely follows immediately from its definition. □

Lemma 2.2 shows that $\mathrm{N} : \mathcal{B}[0, \infty) \times \mathcal{B}(\mathbb{R}\backslash\{0\}) \to \{0, 1, ...\} \cup \{\infty\}$ fulfils the following definition of a Poisson random measure.

Definition 2.3 (Poisson random measure). *In what follows we shall assume that $(S, \mathcal{S}, \eta)$ is an arbitrary σ-finite measure space. Let $\mathrm{N} : \mathcal{S} \to \{0, 1, 2, ...\} \cup \{\infty\}$ in such a way that the family $\{\mathrm{N}(A) : A \in \mathcal{S}\}$ are random variables defined on the probability space $(\Omega, \mathcal{F}, \mathbb{P})$. Then N is called a Poisson random measure on $(S, \mathcal{S}, \eta)$ (or sometimes a Poisson random measure on S with intensity η) if*

(i) for mutually disjoint $A_1, ..., A_n$ in $\mathcal{S}$, the variables $N(A_1), ..., N(A_n)$ are independent,
(ii) for each $A \in \mathcal{S}$, $N(A)$ is Poisson distributed with parameter $\eta(A)$ (here we allow $0 \leq \eta(A) \leq \infty$),
(iii) $\mathbb{P}$-almost surely N is a measure.

In the second condition we note that if $\eta(A) = 0$ then it is understood that $N(A) = 0$ with probability one and if $\eta(A) = \infty$ then $N(A)$ is infinite with probability one.

In the case of (2.6) we have $S = [0,\infty) \times \{\mathbb{R}\backslash\{0\}\}$ and $\mathrm{d}\eta = \lambda \mathrm{d}t \times \mathrm{d}F$. Note also that by construction of the compound Poisson process on some probability space $(\Omega, \mathcal{F}, \mathbb{P})$, all of the random variables $\mathbf{1}_{((T_i,\xi_i)\in A)}$ for each $A \in \mathcal{B}[0,\infty) \times \mathcal{B}(\mathbb{R}\backslash\{0\})$ are $\mathcal{F}$-measurable and hence so are the variables $N(A)$.

We complete this section by proving that a Poisson random measure as defined above exists. This is done in Theorem 2.4, the proof of which has many similarities to the proof of Lemma 2.2.

Theorem 2.4. *There exists a Poisson random measure N as in Definition 2.3.*

Proof. First suppose that S is such that $\eta(S) < \infty$. There exists a standard construction of an infinite product space, say $(\Omega, \mathcal{F}, P)$ on which the independent random variables

$$\mathrm{N} \text{ and } \{\upsilon_1, \upsilon_2, ...\}$$

are collectively defined such that N has a Poisson distribution with parameter $\eta(S)$ and each of the variables υ_i have distribution $\eta(\mathrm{d}x)/\eta(S)$ on S. Define for each $A \in \mathcal{S}$,

$$N(A) = \sum_{i=1}^{\mathrm{N}} \mathbf{1}_{(\upsilon_i \in A)}$$

so that $\mathrm{N} = N(S)$. As for each $A \in \mathcal{S}$ and $i \geq 1$, the random variables $\mathbf{1}_{(\upsilon_i \in A)}$ are $\mathcal{F}$-measurable, then so are the random variables $N(A)$.

When presented with mutually disjoint sets of S, say $A_1, ..., A_k$, a calculation identical to the one given in the proof of Lemma 2.2 shows again that

$$P(N(A_1) = n_1, ..., N(A_k) = n_k) = \prod_{i=1}^{k} \mathrm{e}^{-\eta(A_i)} \frac{\eta(A_i)^{n_i}}{n_i!}$$

for non-negative integers $n_1, n_2, ..., n_k$. Returning to Definition 2.3 it is now clear from the previous calculation together with the assumption that η is non-atomic that conditions (i)–(iii) are met by N. In particular, similar to the case dealt with in Lemma 2.2, the third condition is automatic as N is a counting measure by definition.

Next we turn to the case that $(S, \mathcal{S}, \eta)$ is a σ-finite measure space. The meaning of σ-finite is that there exists a countable disjoint exhaustive sequence of sets $B_1, B_2, ...$ in S such that $0 < \eta(B_i) < \infty$ for each $i \geq 1$. Define the measures $\eta_i(\cdot) = \eta(\cdot \cap B_i)$ for each $i \geq 1$. The first part of this proof shows that for each $i \geq 1$ there exists some probability space $(\Omega_i, \mathcal{F}_i, \mathbb{P}_i)$ on which we can define a Poisson random measure, say N_i, in $(B_i, \mathcal{S} \cap B_i, \eta_i)$ where $\mathcal{S} \cap B_i = \{A \cap B_i : A \in \mathcal{S}\}$ (the reader should verify easily that $\mathcal{S} \cap B_i$ is indeed a sigma algebra on B_i). The idea is now to show that

$$N(\cdot) = \sum_{i \geq 1} N_i(\cdot \cap B_i)$$

is a Poisson random measure on S with intensity η defined on the product space

$$(\Omega, \mathcal{F}, \mathbb{P}) := \prod_{i \geq 1} (\Omega_i, \mathcal{F}_i, \mathbb{P}_i).$$

First note that it is again immediate from its definition that N is $\mathbb{P}$-almost surely a measure. In particular with the help of Fubini's Theorem, for disjoint $A_1, A_2, ...$, we have

$$\begin{aligned} N\left(\bigcup_{j\geq 1} A_j\right) &= \sum_{i\geq 1} N_i \left(\bigcup_{j} A_j \cap B_i\right) = \sum_{i\geq 1}\sum_{j\geq 1} N(A_j \cap B_i) \\ &= \sum_{j\geq 0}\sum_{i\geq 1} N(A_j \cap B_i) \\ &= \sum_{j\geq 0} N(A_j). \end{aligned}$$

Next, for each $i \geq 1$. we have that $N_i(A \cap B_i)$ is Poisson distributed with parameter $\eta_i(A \cap B_i)$; Exercise 2.1 tells us that under $\mathbb{P}$ the random variable $N(A)$ is Poisson distributed with parameter $\eta(A)$. The proof is complete once we show that for disjoint $A_1, ..., A_k$ in $\mathcal{S}$ the variables $N(A_1), ..., N(A_k)$ are all independent under $\mathbb{P}$. However this is obvious since the double array of variables

$$\{N_i(A_j \cap B_i) : i = 1, 2, ... \text{ and } j = 1, ..., k\}$$

is also an independent sequence of variables. □

From the construction of the Poisson random measure, the following two corollaries should be clear.

Corollary 2.5. *Suppose that N is a Poisson random measure on $(S, \mathcal{S}, \eta)$. Then for each $A \in \mathcal{S}$, $N(\cdot \cap A)$ is a Poisson random measure on $(S \cap A, \mathcal{S} \cap A, \eta(\cdot \cap A))$. Further, if $A, B \in \mathcal{S}$ and $A \cap B = \emptyset$ then $N(\cdot \cap A)$ and $N(\cdot \cap B)$ are independent.*

Corollary 2.6. *Suppose that N is a Poisson random measure on $(S, \mathcal{S}, \eta)$, then the support of N is $\mathbb{P}$-almost surely countable. If in addition, η is a finite measure, then the support is $\mathbb{P}$-almost surely finite.*

Finally, note that if η is a measure with an atom at, say, the singleton $s \in S$, then it is intuitively obvious from the construction of N in the proof of Theorem 2.4 that $\mathbb{P}(N(\{s\}) \geq 1) > 0$. Conversely, if η has no atoms then $\mathbb{P}(N(\{s\}) = 0) = 1$ for all singletons $s \in S$. For further discussion on this point, the reader is referred to Kingman (1993).

2.3 Functionals of Poisson Random Measures

Suppose as in Sect. 2.2 that N is a Poisson random measure on the measure space $(S, \mathcal{S}, \eta)$. As N is $\mathbb{P}$-almost surely a measure, the classical theory of Lebesgue integration now allows us to talk of

$$\int_S f(x)N(\mathrm{d}x) \tag{2.7}$$

as a well defined $[0, \infty]$-valued random variable for measurable functions $f : S \to [0, \infty)$. Further (2.7) is still well defined and $[-\infty, \infty]$ valued for signed measurable f provided at most one the integrals of $f^+ = f \vee 0$ and $f^- = (-f) \vee 0$ is infinite. Note however, from the construction of the Poisson random measure in the proof of Theorem 2.4 the integral in (2.7) may be interpreted as equal in law to

$$\sum_{\upsilon \in \Upsilon} f(\upsilon) m_\upsilon,$$

where Υ is the support of N (which from Corollary 2.6) is countable, and m_υ is the multiplicity of points at υ. Recalling the remarks following Corollary 2.6 if η has no atoms then $m_\upsilon = 1$ for all $\upsilon \in N$.

We move to the main theorem of this section which is attributed to Kingman (1967) who himself accredits the earlier work of Campbell (1909, 1910).

Theorem 2.7. *Suppose that N is a Poisson random measure on $(S, \mathcal{S}, \eta)$. Let $f : S \to \mathbb{R}$ be a measurable function.*

(i) Then

$$X = \int f(x)N(\mathrm{d}x)$$

is almost surely absolutely convergent if and only if

$$\int_S (1 \wedge |f(x)|)\eta(\mathrm{d}x) < \infty. \tag{2.8}$$

(ii) When condition (2.8) holds, then (with $\mathbb{E}$ as expectation with respect to $\mathbb{P}$)

$$\mathbb{E}\left(\mathrm{e}^{\mathrm{i}\beta X}\right) = \exp\left\{-\int_S (1 - \mathrm{e}^{\mathrm{i}\beta f(x)})\eta(\mathrm{d}x)\right\} \tag{2.9}$$

for any $\beta \in \mathbb{R}$.

(iii) Further

$$\mathbb{E}(X) = \int_S f(x)\eta(\mathrm{d}x) \text{ when } \int_S |f(x)|\eta(\mathrm{d}x) < \infty \tag{2.10}$$

and

$$\mathbb{E}\left(X^2\right) = \int_S f(x)^2\eta(\mathrm{d}x) + \left(\int_S f(x)\eta(\mathrm{d}x)\right)^2 \text{ when } \int_S f(x)^2\eta(\mathrm{d}x) < \infty. \tag{2.11}$$

Proof. (i) We begin by defining simple functions to be those of the form

$$f(x) = \sum_{i=1}^{n} f_i \mathbf{1}_{A_i}(x),$$

where f_i is constant and $\{A_i : i = 1, ..., n\}$ are disjoint sets in $\mathcal{S}$ and further $\eta(A_1 \cup ... \cup A_n) < \infty$.

For such functions we have

$$X = \sum_{i=1}^{n} f_i N(A_i)$$

which is clearly finite with probability one since each $N(A_i)$ has a Poisson distribution with parameter $\eta(A_i) < \infty$. Now for any $\theta > 0$ we have (using the well known fact that the characteristic function of a Poisson distribution with parameter $\lambda > 0$ is $\exp\{-\lambda(1 - \mathrm{e}^{-\theta})\}$),

$$\begin{aligned}
\mathbb{E}\left(\mathrm{e}^{-\theta X}\right) &= \prod_{i=1}^{n} \mathbb{E}\left(\mathrm{e}^{-\theta f_i N(A_i)}\right) \\
&= \prod_{i=1}^{n} \exp\left\{-(1 - \mathrm{e}^{-\theta f_i})\eta(A_i)\right\} \\
&= \exp\left\{-\sum_{i=1}^{n}(1 - \mathrm{e}^{-\theta f_i})\eta(A_i)\right\}.
\end{aligned}$$

Since $1 - \mathrm{e}^{-\theta f(x)} = 0$ on $S \backslash (A_1 \cup ... \cup A_n)$ we may thus conclude that

$$\mathbb{E}\left(\mathrm{e}^{-\theta X}\right) = \exp\left\{-\int_S (1 - \mathrm{e}^{-\theta f(x)})\eta(\mathrm{d}x)\right\}.$$

Next we establish the above equality for a general positive measurable f. For the latter class of f, there exists a pointwise increasing sequence of positive simple functions $\{f_n : n \geq 0\}$ such that $\lim_{n\uparrow\infty} f_n = f$, where the limit is also understood in the pointwise sense. Since N is an almost surely σ-finite measure, the monotone convergence theorem now implies that

$$\lim_{n\uparrow\infty} \int f_n(x) N(\mathrm{d}x) = \int f(x) N(\mathrm{d}x) = X$$

almost surely. An application of bounded convergence followed by an application of monotone convergence again tells us that for any $\theta > 0$,

$$\begin{aligned}
\mathbb{E}\left(\mathrm{e}^{-\theta X}\right) &= \mathbb{E}\left(\exp\left\{-\theta \int f(x) N(\mathrm{d}x)\right\}\right) \\
&= \lim_{n\uparrow\infty} \mathbb{E}\left(\exp\left\{-\theta \int f_n(x) N(\mathrm{d}x)\right\}\right)
\end{aligned}$$

$$= \lim_{n\uparrow\infty} \exp\left\{-\int_S (1-\mathrm{e}^{-\theta f_n(x)})\eta(\mathrm{d}x)\right\}$$

$$= \exp\left\{-\int_S (1-\mathrm{e}^{-\theta f(x)})\eta(\mathrm{d}x)\right\}. \tag{2.12}$$

Note that the integral on the right-hand side of (2.12) is either infinite for all $\theta > 0$ or finite for all $\theta > 0$ accordingly, with the cases that $X = \infty$ with probability one or $X = \infty$ with probability less than one, respectively. If $\int_S (1-\mathrm{e}^{-\theta f(x)})\eta(\mathrm{d}x) < \infty$ for all $\theta > 0$ then as, for each $x \in S$, $(1-\mathrm{e}^{-\theta f(x)}) \leq (1-\mathrm{e}^{-f(x)})$ for all $0 < \theta < 1$, dominated convergence implies that

$$\lim_{\theta\downarrow 0} \int_S (1-\mathrm{e}^{-\theta f(x)})\eta(\mathrm{d}x) = 0$$

and hence dominated convergence applied again in (2.12) as $\theta \downarrow 0$ tells us that $\mathbb{P}(X = \infty) = 0$.

In conclusion we have that $X < \infty$ if and only if $\int_S (1-\mathrm{e}^{-\theta f(x)})\eta(\mathrm{d}x) < \infty$ for all $\theta > 0$ and it can be checked (see Exercise 2.3) that the latter happens if and only if

$$\int_S (1 \wedge f(x))\eta(\mathrm{d}x) < \infty.$$

Note that both sides of (2.12) may be analytically extended by replacing θ by $\theta - \mathrm{i}\beta$ for $\beta \in \mathbb{R}$. Then taking limits on both sides as $\theta \downarrow 0$ we deduce (2.9).

Now we shall remove the restriction that f is positive. Henceforth assume as in the statement of the theorem that f is a measurable function. We may write $f = f^+ - f^-$ where $f^+ = f \vee 0$ and $f^- = (-f) \vee 0$ are both measurable. The sum X can be written $X_+ - X_-$ where

$$X_+ = \int_S f(x)N_+(\mathrm{d}x) \text{ and } X_- = \int_S f(x)N_-(\mathrm{d}x)$$

and $N_+ = N(\cdot \cap \{x \in S : f(x) \geq 0)$ and $N_- = N(\cdot \cap \{x \in S : f(x) < 0)$. From Corollary 2.5 we know that N_+ and N_- are both Poisson random measures with respective intensities $\eta(\cdot \cap \{f \geq 0\})$ and $\eta(\cdot \cap \{f < 0\})$. Further, they are independent and hence the same is true of X_+ and X_-. It is now clear that X converges absolutely almost surely if and only if X_+ and X_- are convergent. The analysis of the case when f is positive applied to the sums X_+ and X_- now tells us that absolute convergence of X occurs if and only if

$$\int_S (1 \wedge |f(x)|)\eta(\mathrm{d}x) < \infty \tag{2.13}$$

and the proof of (i) is complete.

To complete the proof of (ii), assume that (2.13) holds. Using the independence of X_+ and X_- as well as the conclusion of part (i) we have that for any $\theta \in \mathbb{R}$,

$$\begin{aligned}
\mathbb{E}\left(\mathrm{e}^{\mathrm{i}\theta X}\right) &= \mathbb{E}\left(\mathrm{e}^{\mathrm{i}\theta X_+}\right)\mathbb{E}\left(\mathrm{e}^{-\mathrm{i}\theta X_-}\right) \\
&= \exp\left\{-\int_{\{f>0\}}(1-\mathrm{e}^{\mathrm{i}\theta f^+(x)})\eta(\mathrm{d}x)\right\} \\
&\quad\times \exp\left\{-\int_{\{f<0\}}(1-\mathrm{e}^{-\mathrm{i}\theta f^-(x)})\eta(\mathrm{d}x)\right\} \\
&= \exp\left\{-\int_S(1-\mathrm{e}^{\mathrm{i}\theta f(x)})\eta(\mathrm{d}x)\right\}
\end{aligned}$$

and the proof of (ii) is complete.

Part (iii) is dealt with similarly as in the above treatment; that is first by considering positive, simple f, then extending to positive measurable f and then to a general measurable f by considering its positive and negative parts. The proof is left to the reader. □

2.4 Square Integrable Martingales

We shall predominantly use the identities in the Theorem 2.7 for Poisson random measures on $([0,\infty)\times\mathbb{R}, \mathcal{B}[0,\infty)\times\mathcal{B}(\mathbb{R}), \mathrm{d}t\times\Pi(\mathrm{d}x))$ where Π is a measure concentrated on $\mathbb{R}\backslash\{0\}$. We shall be interested in integrals of the form

$$\int_{[0,t]}\int_B xN(\mathrm{d}s\times\mathrm{d}x), \tag{2.14}$$

where $B\in\mathcal{B}(\mathbb{R})$. The relevant integrals appearing in (2.8)–(2.11) for the above functional of the given Poisson random measure can now be checked to take the form

$$t\int_B(1\wedge|x|)\Pi(\mathrm{d}x),\ t\int_B(1-\mathrm{e}^{\mathrm{i}\beta x})\Pi(\mathrm{d}x),$$
$$t\int_B|x|\Pi(\mathrm{d}x),\quad \text{and } t\int_B x^2\Pi(\mathrm{d}x),$$

with the appearance of the factor t in front of each of the integrals being a consequence of the involvement of Lebesgue measure in the intensity of N. The following two lemmas capture the context in which we use sums of the form (2.14). The first may be considered as a converse to Lemma 2.2 and the second shows the relationship with martingales.

Lemma 2.8. *Suppose that N is a Poisson random measure on $([0,\infty)\times\mathbb{R}, \mathcal{B}[0,\infty)\times\mathcal{B}(\mathbb{R}), \mathrm{d}t\times\Pi(\mathrm{d}x))$ where Π is a measure concentrated on $\mathbb{R}\backslash\{0\}$ and $B\in\mathcal{B}(\mathbb{R})$ such that $0<\Pi(B)<\infty$. Then*

$$X_t := \int_{[0,t]} \int_B x N(\mathrm{d}s \times \mathrm{d}x), \ t \geq 0$$

is a compound Poisson process with arrival rate $\Pi(B)$ and jump distribution $\Pi(B)^{-1}\Pi(\mathrm{d}x)|_B$.

Proof. First note that since $\Pi(B) < \infty$ by assumption, from Corollary 2.6 we know that almost surely, X_t may be written as the sum over finite number of points for each $t > 0$. This explains why $X = \{X_t : t \geq 0\}$ is right continuous with left limits. (One may also see finiteness of X_t from Theorem 2.7 (i)). Next note that for all $0 \leq s < t < \infty$,

$$X_t - X_s = \int_{(s,t]} \int_B x N(\mathrm{d}s \times \mathrm{d}x)$$

which is independent of $\{X_u : u \leq s\}$ as N gives independent counts over disjoint regions. Further, according to Theorem 2.7 (ii) we have that for all $\theta \in \mathbb{R}$,

$$\mathbb{E}(\mathrm{e}^{\mathrm{i}\theta X_t}) = \exp\left\{-t \int_B (1 - \mathrm{e}^{\mathrm{i}\theta x})\Pi(\mathrm{d}x)\right\}. \tag{2.15}$$

When put together with the fact that X has independent increments this shows that

$$\begin{aligned}\mathbb{E}\left(\mathrm{e}^{\mathrm{i}\theta(X_t - X_s)}\right) &= \frac{\mathbb{E}(\mathrm{e}^{\mathrm{i}\theta X_t})}{\mathbb{E}(\mathrm{e}^{\mathrm{i}\theta X_s})} \\ &= \exp\left\{-(t-s)\int_B (1 - \mathrm{e}^{\mathrm{i}\theta x})\Pi(\mathrm{d}x)\right\} \\ &= \mathbb{E}\left(\mathrm{e}^{\mathrm{i}\theta X_{t-s}}\right)\end{aligned}$$

and hence that increments are stationary. Finally, we identify from (2.15) that the Lévy–Khintchine exponent of X corresponds to that of a compound Poisson process with jump distribution and arrival rate given by $\Pi(B)^{-1}\Pi(\mathrm{d}x)|_B$ and $\Pi(B)$, respectively. Since two Lévy processes with the same increment distributions must be equal in law (as all the finite dimensional distributions are determined by the increment distributions and this in turn determines the law of the process) then the Lévy process X must be a compound Poisson process. □

Lemma 2.9. *Suppose that N and B are as in the previous lemma with the additional assumption that $\int_B |x|\Pi(\mathrm{d}x) < \infty$.*

(i) The compound Poisson process with drift

$$M_t := \int_{[0,t]} \int_B x N(\mathrm{d}s \times \mathrm{d}x) - t\int_B x\Pi(\mathrm{d}x), \ t \geq 0$$

is a $\mathbb{P}$-martingale with respect to the filtration

$$\mathcal{F}_t = \sigma\left(N(A) : A \in \mathcal{B}[0,t] \times \mathcal{B}(\mathbb{R})\right), \ t > 0. \tag{2.16}$$

(ii) If further, $\int_B x^2 \Pi(\mathrm{d}x) < \infty$ then it is a square integrable martingale.

Proof. (i) First note that the process $M = \{M_t : t \geq 0\}$ is adapted to the filtration $\{\mathcal{F}_t : t \geq 0\}$. Next note that for each $t > 0$,

$$\mathbb{E}(|M_t|) \leq \mathbb{E}\left(\int_{[0,t]} \int_B |x| N(\mathrm{d}s \times \mathrm{d}x) + t \int_B |x| \Pi(\mathrm{d}x)\right)$$

which, from Theorem 2.7 (iii), is finite because $\int_B |x|\Pi(\mathrm{d}x)$ is. Next use the fact that M has stationary independent increments to deduce that for $0 \leq s \leq t < \infty$,

$$\begin{aligned}
&\mathbb{E}(M_t - M_s|\mathcal{F}_s) \\
&= \mathbb{E}(M_{t-s}) \\
&= \mathbb{E}\left(\int_{[s,t]} \int_B x N(\mathrm{d}s \times \mathrm{d}x)\right) - (t-s) \int_B x \Pi(\mathrm{d}x) \\
&= 0
\end{aligned}$$

where in the final equality we have used Theorem 2.7 (iii) again.

(ii) To see that M is square integrable, we may yet again appeal to Theorem 2.7 (iii) together with the assumption that $\int_B x^2 \Pi(\mathrm{d}x) < \infty$ to deduce that

$$\begin{aligned}
\mathbb{E}\left(\left\{M_t + t\int_B x\Pi(\mathrm{d}x)\right\}^2\right) &= t\int_B x^2 \Pi(\mathrm{d}x) \\
&\quad + t^2 \left(\int_B x \Pi(\mathrm{d}x)\right)^2.
\end{aligned}$$

Recalling from the martingale property that $\mathbb{E}(M_t) = 0$, it follows by developing the left-hand side in the previous display that

$$\mathbb{E}(M_t^2) = t \int_B x^2 \Pi(\mathrm{d}x) < \infty$$

as required. □

The conditions in both Lemmas 2.8 and 2.9 mean that we may consider sets, for example, of the form $B_\varepsilon := (-1, -\varepsilon) \cup (\varepsilon, 1)$ for any $\varepsilon \in (0, 1)$. However it is not necessarily the case that we may consider sets of the form $B = (-1, 0) \cup (0, 1)$. Consider for example the case that $\Pi(\mathrm{d}x) = \mathbf{1}_{(x>0)} x^{-(1+\alpha)} \mathrm{d}x + \mathbf{1}_{(x<0)} |x|^{-(1+\alpha)}$ for $\alpha \in (1, 2)$. In this case, we have that $\int_B |x| \Pi(\mathrm{d}x) = \infty$

whereas $\int_B x^2\Pi(\mathrm{d}x) < \infty$. It will turn out to be quite important in the proof of the Lévy–Itô decomposition to understand the limit of the martingale in Lemma 2.8 for sets of the form B_ε as $\varepsilon \downarrow 0$. For this reason, let us now state and prove the following theorem.

Theorem 2.10. *Suppose that N is as in Lemma 2.8 and $\int_{(-1,1)} x^2\Pi(\mathrm{d}x) < \infty$. For each $\varepsilon \in (0,1)$ define the martingale*

$$M_t^\varepsilon = \int_{[0,t]}\int_{B_\varepsilon} xN(\mathrm{d}s \times \mathrm{d}x) - t\int_{B_\varepsilon} x\Pi(\mathrm{d}x),\ \ t \geq 0$$

and let $\mathcal{F}_t^*$ be equal to the completion of $\bigcap_{s>t}\mathcal{F}_s$ by the null sets of $\mathbb{P}$ where $\mathcal{F}_t$ is given in (2.16). Then there exists a martingale $M = \{M_t : t \geq 0\}$ with the following properties,

(i) for each $T > 0$, there exists a deterministic subsequence $\{\varepsilon_n^T : n = 1, 2, ...\}$ with $\varepsilon_n^T \downarrow 0$ along which

$$\mathbb{P}(\lim_{n\uparrow\infty} \sup_{0\leq s\leq T} (M_s^{\varepsilon_n^T} - M_s)^2 = 0) = 1,$$

(ii) it is adapted to the filtration $\{\mathcal{F}_t^* : t \geq 0\}$,
(iii) it has right continuous paths with left limits almost surely,
(iv) it has, at most, a countable number of discontinuities on $[0,T]$ almost surely and
(v) it has stationary and independent increments.

In short, there exists a Lévy process, which is also a martingale with a countable number of jumps to which, for any fixed $T > 0$, the sequence of martingales $\{M_t^\varepsilon : t \leq T\}$ converges uniformly on $[0,T]$ with probability one along a subsequence in ε which may depend on T.

Before proving Theorem 2.10 we need to remind ourselves of some general facts concerning square integrable martingales. In our account we shall recall a number of well established facts coming from quite standard L^2 theory, measure theory and continuous time martingale theory. The reader is referred to Sects. 2.4, 2.5 and 9.6 of Ash and Doléans-Dade (2000) for a clear account of the necessary background.

Fix a time horizon $T > 0$. Let us assume as in the above theorem that $(\Omega, \mathcal{F}, \{\mathcal{F}_t^* : t \in [0,T]\}, \mathbb{P})$ is a filtered probability space in which the filtration $\{\mathcal{F}_t^* : t \geq 0\}$ is complete with respect to the null sets of $\mathbb{P}$ and right continuous in the sense that $\mathcal{F}_t^* = \bigcap_{s>t}\mathcal{F}_s^*$.

Definition 2.11. *Fix $T > 0$. Define $\mathcal{M}_T^2 = \mathcal{M}_T^2(\Omega, \mathcal{F}, \{\mathcal{F}_t^* : t \in [0,T]\}, \mathbb{P})$ to be the space of real valued, zero mean right-continuous, square integrable $\mathbb{P}$-martingales with respect to the given filtration over the finite time period $[0,T]$.*

One luxury that follows from the assumptions on $\{\mathcal{F}_t^* : t \geq 0\}$ is that any zero mean square integrable martingale with respect to the latter filtration has a right continuous version which is also a member of $\mathcal{M}_T^2$. Recall that $M' = \{M'_t : t \in [0,T]\}$ is a version of M if it is defined on the same probability space and $\{\exists t \in [0,T] : M'_t \neq M_t\}$ is measurable with zero probability.

It is straightforward to deduce that $\mathcal{M}_T^2$ is a vector space over the real numbers with zero element $M_t = 0$ for all $t \in [0,T]$ and all $\omega \in \Omega$. In fact, as we shall shortly see, $\mathcal{M}_T^2$ is a Hilbert space[2] with respect to the inner product

$$\langle M, N \rangle = \mathbb{E}\left(M_T N_T\right),$$

where $M, N \in \mathcal{M}_T^2$. It is left to the reader to verify the fact that $\langle \cdot, \cdot \rangle$ forms an inner product. The only mild technical difficulty in this verification is showing that for $M \in \mathcal{M}_T^2$, $\langle M, M \rangle = 0$ implies that $M = 0$, the zero element. Note however that if $\langle M, M \rangle = 0$ then by Doob's Maximal Inequality, which says that for $M \in \mathcal{M}_T^2$,

$$\mathbb{E}\left(\sup_{0 \le s \le T} M_s^2\right) \le 4\mathbb{E}\left(M_T^2\right),$$

we have that $\sup_{0 \le t \le T} M_t = 0$ almost surely. Since $M \in \mathcal{M}_T^2$ is right continuous it follows necessarily that $M_t = 0$ for all $t \in [0,T]$ with probability one.

As alluded to above, we can show without too much difficulty that $\mathcal{M}_T^2$ is a Hilbert space. To do that we are required to show that given $\{M^{(n)} : n = 1, 2, ...\}$ is a Cauchy sequence of martingales taken from $\mathcal{M}_T^2$ then there exists an $M \in \mathcal{M}_T^2$ such that

$$\left\| M^{(n)} - M \right\| \to 0$$

as $n \uparrow \infty$ where $\|\cdot\| := \langle \cdot, \cdot \rangle^{1/2}$. To this end let us assume that the sequence of processes $\left\{M^{(n)} : n = 1, 2, ...\right\}$ is a Cauchy sequence, in other words,

$$\mathbb{E}\left[\left(M_T^{(m)} - M_T^{(n)}\right)^2\right]^{1/2} \to 0 \text{ as } m, n \uparrow \infty.$$

Necessarily then the sequence of *random variables* $\{M_T^{(k)} : k \ge 1\}$ is a Cauchy sequence in the Hilbert space of zero mean, square integrable random variables defined on $(\Omega, \mathcal{F}_T, \mathbb{P})$, say $L^2(\Omega, \mathcal{F}_T, \mathbb{P})$, endowed with the inner product $\langle M, N \rangle = \mathbb{E}(MN)$. Hence there exists a limiting variable, say M_T in $L^2(\Omega, \mathcal{F}_T, \mathbb{P})$ satisfying

$$\mathbb{E}\left[(M_T^{(n)} - M_T)^2\right]^{1/2} \to 0$$

[2]Recall that $\langle \cdot, \cdot \rangle : L \times L \to \mathbb{R}$ is an inner product on a vector space L over the reals if it satisfies the following properties for $f, g \in L$ and $a, b \in \mathbb{R}$; (i) $\langle af + bg, h \rangle = a\langle f, h \rangle + b\langle g, h \rangle$ for all $h \in L$, (ii) $\langle f, g \rangle = \langle g, f \rangle$, (iii) $\langle f, f \rangle \ge 0$ and (iv) $\langle f, f \rangle = 0$ if and only if $f = 0$.

For each $f \in L$ let $||f|| = \langle f, f \rangle^{1/2}$. The pair $(L, \langle \cdot, \cdot \rangle)$ are said to form a Hilbert space if all sequences, $\{f_n : n = 1, 2, ...\}$ in L that satisfy $||f_n - f_m|| \to 0$ as $m, n \to \infty$, so called Cauchy sequences, have a limit which exists in L.

as $n \uparrow \infty$. Define the martingale M to be the right continuous version[3] of

$$\mathbb{E}\left(M_T | \mathcal{F}_t^*\right) \text{ for } t \in [0, T]$$

and note that by definition

$$\left\| M^{(n)} - M \right\| \to 0$$

as n tends to infinity. Clearly it is an $\mathcal{F}_t^*$-adapted process and by Jensen's inequality

$$\begin{aligned}
\mathbb{E}\left(M_t^2\right) &= \mathbb{E}\left(\mathbb{E}\left(M_T | \mathcal{F}_t^*\right)^2\right) \\
&\leq \mathbb{E}\left(\mathbb{E}\left(M_T^2 | \mathcal{F}_t^*\right)\right) \\
&= \mathbb{E}\left(M_T^2\right)
\end{aligned}$$

which is finite. Hence Cauchy sequences converge in $\mathcal{M}_T^2$ and we see that $\mathcal{M}_T^2$ is indeed a Hilbert space.

We are now ready to return to Theorem 2.10.

Proof (of Theorem 2.10). (i) Choose $0 < \eta < \varepsilon < 1$, fix $T > 0$ and define $M^\varepsilon = \{M_t^\varepsilon : t \in [0, T]\}$. A calculation similar to the one in Lemma 2.9 (ii) gives

$$\begin{aligned}
\mathbb{E}\left((M_T^\varepsilon - M_T^\eta)^2\right) &= \mathbb{E}\left(\left\{\int_{[0,T]} \int_{\eta \leq |x| < \varepsilon} x N(\mathrm{d}s \times \mathrm{d}x)\right\}^2\right) \\
&= T \int_{\eta \leq |x| < \varepsilon} x^2 \Pi(\mathrm{d}x).
\end{aligned}$$

Note however that the left-hand side above is also equal to $||M^\varepsilon - M^\eta||^2$ (where as in the previous discussion, $|| \cdot ||$ is the norm induced by the inner product on $\mathcal{M}_T^2$).

Thanks to the assumption that $\int_{(-1,1)} x^2 \Pi(\mathrm{d}x) < \infty$, we now have that $\lim_{\varepsilon \downarrow 0} ||M^\varepsilon - M^\eta|| = 0$ and hence that $\{M^\varepsilon : 0 < \varepsilon < 1\}$ is a Cauchy family in $\mathcal{M}_T^2$. As $\mathcal{M}_T^2$ is a Hilbert space we know that there exists a martingale $M = \{M_s : s \in [0, T]\} \in \mathcal{M}_T^2$ such that

$$\lim_{\varepsilon \downarrow 0} ||M - M^\varepsilon|| = 0.$$

An application of Doob's maximal inequality tells us that in fact

$$\lim_{\varepsilon \downarrow 0} \mathbb{E}\left[\sup_{0 \leq s \leq T} (M_s - M_s^\varepsilon)^2\right] \leq 4 \lim_{\varepsilon \downarrow 0} ||M - M^\varepsilon|| = 0. \tag{2.17}$$

[3]Here we use the fact that $\{\mathcal{F}_t^* : t \in [0, T]\}$ is complete and right continuous.

From this one may deduce that the limit $\{M_s : s \in [0,T]\}$ does not depend on T. Indeed suppose it did and we adjust our notation accordingly so that $\{M_{s,T} : s \leq T\}$ represents the limit. Then from (2.17) we see that for any $0 < T' < T$,

$$\lim_{\varepsilon \downarrow 0} \mathbb{E}\left[\sup_{0\leq s\leq T'} (M_s^\varepsilon - M_{s,T'})^2\right] = 0$$

as well as

$$\lim_{\varepsilon \downarrow 0} \mathbb{E}\left[\sup_{0\leq s\leq T'} (M_s^\varepsilon - M_{s,T})^2\right] \leq \lim_{\varepsilon \downarrow 0} \mathbb{E}\left[\sup_{0\leq s\leq T} (M_s^\varepsilon - M_{s,T})^2\right] = 0,$$

where the inequality is the result of a trivial upper bound. Hence using that for any two sequences of real numbers $\{a_n\}$ and $\{b_n\}$, $\sup_n a_n^2 = (\sup_n |a_n|)^2$ and $\sup_n |a_n+b_n| \leq \sup_n |a_n|+\sup_n |b_n|$, we have together with an application of Minkowski's inequality that

$$\begin{aligned}\mathbb{E}\left[\sup_{0\leq s\leq T'} (M_{s,T'} - M_{s,T})^2\right]^{1/2} &\leq \lim_{\varepsilon \downarrow 0} \mathbb{E}\left[\sup_{0\leq s\leq T'} (M_s^\varepsilon - M_{s,T'})^2\right]^{1/2} \\ &\quad + \lim_{\varepsilon \downarrow 0} \mathbb{E}\left[\sup_{0\leq s\leq T'} (M_s^\varepsilon - M_{s,T})^2\right]^{1/2} \\ &= 0\end{aligned}$$

thus showing that the processes $M_{\cdot,T}$ and $M_{\cdot,T'}$ are almost surely uniformly equal on $[0,T']$. Since T' and T may be arbitrarily chosen, we may now speak of a well defined martingale limit $M = \{M_t : t \geq 0\}$.

From the limit (2.17) we may also deduce that there exists a deterministic subsequence $\{\varepsilon_n^T : n \geq 0\}$ along which

$$\lim_{\varepsilon_n^T \downarrow 0} \sup_{0\leq s\leq T} (M_s^{\varepsilon_n^T} - M_s)^2 = 0$$

$\mathbb{P}$-almost surely. This follows from the well established fact that L^2 convergence of a sequence of random variables implies almost sure convergence on a deterministic subsequence.

(ii) Fix $0 < t < T$. Clearly as $M_t^{\varepsilon_n^T}$ is $\mathcal{F}_t^*$-measurable, and by part (i) the latter sequence has an almost sure limit as $n \uparrow \infty$, then by standard measure theory its limit is also $\mathcal{F}_t^*$-measurable.

(iii) As the paths of M^ε are right continuous with left limits, almost sure uniform convergence (along a subsequence) on finite time intervals implies that the limiting process, M, also has paths which are right continuous with left limits. We are using here the fact that, if $D[0,1]$ is the space of functions $f : [0,1] \to \mathbb{R}$ which are right continuous with left limits, then $D[0,1]$ is a closed space under the metric $d(f,g) = \sup_{t\in[0,1]} |f(t)-g(t)|$ for $f,g \in D[0,1]$. See Exercise 2.4 for a proof of this fact.

(iv) According to Corollary 2.6 there are at most an almost surely countable number of points in the support of N. Further, as the measure $\mathrm{d}t \times \Pi(\mathrm{d}x)$ has no atoms then N is necessarily $\{0,1\}$ valued on time–space singletons.[4] Hence every discontinuity in $\{M_s : s \geq 0\}$ corresponds to a unique point in the support of N, it follows that M has at most a countable number of discontinuities. Another way to see there are at most a countable number of discontinuities is simply to note that functions in the space $D[0,1]$ have discontinuities of the first kind[5] and hence there are necessarily a countable number of such discontinuities; see Exercise 2.4.

(v) Uniform almost sure convergence along a subsequence also guarantees convergence in distribution for a collection of fixed times along the same subsequence. This shows that M has stationary independent increments since for any $0 \leq u \leq v \leq s \leq t \leq T < \infty$ and $\theta_1, \theta_2 \in \mathbb{R}$, dominated convergence gives

$$\begin{aligned}
&\mathbb{E}(\mathrm{e}^{\mathrm{i}\theta_1(M_v - M_u)}\mathrm{e}^{\mathrm{i}\theta_2(M_t - M_s)}) \\
&= \lim_{n\uparrow\infty} \mathbb{E}(\mathrm{e}^{\mathrm{i}\theta_1(M_v^{\varepsilon_n^T} - M_u^{\varepsilon_n^T})}\mathrm{e}^{\mathrm{i}\theta_2(M_t^{\varepsilon_n^T} - M_s^{\varepsilon_n^T})}) \\
&= \lim_{n\uparrow\infty} \mathbb{E}(\mathrm{e}^{\mathrm{i}\theta_1 M_{v-u}^{\varepsilon_n^T}})\mathbb{E}(\mathrm{e}^{\mathrm{i}\theta_2 M_{t-s}^{\varepsilon_n^T}}) \\
&= \mathbb{E}(\mathrm{e}^{\mathrm{i}\theta_1 M_{v-u}})\mathbb{E}(\mathrm{e}^{\mathrm{i}\theta_2 M_{t-s}})
\end{aligned}$$

thus concluding the proof. □

2.5 Proof of the Lévy–Itô Decomposition

As previously indicated in Sect. 2, we will take $X^{(1)}$ to be the linear Brownian motion (2.2) defined now on some probability space $(\Omega^\#, \mathcal{F}^\#, P^\#)$.

Given Π in the statement of Theorem 2.1 we know from Theorem 2.4 that there exists a probability space, say $(\Omega^*, \mathcal{F}^*, P^*)$, on which may construct a Poisson random measure N on $([0,\infty) \times \mathbb{R}, \mathcal{B}[0,\infty) \times \mathcal{B}(\mathbb{R}), \mathrm{d}t \times \Pi(\mathrm{d}x))$. We may think of the points in the support of N as having a time and space co-ordinate, or alternatively, as points in $\mathbb{R}\backslash\{0\}$ arriving in time.

Now define

$$X_t^{(2)} = \int_{[0,t]} \int_{|x|\geq 1} xN(\mathrm{d}s \times \mathrm{d}x), \ t \geq 0$$

and note from Lemma 2.8 that since $\Pi(\mathbb{R}\backslash(-1,1)) < \infty$ that it is a compound Poisson process with rate $\Pi(\mathbb{R}\backslash(-1,1))$ and jump distribution

[4] If one reconsiders the construction of a general Poisson random measure on $(S, \mathcal{S}, \eta)$ in the proof of Theorem 2.4, then one sees that if η has atoms, it is possible that the points which contribute to the support of N may lie on top of one another. If η has no atoms then the aforementioned points are almost surely distinct.

[5] A discontinuity of the first kind of f at x means that $f(x+)$ and $f(x-)$ both exist but are not equal to one another.

$\Pi(\mathbb{R}\backslash(-1,1))^{-1}\Pi(\mathrm{d}x)|_{\mathbb{R}\backslash(-1,1)}$. (We may assume without loss of generality that $\Pi(\mathbb{R}\backslash(-1,1)) > 0$ as otherwise we may simply take the process $X^{(2)}$ as the process which is identically zero.)

Next we construct a Lévy process having only small jumps. For each $1 > \varepsilon > 0$ define similarly the compound Poisson process with drift

$$X_t^{(3,\varepsilon)} = \int_{[0,t]} \int_{\varepsilon \le |x| < 1} xN(\mathrm{d}s \times \mathrm{d}x) - t \int_{\varepsilon \le |x| < 1} x\Pi(\mathrm{d}x), \ t \ge 0. \tag{2.18}$$

(As in the definition of $X^{(2)}$ we shall assume without loss of generality $\Pi(\{x : |x| < 1\}) > 0$ otherwise the process $X^{(3)}$ may be taken as the process which is identically zero). Using Theorem 2.7 (ii) we can compute its characteristic exponent,

$$\Psi^{(3,\varepsilon)}(\theta) := \int_{\varepsilon \le |x| < 1} (1 - \mathrm{e}^{\mathrm{i}\theta x} + \mathrm{i}\theta x)\Pi(\mathrm{d}x).$$

According to Theorem 2.10 there exists a Lévy process, which is also a square integrable martingale, defined on $(\Omega^*, \mathcal{F}^*, P^*)$, to which $X^{(3,\varepsilon)}$ converges uniformly on $[0,T]$ along an appropriate deterministic subsequence in ε. Note that it is at this point that we are using the assumption that $\int_{(-1,1)} x^2 \Pi(\mathrm{d}x) < \infty$. It is clear that the characteristic exponent of this latter Lévy process is equal to

$$\Psi^{(3)}(\theta) = \int_{|x|<1} (1 - \mathrm{e}^{\mathrm{i}\theta x} + \mathrm{i}\theta x)\Pi(\mathrm{d}x).$$

From Corollary 2.5 we know that for each $t > 0$, N has independent counts over the two domains $[0,t] \times \{\mathbb{R}\backslash(-1,1)\}$ and $[0,t] \times (-1,1)$. It follows that $X^{(2)}$ and $X^{(3)}$ are independent. Further the latter two are independent of $X^{(1)}$ which was defined on another probability space.

To conclude the proof of the Lévy–Itô decomposition in line with the statement of Theorem 2.1, define the process

$$X_t = X_t^{(1)} + X_t^{(2)} + X_t^{(3)}, \ t \ge 0. \tag{2.19}$$

This process is defined on the product space

$$(\Omega, \mathcal{F}, \mathbb{P}) = (\Omega^{\#}, \mathcal{F}^{\#}, P^{\#}) \times (\Omega^*, \mathcal{F}^*, P^*),$$

has stationary independent increments, has paths that are right continuous with left limits and has characteristic exponent

$$\begin{aligned}\Psi(\theta) &= \Psi^{(1)}(\theta) + \Psi^{(2)}(\theta) + \Psi^{(3)}(\theta) \\ &= \mathrm{i}a\theta + \frac{1}{2}\sigma^2\theta^2 + \int_{\mathbb{R}} (1 - \mathrm{e}^{\mathrm{i}\theta x} + \mathrm{i}\theta x \mathbf{1}_{(|x|<1)})\Pi(\mathrm{d}x)\end{aligned}$$

as required. □

Let us conclude this section with some additional remarks on the Lévy–Itô decomposition.

Recall from (2.4) that the exponent $\Psi^{(3)}$ appears to have the form of the infinite sum of characteristic exponents belonging to compound Poisson processes with drift. Thus suggesting that $X^{(3)}$ may be taken as the superposition of such processes. We now see from the above proof that this is exactly the case. Indeed moving ε to zero through the sequence $\{2^{-k} : k \geq 0\}$ shows us that in the appropriate sense of L^2 convergence

$$\begin{aligned}
\lim_{k\uparrow\infty} X_t^{(3,2^{-k})} &= \lim_{k\uparrow\infty} \int_{[0,t]} \int_{2^{-k}<|x|<1} xN(\mathrm{d}s \times \mathrm{d}x) - t \int_{2^{-k}<|x|<1} x\Pi(\mathrm{d}x) \\
&= \lim_{k\uparrow\infty} \sum_{i=0}^{k-1} \left\{ \int_{[0,t]} \int_{2^{-(i+1)}<|x|<2^{-i}} xN(\mathrm{d}s \times \mathrm{d}x) \right. \\
&\qquad \left. -t \int_{2^{-(i+1)}<|x|<2^{-i}} x\Pi(\mathrm{d}x) \right\}.
\end{aligned}$$

It is also worth remarking that the definition of $X^{(2)}$ and $X^{(3)}$ in the proof of the Lévy–Itô decomposition accordingly, with the partition of $\mathbb{R}\backslash\{0\}$ into $\mathbb{R}\backslash(-1,1)$ and $(-1,1)\backslash\{0\}$ is to some extent arbitrary. The point is that one needs to deal with the contribution of N to the path of the Lévy process which comes from points which are arbitrarily close to the origin in a different way to points which are at a positive distance from the origin. In this respect one could have redrafted the proof replacing $(-1,1)$ by (α,β) for any $\alpha < 0$ and $\beta > 0$, in which case one would need to choose a different value of a in the definition of $X^{(1)}$ in order to make terms add up precisely to the expression given in the Lévy–Khintchine exponent. To be more precise, if for example $\alpha < -1$ and $\beta > 1$, then one should take $X_t^{(1)} = a't + \sigma B_t$ where

$$a' = a - \int_{\alpha<|x|\leq -1} x\Pi(\mathrm{d}x) - \int_{1\leq|x|<\beta} x\Pi(\mathrm{d}x).$$

The latter also shows that the Lévy–Khintchine formula (2.1) is not a unique representation and indeed the indicator $\mathbf{1}_{(|x|<1)}$ in (2.1) may be replaced by $\mathbf{1}_{(\alpha<x<\beta)}$ with an appropriate adjustment in the constant a.

2.6 Lévy Processes Distinguished by Their Path Type

As is clear from the proof of the Lévy–Itô decomposition, we should think of the measure Π given in the Lévy–Khintchine formula as characterising a Poisson random measure which encodes the jumps of the associated Lévy process. In this section we shall re-examine elements of the proof of the Lévy–Itô decomposition and show that, with additional assumptions on Π corresponding to restrictions on the way in which jumps occur, we may further identify special classes of Lévy processes embedded within the general class.

2.6.1 Path Variation

It is clear from the Lévy–Itô decomposition that the presence of the linear Brownian motion $X^{(1)}$ would imply that paths of the Lévy process have unbounded variation. On the other hand, should it be the case that $\sigma = 0$, then the Lévy process may or may not have unbounded variation. The term $X^{(2)}$, being a compound Poisson process, has only bounded variation. Hence in the case $\sigma = 0$, understanding whether the Lévy process has unbounded variation is an issue determined by the limiting process $X^{(3)}$; that is to say the process of compensated small jumps.

Reconsidering the definition of $X^{(3)}$ it is natural to ask under what circumstances

$$\lim_{\varepsilon \downarrow 0} \int_{[0,t]} \int_{\varepsilon \leq |x| < 1} xN(\mathrm{d}s \times \mathrm{d}x)$$

exists almost surely without the need for compensation. Once again, the answer is given by Theorem 2.7 (i). Here we are told that

$$\int_{[0,t]} \int_{0<|x|<1} xN(\mathrm{d}s \times \mathrm{d}x) < \infty$$

if and only if $\int_{0<|x|<1} |x| \Pi(\mathrm{d}x) < \infty$. In that case we may identify $X^{(3)}$ directly via

$$X_t^{(3)} = \int_{[0,t]} \int_{0<|x|<1} xN(\mathrm{d}s \times \mathrm{d}x) - t \int_{0<|x|<1} x\Pi(\mathrm{d}x), \ t \geq 0.$$

This also tells us that $X^{(3)}$ will be of bounded variation if and only if $\int_{0<|x|<1} |x|\Pi(\mathrm{d}x) < \infty$. The latter combined with the general integral condition $\int_{\mathbb{R}} (1 \wedge x^2)\Pi(\mathrm{d}x) < \infty$ in the light of the Lévy–Itô decomposition yields the following Lemma.

Lemma 2.12. *A Lévy process with Lévy–Khintchine exponent corresponding to the triple (a, σ, Π) has paths of bounded variation if and only if*

$$\sigma = 0 \ \textit{and} \ \int_{\mathbb{R}} (1 \wedge |x|)\Pi(\mathrm{d}x) < \infty. \tag{2.20}$$

Note that the finiteness of the integral in (2.20) also allows for the Lévy–Khintchine exponent of any such bounded variation process to be re-written as follows

$$\Psi(\theta) = -\mathrm{i}d\theta + \int_{\mathbb{R}} (1 - \mathrm{e}^{\mathrm{i}\theta x})\Pi(\mathrm{d}x), \tag{2.21}$$

where the constant $d \in \mathbb{R}$ relates to the constant a and Π via

$$d = -\left(a + \int_{|x|<1} x\Pi(\mathrm{d}x) \right).$$

In this case, we may write the Lévy process in the form

$$X_t = dt + \int_{[0,t]} \int_{\mathbb{R}} xN(\mathrm{d}s \times \mathrm{d}x),\ t \geq 0. \tag{2.22}$$

In view of the decomposition of the Lévy–Khintchine formula for a process of bounded variation and the corresponding representation (2.22), the term d is often referred to as the *drift*. Strictly speaking, one should not talk of drift in the case of a Lévy process whose jump part is a process of unbounded variation. If drift is to be understood in terms of a purely deterministic trend, then it is ambiguous on account of the "infinite limiting compensation" that one sees in $X^{(3)}$ coming from the second term on the right hand side of (2.18).

From the expression given in (1.3) of the Chap. 1 we see that if X is a compound Poisson process with drift then its characterisitic exponent takes the form of (2.21) with $\Pi(\mathbb{R}) < \infty$. Conversely, suppose that $\sigma = 0$ and Π has finite total mass, then we know from Lemma 2.8 that (2.22) is a compound Poisson process with drift d. In conclusion we have the following lemma.

Lemma 2.13. *A Lévy process is a compound Poisson process with drift if and only if $\sigma = 0$ and $\Pi(\mathbb{R}) < \infty$.*

Processes which are of bounded variation but which are not compound Poisson processes with drift are sometimes referred to as *generalised compound Poisson processes (with drift)*. This is because they are structurally the same as compound Poisson processes in that they may be represented in the form of (2.22). This is of course not true for Lévy processes of unbounded variation, even if the Gaussian component is zero.

2.6.2 One-Sided Jumps

Suppose now that $\Pi(-\infty, 0) = 0$. From the proof of the Lévy–Itô decomposition, we see that this implies that the corresponding Lévy process has no negative jumps. If further we have that $\int_{(0,\infty)}(1 \wedge x)\Pi(\mathrm{d}x) < \infty$, $\sigma = 0$ and, in the representation (2.21) of the characteristic exponent, $d \geq 0$, then from the representation (2.22) it becomes clear that the Lévy process has non-decreasing paths. Conversely, suppose that a Lévy process has non-decreasing paths then necessarily it has bounded variation. Hence $\int_{(0,\infty)}(1 \wedge x)\Pi(\mathrm{d}x) < \infty$, $\sigma = 0$ and then it is easy to see that in the representation (2.21) of the characteristic exponent, we necessarily have $d \geq 0$. Examples of such a process were given in Chap. 1 (the gamma process and the inverse Gaussian process) and were named subordinators. Summarising we have the following.

Lemma 2.14. *A Lévy process is a subordinator if and only if $\Pi(-\infty, 0) = 0$, $\int_{(0,\infty)}(1 \wedge x)\Pi(\mathrm{d}x) < \infty$, $\sigma = 0$ and $d = -\left(a + \int_{(0,1)} x\Pi(\mathrm{d}x)\right) \geq 0$.*

For the sake of clarity, we note that when X is a subordinator, further to (2.21), its Lévy-Khinchine formula may be written

$$\Psi(\theta) = -id\theta + \int_{(0,\infty)} (1 - e^{i\theta x})\Pi(dx), \quad (2.23)$$

where necessarily $d \geq 0$.

If $\Pi(-\infty, 0) = 0$ and X is not a subordinator, then it is referred to in general as a *spectrally positive* Lévy process. A Lévy process, X, will then be referred to as *spectrally negative* if $-X$ is spectrally positive. Together the latter two classes of processes are called *spectrally one-sided.* Spectrally one-sided Lévy processes may be of bounded or unbounded variation and, in the latter case, may or may not possess a Gaussian component. Note in particular that when $\sigma = 0$ it is still possible to have paths of unbounded variation. When a spectrally positive Lévy process has bounded variation then it must take the form

$$X_t = -dt + S_t, \ t \geq 0$$

where $\{S_t : t \geq 0\}$ is a pure jump subordinator and necessarily $d > 0$ as otherwise X would conform to the definition of a subordinator. Note that the above decomposition implies that if $\mathbb{E}(X_1) \leq 0$ then $\mathbb{E}(S_1) < \infty$ as opposed to the case that $\mathbb{E}(X_1) > 0$ in which case it is possible that $\mathbb{E}(S_1) = \infty$.

A special feature of spectrally positive processes is that, if $\tau_x^- = \inf\{t > 0 : X_t < x\}$ where $x < 0$, then $\mathbb{P}(\tau_x^- < \infty) > 0$ and hence, as there can be no downwards jumps,

$$\mathbb{P}(X_{\tau_x^-} = x | \tau_x^- < \infty) = 1$$

with a similar property for first passage upwards being true for spectrally negative processes. A rigorous proof of the first of the above two facts will be given in Corollary 3.13 at the end of Sect. 3.3. It turns out that the latter fact plays a very important role in the simplification of a number of theorems we shall encounter later on in this text which concern the fluctuations of general Lévy processes.

2.7 Interpretations of the Lévy–Itô Decomposition

Let us return to some of the models considered in Chap. 1 and consider how the understanding of the Lévy–Itô decomposition helps to justify working with more general classes of Lévy processes.

2.7.1 The Structure of Insurance Claims

Recall from Sect. 1.3.1 that the Cramér–Lundberg model corresponds to a Lévy process with characteristic exponent given by

$$\Psi(\theta) = -\mathrm{i}c\theta + \lambda \int_{(-\infty,0)} (1 - \mathrm{e}^{\mathrm{i}\theta x}) F(\mathrm{d}x),$$

for $\theta \in \mathbb{R}$. In other words, a compound Poisson process with arrival rate $\lambda > 0$ and negative jumps, corresponds to claims, having common distribution F as well as a drift $c > 0$ corresponding to a steady income due to premiums. Suppose instead we work with a general spectrally negative Lévy process. That is a process for which $\Pi(0,\infty) = 0$ (but not the negative of a subordinator). In this case, the Lévy–Itô decomposition offers an interpretation for large scale insurance companies as follows. The Lévy–Khintchine exponent may be written in the form

$$\Psi(\theta) = \left\{\frac{1}{2}\sigma^2\theta^2\right\} + \left\{-\mathrm{i}\theta c + \int_{(-\infty,-1]} (1 - \mathrm{e}^{\mathrm{i}\theta x}) \Pi(\mathrm{d}x)\right\} + \left\{\int_{(-1,0)} (1 - \mathrm{e}^{\mathrm{i}\theta x} + \mathrm{i}\theta x) \Pi(\mathrm{d}x)\right\} \quad (2.24)$$

for $\theta \in \mathbb{R}$. Assume that $\Pi(-\infty, 0) = \infty$ so Ψ is genuinely different from the characteristic of a Cramér–Lundberg model. The third bracket in (2.24) we may understand as a Lévy process representing a countably infinite number of arbitrarily small claims compensated by a deterministic positive drift (which may be infinite in the case that $\int_{(-1,0)} |x|\Pi(\mathrm{d}x) = \infty$) corresponding to the accumulation of premiums over an infinite number of contracts. Roughly speaking, the way in which claims occur is such that in any arbitrarily small period of time $\mathrm{d}t$, a claim of size $|x|$ (for $x < 0$) is made independently with probability $\Pi(\mathrm{d}x)\mathrm{d}t + o(\mathrm{d}t)$. The insurance company thus counterbalances such claims by ensuring that it collects premiums in such a way that in any $\mathrm{d}t$, $|x|\Pi(\mathrm{d}x)\mathrm{d}t$ of its income is devoted to the compensation of claims of size $|x|$. The second bracket in (2.24) we may understand as coming from large claims which occur occasionally and are compensated against by a steady income at rate $c > 0$ as in the Cramér–Lundberg model. Here "large" is taken to mean claims of size one or more and $c = -a$ in the terminology of the Lévy–Khintchine formula given in Theorem 1.6. Finally the first bracket in (2.24) may be seen as a stochastic perturbation of the system of claims and premium income.

Since the first and third brackets in (2.24) correspond to martingales, the company may guarantee that its revenues drift to infinity over an infinite time horizon by assuming the latter behaviour applies to the compensated process of large claims corresponding to the second bracket in (2.24).

2.7.2 General Storage Models

The workload of the $M/G/1$ queue was presented in Sect. 1.3.2 as a spectrally negative compound Poisson process with rate $\lambda > 0$ and jump distribution F

with positive unit drift reflected in its supremum. In other words, the underlying Lévy process has characteristic exponent

$$\Psi(\theta) = -\mathrm{i}\theta + \lambda \int_{(-\infty,0)} (1 - \mathrm{e}^{\mathrm{i}\theta x}) F(\mathrm{d}x)$$

for all $\theta \in \mathbb{R}$. A general storage model, described for example in the classic books of Prabhu (1998) and Takács (1966), consists of working with a general Lévy process which is the difference of a positive drift and a subordinator and then reflected in its supremum. Its Lévy–Khintchine exponent thus takes the form

$$\Psi(\theta) = -\mathrm{i}d\theta + \int_{(-\infty,0)} (1 - \mathrm{e}^{\mathrm{i}\theta x}) \Pi(\mathrm{d}x),$$

where $d > 0$ and $\int_{(-\infty,0)} (1 \wedge |x|) \Pi(\mathrm{d}x) < \infty$. As with the case of the $M/G/1$ queue, the reflected process

$$W_t = (w \vee \overline{X}_t) - X_t, \ t \geq 0$$

may be thought of the stored volume or workload of some system where X is the Lévy process with characteristic exponent Ψ given above and $\overline{X}$ is its running supremum. The Lévy–Itô decomposition tells us that during the periods of time that X is away from its supremum, there is a natural "drainage" of volume or "processing" of workload, corresponding to the downward movement of W in a linear fashion with rate d. At the same time new "volume for storage" or equivalently new "jobs" arrive independently so that in each $\mathrm{d}t$, one arrives of size $|x|$ (where $x < 0$) with probability $\Pi(\mathrm{d}x)\mathrm{d}t + o(\mathrm{d}t)$ (thus giving similar interpretation to the occurrence of jumps in the insurance risk model described above). When $\Pi(-\infty, 0) = \infty$ the number of jumps are countably infinite over any finite time interval thus indicating that our model is processing with "infinite frequency" in comparison to the finite activity of the workload of the $M/G/1$ process.

Of course one may also envisage working with a jump measure which has some mass on the positive half line. This would correspond negative jumps in the process W. This, in turn, can be interpreted as follows. Over and above the natural drainage or processing at rate d, in each $\mathrm{d}t$ there is independent removal of a "volume" or "processing time of job" of size $y > 0$ with probability $\Pi(\mathrm{d}y)\mathrm{d}t + o(\mathrm{d}t)$. One may also consider moving to models of unbounded variation. However, in this case, the interpretation of drift is lost.

2.7.3 Financial Models

Financial mathematics has become a field of applied probability which has embraced the use of Lévy processes, in particular, for the purpose of modelling the evolution of risky assets. We shall not attempt to give anything like a comprehensive exposure of this topic here, nor elsewhere in this book. Especially since the existing text books of Boyarchenko and Levendorskii (2002b), Schoutens (2003), Cont and Tankov (2004) and Barndorff–Nielsen and Shephard (2005) already offer a clear and up-to-date overview between them. It is however worth mentioning briefly some of the connections between path properties of Lévy processes seen above and modern perspectives within financial modelling.

One may say that financial mathematics proper begins with the thesis of Louis Bachellier who proposed the use of linear Brownian motion to model the value of a risky asset, say the value of a stock (See Bachelier (1900, 1901)). The *classical* model, proposed by Samuleson (1965), for the evolution of a risky asset however is generally accepted to be that of an exponential linear Brownian motion with drift;

$$S_t = s e^{\sigma B_t + \mu t}, \quad t \geq 0 \tag{2.25}$$

where $s > 0$ is the initial value of the asset, $B = \{B_t : t \geq 0\}$ is a standard Brownian motion, $\sigma > 0$ and $\mu \in \mathbb{R}$. This choice of model offers the feature that asset values have multiplicative stationarity and independence in the sense that for any $0 \leq u < t < \infty$,

$$S_t = S_u \times \widetilde{S}_{t-u} \tag{2.26}$$

where $\widetilde{S}_{t-u}$ is independent of S_u and has the same distribution as S_{t-u}. Whether this is a realistic assumption in terms of temporal correlations in financial markets is open to debate. Nonetheless, for the purpose of a theoretical framework in which one may examine the existence or absence of certain economic mechanisms, such as risk-neutrality, hedging and arbitrage as well as giving sense to the value of certain financial products such as option contracts, exponential Brownian motion has proved to be the right model to capture the imagination of mathematicians, economists and financial practitioners alike. Indeed, what makes (2.25) "classical" is that Black and Scholes (1973) and Merton (1973) demonstrated how one may construct rational arguments leading to the pricing of a call option on a risky asset driven by exponential Brownian motion.

Two particular points (of the many) where the above model of a risky asset can be shown to be inadequate concern the continuity of the paths and the distribution of log-returns of the value of a risky asset. Clearly (2.25) has continuous paths and therefore cannot accommodate for jumps which arguably are present in observed historical data of certain risky assets due to shocks in the market. The feature (2.26) suggests that if one would choose a fixed period of time Δ, then for each $n \geq 1$, the innovations $\log(S_{(n+1)\Delta}/S_{n\Delta})$ are independent and normally distributed with mean $\mu\Delta$ and variance $\sigma\sqrt{\Delta}$. Empirical data suggests however that the tails of the distribution of the log-returns are asymmetric as well as having heavier tails than those of normal distributions. The tails of the latter being particularly light as they decay like $\exp\{-x^2\}$ for large $|x|$. See for example the discussion in Schoutens (2003).

Recent literature suggests that a possible remedy for these two points is to work with

$$S_t = s\mathrm{e}^{X_t},\ t \geq 0$$

instead of (2.25) where again $s > 0$ is the initial value of the risky asset and $X = \{X_t : t \geq 0\}$ is now a Lévy process. This preserves multiplicative stationary and independent increments as well as allowing for jumps, distributional asymmetry and the possibility of heavier tails than the normal distribution can offer. A rather unsophisticated example of how the latter may happen is simply to take for X a compound Poisson process whose jump distribution is asymmetric and heavy tailed. A more sophisticated example however, and indeed quite a popular model in the research literature, is the so-called *variance gamma* process, introduced by Madan and Seneta (1990). This Lévy process is pure jump, that is to say $\sigma = 0$, and has Lévy measure given by

$$\Pi(\mathrm{d}x) = \mathbf{1}_{(x<0)} \frac{C}{|x|} \mathrm{e}^{Gx} \mathrm{d}x + \mathbf{1}_{(x>0)} \frac{C}{x} \mathrm{e}^{-Mx} \mathrm{d}x,$$

where $C, G, M > 0$. It is easily seen by computing explicitly the integral $\int_{\mathbb{R}\setminus\{0\}} (1 \wedge |x|) \Pi(\mathrm{d}x)$ and the total mass $\Pi(\mathbb{R})$ that the variance gamma process has paths of bounded variation and further is not a compound Poisson process. It turns out that the exponential weighting in the Lévy measure ensures that the distribution of the variance gamma process at a fixed time t has exponentially decaying tails (as opposed to the much lighter tails of the Gaussian distribution).

Working with pure jump processes implies that there is no diffusive nature to the evolution of risky assets. Diffusive behaviour is often found attractive for modelling purposes as it has the taste of a physical interpretation in which increments in infinitesimal periods of time are explained as the aggregate effect of many simultaneous conflicting external forces. Geman et al. (2001) argue however the case for modelling the value of risky assets with Lévy processes which have paths of bounded variation which are not compound Poisson processes. In their reasoning, the latter has a countable number of jumps over finite periods of time which correspond to the countable, but nonetheless infinite number of purchases and sales of the asset which collectively dictate its value as a net effect. In particular being of bounded variation means the Lévy process can be written as the difference to two independent subordinators (see Exercise 2.8). The latter two should be thought of the total prevailing price buy orders and total prevailing price sell orders on the logarithmic price scale.

Despite the fundamental difference between modelling with bounded variation Lévy processes and exponential Brownian motion, Geman et al. (2001) also provide an interesting link to the classical model (2.25) via time change. The basis of their ideas lies with the following lemma.

Lemma 2.15. *Suppose that* $X = \{X_t : t \geq 0\}$ *is a Lévy process with characteristic exponent* Ψ *and* $\tau = \{\tau_s : s \geq 0\}$ *is an independent subordinator with characteristic* Ξ. *Then* $Y = \{X_{\tau_s} : s \geq 0\}$ *is again a Lévy process with characteristic exponent* $\Xi \circ \Psi$.

Proof. First let us make some remarks about Ξ. We already know that the formula

$$\mathbb{E}(\mathrm{e}^{\mathrm{i}\theta\tau_s}) = \mathrm{e}^{-\Xi(\theta)s}$$

holds for all $\theta \in \mathbb{R}$. However, since τ is a non-negative valued process, via analytical extension, we may claim that the previous equality is still valid for $\theta \in \{z \in \mathbb{C} : \Im z \geq 0\}$. Note in particular then that since

$$\Re\Psi(u) = \frac{1}{2}\sigma^2 u^2 + \int_{\mathbb{R}} (1 - \cos(ux))\Pi(\mathrm{d}x) > 0$$

for all $u \in \mathbb{R}$, the equality

$$\mathbb{E}(\mathrm{e}^{-\Psi(u)\tau_s}) = \mathrm{e}^{-\Xi(\Psi(u))s} \tag{2.27}$$

holds.

Since X and τ have right continuous paths, then so does Y. Next consider $0 \leq u \leq v \leq s \leq t < \infty$ and $\theta_1, \theta_2 \in \mathbb{R}$. Then by first conditioning on τ and noting that $0 \leq \tau_u \leq \tau_v \leq \tau_s \leq \tau_t < \infty$ we have

$$\begin{aligned}\mathbb{E}\left(\mathrm{e}^{\mathrm{i}\theta_1(Y_v - Y_u) + \mathrm{i}\theta_2(Y_t - Y_s)}\right) &= \mathbb{E}\left(\mathrm{e}^{-\Psi(\theta_1)(\tau_v - \tau_u) - \Psi(\theta_2)(\tau_t - \tau_s)}\right)\\ &= \mathbb{E}\left(\mathrm{e}^{-\Psi(\theta_1)\tau_{v-u} - \Psi(\theta_2)\tau_{t-s}}\right)\\ &= \mathrm{e}^{-\Xi(\Psi(\theta_1))(v-u) - \Xi(\Psi(\theta_2))(t-s)}\end{aligned}$$

where in the final equality we have used the fact that τ has stationary independent increments together with (2.27). This shows that Y has stationary and independent increments. □

Suppose in the above lemma we take for X a linear Brownian motion with drift as in the exponent of (2.25). By sampling this continuous path process along the range of an independent subordinator, one recovers another Lévy process. Geman et al. (2001) suggest that one may consider the value of a risky asset to evolve as the process (2.25) on an abstract time scale suitable to the rate of business transactions called *business time*. The link between business time and real time is given by the subordinator τ. That is to say, one assumes that the value of a given risky asset follows the process $Y = X \circ \tau$ because at real time $s > 0$, τ_s units of business time have passed and hence the value of the risky asset is positioned at X_{τ_s}.

Returning to the example of the variance gamma process given above, it turns out that one may recover it from a linear Brownian motion by applying a time change using a gamma subordinator. See Exercise 2.9 for more details on the facts mentioned here concerning the variance gamma process as well as Exercise 2.12 for more examples of Lévy processes which may be written in terms of a subordinated Brownian motion with drift.

Exercises

2.1. The object of this exercise is to give a reminder of the additive property of Poisson distributions (which is also the reason why they belong to the class of infinite divisible distributions). Suppose that $\{N_i : i = 1, 2, ...\}$ is an independent sequence of random variables defined on $(\Omega, \mathcal{F}, P)$ which are Poisson distributed with parameters λ_i for $i = 1, 2, ...$, respectively. Let $S = \sum_{i\geq 1} N_i$. Show that

(i) if $\sum_{i\geq 1} \lambda_i < \infty$ then S is Poisson distributed with parameter $\sum_{i\geq 1} \lambda_i$ and hence in particular $P(S < \infty) = 1$,
(ii) if $\sum_{i\geq 1} \lambda_i = \infty$ then $P(S = \infty) = 1$.

2.2. Denote by $\{T_i : i \geq 1\}$ the arrival times in the Poisson process $N = \{N_t : t \geq 0\}$ with parameter λ.

(i) By recalling that inter-arrival times are independent and exponential, show that for any $A \in \mathcal{B}([0, \infty)^n)$,

$$P((T_1, ..., T_n) \in A | N_t = n) = \int_A \frac{n!}{t^n} \mathbf{1}_{(0\leq t_1 \leq ... \leq t_n \leq t)} \mathrm{d}t_1 \times ... \times \mathrm{d}t_n.$$

(ii) Deduce that the distribution of $(T_1, ..., T_n)$ conditional on $N_t = n$ has the same law as the distribution of an ordered independent sample of size n taken from the uniform distribution on $[0, t]$.

2.3. If η is a measure on $(S, \mathcal{S})$ and $f : S \to [0, \infty)$ is measurable then show that $\int_S (1 - \mathrm{e}^{-\phi f(x)}) \eta(\mathrm{d}x) < \infty$ for all $\phi > 0$ if and only if $\int_S (1 \wedge f(x)) \eta(\mathrm{d}x) < \infty$.

2.4. Recall that $D[0, 1]$ is the space of functions $f : [0, 1] \to \mathbb{R}$ which are right continuous with left limits.

(i) Define the norm $||f|| = \sup_{x\in[0,1]} |f(x)|$. Use the triangle inequality to deduce that $D[0, 1]$ is closed under uniform convergence with respect to the norm $||\cdot||$. That is to say, show that if $\{f_n : n \geq 1\}$ is a sequence in $D[0, 1]$ and $f : [0, 1] \to \mathbb{R}$ such that $\lim_{n\uparrow\infty} ||f_n - f|| = 0$ then $f \in D[0, 1]$.
(ii) Suppose that $f \in D[0, 1]$ and let $\Delta = \{t \in [0, 1] : |f(t) - f(t-)| \neq 0\}$ (the set of discontinuity points). Show that Δ is countable if Δ_c is countable for all $c > 0$ where $\Delta_c = \{t \in [0, 1] : |f(t) - f(t-)| > c\}$. Next fix $c > 0$. By supposing for contradiction that Δ_c has an accumulation point, say x, show that the existence of either a left or right limit at x fails as it would imply that there is no left or right limit of f at x. Deduce that Δ_c and hence Δ is countable.

2.5. The explicit construction of a Lévy process given in the Lévy–Itô decomposition begs the question as to whether one may construct examples of *deterministic* functions which have similar properties to those of the paths of Lévy processes. The objective of this exercise is to do precisely that. The reader is warned however, that this is purely an analytical exercise and one should not necessarily think of the paths of Lévy processes as being entirely similar to the functions constructed below in all respects.

(i) Let us recall the definition of the Cantor function which we shall use to construct a deterministic function which has bounded variation and that is right continuous with left limits. Take the interval $C_0 := [0,1]$ and perform the following iteration. For $n \geq 0$ define C_n as the union of intervals which remain when removing the middle third of each of the intervals which make up C_{n-1}. The Cantor set C is the limiting object, $\bigcap_{n\geq 0} C_n$ and can be described by

$$C = \{x \in [0,1] : x = \sum_{k\geq 1} \frac{\alpha_k}{3^k} \text{ such that } \alpha_k \in \{0,2\} \text{ for each } k \geq 1\}.$$

One sees then that the Cantor set is simply the points in $[0,1]$ which omits numbers whose tertiary expansion contain the digit 1. To describe the Cantor function, for each $x \in [0,1]$ let $j(x)$ be the smallest j for which $\alpha_j = 1$ in the tertiary expansion of $\sum_{k\geq 1} \alpha_k/3^k$ of x. If $x \in C$ then $j(x) = \infty$ and otherwise if $x \in [0,1]\backslash C$ then $1 \leq j(x) < \infty$. The Cantor function is defined as follows

$$f(x) = \frac{1}{2^{j(x)}} + \sum_{i=1}^{j(x)-1} \frac{\alpha_j}{2^{i+1}} \text{ for } x \in [0,1].$$

Now consider the function $g : [0,1] \to [0,1]$ given by $g(x) = f^{-1}(x) - at$ for $a \in \mathbb{R}$. Here we understand $f^{-1}(x) = \inf\{\theta : f(\theta) > x\}$. Note that g is monotone if and only if $a \leq 0$. Show that g has only positive jumps and the value of x for which g jumps form a dense set in $[0,1]$. Show further that g has bounded variation on $[0,1]$.

(ii) Now let us construct an example of a deterministic function which has unbounded variation and that is right continuous with left limits. Denote by $\mathbb{Q}_2$ the dyadic rationals. Consider a function $J : [0,\infty) \to \mathbb{R}$ as follows. For all $x \geq 0$ which are not in $\mathbb{Q}_2$, set $J(x) = 0$. It remains to assign a value for each $x = a/2^n$ where $a = 1,3,5,...$ (even values of a cancel). Let

$$J(a/2^n) = \begin{cases} 2^{-n} & \text{if } a = 1,5,9,... \\ -2^{-n} & \text{if } a = 3,7,11,... \end{cases}$$

and define

$$f(x) = \sum_{s\in[0,x]\cap\mathbb{Q}_2} J(x).$$

Show that f is uniformly bounded on $[0,1]$, is right continuous with left limits and has unbounded variation over $[0,1]$.

2.6. Suppose that X is a Lévy process with Lévy measure Π.

(i) For each $n \geq 2$ show that for each $t > 0$,

$$\int_{[0,t]} \int_{\mathbb{R}} x^n N(\mathrm{d}s \times \mathrm{d}x) < \infty$$

almost surely if and only if

$$\int_{|x| \geq 1} |x|^n \Pi(\mathrm{d}x) < \infty.$$

(ii) Suppose now that Π satisfies $\int_{|x|\geq 1} |x|^n \Pi(\mathrm{d}x) < \infty$ for $n \geq 2$. Show that

$$\int_{[0,t]} \int_{\mathbb{R}} x^n N(\mathrm{d}s \times \mathrm{d}x) - t \int_{\mathbb{R}} x^n \Pi(\mathrm{d}x), t \geq 0$$

is a martingale.

2.7. Let X be a Lévy process with Lévy measure Π. Denote by N the Poisson random measure associated with its jumps.

(i) Show that

$$\mathbb{P}(\sup_{0<s\leq t} |X_s - X_{s-}| > a) = 1 - \mathrm{e}^{-t\Pi(\mathbb{R}\backslash(-a,a))}$$

for $a > 0$.

(ii) Show that the paths of X are continuous if and only if $\Pi = 0$.

(iii) Show that the paths of X are piece-wise linear if and only if it is a compound Poisson process with drift if and only if $\sigma = 0$ and $\Pi(\mathbb{R}) < \infty$. [Recall that a function $f : [0, \infty) \to \mathbb{R}$ is right continuous and piece-wise linear if there exist sequence of times $0 = t_0 < t_1 < ... < t_n < ...$ with $\lim_{n\uparrow\infty} t_n = \infty$ such that on $[t_{j-1}, t_j)$ the function f is linear].

(iv) Now suppose that $\Pi(\mathbb{R}) = \infty$. Argue by contradiction that for each positive rational $q \in \mathbb{Q}$ there exists a decreasing sequence of jump times for X, say $\{T_n(\omega) : n \geq 0\}$, such that $\lim_{n\uparrow\infty} T_n = q$. Hence deduce that the set of jump times are dense in $[0, \infty)$.

2.8. Show that any Lévy process of bounded variation may be written as the difference of two independent subordinators.

2.9. This exercise gives another explicit example of a Lévy process; the variance gamma process, introduced by Madan and Seneta (1990) for modelling financial data.

(i) Suppose that $\Gamma = \{\Gamma_t : t \geq 0\}$ is a gamma subordinator with parameters α, β and that $B = \{B_t : t \geq 0\}$ is a standard Brownian motion. Show that for $c \in \mathbb{R}$ and $\sigma > 0$, the variance gamma process

$$X_t := c\Gamma_t + \sigma B_{\Gamma_t},\ t \geq 0$$

is a Lévy process with characteristic exponent

$$\Psi(\theta) = \beta \log(1 - \mathrm{i}\frac{\theta c}{\alpha} + \frac{\sigma^2\theta^2}{2\alpha}).$$

(ii) Show that the variance gamma process is equal in law to the Lévy process

$$\Gamma^{(1)} - \Gamma^{(2)} = \{\Gamma_t^{(1)} - \Gamma_t^{(2)} : t \geq 0\},$$

where $\Gamma^{(1)}$ is a Gamma subordinator with parameters

$$\alpha^{(1)} = \left(\sqrt{\frac{1}{4}\frac{c^2}{\alpha^2} + \frac{1}{2}\frac{\sigma^2}{\alpha}} + \frac{1}{2}\frac{c}{\alpha}\right)^{-1} \text{ and } \beta^{(1)} = \beta$$

and $\Gamma^{(2)}$ is a Gamma subordinator, independent of $\Gamma^{(1)}$, with parameters

$$\alpha^{(2)} = \left(\sqrt{\frac{1}{4}\frac{c^2}{\alpha^2} + \frac{1}{2}\frac{\sigma^2}{\alpha}} - \frac{1}{2}\frac{c}{\alpha}\right)^{-1} \text{ and } \beta^{(2)} = \beta.$$

2.10. Suppose that d is an integer greater than one. Choose $\mathbf{a} \in \mathbb{R}^d$ and let Π be a measure concentrated on $\mathbb{R}^d \backslash \{0\}$ satisfying

$$\int_{\mathbb{R}^d} (1 \wedge |\mathbf{x}|^2)\Pi(\mathrm{d}\mathbf{x})$$

where $|\cdot|$ is the standard Euclidian norm. Show that it is possible to construct a d-dimensional process $\mathbf{X} = \{\mathbf{X}_t : t \geq 0\}$ on a probability space $(\Omega, \mathcal{F}, \mathbb{P})$ having the following properties.

(i) The paths of $\mathbf{X}$ are right continuous with left limits $\mathbb{P}$-almost surely in the sense that for each $t \geq 0$, $\mathbb{P}(\lim_{s\downarrow t} \mathbf{X}_s = \mathbf{X}_t) = 1$ and $\mathbb{P}(\lim_{s\uparrow t} \mathbf{X}_s \text{ exists}) = 1$.
(ii) $\mathbb{P}(\mathbf{X}_0 = \mathbf{0}) = 1$, the zero vector in $\mathbb{R}^d$.
(iii) For $0 \leq s \leq t$, $\mathbf{X}_t - \mathbf{X}_s$ is independent of $\sigma(\mathbf{X}_u : u \leq s)$.
(iv) For $0 \leq s \leq t$, $\mathbf{X}_t - \mathbf{X}_s$ is equal in distribution to $\mathbf{X}_{t-s}$.
(v) For any $t \geq 0$ and $\theta \in \mathbb{R}^d$,

$$\mathbb{E}(\mathrm{e}^{\mathrm{i}\theta\cdot\mathbf{X}_t}) = \mathrm{e}^{-\Psi(\theta)t}$$

and

$$\Psi(\theta) = \mathrm{i}\mathbf{a}\cdot\theta + \frac{1}{2}\theta\cdot\mathbf{A}\theta + \int_{\mathbb{R}^d} (1 - \mathrm{e}^{\mathrm{i}\theta\cdot\mathbf{x}} + \mathrm{i}(\theta\cdot\mathbf{x})\mathbf{1}_{(|\mathbf{x}|<1)})\Pi(\mathrm{d}\mathbf{x}), \quad (2.28)$$

where for any two vectors $\mathbf{x}$ and $\mathbf{y}$ in $\mathbb{R}^d$, $\mathbf{x}\cdot\mathbf{y}$ is the usual inner product and $\mathbf{A}$ is a $d \times d$ matrix whose eigenvalues are all non-negative.

2.11. Suppose that X is a subordinator. Show that it has a Laplace exponent given by

$$-\log \mathbb{E}(\mathrm{e}^{-qX_1}) = \Phi(q) = dq + \int_{(0,\infty)} (1 - \mathrm{e}^{-qx})\Pi(\mathrm{d}x)$$

for $q \geq 0$. Show using integration by parts that

$$\Phi(q) = dq + q\int_0^\infty \mathrm{e}^{-qx}\Pi(x,\infty)\mathrm{d}x$$

and hence that the drift term d may be recovered from the limit

$$\lim_{q\uparrow\infty} \frac{\Phi(q)}{q} = d.$$

2.12. Here are some more examples of Lévy processes which may be written as a subordinated Brownian motion.

(i) Let $\alpha \in (0,2)$. Show that a Brownian motion subordinated by a stable process of index $\alpha/2$ is a symmetric stable process of index α.
(ii) Suppose that $X = \{X_t : t \geq 0\}$ is a compound Poisson process with Lévy measure given by

$$\Pi(\mathrm{d}x) = \left\{\mathbf{1}_{(x<0)}\mathrm{e}^{-a|x|} + \mathbf{1}_{(x>0)}\mathrm{e}^{-ax}\right\}\mathrm{d}x$$

for $a > 0$. Now let $\tau = \{\tau_s : s \geq 0\}$ be a pure jump subordinator with Lévy measure

$$\pi(\mathrm{d}x) = \mathbf{1}_{(x>0)}2a\mathrm{e}^{-a^2x}\mathrm{d}x.$$

Show that $\{\sqrt{2}B_{\tau_s} : s \geq 0\}$ has the same law as X where $B = \{B_t : t \geq 0\}$ is a standard Brownian motion independent of τ.
(iii) Suppose now that $X = \{X_t : t \geq 0\}$ is a compound Poisson process with Lévy measure given by

$$\Pi(\mathrm{d}x) = \frac{\lambda\sqrt{2}}{\sigma\sqrt{\pi}}\mathrm{e}^{-x^2/2\sigma^2}\mathrm{d}x$$

for $x \in \mathbb{R}$. Show that $\{\sigma B_{N_t} : t \geq 0\}$ has the same law as X where B is as in part (ii) and $\{N_s : s \geq 0\}$ is a Poisson process with rate 2λ independent of B.

Further, the final part gives a simple example of Lévy processes which may be written as a subordinated Lévy process.

(iv) Suppose that X is a symmetric stable process of index $\alpha \in (0,2)$. Show that X can be written as a symmetric stable process of index α/β subordinated by an independent stable subordinator of index $\beta \in (0,1)$.

3

More Distributional and Path-Related Properties

In this chapter we consider some more distributional and path-related properties of general Lévy processes. Specifically we examine the Strong Markov Property, duality, moments and exponential change of measure.

In Chap. 1 it was mentioned that any Lévy process $X = \{X_t : t \geq 0\}$ is assumed to be defined on a probability space $(\Omega, \mathcal{F}, \mathbb{P})$. We now broaden this assumption.

Henceforth any Lévy process X is assumed to be defined on a filtered probability space $(\Omega, \mathcal{F}, \mathbb{F}, \mathbb{P})$ where now the filtration $\mathbb{F} = \{\mathcal{F}_t : t \geq 0\}$, defined by $\mathcal{F}_t = \sigma(\mathcal{F}_t^0, \mathcal{O})$, $\mathcal{F}_t^0 = \sigma(X_s : s \leq t)$ and $\mathcal{O}$ is the set of null sets of $\mathbb{P}$.

Principally what will be of repeated use in making this assumption is the consequence that the filtration $\mathbb{F}$ becomes right continuous.[1] That is to say for each $t \geq 0$,

$$\mathcal{F}_t = \bigcap_{s>t} \mathcal{F}_s.$$

The proof of this fact is given in Exercise 3.1.

3.1 The Strong Markov Property

Recall that $X = \{X_t : t \geq 0\}$ defined on the filtered space $(\Omega, \mathcal{F}, \mathbb{F}, \mathbb{P})$ is a Markov Process if for each $B \in \mathcal{B}(\mathbb{R})$ and $s, t \geq 0$,

$$\mathbb{P}(X_{t+s} \in B | \mathcal{F}_t) = \mathbb{P}(X_{t+s} \in B | \sigma(X_t)). \tag{3.1}$$

[1] It is important to stress that, in general, completing the filtration of a stochastic process by the null sets of its measure is not sufficient to induce right continuity in the modified filtration.

In addition, recall that a non-negative random variable, say τ, defined on the same filtered probability space $(\Omega, \mathcal{F}, \mathbb{F}, \mathbb{P})$ is called a stopping time if

$$\{\tau \leq t\} \in \mathcal{F}_t$$

for all $t > 0$. It is possible that a stopping time may have the property that $\mathbb{P}(\tau = \infty) > 0$. In addition, for any stopping time τ,

$$\{\tau < t\} = \bigcup_{n \geq 1} \{\tau \leq t - 1/n\} \in \bigcup_{n \geq 1} \mathcal{F}_{t-1/n} \subseteq \mathcal{F}_t.$$

However, given the right continuity of $\mathbb{F}$, conversely we also have that any random time τ which has the property that $\{\tau < t\} \in \mathcal{F}_t$ for all $t \geq 0$, must also be a stopping time since

$$\{\tau \leq t\} = \bigcap_{n \geq 1} \{\tau < t + 1/n\} \in \bigcap_{n \geq 1} \mathcal{F}_{t+1/n} = \mathcal{F}_{t+} = \mathcal{F}_t,$$

where in the last equality we use the right continuity of the filtration. In other words, for a Lévy process whose filtration is right continuous we may also say that τ is a stopping time if and only if $\{\tau < t\} \in \mathcal{F}_t$ for all $t \geq 0$.

Associated with a given stopping time is the sigma algebra

$$\mathcal{F}_\tau := \{A \in \mathcal{F} : A \cap \{\tau \leq t\} \in \mathcal{F}_t \text{ for all } t \geq 0\}.$$

(Note, it is a simple exercise to verify that the above set of sets is a sigma algebra). A Strong Markov Process satisfies the analog of relation (3.1) in which the fixed time t may be replaced by any stopping time τ with respect to $\mathbb{F}$;

$$\mathbb{P}(X_{\tau+s} \in B | \mathcal{F}_\tau) = \mathbb{P}(X_{\tau+s} \in B | \sigma(X_\tau)) \text{ on } \{\tau < \infty\}.$$

It is easy to see that all Lévy processes are Markovian. Indeed Lévy processes satisfy the stronger condition that the law of $X_{t+s} - X_t$ is independent of $\mathcal{F}_t$. The next theorem indicates that the continuation of X beyond an $\mathbb{F}$-stopping time relative to its stopped position is completely independent of the history of the process up to that stopping time and is equal in law to X. Hence in particular, all Lévy processes exhibit the strong Markov property.

Theorem 3.1. *Suppose that τ is a stopping time. Define on $\{\tau < \infty\}$ the process $\widetilde{X} = \{\widetilde{X}_t : t \geq 0\}$ where*

$$\widetilde{X}_t = X_{\tau+t} - X_\tau, \ t \geq 0.$$

Then on the event $\{\tau < \infty\}$ the process $\widetilde{X}$ is independent of $\mathcal{F}_\tau$ and has the same law as X and hence in particular is a Lévy process.

Proof. As finite dimensional distributions determine the law of a stochastic process, it would suffice to prove that for any $0 \leq v \leq u \leq s \leq t < \infty$, $H \in \mathcal{F}_\tau$ and $\theta_1, \theta_2 \in \mathbb{R}$,

$$\mathbb{E}\left(\mathrm{e}^{\mathrm{i}\theta_1(X_{\tau+t}-X_{\tau+s})+\mathrm{i}\theta_2(X_{\tau+u}-X_{\tau+v})}; H \cap \{\tau < \infty\}\right)$$
$$= \mathrm{e}^{-\Psi(\theta_1)(t-s)-\Psi(\theta_2)(u-v)}\mathbb{P}\left(H \cap \{\tau < \infty\}\right),$$

where Ψ is the characteristic exponent of X.

To this end, define a sequence of stopping times $\{\tau^{(n)} : n \geq 1\}$ by

$$\tau^{(n)} = \begin{cases} k2^{-n} \text{ if } (k-1)\,2^{-n} < \tau \leq k2^{-n} \text{ for } k = 1, 2, ... \\ 0 \qquad\qquad\qquad\qquad \text{if } \tau = 0. \end{cases} \tag{3.2}$$

Stationary independent increments together with the fact that $H \cap \{\tau^{(n)} = k2^{-n}\} \in \mathcal{F}_{k2^{-n}}$ gives

$$\mathbb{E}\left(\mathrm{e}^{\mathrm{i}\theta_1(X_{\tau^{(n)}+t}-X_{\tau^{(n)}+s})+\mathrm{i}\theta_2(X_{\tau^{(n)}+u}-X_{\tau^{(n)}+v})}; H \cap \{\tau^{(n)} < \infty\}\right)$$
$$= \sum_{k\geq 0}\mathbb{E}\left(\mathrm{e}^{\mathrm{i}\theta_1(X_{\tau^{(n)}+t}-X_{\tau^{(n)}+s})+\mathrm{i}\theta_2(X_{\tau^{(n)}+u}-X_{\tau^{(n)}+v})}; H \cap \{\tau^{(n)} = k2^{-n}\}\right)$$
$$= \sum_{k\geq 0}\mathbb{E}\Big[\mathbf{1}_{(H\cap\{\tau^{(n)}=k2^{-n}\})}$$
$$\cdot\, \mathbb{E}\left(\mathrm{e}^{\mathrm{i}\theta_1(X_{k2^{-n}+t}-X_{k2^{-n}+s})+\mathrm{i}\theta_2(X_{k2^{-n}+u}-X_{k2^{-n}+v})}\Big| \mathcal{F}_{k2^{-n}}\right)\Big]$$
$$= \sum_{k\geq 0}\mathbb{E}\left(\mathrm{e}^{\mathrm{i}\theta_1 X_{t-s}+\mathrm{i}\theta_2 X_{u-v}}\right)\mathbb{P}(H \cap \{\tau^{(n)} = k2^{-n}\})$$
$$= \mathrm{e}^{-\Psi(\theta_1)(t-s)-\Psi(\theta_2)(u-v)}\mathbb{P}(H \cap \{\tau^{(n)} < \infty\}).$$

The paths of X are almost surely right continuous and $\tau^{(n)} \downarrow \tau$ as n tends to infinity, hence $X_{\tau^{(n)}+s} \to X_{\tau+s}$ almost surely for all $s \geq 0$ as n tends to infinity. It follows by the dominated convergence theorem that

$$\mathbb{P}\left(\mathrm{e}^{\mathrm{i}\theta_1(X_{\tau+t}-X_{\tau+s})+\mathrm{i}\theta_2(X_{\tau+u}-X_{\tau+v})}; H \cap \{\tau < \infty\}\right)$$
$$= \lim_{n\uparrow\infty}\mathbb{E}\left(\mathrm{e}^{\mathrm{i}\theta_1(X_{\tau^{(n)}+t}-X_{\tau^{(n)}+s})+\mathrm{i}\theta_2(X_{\tau^{(n)}+u}-X_{\tau^{(n)}+v})}; H \cap \{\tau^{(n)} < \infty\}\right)$$
$$= \lim_{n\uparrow\infty}\mathrm{e}^{-\Psi(\theta_1)(t-s)-\Psi(\theta_2)(u-v)}\mathbb{P}(H \cap \{\tau^{(n)} < \infty\})$$
$$= \mathrm{e}^{-\Psi(\theta_1)(t-s)-\Psi(\theta_2)(u-v)}\mathbb{P}\left(H \cap \{\tau < \infty\}\right)$$

showing that $\widetilde{X}$ is independent of $\mathcal{F}_\tau$ on $\{\tau < \infty\}$ and has the same law as X. □

Examples of $\mathbb{F}$-stopping times which will repeatedly occur in the remaining text are those of the *first entrance time* and *first hitting time* of a given open or closed set $B \subseteq \mathbb{R}$. They are defined, respectively, as

$$T^B = \inf\{t \geq 0 : X_t \in B\} \text{ and } \tau^B = \inf\{t > 0 : X_t \in B\}.$$

We take the usual definition $\inf \emptyset = \infty$ here. Typically throughout this book we shall work with the choice of B equal to (x, ∞), $[x, \infty)$, $(-\infty, x)$, $(-\infty, x]$ and $\{x\}$ where $x \in \mathbb{R}$. The two times T^B and τ^B are very closely related. They are equal when $X_0 \notin B$, however, they may possibly differ in value when $X_0 \in B$. Consider for example the case that $B = [0, \infty)$ and X is a compound Poisson process with strictly negative drift. When $X_0 = 0$ we have $\mathbb{P}(T^B = 0) = 1$ where as $\mathbb{P}(\tau^B > 0) = 1$.[2]

To some extent it is intuitively obvious why T^B and τ^B are stopping times. Nonetheless, we complete this section by justifying this claim. The justification comes in the form of a supporting lemma and a theorem establishing the claim. The lemma illustrates that there exists a sense of left continuity of Lévy processes when appropriately sampling the path with an increasing sequence of stopping times; so called *quasi-left-continuity*. The proofs of the forthcoming lemma and theorem are quite technical and it will do no harm if the reader chooses to bypass their proofs and continue reading on to the next section at this point in time. The arguments given are rooted in the works of Dellacherie and Meyer (1975–1993) and Blumenthal and Getoor (1968) who give a comprehensive and highly detailed account of the theory of Markov processes in general.

Lemma 3.2 (Quasi-Left-Continuity). *Suppose that T is a $\mathbb{F}$-stopping time and that $\{T_n : n \geq 1\}$ is an increasing sequence of $\mathbb{F}$-stopping times such that $\lim_{n\uparrow\infty} T_n = T$ almost surely, then $\lim_{n\uparrow\infty} X_{T_n} = X_T$ on $\{T < \infty\}$. Hence if $T_n < T$ almost surely for each $n \geq 1$, then X is left continuous at T on $\{T < \infty\}$.*

Note that for any fixed $t > 0$, the probability that X jumps at time t is zero. Hence if $\{t_n : n = 1, 2, ...\}$ is a sequence of deterministic times satisfying $t_n \to t$ as $n \uparrow \infty$ then with probability one $X_{t_n} \to X_t$; in other words, t is a point of continuity of X.[3] The statement in the above lemma thus asserts that this property extends to the case of stopping times.

Proof (of Lemma 3.2). First suppose that $\mathbb{P}(T < \infty) = 1$. As the sequence $\{T_n : n \geq 1\}$ is almost surely increasing we can identify the limit of $\{X_{T_n} : n \geq 0\}$ by

[2] As we shall see later, this is a phenomenon which is not exclusive to compound Poisson processes with strictly negative drift. The same behaviour is experienced by Lévy processes of bounded variation with strictly negative drift.

[3] It is worth reminding oneself for the sake of clarity that $X_{t_n} \to X_t$ as $n \uparrow \infty$ means that for all $\varepsilon > 0$ there exists a $N > 0$ such that $|X_{t_n} - X_t| < \varepsilon$ for all $n > N$ and this does not contradict the fact that there might be an infinite number of discontinuities in the path of X in an arbitrary small neighbourhood of t.

$$Z = \mathbf{1}_A X_T + \mathbf{1}_{A^c} X_{T-}$$

where $A = \bigcup_{n\geq 1} \bigcap_{k\geq n} \{T_k = T\} = \{T_k = T \text{ eventually}\}$. Suppose that f and g are two continuous functions with compact supports. Appealing to bounded convergence and then to bounded convergence again together with the right continuity and left limits of paths we have

$$\begin{aligned}\lim_{t\downarrow 0}\lim_{n\uparrow\infty} \mathbb{E}(f(X_{T_n})g(X_{T_n+t})) &= \lim_{t\downarrow 0} \mathbb{E}(f(Z)g(X_{(T+t)-})) \\ &= \mathbb{E}(f(Z)g(X_T)). \qquad (3.3)\end{aligned}$$

Now write for short $P_t g(x) = \mathbb{E}(g(x+X_t)) = \mathbb{E}_x(g(X_t))$ which is uniformly bounded in x and t and, by bounded convergence, continuous in x. Note that right continuity of X together with bounded convergence also implies that $\lim_{t\downarrow 0} P_t g(x) = g(x)$ for each $x \in \mathbb{R}$. These facts together with the Markov property applied at time T_n and bounded convergence imply that

$$\begin{aligned}\lim_{t\downarrow 0}\lim_{n\uparrow\infty} \mathbb{E}(f(X_{T_n})g(X_{T_n+t})) &= \lim_{t\downarrow 0}\lim_{n\uparrow\infty} \mathbb{E}(f(X_{T_n})P_t g(X_{T_n})) \\ &= \lim_{t\downarrow 0} \mathbb{E}(f(Z)P_t g(Z)) \\ &= \mathbb{E}(f(Z)g(Z)). \qquad (3.4)\end{aligned}$$

Equating (3.3) and (3.4) we see that for all uniformly bounded continuous functions f and g,

$$\mathbb{E}(f(Z)g(X_T)) = \mathbb{E}(f(Z)g(Z)).$$

From this equality we may deduce (using standard measure theory) that

$$\mathbb{E}(h(Z, X_T)) = \mathbb{E}(h(Z, Z))$$

for any bounded measurable function h. In particular, if we take $h(x, y) = \mathbf{1}_{(x=y)}$ then we deduce that $Z = X_T$ almost surely.

When $T_n < T$ almost surely for all $n \geq 1$ it is clear that $Z = X_{T-}$ and the concluding sentence in the statement of the lemma follows for the case that $\mathbb{P}(T < \infty) = 1$.

To remove the requirement that $\mathbb{P}(T < \infty) = 1$ recall that for each $t > 0$, $T \wedge t$ is a finite stopping time. We have that $T_n \wedge t \uparrow T \wedge t$ as $n \uparrow \infty$ and hence, from the previous part of the proof, $\lim_{n\uparrow\infty} X_{T_n \wedge t} = X_{T\wedge t}$ almost surely. In other words, $\lim_{n\uparrow\infty} X_{T_n} = X_T$ on $\{T \leq t\}$. Since we may take t arbitrarily large the result follows. □

Theorem 3.3. *Suppose that B is open or closed. Then,*

(i) T^B is a stopping time and $X_{T^B} \in \overline{B}$ on $\{T^B < \infty\}$ and
(ii) τ^B is a stopping time and $X_{\tau^B} \in \overline{B}$ on $\{\tau^B < \infty\}$

(note that $\overline{B} = B$ when B is closed).

Proof. (i) First we deal with the case that B is open. Since any Lévy process $X = \{X_t : t \geq 0\}$ has right continuous paths and B is open we may describe the event $\{T^B < t\}$ in terms of the path of X at rational times. That is to say,

$$\{T^B < t\} = \bigcup_{s \in \mathbb{Q} \cap [0,t)} \{X_s \in B\}. \tag{3.5}$$

Since each of the sets in the union are $\mathcal{F}_t$-measurable and sigma algebras are closed under countable set operations, then $\{T^B < t\}$ is also $\mathcal{F}_t$-measurable. Recalling that $\mathbb{F}$ is right continuous we have that $\{T^B < t\}$ is $\mathcal{F}_t$-measurable if and only if $\{T^B \leq t\}$ is $\mathcal{F}_t$-measurable and hence T^B fulfills the definition of an $\mathbb{F}$-stopping time. Now note that on $\{T^B < \infty\}$ we have that either $X_{T^B} \in B$ or that at the time T^B, X is at the boundary of B and at the next instant moves into B; that is to say, X moves continuously into B by first approaching its boundary arbitrarily closely. In the latter case, right continuity of paths implies that $X_{T^B} \in \overline{B}$. To illustrate this latter case, consider the example where $B = (x, \infty)$ for some $x > 0$ and X is any compound Poisson process with strictly positive drift and negative jumps. It is clear that $\mathbb{P}(X_{T^{(x,\infty)}} = x) > 0$ as the process may drift up to the boundary point $\{x\}$ and then continue into (x, ∞) before, for example, the first jump occurs.

For the case of closed B, the argument given above does not work. The reason why lies with the possibility that X may enter B simply by touching its boundary which is now included in B. Further, this may occur in a way that cannot be described in terms of a countable sequence of events.

We thus employ another technique for the proof of (i) when B is closed. Suppose then that $\{B_n : n \geq 1\}$ is a sequence of open sets given by

$$B_n = \{x \in \mathbb{R} : |x - y| < 1/n \text{ for some } y \in B\}.$$

Note that $B \subset B_n$ for all $n \geq 1$ and $\bigcap_{n\geq 1} \overline{B}_n = B$. From the previous paragraph we have that T^{B_n} are $\mathbb{F}$-stopping times and clearly they are increasing. Denote their limit by T. Since for all $t \geq 0$,

$$\{T \leq t\} = \{\sup_{n\geq 1} T^{B_n} \leq t\} = \bigcap_{n\geq 1} \{T^{B_n} \leq t\} \in \mathcal{F}_t,$$

we see that T is an $\mathbb{F}$-stopping time. Obviously $T^{B_n} \leq T^B$ for all n and hence $T \leq T^B$. On the other hand, according to quasi-left-continuity described in the previous lemma, $\lim_{n\uparrow\infty} X_{T^{B_n}} = X_T$ on the event $\{T < \infty\}$ showing that $X_T \in \overline{B} = B$ and hence that $T \geq T^B$ on $\{T < \infty\}$. In conclusion we have that $T = T^B$ and $X_{T^B} \in B$ on $\{T^B < \infty\}$.

(ii) Suppose now that B is open. Let $T^B_\varepsilon = \inf\{t \geq \varepsilon : X_t \in B\}$. Note that $\{T^B_\varepsilon < t\} = \emptyset \in \mathcal{F}_t$ for all $t < \varepsilon$ and for $t \geq \varepsilon$,

$$\{T_\varepsilon^B < t\} = \bigcup_{s \in \mathbb{Q} \cap [\varepsilon, t)} \{X_s \in B\},$$

which is $\mathcal{F}_t$ measurable and hence by right continuity of $\mathbb{F}$, T_ε^B is an $\mathbb{F}$-stopping time. Now suppose that B is closed. Following the arguments in part (i) but with $T_\varepsilon^{B_n} := \inf\{t \geq \varepsilon : X_t \in B_n\}$, we conclude for closed B that T_ε^B is again an $\mathbb{F}$-stopping time. In both cases, when B is open or closed, we also see as in part (i) that $X_{T_\varepsilon^B} \in \overline{B}$ on $\{T_\varepsilon^B < \infty\}$.

Now suppose that B is open or closed. The sequence of stopping times $\{T_\varepsilon^B : \varepsilon > 0\}$ forms a decreasing sequence as $\varepsilon \downarrow 0$ and hence has an almost sure limit which is equal to τ^B by definition. Note also that $\{T_\varepsilon^B < \infty\}$ increases to $\{\tau^B < \infty\}$ as $\varepsilon \downarrow 0$. Since for all $t \geq 0$ and decreasing sequences $\varepsilon \downarrow 0$,

$$\{\tau^B \leq t\}^c = \{\inf_{n \geq 1} T_{\varepsilon_n}^B > t\} = \bigcap_{n \geq 1} \{T_{\varepsilon_n}^B > t\} \in \mathcal{F}_t,$$

we see that τ^B is an $\mathbb{F}$-stopping time. Right continuity of the paths of X tell us that $\lim_{\varepsilon \downarrow 0} X_{T_\varepsilon^B} = X_{\tau^B}$ on $\{\tau^B < \infty\}$ where the limit is taken in $\mathbb{Q} \cap [0, 1]$. Hence $X^{\tau^B} \in \overline{B}$ whenever $\{\tau^B < \infty\}$. □

3.2 Duality

In this section we discuss a simple feature of all Lévy processes which follows as a direct consequence of stationary independent increments. That is, when the path of a Lévy process over a finite time horizon is time reversed (in an appropriate sense) the new path is equal in law to the process reflected about the origin. This property will prove to be of crucial importance in a number of fluctuation calculations later on.

Lemma 3.4 (Duality Lemma). *For each fixed $t > 0$, define the reversed process*

$$\{X_{(t-s)-} - X_t : 0 \leq s \leq t\}$$

and the dual process,

$$\{-X_s : 0 \leq s \leq t\}.$$

Then the two processes have the same law under $\mathbb{P}$.

Proof. Define the time reversed process $Y_s = X_{(t-s)-} - X_t$ for $0 \leq s \leq t$ and note that under $\mathbb{P}$ we have $Y_0 = 0$ almost surely as in the Poisson random measure describing the jumps $\mathbb{P}(N(\{t\} \times \mathbb{R}) = 0) = 1$ and hence t is a jump time with probability zero. As can be seen from Fig. 3.1, the paths of Y are obtained from those of X by a reflection about the vertical axis with an

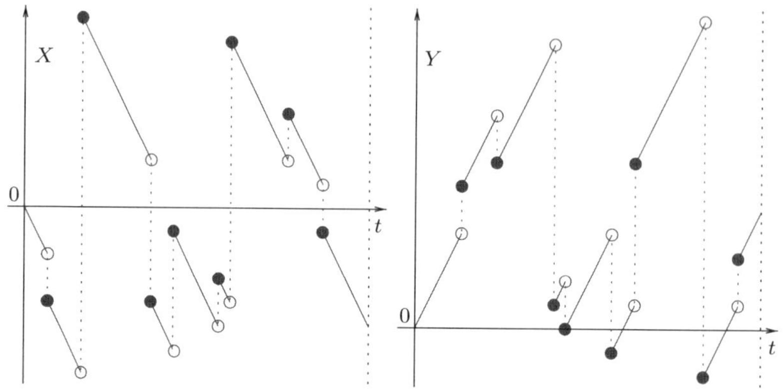

Fig. 3.1. Duality of the processes $X = \{X_s : s \leq t\}$ and $Y = \{X_{(t-s)-} - X_t : s \leq t\}$. The path of Y is a reflection of the path of X with an adjustment of continuity at jump times.

adjustment of the continuity at the jump times so that its paths are almost surely right continuous with left limits. Further, the stationary independent increments of X imply directly the same as is true of Y. Further, for each $0 \leq s \leq t$, the distribution of $X_{(t-s)-} - X_t$ is identical to that of $-X_s$ and hence, since the finite time distributions of Y determine its law, the proof is complete. □

The Duality Lemma is also well known for (and in fact originates from) random walks, the discrete time analogue of Lévy processes, and is justified using an identical proof. See for example Feller (1971).

One interesting feature that follows as a consequence of the Duality Lemma is the relationship between the running supremum, the running infimum, the process reflected in its supremum and the process reflected in its infimum. The last four objects are, respectively,

$$\overline{X}_t = \sup_{0\leq s\leq t} X_s, \qquad \underline{X}_t = \inf_{0\leq s\leq t} X_s$$

$$\{\overline{X}_t - X_t : t \geq 0\} \text{ and } \{X_t - \underline{X}_t : t \geq 0\}.$$

Lemma 3.5. *For each fixed* $t > 0$, *the pairs* $(\overline{X}_t, \overline{X}_t - X_t)$ *and* $(X_t - \underline{X}_t, -\underline{X}_t)$ *have the same distribution under* $\mathbb{P}$.

Proof. Define $\widetilde{X}_s = X_t - X_{(t-s)}$ for $0 \leq s \leq t$ and write $\widetilde{\underline{X}}_t = \inf_{0\leq s\leq t} \widetilde{X}_s$. Using right continuity and left limits of paths we may deduce that

$$(\overline{X}_t, \overline{X}_t - X_t) = (\widetilde{X}_t - \widetilde{\underline{X}}_t, -\widetilde{\underline{X}}_t)$$

almost surely. One may visualise this in Fig. 3.2. By rotating the picture about by 180° one sees the almost sure equality of the pairs $(\overline{X}_t, \overline{X}_t - X_t)$ and

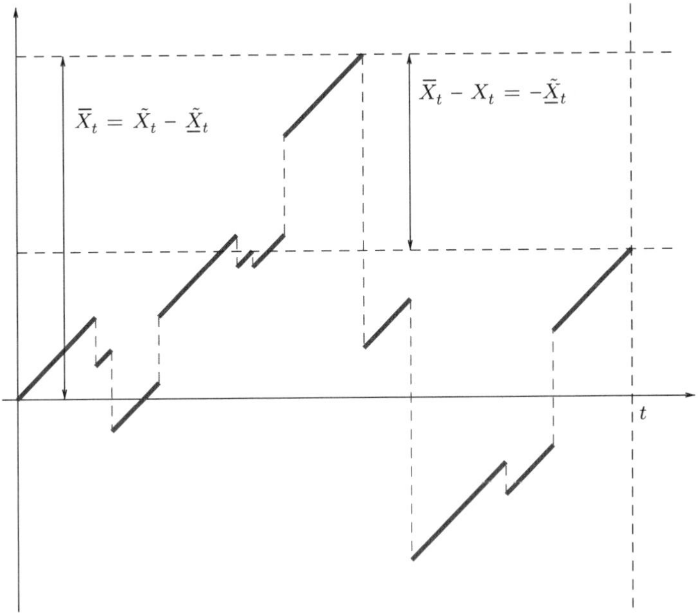

Fig. 3.2. Duality of the pairs $(\overline{X}_t, \overline{X}_t - X_t)$ and $(X_t - \underline{X}_t, -\underline{X}_t)$.

$(\widetilde{X}_t - \underline{\widetilde{X}}_t, -\underline{\widetilde{X}})$. Now appealing to the Duality Lemma we have that $\{\widetilde{X}_s : 0 \leq s \leq t\}$ is equal in law to $\{X_s : 0 \leq s \leq t\}$ under $\mathbb{P}$. The result now follows. □

3.3 Exponential Moments and Martingales

It is well known that the position of a Brownian motion at a fixed time has moments of all orders. It is natural therefore to cast an eye on similar issues for Lévy processes. In general the picture is not so straightforward. One needs only to consider compound Poisson processes to see how things can difffer. Suppose we write the latter in the form

$$X_t = \sum_{i=1}^{N_t} \xi_i$$

where $N = \{N_t \colon t \geq 0\}$ is a Poisson process and $\{\xi_i \colon i \geq 0\}$ are independent and identically distributed. By choosing the jump distribution of each ξ_i in such a way that it has infinite first moment (for example any stable distribution on $(0, \infty)$ with index $\alpha \in (0, 1)$) it is clear that

$$\mathbb{E}(X_t) = \lambda t \mathbb{E}(\xi_1) = \infty$$

for all $t > 0$.

As one might suspect, there is an intimate relationship between the moments of the Lévy measure and the moments of the distribution of the associated Lévy process at any fixed time. We have the following theorem.

Theorem 3.6. *Let $\beta \in \mathbb{R}$, then*

$$\mathbb{E}\left(\mathrm{e}^{\beta X_t}\right) < \infty \text{ for all } t \geq 0 \text{ if and only if } \int_{|x|\geq 1} \mathrm{e}^{\beta x} \Pi(\mathrm{d}x) < \infty.$$

Proof. First suppose that $\mathbb{E}\left(e^{\beta X_t}\right) < \infty$ for some $t > 0$. Recall $X^{(1)}$, $X^{(2)}$ and $X^{(3)}$ given in the Lévy–Itô decomposition. Note in particular that $X^{(2)}$ is a compound Poisson process with arrival rate $\lambda := \Pi(\mathbb{R}\backslash(-1,1))$ and jump distribution $F(\mathrm{d}x) := \mathbf{1}_{(|x|\geq 1)}\Pi(\mathrm{d}x)/\Pi(\mathbb{R}\backslash(-1,1))$ and $X^{(1)} + X^{(3)}$ is a Lévy process with Lévy measure $\mathbf{1}_{(|x|\leq 1)}\Pi(\mathrm{d}x)$. Since

$$\mathbb{E}\left(\mathrm{e}^{\beta X_t}\right) = \mathbb{E}\left(\mathrm{e}^{\beta X_t^{(2)}}\right)\mathbb{E}\left(\mathrm{e}^{\beta(X_t^{(1)}+X_t^{(3)})}\right),$$

it follows that

$$\mathbb{E}\left(\mathrm{e}^{\beta X_t^{(2)}}\right) < \infty, \tag{3.6}$$

and hence as $X^{(2)}$ is a compound Poisson process,

$$\begin{aligned}\mathbb{E}(\mathrm{e}^{\beta X_t^{(2)}}) &= \mathrm{e}^{-\lambda t}\sum_{k\geq 0}\frac{(\lambda t)^k}{k!}\int_{\mathbb{R}} \mathrm{e}^{\beta x} F^{*k}(\mathrm{d}x) \\ &= \mathrm{e}^{-\Pi(\mathbb{R}\backslash(-1,1))t}\sum_{k\geq 0}\frac{t^k}{k!}\int_{\mathbb{R}} e^{\beta x}(\Pi|_{\mathbb{R}\backslash(-1,1)})^{*k}(\mathrm{d}x) < \infty, \end{aligned} \tag{3.7}$$

where F^{*n} and $(\Pi|_{\mathbb{R}\backslash(-1,1)})^{*n}$ are the n-fold convolution of F and $\Pi|_{\mathbb{R}\backslash(-1,1)}$, the restriction of Π to $\mathbb{R}\backslash(-1,1)$, respectively. In particular the summand corresponding to $k = 1$ must be finite; that is

$$\int_{|x|\geq 1} \mathrm{e}^{\beta x}\Pi(\mathrm{d}x) < \infty.$$

Now suppose that $\int_{\mathbb{R}} \mathrm{e}^{\beta x}\mathbf{1}_{(|x|\geq 1)}\Pi(\mathrm{d}x) < \infty$ for some $\beta \in \mathbb{R}$. Since $(\Pi|_{\mathbb{R}\backslash(-1,1)})^{*n}(dx)$ is a finite measure, we have

$$\int_{\mathbb{R}} \mathrm{e}^{\beta x}(\Pi|_{\mathbb{R}\backslash(-1,1)})^{*n}(\mathrm{d}x) = \left(\int_{|x|\geq 1} \mathrm{e}^{\beta x}\Pi(\mathrm{d}x)\right)^n,$$

and hence (3.7) and (3.6) hold for all $t > 0$. The proof is thus complete once we show that for all $t > 0$,

$$\mathbb{E}\left(\mathrm{e}^{\beta(X_t^{(1)}+X_t^{(3)})}\right) < \infty. \tag{3.8}$$

However, since $X^{(1)} + X^{(3)}$ has a Lévy measure with bounded support, it follows that its characteristic exponent,

$$-\frac{1}{t}\log \mathbb{E}\left(\mathrm{e}^{\mathrm{i}\theta(X_t^{(1)}+X_t^{(3)})}\right)$$
$$= \mathrm{i}a\theta + \frac{1}{2}\sigma^2\theta^2 + \int_{(-1,1)} (1 - \mathrm{e}^{\mathrm{i}\theta x} + \mathrm{i}\theta x)\Pi\,(\mathrm{d}x)\,,\ \theta \in \mathbb{R}, \tag{3.9}$$

can be extended to an entire function (analytic on the whole of $\mathbb{C}$). To see this, note that

$$\int_{(-1,1)} (1 - \mathrm{e}^{\mathrm{i}\theta x} + \mathrm{i}\theta x)\Pi\,(\mathrm{d}x) = \int_{(-1,1)} \sum_{k\geq 0} \frac{(\mathrm{i}\theta x)^{k+2}}{(k+2)!}\Pi\,(\mathrm{d}x)\,.$$

The sum and the integral may be exchanged in the latter using Fubini's Theorem and the estimate

$$\sum_{k\geq 0}\int_{(-1,1)} \frac{(|\theta| x)^{k+2}}{(k+2)!}\Pi\,(\mathrm{d}x) \leq \sum_{k\geq 0}\frac{(|\theta|)^{k+2}}{(k+2)!}\int_{(-1,1)} x^2\Pi\,(\mathrm{d}x) < \infty.$$

Hence the right–hand side of (3.9) can be written as a power series for all $\theta \in \mathbb{C}$ and is thus entire. We have then in particular that (3.8) holds. □

In general the conclusion of the previous theorem can be extended to a larger class of functions than just the exponential functions.

Definition 3.7. *A function $g : \mathbb{R} \to [0,\infty)$ is called submultiplicative if there exists a constant $c > 0$ such that $g(x + y) \leq cg(x)g(y)$ for all $x, y \in \mathbb{R}$.*

It follows easily from the definition that, for example, the product of two submultiplicative functions is submultiplicative. Again working directly with the definition it is also easy to show that if $g(x)$ is submultiplicative, then so is $\{g(cx + \gamma)\}^\alpha$ where $c \in \mathbb{R}$, $\gamma \in \mathbb{R}$ and $\alpha > 0$. An easy way to see this is first to prove the statement for $g(cx)$, then $g(x+\gamma)$ and finally for $g(x)^\alpha$, and then combine the conclusions.

Theorem 3.8. *Suppose that g is measurable, submultiplicative and bounded on compacts. Then*

$$\int_{|x|\geq 1} g(x)\Pi(\mathrm{d}x) < \infty \textit{ if and only if } \mathbb{E}\left(g(X_t)\right) < \infty \textit{ for all } t > 0.$$

The proof is essentially the same once one has established that for each submultiplicative function, g, there exist constants $a_g > 0$ and $b_g > 0$ such that $g\,(x) \leq a_g \exp\{b_g|x|\}$. See Exercise 3.3 where examples of submultiplicative functions other than the exponential can be found.

Theorem 3.6 gives us a criterion under which we can perform an exponential change of measure. Define the Laplace exponent

$$\psi(\beta) = \frac{1}{t}\log \mathbb{E}(\mathrm{e}^{\beta X_t}) = -\Psi(-\mathrm{i}\beta), \tag{3.10}$$

defined for all β for which it exits. We now know that the Laplace exponent is finite if and only if $\int_{|x|\geq 1} \mathrm{e}^{\beta x}\Pi(\mathrm{d}x) < \infty$. Following Exercise 1.5 it is easy to deduce under the latter assumption that $\mathcal{E}(\beta) = \{\mathcal{E}_t(\beta)\colon t \geq 0\}$ is a $\mathbb{P}$-martingale with respect to $\mathbb{F}$ where

$$\mathcal{E}_t(\beta) = \mathrm{e}^{\beta X_t - \psi(\beta)t},\ t \geq 0. \tag{3.11}$$

Since it has mean one, it may be used to perform a change of measure via

$$\left.\frac{\mathrm{d}\mathbb{P}^\beta}{\mathrm{d}\mathbb{P}}\right|_{\mathcal{F}_t} = \mathcal{E}_t(\beta).$$

The change of measure above, known as the *Esscher transform*, is a natural generalisation of the Cameron–Martin–Girsanov change of measure. As the next theorem shows, it has the important property that the process X under $\mathbb{P}^\beta$ is still a Lévy process. This fact will play a crucial role in the analysis of risk insurance models and spectrally negative Lévy processes later on in this text.

Theorem 3.9. *Suppose that X is Lévy process with characteristic triple (a, σ, Π) and that $\beta \in \mathbb{R}$ is such that*

$$\int_{|x|\geq 1} \mathrm{e}^{\beta x}\Pi(\mathrm{d}x) < \infty.$$

Under the change of measure $\mathbb{P}^\beta$ the process X is still a Lévy process with characteristic triple (a^*, σ^*, Π^*) where

$$a^* = a - \beta\sigma^2 + \int_{|x|<1}(1-\mathrm{e}^{\beta x})x\Pi(\mathrm{d}x),\ \sigma^* = \sigma \text{ and } \Pi^*(\mathrm{d}x) = \mathrm{e}^{\beta x}\Pi(\mathrm{d}x)$$

so that

$$\psi_\beta(\theta) = -\theta\left(a - \beta\sigma^2 + \int_{|x|<1}(1-\mathrm{e}^{\beta x})x\Pi(\mathrm{d}x)\right) + \frac{1}{2}\sigma^2\theta^2$$
$$\int_{\mathbb{R}}(\mathrm{e}^{\theta x} - 1 - \theta x\mathbf{1}_{(|x<1|)})\mathrm{e}^{\beta x}\Pi(\mathrm{d}x).$$

Proof. Suppose, without loss of generality that $\beta > 0$. In the case that $\beta < 0$ simply consider the forthcoming argument for $-X$ and for $\beta = 0$ the statement of the theorem is trivial. Begin by noting from Hölder's inequality that for any $\theta \in [0, \beta]$ and all $t \geq 0$,

$$\mathbb{E}(\mathrm{e}^{\theta X_t}) \leq \mathbb{E}(\mathrm{e}^{\beta X_t})^{\theta/\beta} < \infty.$$

Hence $\psi(\theta) < \infty$ for all $\theta \in [0, \beta]$. (In fact the inequality shows that ψ is convex in this interval). This in turn implies that $\mathbb{E}(\mathrm{e}^{\mathrm{i}\theta X_t}) < \infty$ for all θ such that $-\Im\theta \in [0, \beta]\}$ and $t \geq 0$. By analytic extension the characteristic exponent Ψ of X is thus finite on the same region of the complex plane.

Fix a time horizon $t > 0$ and note that the density $\exp\{\beta X_t - \psi(\beta)t\}$ is positive with probability one and hence $\mathbb{P}$ and $\mathbb{P}^\beta$ are equivalent measures on $\mathcal{F}_t$. For each $t > 0$, let

$$A_t = \{\forall s \in (0, t],\ \exists \lim_{u \uparrow s} X_u \text{ and } \forall s \in [0, t),\ \exists \lim_{u \downarrow s} X_u = X_s\}.$$

Then, since $\mathbb{P}(A_t) = 1$ for all $t > 0$ it follows that $\mathbb{P}^\beta(A_t) = 1$ for all $t > 0$. That is to say, under $\mathbb{P}^\beta$, the process X still has paths which are almost surely continuous from the right with left limits.

Next let $0 \leq v \leq u \leq s \leq t < \infty$ and $\theta_1, \theta_2 \in \mathbb{R}$. Write

$$\begin{aligned}
&\mathbb{E}^\beta\left(\mathrm{e}^{\mathrm{i}\theta_1(X_t - X_s) + \mathrm{i}\theta_2(X_u - X_v)}\right) \\
&= \mathbb{E}\left(\mathrm{e}^{\beta X_v - \psi(\beta)v}\mathrm{e}^{(\mathrm{i}\theta_2 + \beta)(X_u - X_v) - \psi(\beta)(u-v)}\right. \\
&\qquad\qquad \left.\times\ \mathrm{e}^{\beta(X_s - X_u) - \psi(\beta)(s-u)}\mathrm{e}^{(\mathrm{i}\theta_1 + \beta)(X_t - X_s) - \psi(\beta)(t-s)}\right).
\end{aligned}$$

Using the martingale property of the change of measure and stationary independent increments of X under $\mathbb{P}$, by first conditioning on $\mathcal{F}_s$, then $\mathcal{F}_u$ and finally $\mathcal{F}_v$, we find from the previous equality that

$$\mathbb{E}^\beta\left(\mathrm{e}^{\mathrm{i}\theta_1(X_t - X_s) + \mathrm{i}\theta_2(X_u - X_v)}\right) = \mathrm{e}^{(\Psi(-\mathrm{i}\beta) - \Psi(\theta_1 - \mathrm{i}\beta))(t-s)}\mathrm{e}^{(\Psi(-\mathrm{i}\beta) - \psi(\theta_2 - \mathrm{i}\beta))(u-v)}.$$

Hence under $\mathbb{P}^\beta$, X has stationary independent increments with characteristic exponent given by

$$\Psi(\theta - \mathrm{i}\beta) - \Psi(-\mathrm{i}\beta),\ \theta \in \mathbb{R}.$$

By writing out the latter exponent in terms of the triple (a, σ, Π) associated with X we have

$$\mathrm{i}\theta\left(a - \beta\sigma^2 + \int_{|x|<1}(1 - \mathrm{e}^{\beta x})x\Pi(\mathrm{d}x)\right) + \frac{1}{2}\theta^2\sigma^2, \quad (3.12)$$

$$+ \int_{\mathbb{R}}(1 - e^{\mathrm{i}\theta x} + \mathrm{i}\theta x\mathbf{1}_{|x|<1})\mathrm{e}^{\beta x}\Pi(\mathrm{d}x). \quad (3.13)$$

We thus identify from the Lévy–Khintchine formula for X under $\mathbb{P}^\beta$, that the triple (a^*, σ^*, Π^*) are as given. The Laplace exponent is obtained by applying (3.10)–(3.13). □

Seemingly then, the effect of the Esscher transform is to exponentially tilt the Lévy measure, introduce an additional linear drift and leave the Gaussian contribution untouched.

Note that in the case of a spectrally negative Lévy process the Laplace exponent satisfies $|\psi(\theta)| < \infty$ for $\theta \geq 0$. This follows as a consequence of Theorem 3.6 together with the fact that $\Pi(-\infty, 0) = 0$.

Corollary 3.10. *The Esscher transform may be applied for all $\beta \geq 0$ when X is a spectrally negative Lévy process. Further, under $\mathbb{P}^\beta$, X remains within the class of spectrally negative Lévy process. The Laplace exponent ψ_β of X under $\mathbb{P}^\beta$ satisfies*

$$\psi_\beta(\theta) = \psi(\theta + \beta) - \psi(\beta)$$

for all $\theta \geq -\beta$.

Proof. The Esscher change of measure has the effect of exponentially tilting the original Lévy measure and therefore does not have any influence on the support of the Lévy measure. Compute as before, for $\theta \geq -\beta$,

$$\mathrm{e}^{-\psi_\beta(\theta)} = \mathbb{E}^\beta(\mathrm{e}^{\theta X_1}) = \mathbb{E}(\mathrm{e}^{(\theta+\beta)X_1 - \psi(\beta)}) = \mathbb{E}(\mathrm{e}^{(\theta+\beta) - \psi(\beta)}),$$

which establishes the final statement of the corollary. □

Corollary 3.11. *Under the conditions of Theorem 3.9, if τ is an $\mathbb{F}$-stopping time then on $\{\tau < \infty\}$,*

$$\left.\frac{\mathrm{d}\mathbb{P}^\beta}{\mathrm{d}\mathbb{P}}\right|_{\mathcal{F}_\tau} = \mathcal{E}_\tau(\beta).$$

Proof. By definition if $A \in \mathcal{F}_\tau$, then $A \cap \{\tau \leq t\} \in \mathcal{F}_t$. Hence

$$\begin{aligned}\mathbb{P}^\beta(A \cap \tau \leq t) &= \mathbb{E}(\mathcal{E}_t(\beta)\mathbf{1}_{(A,\tau\leq t)}) \\ &= \mathbb{E}(\mathbb{E}(\mathcal{E}_t(\beta)\mathbf{1}_{(A,\tau\leq t)}|\mathcal{F}_\tau)) \\ &= \mathbb{E}(\mathcal{E}_\tau(\beta)\mathbf{1}_{(A,\tau\leq t)}),\end{aligned}$$

where in the third equality we have used the Strong Markov Property as well as the martingale property for $\mathcal{E}(\beta)$. Now taking limits as $t \uparrow \infty$, the result follows with the help of the Monotone Convergence Theorem. □

We conclude this section, remaining with spectrally negative Lévy processes, by giving another application of the exponential martingale $\mathcal{E}(\alpha)$. Recall the stopping times

$$\tau_x^+ = \inf\{t > 0 : X_t > x\}, \tag{3.14}$$

for $x \geq 0$; also called *first passage times*.

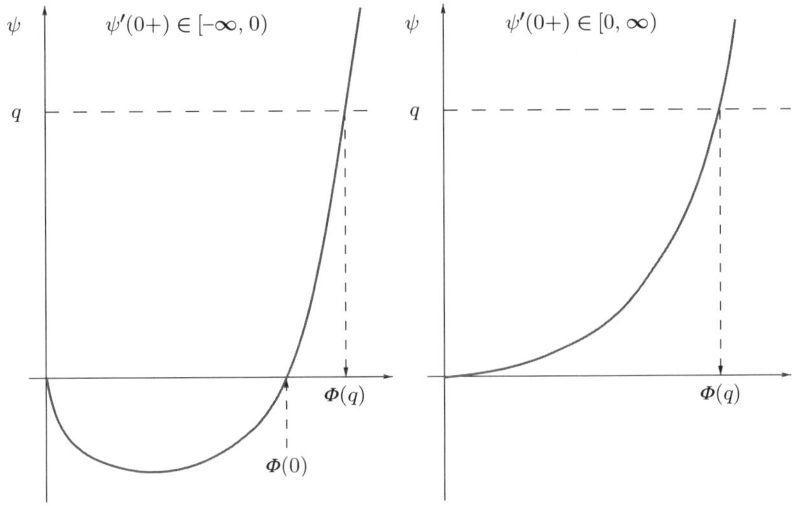

Fig. 3.3. Two examples of ψ, the Laplace exponent of a spectrally negative Lévy process, and the relation to Φ.

Theorem 3.12. *For any spectrally negative Lévy process, with* $q \geq 0$,

$$\mathbb{E}(\mathrm{e}^{-q\tau_x^+}\mathbf{1}_{(\tau_x^+<\infty)}) = \mathrm{e}^{-\Phi(q)x},$$

where $\Phi(q)$ *is the largest root of the equation* $\psi(\theta) = q$.

Before proceeding to the proof, let us make some remarks about the function

$$\Phi(q) = \sup\{\theta \geq 0 : \psi(\theta) = q\}, \tag{3.15}$$

also known as the *right inverse* of ψ. Exercise 3.5 shows that on $[0,\infty)$, ψ is infinitely differentiable, strictly convex and that $\psi(0) = 0$ whilst $\psi(\infty) = \infty$. As a particular consequence of these facts, it follows that $\mathbb{E}(X_1) = \psi'(0+) \in [-\infty,\infty)$. In the case that $\mathbb{E}(X_1) \geq 0$, $\Phi(q)$ is the unique solution to $\psi(\theta) = q$ in $[0,\infty)$. When $\mathbb{E}(X_1) < 0$ the latter statement is true only when $q > 0$ and when $q = 0$ there are two roots to the equation $\psi(\theta) = 0$, one of them being $\theta = 0$ and the other being $\Phi(0) > 0$. See Fig. 3.3 for further clarification.

Proof (of Theorem 3.12). Fix $q > 0$. Using spectral negativity to write $x = X_{\tau_x^+}$ on $\{\tau_x^+ < \infty\}$, note with the help of the Strong Markov Property that

$$\begin{aligned}
&\mathbb{E}(\mathrm{e}^{\Phi(q)X_t - qt}|\,\mathcal{F}_{\tau_x^+}) \\
&= \mathbf{1}_{(\tau_x^+\geq t)}\mathrm{e}^{\Phi(q)X_t-qt} + \mathbf{1}_{(\tau_x^+<t)}\mathrm{e}^{\Phi(q)x-q\tau_x^+}\mathbb{E}(\mathrm{e}^{\Phi(q)(X_t-X_{\tau_x^+})-q(t-\tau_x^+)}|\mathcal{F}_{\tau_x^+}), \\
&= \mathrm{e}^{\Phi(q)X_{t\wedge\tau_x^+}-q(t\wedge\tau_x^+)}
\end{aligned}$$

where in the final equality we have used the fact that $\mathbb{E}(\mathcal{E}_t(\Phi(q))) = 1$ for all $t \geq 0$. Taking expectations again we have

$$\mathbb{E}(\mathrm{e}^{\Phi(q)X_{t\wedge\tau_x^+} - q(t\wedge\tau_x^+)}) = 1.$$

Noting that the expression in the latter expectation is bounded above by $\mathrm{e}^{\Phi(q)x}$, an application of dominated convergence yields

$$\mathbb{E}(\mathrm{e}^{\Phi(q)x - q\tau_x^+}\mathbf{1}_{(\tau_x^+<\infty)}) = 1$$

which is equivalent to the statement of the theorem. □

The following two corollaries are worth recording for later.

Corollary 3.13. *From the previous theorem we have that* $\mathbb{P}(\tau_x^+ < \infty) = \mathrm{e}^{-\Phi(0)x}$ *which is one, if and only if* $\Phi(0) = 0$*, if and only if* $\psi'(0+) \geq 0$*, if and only if* $\mathbb{E}(X_1) \geq 0$*.*

For the next corollary we define a killed subordinator to be a subordinator which is sent to an additional "cemetery" state at an independent and exponentially distributed time.

Corollary 3.14. *If* $\mathbb{E}(X_1) \geq 0$ *then the process* $\{\tau_x^+ : x \geq 0\}$ *is a subordinator and otherwise it is equal in law to a subordinator killed at an independent exponential time with parameter* $\Phi(0)$*.*

Proof. First we claim that $\Phi(q) - \Phi(0)$ is the Laplace exponent of a non-negative infinitely divisible random variable. To see this, note that for all $x \geq 0$,

$$\mathbb{E}(\mathrm{e}^{-q\tau_x^+}|\tau_x^+ < \infty) = \mathrm{e}^{-(\Phi(q)-\Phi(0))x} = \mathbb{E}(\mathrm{e}^{-q\tau_1^+}|\tau_1^+ < \infty)^x,$$

and hence in particular

$$\mathbb{E}(\mathrm{e}^{-q\tau_1^+}|\tau_1^+ < \infty) = \mathbb{E}(\mathrm{e}^{-q\tau_{1/n}^+}|\tau_{1/n}^+ < \infty)^n$$

showing that $\mathbb{P}(\tau_1^+ \in \mathrm{d}z|\tau_1^+ < \infty)$ for $z \geq 0$ is the law of an infinitely divisible random variable. Next, using the Strong Markov Property, spatial homogeneity and again the special feature of spectral negativity that $\{X_{\tau_x^+} = x\}$ on the event $\{\tau_x^+ < \infty\}$, we have for $x, y \geq 0$ and $q \geq 0$,

$$\begin{aligned}
&\mathbb{E}(\mathrm{e}^{-q(\tau_{x+y}^+ - \tau_x^+)}\mathbf{1}_{(\tau_{x+y}^+<\infty)}|\mathcal{F}_{\tau_x^+})\mathbf{1}_{(\tau_x^+<\infty)} \\
&= \mathbb{E}_x(\mathrm{e}^{-q\tau_y^+}\mathbf{1}_{(\tau_y^+<\infty)})\mathbf{1}_{(\tau_x^+<\infty)} \\
&= \mathrm{e}^{-(\Phi(q)-\Phi(0))y}\mathrm{e}^{-\Phi(0)y}\mathbf{1}_{(\tau_x^+<\infty)}.
\end{aligned}$$

In the first equality we have used standard notation for Markov processes, $\mathbb{E}_x(\cdot) = \mathbb{E}(\cdot|X_0 = x)$. We see then that the increment $\tau^+_{x+y} - \tau^+_x$ is independent of $\mathcal{F}_{\tau^+_x}$ on $\{\tau^+_x < \infty\}$ and has the same law as the subordinator with Laplace exponent $\Phi(q) - \Phi(0)$ but killed at an independent exponential time with parameter $\Phi(0)$.

When $\mathbb{E}(X_1) \geq 0$ we have that $\Phi(0) = 0$ and hence the concluding statement of the previous paragraph indicates that $\{\tau^+_x : x \geq 0\}$ is a subordinator (without killing). On the other hand, if $\mathbb{E}(X_1) < 0$, or equivalently $\Phi(0) > 0$ then the second statement of the corollary follows. □

Note that embedded in the previous corollary is the same reasoning which lies behind the justification of the fact that an inverse Gaussian process is a Lévy process. See Sect. 1.2.5 and Exercise 1.6.

Exercises

3.1. Suppose that X is a Lévy process defined on $(\Omega, \mathcal{F}, \mathbb{P})$ and that $\mathcal{F}_t$ is the sigma algebra obtained by completing $\sigma(X_s : s \leq t)$ by the null sets of $\mathbb{P}$. The aim of this exercise is to show that $\mathbb{F} := \{\mathcal{F}_t : t \geq 0\}$ is automatically right continuous; that is to say that for all $t \geq 0$,

$$\mathcal{F}_t = \bigcap_{s>t} \mathcal{F}_s.$$

(i) Fix $t_2 > t_1 \geq 0$ and show that for any $t \geq 0$,

$$\lim_{u \downarrow t} \mathbb{E}(e^{i\theta_1 X_{t_1} + i\theta_2 X_{t_2}} | \mathcal{F}_u) = \mathbb{E}(e^{i\theta_1 X_{t_1} + i\theta_2 X_{t_2}} | \mathcal{F}_t)$$

almost surely where $\theta_1, \theta_2 \in \mathbb{R}$.

(ii) Deduce that for any sequence of times $t_1, ..., t_n \geq 0$,

$$\mathbb{E}(g(X_{t_1}, ..., X_{t_n})|\mathcal{F}_t) = \mathbb{E}(g(X_{t_1}, ..., X_{t_n})|\mathcal{F}_{t+})$$

almost surely for all functions g satisfying $\mathbb{E}(|g(X_{t_1}, ..., X_{t_n})|) < \infty$.

(iii) Conclude that for each $A \in \mathcal{F}_{t+}$, $\mathbb{E}(\mathbf{1}_A|\mathcal{F}_t) = \mathbf{1}_A$ and hence that $\mathcal{F}_t = \mathcal{F}_{t+}$.

3.2. Show that for any $y \geq 0$,

$$\{(y \vee \overline{X}_t) - X_t : t \geq 0\} \text{ and } \{X_t - (\underline{X}_t \wedge (-y)) : t \geq 0\}$$

are $[0, \infty)$-valued strong Markov process.

3.3 (Proof of Theorem 3.8 and examples).

(i) Use the comments under Theorem 3.8 to prove it.

(ii) Prove that the following functions are submultiplicative: $x \vee 1$, $x^\alpha \vee 1$ $|x| \vee 1$, $|x|^\alpha \vee 1$, $\exp(|x|^\beta)$, $\log(|x| \vee \mathrm{e})$, $\log\log(|x| \vee \mathrm{e}^{\mathrm{e}})$ where $\alpha > 0$ and $\beta \in (0, 1]$.
(iii) Suppose that X is a stable process of index $\alpha \in (0, 2)$. Show that $\mathbb{E}(|X_t|^\eta) < \infty$ for all $t \geq 0$ if and only if $\eta \in (0, \alpha)$.

3.4. A generalised tempered stable process is a Lévy process with no Gaussian component and Lévy measure given by

$$\Pi(\mathrm{d}x) = \mathbf{1}_{(x>0)} \frac{c^+}{x^{1+\alpha^+}} \mathrm{e}^{-\gamma^+ x} \, \mathrm{d}x + \mathbf{1}_{(x<0)} \frac{c^-}{|x|^{1+\alpha^-}} \mathrm{e}^{\gamma^- x} \, \mathrm{d}x,$$

where $c^\pm > 0$, $\alpha^\pm \in (-\infty, 2)$ and $\gamma^\pm > 0$. Show that if X is a generalized stable process, then X may always be written in the form $X = X^+ - X^-$ where $X^+ = \{X_t^+ : t \geq 0\}$ and $X^- = \{X_t^- : t \geq 0\}$ satisfy the following:

(i) If $\alpha^\pm < 0$ then $X^\pm$ is a compound Poisson process with drift.
(ii) If $\alpha^\pm = 0$ then $X^\pm$ is a gamma process with drift.
(iii) If $\alpha^\pm = 1$ then $X^\pm$ is a one-sided Cauchy process.
(iv) If $\alpha^\pm \in (0, 1) \cup (1, 2)$ then up to the addition of a linear drift $X^\pm$ has the same law as a spectrally positive stable process with index $\alpha^\pm$ but considered under the change of measure $\mathbb{P}^{-\gamma^\pm}$.

3.5. Suppose that ψ is the Laplace exponent of a spectrally negative Lévy process. By considering explicitly the formula

$$\psi(\beta) = -a\beta + \frac{1}{2}\sigma^2\beta^2 + \int_{(-\infty,0)} (\mathrm{e}^{\beta x} - 1 - \beta x \mathbf{1}_{(x>-1)}) \Pi(\mathrm{d}x)$$

show that on $[0, \infty)$, ψ is infinitely differentiable, strictly convex and that $\psi(0) = 0$ whilst $\psi(\infty) = \infty$.

3.6. Suppose that X is a spectrally negative Lévy process with Lévy–Khintchine exponent Ψ. Here we give another proof of the existence of a finite Laplace exponent for all spectrally negative Lévy processes.

(i) Use spectral negativity together with the lack of memory property to show that for $x, y > 0$,

$$\mathbb{P}(\overline{X}_{\mathbf{e}_q} > x + y) = \mathbb{P}(\overline{X}_{\mathbf{e}_q} > x)\mathbb{P}(\overline{X}_{\mathbf{e}_q} > y)$$

where $\mathbf{e}_q$ is an exponentially distributed random variable[4] independent of X and $\overline{X}_t = \sup_{s \leq t} X_s$.

[4] We are making an abuse of notation in the use of the measure $\mathbb{P}$. Strictly speaking we should work with the measure $\mathbb{P} \times \mathcal{P}$ where $\mathcal{P}$ is the probability measure on the space in which the random variable $\mathbf{e}_q$ is defined. This abuse of notation will be repeated for the sake of convenience at various points throughout this text.

(ii) Deduce that $\overline{X}_{\mathbf{e}_q}$ is exponentially distributed and hence the Laplace exponent $\psi(\beta) = -\Psi(-\mathrm{i}\beta)$ exists and is finite for all $\beta \geq 0$.
(iii) By considering the Laplace transform of the first passage time τ_x^+ as in Sect. 3.3, show that one may also deduce via a different route that $\overline{X}_{\mathbf{e}_q}$ is exponentially distributed with parameter $\Phi(q)$. In particular show that $\overline{X}_\infty$ is either infinite with probability one or is exponentially distributed accordingly as $\mathbb{E}(X_1) \geq 0$ or $\mathbb{E}(X_1) < 0$. [Hint: reconsider Exercise 3.5].

3.7. For this exercise it will be useful to refer to Sect. 1.2.6. Suppose that X is a Stable Lévy process with index $\beta = 1$; that is to say $\Pi(-\infty, 0) = 0$.

(i) Show that if $\alpha \in (0,1)$ then X is a *driftless* subordinator with Laplace exponent satisfying

$$-\log\ \mathbb{E}(\mathrm{e}^{-\theta X_1}) = c\theta^\alpha,\ \theta \geq 0$$

for some $c > 0$.
(ii) Show that if $\alpha \in (1,2)$, then X has a Laplace exponent satisfying

$$-\log \mathbb{E}(\mathrm{e}^{-\theta X_1}) = -C\theta^\alpha,\ \theta \geq 0$$

for some $C > 0$. Confirm that X has no integer moments of order 2 and above as well as being a process of unbounded variation.

4

General Storage Models and Paths of Bounded Variation

In this chapter we return to the queuing and general storage models discussed in Sects. 1.3.2 and 2.7.2. Predominantly we shall concentrate on the asymptotic behaviour of the two quantities that correspond to the workload process and the idle time in the $M/G/1$ queue but now in the general setting described in Sect. 2.7.2. Along the way we will introduce some new tools which will be of help, both in this chapter and in later chapters. Specifically we shall spend some additional time looking at the change of variable and compensation formulae. We also spend some time discussing similarities between the mathematical description of the limiting distribution of workload process (when it is non-trivial) and the Pollaczek–Khintchine formula which requires a study of the small scale behaviour of Lévy processes of bounded variation. We start however by briefly recalling and expanding a little on the mathematical background of general storage models.

4.1 General Storage Models

A general storage model may be considered as consisting of two processes; $\{A_t : t \geq 0\}$, the volume of incoming work and $\{B_t : t \geq 0\}$, the total amount of work that can potentially exit from the system as a result of processing work continuously. In the case of the $M/G/1$ queue we have $A_t = \sum_{i=1}^{N_t} \xi_i$ where $\{N_t : t \geq 0\}$ is a Poisson process and $\{\xi_i : i = 1, 2, ...\}$ are the independent service times of the ordered customers. Further, as the server processes at a constant unit rate, we have simply that $B_t = t$. For all $t \geq 0$ let $D_t = A_t - B_t$. The process $D = \{D_t : t \geq 0\}$ is clearly related to the workload of the system, although it is itself *not* a suitable candidate to model the workload as in principle D may become negative and the workload is clearly a non-negative quantity. The work stored in the system, $W = \{W_t : t \geq 0\}$, is instead defined by

$$W_t = D_t + L_t,$$

where $L = \{L_t : t \geq 0\}$ is increasing with paths that are right continuous (and left limits are of course automatic by monotonicity) and is added to the process D to ensure that $W_t \geq 0$ for all $t \geq 0$. The process L must only increase when $W = 0$ so in particular

$$\int_0^\infty \mathbf{1}_{(W_t>0)} \mathrm{d}L_t = 0.$$

It is easy to check that we may take $L_t = -(\inf_{s\leq t} D_s \wedge 0)$. Indeed with this choice of L we have that $\{W_t = 0\}$ if and only if $\{D_t = \inf_{s\leq t} D_s \wedge 0\}$ if and only if t is in the support of the measure $\mathrm{d}L$. It can also be proved that there is no other choice of L fulfilling these requirements (see for example Kella and Whitt 1996).

We are concerned with the case that the process A is a pure jump subordinator and B is a linear trend. Specifically, $D_t = x - X_t$ where $w \geq 0$ is the workload already in the system at time $t = 0$ and X is a spectrally negative Lévy process of bounded variation. A little algebra with the given expressions for D and L shows that

$$W_t = (w \vee \overline{X}_t) - X_t, \ t \geq 0,$$

where $\overline{X}_t = \sup_{s\leq t} X_s$.

We know from the discussion in Sect. 3.3 (see also Exercise 3.6) that the process X has Laplace exponent $\psi(\theta) = \log \mathbb{E}(\mathrm{e}^{\theta X_1})$ which is finite for all $\theta \geq 0$. Writing X in the form $dt - S_t$ where $d > 0$ and $S = \{S_t : t \geq 0\}$ is a pure jump subordinator, it is convenient to write the Laplace exponent of X in the form

$$\psi(\theta) = d\theta - \int_{(0,\infty)} (1 - \mathrm{e}^{-\theta x})\nu(\mathrm{d}x),$$

where ν is the Lévy measure of the subordinator S and necessarily $\int_{(0,\infty)}(1 \wedge x)\nu(\mathrm{d}x) < \infty$. We interpret $\nu(\mathrm{d}x)\mathrm{d}t + o(\mathrm{d}t)$ as the probability a job or storage bulk of size x arrives independently in each $\mathrm{d}t$.

4.2 Idle Times

We start by introducing the parameter

$$\rho := \frac{d - \psi'(0+)}{d}.$$

Note that regimes $0 < \rho < 1$, $\rho = 1$ and $\rho > 1$ correspond precisely to the regimes $\psi'(0+) > 0$, $\psi'(0+) = 0$ and $\psi'(0+) < 0$, respectively. The first two of these cases thus imply that $\Phi(0) = 0$ and the third case implies $\Phi(0) > 0$ where Φ is the right inverse of ψ defined in (3.15). When $d = 1$ and $\nu = \lambda F$ where F is a distribution function and $\lambda > 0$ is the arrival rate, the process W is the workload of an $M/G/1$ queue. In that case $\rho = \lambda \mathbb{E}(\xi)$ where ξ is a random variable with distribution F and is called the *traffic intensity.*

The main purpose of this section is to prove the following result which generalises Theorem 1.11.

Theorem 4.1. *Suppose that $\rho > 1$ then the total time that the storage process spends idle*

$$I := \int_0^\infty \mathbf{1}_{(W_t=0)} \mathrm{d}t,$$

has the following distribution

$$\mathbb{P}(I \in \mathrm{d}x) = (1 - \mathrm{e}^{-\Phi(0)w})\delta_0(\mathrm{d}x) + \Phi(0)\mathrm{e}^{-\Phi(0)(w+xd)}\mathrm{d}x.$$

Otherwise if $0 < \rho \leq 1$ then I is infinite with probability one.

Proof (of Theorem 4.1). Essentially the proof mimics the steps of Exercise 1.9. As one sees for the case of the $M/G/1$ queue, a key ingredient to the proof is that one may identify the process $\{d\int_0^t \mathbf{1}_{(W_s=0)}\mathrm{d}s : t \geq 0\}$ as the process $\{\overline{X}_t : t \geq 0\}$. To see why this is true in the general storage model, recall from the Lévy–Itô decomposition that X has a countable number of jumps over finite intervals of time and hence the same is true of W. Further since X has negative jumps, $W_s = 0$ only if there is no jump at time s. Hence, given that X is the difference of a linear drift with rate d and a subordinator S, it follows that for each $t \geq 0$,

$$\begin{aligned}\overline{X}_t &= \int_0^t \mathbf{1}_{(\overline{X}_s=X_s)}\mathrm{d}X_s\\ &= d\int_0^t \mathbf{1}_{(\overline{X}_s=X_s)}\mathrm{d}s - \int_0^t \mathbf{1}_{(\overline{X}_s=X_s)}\mathrm{d}S_s\\ &= d\int_0^t \mathbf{1}_{(\overline{X}_s=X_s)}\mathrm{d}s\end{aligned}$$

almost surely where the final equality follows as a consequence of the fact that

$$\int_0^t \mathbf{1}_{(\overline{X}_s=X_s)}\mathrm{d}S_s \leq \int_0^t \mathbf{1}_{(\Delta S_s=0)}\mathrm{d}S_s = 0.$$

It is important to note that the latter calculation only works for spectrally negative Lévy processes of bounded variation on account of the Lévy–Itô decomposition.

Now following Exercise 3.6 (iii) we can use the equivalence of the events $\{\overline{X}_\infty > x\}$ and $\{\tau_x^+ < \infty\}$, where τ_x^+ is the first hitting time of (x, ∞) defined in (3.14), to deduce that $\overline{X}_\infty$ is exponentially distributed with parameter $\Phi(0)$. When $\Phi(0) = 0$ then the previous statement is understood to mean that $\mathbb{P}(\overline{X}_\infty = \infty) = 1$. When $w = 0$ we have precisely that

$$\frac{\overline{X}_\infty}{d} = \int_0^\infty \mathbf{1}_{(\overline{X}_s=X_s)}\mathrm{d}s - \int_0^\infty \mathbf{1}_{(W_s=0)}\mathrm{d}s. \tag{4.1}$$

Hence we see that I is exponentially distributed with parameter $d\Phi(0)$. Recalling which values of ρ imply that $\Phi(0) > 0$ we see that the statement of the theorem follows for the case $w = 0$.

In general however, when $w > 0$ the equality (4.1) is not valid. Instead we have that

$$\begin{aligned}
\int_0^\infty \mathbf{1}_{(W_s=0)}\mathrm{d}s &= \int_0^{\tau_w^+} \mathbf{1}_{(W_s=0)}\mathrm{d}s + \int_{\tau_w^+}^\infty \mathbf{1}_{(W_s=0)}\mathrm{d}s \\
&= \mathbf{1}_{(\tau_w^+<\infty)} \int_{\tau_w^+}^\infty \mathbf{1}_{(W_s=0)}\mathrm{d}s \\
&= \mathbf{1}_{(\overline{X}_\infty \geq w)} I^*, \qquad (4.2)
\end{aligned}$$

where I^* is independent of $\mathcal{F}_{\tau_w^+}$ on $\{\tau_w^+ < \infty\}$ and equal in distribution to $\int_0^\infty \mathbf{1}_{(W_s=0)}\mathrm{d}s$ when $w = 0$. Note that the first integral in the right-hand side of the first equality disappears on account of the fact that $W_s > 0$ for all $s < \tau_w^+$. The statement of the theorem now follows for $0 \leq \rho \leq 1$ by once again recalling that in this regime $\Phi(0) = 0$ and hence from (4.2) $\overline{X}_\infty = \infty$ with probability one and hence $I = I^*$. The latter has previously been shown to be infinite with probability one. On the other hand, when $\rho > 1$, we see from (4.2) that there is an atom at zero corresponding to the event $\{\overline{X}_\infty < w\}$ with probability $1 - \mathrm{e}^{-\Phi(0)w}$. Otherwise, with independent probability $\mathrm{e}^{-\Phi(0)w}$, the integral I has the same distribution as I^*. Again from previous calculations for the case $w = 0$ we have seen that this is exponential with parameter $d\Phi(0)$ and the proof is complete. □

4.3 Change of Variable and Compensation Formulae

Next we spend a little time introducing the change of variable formula and the compensation formula. Both formulae pertain to a form of stochastic calculus. The theory of stochastic calculus is an avenue which we choose not to pursue in full generality, choosing instead to make some brief remarks. Our exposition will suffice to study in more detail the storage processes discussed in Chap. 1 as well as a number of other applications in later Chapters.

4.3.1 The Change of Variable Formula

We assume that $X = \{X_t : t \geq 0\}$ is a Lévy process of bounded variation. Referring back to Chap. 2, (2.21) and (2.22) we recall that we may always write its Lévy–Khintchine exponent as

$$\Psi(\theta) = -\mathrm{i}d\theta + \int_{\mathbb{R}} (1 - \mathrm{e}^{\mathrm{i}\theta x})\Pi(\mathrm{d}x)$$

and correspondingly identify X path-wise in the form of a generalised compound Poisson process

$$X_t = dt + \int_{[0,t]} \int_{\mathbb{R}} xN(\mathrm{d}s \times \mathrm{d}x),$$

where $d \in \mathbb{R}$ and, as usual, N is the Poisson random measure associated with the jumps of X.

Our goal in this section is to prove the following change of variable formula.

Theorem 4.2. *Let $C^{1,1}([0,\infty)\times\mathbb{R})$ be the space of functions $f:[0,\infty)\times\mathbb{R}\to\mathbb{R}$ which are continuously differentiable in each variable (in the case of the derivative in the first variable at the origin a right derivative is understood). If $f(s,x)$ belongs to the class $C^{1,1}([0,\infty)\times\mathbb{R})$ then*

$$f(t,X_t) = f(0,X_0) + \int_0^t \frac{\partial f}{\partial s}(s,X_s)\mathrm{d}s + d\int_0^t \frac{\partial f}{\partial x}(s,X_s)\mathrm{d}s + \int_{[0,t]}\int_{\mathbb{R}} (f(s,X_{s-}+x) - f(s,X_{s-}))N(\mathrm{d}s\times\mathrm{d}x).$$

It will become apparent from the proof of this theorem that the final integral with respect to N in the change of variable formula is well defined.

It is worth mentioning that the change of variable formula exists in a much more general form. For example it is known (cf. Protter 2004 that if $V=\{V_t : t\geq 0\}$ is any mapping from $[0,\infty)$ to $\mathbb{R}$ (random or deterministic) of bounded variation with paths that are right continuous and $f(s,x)\in C^{1,1}([0,\infty)\times\mathbb{R})$ is continuously differentiable, then $\{f(t,V_t): t\geq 0\}$ is a mapping from $[0,\infty)$ to $\mathbb{R}$ of bounded variation which satisfies

$$f(t,V_t) = f(0,V_0) + \int_0^t \frac{\partial f}{\partial s}(s,V_s)\mathrm{d}s + \int_0^t \frac{\partial f}{\partial x}(s,V_{s-})\mathrm{d}V_s + \sum_{s\leq t}\{f(s,V_s) - f(s,V_{s-}) - \Delta V_s\frac{\partial f}{\partial x}(s,V_s)\}, \tag{4.3}$$

where $\Delta V_s = V_s - V_{s-}$. Note also that since V is of bounded variation then it has a minimal decomposition as the difference of two increasing functions mapping $[0,\infty)$ to $[0,\infty)$ and hence left continuity of V is automatically guaranteed. This means that V has a countable number of jumps (see Exercise 2.4). One therefore may understand the final term on the right-hand side of (4.3) as a convergent sum over the countable discontinuities of V. In the case that V is a Lévy process of bounded variation, it is a straightforward exercise to deduce that when one represents the discontinuities of V via a Poisson random measure then (4.3) and the conclusion of Theorem 4.2 agree.

Proof (of Theorem 4.2). Define for all $\varepsilon>0$,

$$X_t^\varepsilon = dt + \int_{[0,t]}\int_{\{|x|\geq\varepsilon\}} xN(\mathrm{d}s\times\mathrm{d}x),\ t\geq 0.$$

As $\Pi(\mathbb{R}\backslash(-\varepsilon,\varepsilon))<\infty$ it follows that N counts an almost surely finite number of jumps over $[0,t]\times\{\mathbb{R}\backslash(-\varepsilon,\varepsilon)\}$ and $X^\varepsilon = \{X_t^\varepsilon : t\geq 0\}$ is a compound Poisson process with drift. Suppose the collection of jumps of X^ε up to time

t are described by the time-space points $\{(T_i, \xi_i) : i = 1, ..., N\}$ where $N = N([0,t] \times \{\mathbb{R}\backslash(-\varepsilon, \varepsilon)\})$. Let $T_0 = 0$. Then a simple telescopic sum gives

$$f(t, X_t^\varepsilon) = f(0, X_0^\varepsilon) + \sum_{i=1}^{N}(f(T_i, X_{T_i}^\varepsilon) - f(T_{i-1}, X_{T_{i-1}}^\varepsilon)) \\ +(f(t, X_t^\varepsilon) - f(T_N, X_{T_N}^\varepsilon)).$$

Now noting that X^ε is piece-wise linear we have,

$$\begin{aligned}
&f(t, X_t^\varepsilon) \\
&= f(0, X_0^\varepsilon) \\
&+\sum_{i=1}^{N}\left(\int_{T_{i-1}}^{T_i} \frac{\partial f}{\partial s}(s, X_s^\varepsilon) + d\frac{\partial f}{\partial x}(s, X_s^\varepsilon)\mathrm{d}s + (f(T_i, X_{T_i-}^\varepsilon + \xi_i) - f(T_i, X_{T_i-}^\varepsilon))\right) \\
&\qquad + \int_{T_N}^{t} \frac{\partial f}{\partial s}(s, X_s^\varepsilon) + d\frac{\partial f}{\partial x}(s, X_s^\varepsilon)\mathrm{d}s \\
&= f(0, X_0^\varepsilon) + \int_0^t \frac{\partial f}{\partial s}(s, X_s^\varepsilon) + d\frac{\partial f}{\partial x}(s, X_s^\varepsilon)\mathrm{d}s \\
&\qquad + \int_{[0,t]}\int_{\mathbb{R}\backslash\{0\}} (f(s, X_{s-}^\varepsilon + x) - f(x, X_{s-}^\varepsilon))\mathbf{1}_{(|x|\geq\varepsilon)}N(\mathrm{d}s \times \mathrm{d}x). \qquad (4.4)
\end{aligned}$$

(Note that the smoothness of f has been used here).

From Exercise 2.8 we know that any Lévy process of bounded variation may be written as the difference of two independent subordinators. In this spirit write $X_t = X_t^{(1)} - X_t^{(2)}$ where

$$X_t^{(1)} = (d \vee 0)t + \int_{[0,t]}\int_{(0,\infty)} xN(\mathrm{d}s \times \mathrm{d}x), \quad t \geq 0$$

and

$$X_t^{(2)} = |d \wedge 0|t - \int_{[0,t]}\int_{(-\infty,0)} xN(\mathrm{d}s \times \mathrm{d}x), \quad t \geq 0.$$

Now let

$$X_t^{(1,\varepsilon)} = (d \vee 0)t + \int_{[0,t]}\int_{[\varepsilon,\infty)} xN(\mathrm{d}s \times \mathrm{d}x), \quad t \geq 0$$

and

$$X_t^{(2,\varepsilon)} = |d \wedge 0|t - \int_{[0,t]}\int_{(-\infty,-\varepsilon]} xN(\mathrm{d}s \times \mathrm{d}x), \quad t \geq 0$$

and note by almost sure monotone convergence that for each fixed $t \geq 0$, $X_t^{(i,\varepsilon)} \uparrow X_t^{(i)}$ almost surely for $i = 1, 2$ as $\varepsilon \downarrow 0$. Since $X_t^\varepsilon = X_t^{(1,\varepsilon)} - X_t^{(2,\varepsilon)}$ we see that for each fixed $t > 0$ we have $\lim_{\varepsilon\downarrow 0} X_t^\varepsilon = X_t$ almost surely. By replacing $[0,t]$ by $[0,t)$ in the delimiters of the definitions above it is also clear that for each fixed $t > 0$, $\lim_{\varepsilon\downarrow 0} X_{t-}^\varepsilon = X_{t-}$ almost surely.

Now define the random region $B = \{0 \le x \le |X_s^\varepsilon| : s \le t \text{ and } \varepsilon > 0\}$ and note that B is almost surely bounded in $\mathbb{R}$ since it is contained in

$$\{0 \le x \le X_s^{(1)} : s \le t\} \cup \{0 \ge x \ge -X_s^{(2)} : s \le t\}$$

and the latter two are almost surely bounded sets on account of right continuity of paths. Due to the assumed smoothness of f, both derivatives of f are uniformly bounded (by a random value) on $[0,t] \times \overline{B}$, where $\overline{B}$ is the closure of the set B. Using the limiting behaviour of X^ε in ε and boundedness of the derivatives of f on $[0,t] \times \overline{B}$ together with almost sure dominated convergence we see that

$$\lim_{\varepsilon \downarrow 0} \int_0^t \frac{\partial f}{\partial s}(s, X_s^\varepsilon) + d\frac{\partial f}{\partial x}(s, X_s^\varepsilon)\mathrm{d}s = \int_0^t \frac{\partial f}{\partial s}(s, X_s) + d\frac{\partial f}{\partial x}(s, X_s)\mathrm{d}s.$$

Again using uniform boundedness of $\partial f/\partial x$ but this time on $[0,t] \times \{x + \overline{B} : |x| \le 1\}$ we note with the help of the Mean Value Theorem that for all $\varepsilon > 0$ and $s \in [0,t]$,

$$\left|(f(s, X_{s-}^\varepsilon + x) - f(s, X_{s-}^\varepsilon))\mathbf{1}_{(\varepsilon \le |x| < 1)}, \right| \le C\,|x|\,\mathbf{1}_{(|x|<1)},$$

where $C > 0$ is some random variable, independent of s, ε and x. The function $|x|$ integrates against N on $[0,t] \times \{(-1,1)\}$ thanks to the assumption that X has bounded variation. Now appealing to almost sure dominated convergence again we have that

$$\lim_{\varepsilon \downarrow 0} \int_{[0,t]} \int_{(-1,1)} (f(s, X_{s-}^\varepsilon + x) - f(s, X_{s-}^\varepsilon))\mathbf{1}_{(|x| \ge \varepsilon)} N(\mathrm{d}s \times \mathrm{d}x)$$

$$= \int_{[0,t]} \int_{(-1,1)} f(s, X_{s-} + x) - f(s, X_{s-}) N(\mathrm{d}s \times \mathrm{d}x).$$

A similar limit holds when the delimiters in the double integrals above are replaced by $[0,t] \times \{\mathbb{R}\backslash(-1,1)\}$ as there are at most a finite number of atoms in the support of N in this domain. Now taking limits on both sides of (4.4) the statement of the theorem follows. □

It is clear from the above proof that one could not expect such a formula to be valid for a general Lévy process. In order to write down a change of variable formula for a general Lévy process, X, one must first progress to the construction of stochastic integrals with respect to the X; at the very least, integrals of the form

$$\int_0^t g(s, X_{s-})\mathrm{d}X_s \tag{4.5}$$

for continuous functions g. Roughly speaking the latter integral may be understood as the limit

$$\lim_{||\mathcal{P}||\downarrow 0}\sum_{i\geq 1} g(t_{i-1}, X_{t_{i-1}})(X_{t\wedge t_i} - X_{t\wedge t_{i-1}}),$$

where $\mathcal{P} = \{0 = t_0 \leq t_1 \leq t_2 \leq ...\}$ is a partition of $[0,\infty)$, $||\mathcal{P}|| = \sup_{i\geq 1}(t_i - t_{i-1})$ and the limit is taken in probability uniformly in t on $[0,T]$ where $T > 0$ is some finite time horizon. This is not however the only way to make sense of (4.5) although all definitions must be equivalent; see for example Exercise 4.5. In the case that X has bounded variation the integral (4.5) takes the recognisable form

$$\int_0^t g(s, X_{s-})\mathrm{d}X_s = d\int_0^t g(s, X_s)\mathrm{d}s + \int_{[0,t]}\int_{\mathbb{R}} g(s, X_{s-})N(\mathrm{d}s\times\mathrm{d}x). \quad (4.6)$$

Establishing these facts is of course non-trivial and in keeping with the title of this book we shy away from them. The reader is otherwise directed to Applebaum (2004) for a focused account of the necessary calculations. Protter (2004) also gives the much broader picture for integration with respect to a general semi-martingale. (A Lévy process is an example of a broader family of stochastic processes called semi-martingales which form a natural class from which to construct a theory of stochastic integration). We finish this section however by simply stating Itô's formula for a general Lévy process[1] which functions as a change of variable for the cases not covered by Theorem 4.2.

Theorem 4.3. *Let $C^{1,2}([0,\infty)\times\mathbb{R})$ be the space of functions $f : [0,\infty)\times\mathbb{R}$ which are continuously differentiable in the first variable (where the right derivative is understood at the origin) and twice continuously differentiable in the second variable. Then for a general Lévy process X with Gaussian coefficient $\sigma \geq 0$ and $f \in C^{1,2}([0,\infty)\times\mathbb{R})$ we have*

$$\begin{aligned} f(t, X_t) = f(0, X_0) &+ \int_0^t \frac{\partial f}{\partial s}(s, X_s)\mathrm{d}s + \int_0^t \frac{\partial f}{\partial x}(s, X_{s-})\mathrm{d}X_s \\ &+ \int_0^t \frac{1}{2}\sigma^2\frac{\partial^2 f}{\partial x^2}(s, X_s)\mathrm{d}s \\ &+ \int_{[0,t]}\int_{\mathbb{R}}\left(f(s, X_{s-}+x) - f(s, X_{s-}) - x\frac{\partial f}{\partial x}(s, X_{s-})\right)N(\mathrm{d}s\times\mathrm{d}x). \end{aligned}$$

4.3.2 The Compensation Formula

Although it was indicated that this chapter principally concerns processes of bounded variation, the compensation formula, which we will shortly establish, is applicable to all Lévy processes. Suppose then that X is a general Lévy process with Lévy measure Π. Recall our running assumption that X is defined on the filtered probability space $(\Omega, \mathcal{F}, \mathbb{F}, \mathbb{P})$ where $\mathbb{F} = \{\mathcal{F}_t : t \geq 0\}$

[1] As with the change of variable formula, a more general form of Itô's formula exists which includes the statement of Theorem 4.3. The natural setting as indicated above is the case that X is a semi-martingale.

is assumed to is completed by null sets (and hence is right continuous as X is a Lévy process). As usual, N will denote Poisson random measure with intensity $\mathrm{d}s \times \mathrm{d}\Pi$ describing the jumps of X. The main result of this section may be considered as a generalisation of the results in Theorem 2.7.

Theorem 4.4. *Suppose $\phi : [0,\infty) \times \mathbb{R} \times \Omega \to [0,\infty)$ is a random time-space function such that*

(i) as a trivariate function $\phi = \phi(t,x)[\omega]$ is measurable,
(ii) for each $t \geq 0$ $\phi(t,x)[\omega]$ is $\mathcal{F}_t \times \mathcal{B}(\mathbb{R})$-measurable and
(iii) for each $x \in \mathbb{R}$, with probability one, $\{\phi(t,x)[\omega] : t \geq 0\}$ is a left continuous process.

Then for all $t \geq 0$,

$$\mathbb{E}\left(\int_{[0,t]}\int_{\mathbb{R}} \phi(s,x)N(\mathrm{d}s \times \mathrm{d}x)\right) = \mathbb{E}\left(\int_0^t \int_{\mathbb{R}} \phi(s,x)\mathrm{d}s\Pi(\mathrm{d}x)\right) \tag{4.7}$$

with the understanding that the right-hand side is infinite if and only if the left-hand side is.

Note that for each $\varepsilon > 0$,

$$\int_{[0,t]}\int_{\mathbb{R}\backslash(-\varepsilon,\varepsilon)} \phi(s,x)N(\mathrm{d}s \times \mathrm{d}x)$$

is nothing but the sum over a finite number of terms of positive random objects and hence, under the first assumption on ϕ, is measurable in ω. By (almost sure) monotone convergence the integral $\int_{[0,t]}\int_{\mathbb{R}} \phi(s,x)N(\mathrm{d}s \times \mathrm{d}x)$ is well defined as

$$\lim_{\varepsilon \downarrow 0}\int_{[0,t]}\int_{\mathbb{R}\backslash(-\varepsilon,\varepsilon)} \phi(s,x)N(\mathrm{d}s \times \mathrm{d}x)$$

and is measurable in ω (recall when the limit of a sequence of measurable functions exists it is also measurable). Hence the left-hand side of (4.7) is well defined even if infinite in value.

On the other hand, under the first assumption on ϕ, Fubini's Theorem implies that, as a random variable,

$$\int_0^t \int_{\mathbb{R}} \phi(s,x)\mathrm{d}s\Pi(\mathrm{d}x)$$

is measurable in ω. Hence the expression on the right-hand side of (4.7) is also well defined even when infinite in value.

Proof (of Theorem 4.4). Suppose initially that in addition to the assumptions of the theorem, ϕ is uniformly bounded by $C(1 \wedge x^2)$ for some $C > 0$. Note that this ensures the finiteness of the expressions on the left-hand and right-hand side of (4.7). Write

$$\phi^n(t,x) = \phi(0,x)\mathbf{1}_{(t=0)} + \sum_{k\geq 0} \phi(k/2^n, x)\mathbf{1}_{(t\in(k/2^n,(k+1)/2^n])} \tag{4.8}$$

noting that ϕ^n also satisfies the assumptions (i)–(iii) of the theorem. Hence, as remarked above, for each $\varepsilon > 0$,

$$\int_{[0,t]}\int_{\mathbb{R}\backslash(-\varepsilon,\varepsilon)} \phi^n(s,x)N(\mathrm{d}s\times\mathrm{d}x)$$

is well defined and measurable in ω. We have

$$\begin{aligned}
&\mathbb{E}\left(\int_{[0,t]}\int_{\mathbb{R}\backslash(-\varepsilon,\varepsilon)} \phi^n(s,x)N(\mathrm{d}s\times\mathrm{d}x)\right)\\
&= \mathbb{E}\left(\sum_{k\geq 0}\int_{(\frac{k}{2^n},\frac{k+1}{2^n}]}\int_{\mathbb{R}\backslash(-\varepsilon,\varepsilon)} \phi(k/2^n,x)N(\mathrm{d}s\times\mathrm{d}x)\right)\\
&= \mathbb{E}\left(\sum_{k\geq 0}\mathbb{E}\left(\int_{(\frac{k}{2^n}\wedge t,\frac{k+1}{2^n}\wedge t]}\int_{\mathbb{R}\backslash(-\varepsilon,\varepsilon)} \phi(k/2^n,x)N(\mathrm{d}s\times\mathrm{d}x)\middle|\mathcal{F}_{\frac{k}{2^n}\wedge t}\right)\right)\\
&= \mathbb{E}\left(\sum_{k\geq 0}\int_{(\frac{k}{2^n}\wedge t,\frac{k+1}{2^n}\wedge t]}\int_{\mathbb{R}\backslash(-\varepsilon,\varepsilon)} \phi(k/2^n,x)\mathrm{d}s\Pi(\mathrm{d}x)\right)\\
&= \mathbb{E}\left(\int_{[0,t]}\int_{\mathbb{R}\backslash(-\varepsilon,\varepsilon)} \phi^n(s,x)\mathrm{d}s\Pi(\mathrm{d}x)\right),
\end{aligned} \tag{4.9}$$

where in the third equality we have used the fact that N has independent counts on disjoint domains, the measurability of $\phi^n(k/2^n,x)$ together with an application of Theorem 2.7 (iii). Since it is assumed that ϕ is uniformly bounded by $C(1\wedge x^2)$ we may apply dominated convergence on both sides of (4.9) as $n\uparrow\infty$ together with the fact that $\lim_{n\uparrow\infty}\phi^n(t,x) = \phi(t-,x) = \phi(t,x)$ almost surely (by the assumed left continuity of ϕ) to conclude that

$$\mathbb{E}\left(\int_{[0,t]}\int_{\mathbb{R}\backslash(-\varepsilon,\varepsilon)} \phi(s,x)N(\mathrm{d}s\times\mathrm{d}x)\right) = \mathbb{E}\left(\int_0^t\int_{\mathbb{R}\backslash(-\varepsilon,\varepsilon)} \phi(s,x)\mathrm{d}s\Pi(\mathrm{d}x)\right).$$

Now take limits as $\varepsilon\downarrow 0$ and apply the Monotone Convergence theorem on each side of the above equality to deduce (4.7) for the case that ϕ is uniformly bounded by $C(1\wedge x^2)$.

To remove the latter condition, note that it has been established that (4.7) holds for $\phi\wedge C(1\wedge x^2)$ where ϕ is given in the statement of the theorem. By taking limits as $C\uparrow\infty$ in the aforementioned equality, again with the help of the Monotone Convergence theorem, the required result follows. □

Reviewing the proof of this result, there is a rather obvious corollary which follows. We leave its proof to the reader as an exercise.

Corollary 4.5. *Under the same conditions as Theorem 4.4 we have for all* $0 \le u \le t < \infty$,

$$\mathbb{E}\left(\int_{(u,t]} \int_{\mathbb{R}} \phi(s,x) N(\mathrm{d}s \times \mathrm{d}x) \middle| \mathcal{F}_u \right) = \mathbb{E}\left(\int_u^t \int_{\mathbb{R}} \phi(s,x) \mathrm{d}s \Pi(\mathrm{d}x) \middle| \mathcal{F}_u \right).$$

The last corollary also implies the martingale result below.

Corollary 4.6. *Under the same conditions as Theorem 4.4 with the additional assumption that*

$$\mathbb{E}\left(\int_{[0,t]} \int_{\mathbb{R}} |\phi(s,x)| \mathrm{d}s \Pi(\mathrm{d}x) \right) < \infty,$$

we have that

$$M_t := \int_{[0,t]} \int_{\mathbb{R}} \phi(s,x) N(\mathrm{d}s \times \mathrm{d}x) - \int_{[0,t]} \int_{\mathbb{R}} \phi(s,x) \mathrm{d}s \Pi(\mathrm{d}x), \ t \ge 0,$$

is a martingale.

Proof. The additional integrability condition on ϕ and Theorem 4.4 implies that for each $t \ge 0$,

$$\mathbb{E}|M_t| \le 2\mathbb{E}\left(\int_{[0,t]} \int_{\mathbb{R}} |\phi(s,x)| \mathrm{d}s \Pi(\mathrm{d}x) \right) < \infty.$$

For $0 \le u \le t$ we see that

$$\begin{aligned} \mathbb{E}(M_t|\mathcal{F}_u) &= M_u + \mathbb{E}\left(\int_{(u,t]} \int_{\mathbb{R}} \phi(s,x) N(\mathrm{d}s \times \mathrm{d}x) \middle| \mathcal{F}_u \right) \\ &\quad - \mathbb{E}\left(\int_u^t \int_{\mathbb{R}} \phi(s,x) \mathrm{d}s \Pi(\mathrm{d}x) \middle| \mathcal{F}_u \right) \\ &= M_u \end{aligned}$$

where the last equality is a consequence of Corollary 4.5. □

4.4 The Kella–Whitt Martingale

In this section we introduce a martingale, the Kella–Whitt martingale, which will prove to be useful for the analysis concerning the existence of a stationary distribution of the process W. The martingale itself is of implicit interest as far as fluctuation theory of general spectrally negative Lévy processes are concerned since one may derive a number of important identities from it. These identities also appear later in this text as a consequence of other techniques centred around the Wiener–Hopf factorisation. See in particular Exercise 4.8.

The Kella–Whitt martingale takes its name from Kella and Whitt (1992) and is presented in the theorem below.

Theorem 4.7. *Suppose that X is a spectrally negative Lévy process of bounded variation as described in the introduction. For each $\alpha \geq 0$, the process*

$$\psi(\alpha)\int_0^t \mathrm{e}^{-\alpha(\overline{X}_s - X_s)}\mathrm{d}s + 1 - \mathrm{e}^{-\alpha(\overline{X}_t - X_t)} - \alpha\overline{X}_t, \quad t \geq 0$$

is a $\mathbb{P}$-martingale with respect to $\mathbb{F}$.

Proof. The proof of this theorem will rely on the change of variable and compensation formulae. To be more precise, we will make use of the slightly more general version of the change of variable formula given in Exercise 4.2 which looks like

$$f(\overline{X}_t, X_t) = f(\overline{X}_0, X_0) + d\int_0^t \frac{\partial f}{\partial x}(\overline{X}_s, X_s)\mathrm{d}s + \int_0^t \frac{\partial f}{\partial s}(\overline{X}_s, X_s)\mathrm{d}\overline{X}_s$$
$$+ \int_{[0,t]}\int_{\mathbb{R}} (f(\overline{X}_s, X_{s-} + x) - f(\overline{X}_s, X_{s-}))N(\mathrm{d}s \times \mathrm{d}x).$$

for $f(s, x) \in C^{1,1}([0, \infty) \times \mathbb{R})$. From the latter we have that

$$\begin{aligned}
\mathrm{e}^{-\alpha(\overline{X}_t - X_t)} &= 1 + \alpha d\int_0^t \mathrm{e}^{-\alpha(\overline{X}_s - X_s)}\mathrm{d}s - \alpha\int_0^t \mathrm{e}^{-\alpha(\overline{X}_s - X_s)}\mathrm{d}\overline{X}_s \\
&\quad + \int_{[0,t]}\int_{(-\infty,0)} (\mathrm{e}^{-\alpha(\overline{X}_s - X_{s-} - x)} - \mathrm{e}^{-\alpha(\overline{X}_s - X_{s-})})N(\mathrm{d}s \times \mathrm{d}x) \\
&= 1 + \alpha d\int_0^t \mathrm{e}^{-\alpha(\overline{X}_s - X_s)}\mathrm{d}s - \alpha\int_0^t \mathrm{e}^{-\alpha(\overline{X}_s - X_s)}\mathrm{d}\overline{X}_s \\
&\quad + \int_0^t\int_{(0,\infty)} (\mathrm{e}^{-\alpha(\overline{X}_s - X_{s-} + x)} - \mathrm{e}^{-\alpha(\overline{X}_s - X_{s-})})\mathrm{d}s\nu(\mathrm{d}x) \\
&\quad + M_t
\end{aligned} \tag{4.10}$$

(recall that ν is the Lévy measure of $-X$, defined at the end of Sect. 4.1) where for each $t \geq 0$,

$$\begin{aligned}
M_t = &\int_{[0,t]}\int_{(-\infty,0)} (\mathrm{e}^{-\alpha(\overline{X}_s - X_{s-} - x)} - \mathrm{e}^{-\alpha(\overline{X}_s - X_{s-})})N(\mathrm{d}s \times \mathrm{d}x) \\
&- \int_0^t\int_{(0,\infty)} (\mathrm{e}^{-\alpha(\overline{X}_s - X_{s-} + x)} - \mathrm{e}^{-\alpha(\overline{X}_s - X_{s-})})\mathrm{d}s\nu(\mathrm{d}x).
\end{aligned}$$

Note that the second integral on the right-hand side of (4.10) can be replaced by $\overline{X}_t$ since the process $\overline{X}$ increases if and only if the integrand is equal to one. Note also that the final integral on the right-hand side of (4.10) also combines with the first integral to give

$$\begin{aligned}
&\alpha d\int_0^t \mathrm{e}^{-\alpha(\overline{X}_s - X_s)}\mathrm{d}s + \int_0^t \mathrm{e}^{-\alpha(\overline{X}_s - X_s)}\mathrm{d}s\int_{(0,\infty)} (\mathrm{e}^{-\alpha x} - 1)\nu(\mathrm{d}x) \\
&= \psi(\alpha)\int_0^t \mathrm{e}^{-\alpha(\overline{X}_s - X_s)}\mathrm{d}s.
\end{aligned}$$

The theorem is thus proved once we show that $M = \{M_t : t \geq 0\}$ is a martingale. However, this is a consequence of Corollary 4.6. □

For the reader who is more familiar with stochastic calculus and Itô's formula for a general Lévy process, the conclusion of the previous theorem is still valid when we replace X by a general spectrally negative Lévy process. See Exercise 4.7. The reader is also encouraged to consult Kella and Whitt (1992) where general complex valued martingales of this type are derived. Similarly in this vein one may consult Kennedy (1976), Jacod and Shiryaev (1987) and Nugyen-Ngoc (2005).

The theorem below, taken from Kyprianou and Palmowski (2005), is an example of how one may use the Kella–Whitt martingale to study the distribution of the running infimum $\underline{X} = \{\underline{X}_t : t \geq 0\}$ where $\underline{X}_t := \inf_{s \leq t} X_s$.

Theorem 4.8. *Suppose that X is a general spectrally negative Lévy process with Laplace exponent ψ and that $\mathbf{e}_q$ is a random variable which is exponentially distributed with parameter q which is independent of X. Then for all $\beta \geq 0$ and $q > 0$,*

$$\mathbb{E}(\mathrm{e}^{-\beta(\overline{X}_{\mathbf{e}_q} - X_{\mathbf{e}_q})}) = \frac{q}{\Phi(q)} \frac{\beta - \Phi(q)}{\psi(\beta) - q}. \tag{4.11}$$

Proof. As indicated in the remarks following the proof of Theorem 4.7, the statement of the latter theorem is still valid when X is a general spectrally negative Lévy process. We will assume this fact without proof here (otherwise refer to Exercise 4.7).

Since M is a martingale, it follows that $\mathbb{E}(M_{\mathbf{e}_q}) = 0$. That is to say, for all $\alpha \geq 0$,

$$\psi(\alpha)\mathbb{E}\left(\int_0^{\mathbf{e}_q} \mathrm{e}^{-\alpha(\overline{X}_s - X_s)} \mathrm{d}s\right) + 1 - \mathbb{E}(\mathrm{e}^{-\alpha(\overline{X}_{\mathbf{e}_q} - X_{\mathbf{e}_q})}) - \alpha\mathbb{E}(\overline{X}_{\mathbf{e}_q}) = 0. \tag{4.12}$$

Taking the first of the three expectations note that

$$\begin{aligned}
\mathbb{E}\left(\int_0^{\mathbf{e}_q} \mathrm{e}^{-\alpha(\overline{X}_s - X_s)} \mathrm{d}s\right) &= \mathbb{E}\left(\int_0^\infty \mathrm{d}u \cdot q\mathrm{e}^{-qu} \int_0^\infty \mathbf{1}_{(s \leq u)} \mathrm{e}^{-\alpha(\overline{X}_s - X_s)} \mathrm{d}s\right) \\
&= \frac{1}{q}\mathbb{E}\left(\int_0^\infty q\mathrm{e}^{-qs} \mathrm{e}^{-\alpha(\overline{X}_s - X_s)} \mathrm{d}s\right) \\
&= \frac{1}{q}\mathbb{E}\left(\mathrm{e}^{-\alpha(\overline{X}_{\mathbf{e}_q} - X_{\mathbf{e}_q})}\right).
\end{aligned}$$

To compute the third expectation of (4.12) we recall from Exercise 3.6 that $\overline{X}_{\mathbf{e}_q}$ is exponentially distributed with parameter $\Phi(q)$. Hence the latter expectation is equal to $1/\Phi(q)$.

Now returning to (4.12) and using the previous two main observations we may re-write it as (4.11). □

Corollary 4.9. *In the previous theorem, by taking limits as $q \downarrow 0$ on both sides of the above equality, we obtain*

$$\mathbb{E}(e^{-\beta(\overline{X}_\infty - X_\infty)}) = (0 \vee \psi'(0+))\frac{\beta}{\psi(\beta)}.$$

In particular this shows that $\mathbb{P}(\overline{X}_\infty - X_\infty = \infty) = 1$ if and only if $\psi'(0+) \leq 0$ and otherwise $\mathbb{P}(\overline{X}_\infty - X_\infty < \infty) = 1$.

Proof. Note that when $\Phi(0) = 0$, equivalently $\psi'(0+) \geq 0$,

$$\psi'(0+) = \lim_{\theta \downarrow 0} \frac{\psi(\theta)}{\theta} = \lim_{q \downarrow 0} \frac{q}{\Phi(q)}.$$

On the other hand, when $\Phi(0) > 0$, equivalently $\psi'(0+) < 0$,

$$\lim_{q \downarrow 0} \frac{q}{\Phi(q)} = 0.$$

Using these limits in (4.11) the statement of the Corollary follows. □

4.5 Stationary Distribution of the Workload

In this section we turn to the stationary distribution of the workload process W, making use of the conclusion in Corollary 4.9 which itself is drawn from the Kella–Whitt martingale. The setting is as in the introduction to this chapter.

Theorem 4.10. *Suppose that $0 < \rho < 1$. Then for all $w \geq 0$ the workload has a stationary distribution,*

$$\lim_{t \uparrow \infty} \mathbb{P}(W_t \in dx | W_0 = w) = (1 - \rho) \sum_{k=0}^{\infty} \rho^k \eta^{*k}(dx), \tag{4.13}$$

where

$$\eta(dx) = \frac{1}{d\rho} \nu(x, \infty) dx.$$

Here we understand $\eta^{*0}(dx) = \delta_0(dx)$ so that the distribution of W_∞ has an atom at zero. Otherwise if $\rho \geq 1$ then there is no non-trivial stationary distribution.

Proof. First suppose that $\rho \geq 1$. In this case we know that $\psi'(0+) \leq 0$ and then from Corollary 4.9 $\mathbb{P}(\overline{X}_\infty - X_\infty = \infty) = 1$. Next note that $W_t = (w \vee \overline{X}_t) - X_t \geq \overline{X}_t - X_t$ thus showing that $\mathbb{P}(W_\infty = \infty) = 1$.

Now suppose that $0 < \rho < 1$. In this case $\psi'(0+) > 0$ and hence from Corollary 3.14 we know that $\mathbb{P}(\tau_w^+ < \infty) = 1$. It follows that for all $t \geq \tau_w^+$, $W_t = \overline{X}_t - X_t$ and so from Corollary 4.9 we see that for all $\beta > 0$,

$$\lim_{t\uparrow\infty} \mathbb{E}(\mathrm{e}^{-\beta W_t}) = \psi'(0+)\frac{\beta}{\psi(\beta)}. \tag{4.14}$$

The remainder of the proof thus requires us to show that the right-hand side of (4.13) has Laplace–Stieltjes transform equal to the right-hand side of (4.14).

To this end, note using integration by parts in the definition of ψ that

$$\frac{\psi(\beta)}{\beta} = d - \int_0^\infty \mathrm{e}^{-\beta x}\nu(x,\infty)\mathrm{d}x. \tag{4.15}$$

As $\psi'(0+) > 0$ we have that $d^{-1}\int_0^\infty \nu(x,\infty)\mathrm{d}x < 1$; indeed, for all $\beta \geq 0$ we have that $d^{-1}\int_0^\infty e^{-\beta x}\nu(x,\infty)\mathrm{d}x < 1$. We may thus develop the right-hand side of (4.14) as follows

$$\psi'(0+)\frac{\beta}{\psi(\beta)} = \frac{\psi'(0+)}{d}\sum_{k\geq 0}\left(\frac{1}{d}\int_0^\infty \mathrm{e}^{-\beta x}\nu(x,\infty)\mathrm{d}x\right)^k.$$

Now define the measure $\eta(\mathrm{d}x) = (\mathrm{d}\rho)^{-1}\nu(x,\infty)\mathrm{d}x$. We have

$$\psi'(0+)\frac{\beta}{\psi(\beta)} = \frac{\psi'(0+)}{d}\sum_{k\geq 0}\rho^k\int_0^\infty \mathrm{e}^{-\beta x}\eta^{*k}(\mathrm{d}x) \tag{4.16}$$

with the understanding that $\eta^{*0}(\mathrm{d}x) = \delta_0(\mathrm{d}x)$. Noting that $\psi'(0+)/d = 1-\rho$ the result now follows by comparing (4.16) against (4.14). Note in particular that the stationary distribution, as one would expect, is independent of the initial value of the workload. □

Theorem 4.10 contains Theorem 1.12. To see this simply set $d = 1$, $\nu = \lambda F$ where F is the distribution with mean μ.

As noted earlier in Sect. 1.3.2 for the case of the $M/G/1$ queue, the expression for the stationary distribution given in statement Theorem 4.10 for the case $0 < \rho < 1$ is remarkably similar to the expression for the Pollaczek-Khintchine formula given in Theorem 1.8. The similarity of these two can be explained simply using Duality Lemma 3.4. Duality implies that for each fixed $t \geq 0$, $\overline{X}_t - X_t$ is equal in distribution to $-\underline{X}_t$. As was noted in the proof of Theorem 4.10, when $0 < \rho < 1$, the limit in distribution of W is independent of w and equal to the distributional limit of $\overline{X} - X$ and hence by the previous remarks, is also equal to the distribution of $-\underline{X}_\infty$. Noting further that

$$\mathbb{P}(-\underline{X}_\infty \leq x) = \mathbb{P}_x(\tau_0^- = \infty),$$

where $\tau_0^- = \inf\{t > 0 : X_t < 0\}$, we see that Theorem 4.10 also reads for all $x > 0$,

$$\mathbb{P}_x(\tau_0^- = \infty) = (1-\rho)\sum_{k=0}^\infty \rho^k\eta^{*k}(x)$$

where now $\eta^{*0}(x) = 1$. However, this is precisely the combined statements of Theorems 1.8 and 1.9, but now for a general spectrally negative Lévy process of bounded variation.

4.6 Small-Time Behaviour and the Pollaczek–Khintchine Formula

Either within the context of the stationary distribution of the workload process or the ruin problem, the reason for the appearance of a geometric-type sum in both cases is related to how spectrally negative Lévy processes of bounded variation behave at arbitrarily small times and consequently, how the entire path of the process X decomposes into objects called *excursions*. This section is dedicated to explaining this phenomenon.

We start the discussion with a Lemma, essentially due to Shtatland (1965); see also Chap. IV of Gikhman and Skorokhod (1975).

Lemma 4.11. *Suppose that X is a spectrally negative Lévy process of bounded variation. Then*

$$\lim_{t\downarrow 0}\frac{X_t}{t}=d$$

almost surely.

Proof. Recall from the Lévy–Itô decomposition that jumps of Lévy processes are described by a Poisson random measure with intensity $\mathrm{d}t\times\nu(\mathrm{d}x)$. From this it follows that the first jump of X of magnitude greater than δ appears after a length of time which is exponentially distributed with parameter $\nu(\delta,\infty)$. Since we are interested in small-time behaviour, it therefore is of no consequence if we assume that ν is concentrated on $(0,\delta)$. That is to say, there are no negative jumps of magnitude greater than δ.

Recall that as X is written in the form $X_t=\mathrm{d}t-S_t$ for $t\geq 0$ where $S=\{S_t:t\geq 0\}$ is a pure jump subordinator with Lévy measure ν. The proof is then completed by showing that

$$\lim_{t\downarrow 0}\frac{S_t}{t}=0.$$

Note however that since S has non-decreasing paths, it follows that when $t\in[2^{-(n-1)},2^{-n})$, $S_t\leq S_{2^{-n}}$ and hence it suffices to prove that

$$\lim_{n\uparrow\infty}\frac{S_{2^{-n}}}{2^{-(n-1)}}=0.$$

To achieve the latter, set $M_n=S_{2^{-n}}/2^{-n}$ and compute on the one hand

$$\mathbb{E}(M_{n+1}|M_1,\ ...,\ M_n)=2M_n-2^{n+1}\mathbb{E}(S_{2^{-n}}-S_{2^{-(n+1)}}|M_1,\ ...,\ M_n).\quad(4.17)$$

On the other hand, time reversing the path $\{S_t:t\leq 2^{-n}\}$ and using the stationarity and independence of increments we have that the law of $S_{2^{-n}/2}-S_0$ given $\{S_{2^{-n}},S_{2^{-(n-1)}},\ ...,\ S_{1/2}\}$ is equal in law to the law of $S_{2^{-n}}-S_{2^{-n}/2}$ given $\{S_{2^{-n}},S_{2^{-(n-1)}},\ ...,\ S_{1/2}\}$ and hence

$$\mathbb{E}(S_{2^{-n}}-S_{2^{-(n+1)}}|M_1,\ ...,\ M_n)=\mathbb{E}(S_{2^{-(n+1)}}|M_1,\ ...,\ M_n).$$

Substituting back into (4.17) we see that $\mathbb{E}(M_{n+1}|M_1, ..., M_n) = M_n$ and hence the sequence $M = \{M_n : n \geq 1\}$ is a positive $\mathbb{P}$-martingale with respect to the filtration generated by M. The Martingale Convergence Theorem implies that $M_\infty := \lim_{n\uparrow\infty} M_n$ exists and Fatou's Lemma implies that

$$\mathbb{E}(M_\infty) \leq \mathbb{E}(M_1) = \int_{(0,\delta)} x\nu(\mathrm{d}x).$$

We have then that

$$\mathbb{E}\left(\limsup_{t\downarrow 0} \frac{S_t}{t}\right) \leq \frac{1}{2}\mathbb{E}(\limsup_{n\uparrow\infty} M_n) = \frac{1}{2}\mathbb{E}(M_\infty) \leq \frac{1}{2}\int_{(0,\delta)} x\nu(\mathrm{d}x). \tag{4.18}$$

Since $\int_{(0,1)} x\nu(\mathrm{d}x) < \infty$ the right-hand side above can be made arbitrarily small by letting $\delta \downarrow 0$. This shows that the expectation on the left-hand side of (4.18) is equal to zero and hence so is the limsup in the expectation with probability one. □

The Lemma shows that for all sufficiently small times, $X_t > 0$ and hence $\mathbb{P}(\tau_0^- > 0) = 1$. That is to say, when starting from zero, it takes a strictly positive amount of time before X visits $(-\infty, 0)$. Compare this with, for example, the situation of a Brownian motion. It is intuitively clear that it will visit both sides of the origin immediately. To be rigorous about this, recall from Exercise 1.7 that the first passage process of a Brownian motion is a stable-$\frac{1}{2}$ subordinator. Since the latter subordinator is not a compound Poisson process, and hence does not remain at the origin for an initial almost surely strictly positive period of time, first passage strictly above level zero of B occurs immediately. By symmetry, the same can be said about first passage strictly below the level zero.

In order to complete our explanation of the geometric-type sum appearing in the Pollaczek–Khinchine formula let us proceed for the sake of convenience by showing that $\mathbb{P}(\sigma_x^+ = \infty)$ takes the form given in the right-hand side of (4.13) where now we take $Y = -X$ and for each $x \geq 0$, $\sigma_x^+ = \inf\{t > 0 : Y_t > x\}$. Lemma 4.11 shows that $\mathbb{P}(\sigma_0^+ > 0) = 1$. This information allows us to make the following path decomposition.

Define $T_0 = 0$ and $H_0 = 0$. Let $T_1 := \sigma_0^+$ and

$$H_1 = \begin{cases} X_{T_1} & \text{if } T_1 < \infty \\ \infty & \text{if } T_1 = \infty. \end{cases}$$

Next we construct iteratively the variables $T_1, T_2, ...$ and $H_1, H_2, ...$ in such a way that

$$T_n := \begin{cases} \inf\{t > T_{n-1} : X_t > H_{n-1}\} & \text{if } T_{n-1} < \infty \\ \infty & \text{if } T_{n-1} = \infty \end{cases}$$

and

$$H_n := \begin{cases} X_{T_n} & \text{if } T_n < \infty \\ \infty & \text{if } T_n = \infty. \end{cases}$$

Note in particular that $T_1 = \sigma_0^+$ is a stopping time and that for each $n \geq 1$, $T_{n+1} - T_n$ is equal in distribution to T_1. The Strong Markov Property and stationary independent increments imply that in the event that $\{T_{n-1} < \infty\}$ the path

$$\epsilon_n = \{X_t : T_{n-1} < t \leq T_n\} \tag{4.19}$$

is independent of $\mathcal{F}_{T_{n-1}}$ and has the same law as

$$\{X_t : 0 < t \leq T_1\}.$$

In particular, the pair $(T_n - T_{n-1}, H_n - H_{n-1})$ are independent of $\mathcal{F}_{T_{n-1}}$ and have the same distribution as $(\sigma_0^+, X_{\sigma_0^+})$ under law $\mathbb{P}$.

The sequence of independent and identically distributed sections of path $\{\epsilon_n : n \geq 1\}$ are called the excursions of X from its maximum. The sequence of pairs $\{(T, H) := (T_n, H_n) : n \geq 0\}$ are nothing more than the jump times and the consecutive heights of the new maxima of X so long as they are finite. The assumption that X drifts to infinity implies that the distribution of σ_0^+ (and hence $X_{\sigma_0^+}$) under $\mathbb{P}$ is defective. To see this, recall that $\overline{X}_\infty - X_\infty$ is equal in distribution to $-\underline{X}_\infty$ which in turn is equal to $\overline{Y}_\infty$. From (4.15) we see that as $\lim_{\beta\uparrow\infty} \psi(\beta)/\beta = d$. Hence, as it is assumed that $0 < \rho < 1$, or equivalently that $\psi'(0+) > 0$, we see from Corollary 4.9 that

$$1 - \rho = \frac{\psi'(0+)}{d} = \lim_{\beta\uparrow\infty} \mathbb{E}(\mathrm{e}^{-\beta \overline{Y}_\infty}) = \mathbb{P}(\overline{Y}_\infty = 0) = \mathbb{E}(\sigma_0^+ = \infty).$$

It follows then that there exists an almost surely finite N for which (T_n, H_n) are all finite for all $n \leq N$ and infinite for all $n > N$. We say that the excursion ϵ_n is *infinite* if $T_n - T_{n-1} = \infty$ and otherwise *finite*. Since excursions are independent and identically distributed, the total number of excursions $N + 1$ is the first time to failure in Bernoulli trials where "failure" means the occurrence of an infinite excursion and, as noted above, failure has probability $1 - \rho$. That is to say $N + 1$ is geometrically distributed with parameter $1 - \rho$. As the process Y is assumed to drift to ∞ the structure of the path of Y must correspond to the juxtaposition of N finite excursions followed by a final infinite excursion. Figure 4.1 gives a symbolic impression of this decomposition leaving out details of the path within excursions.

Using the decomposition of the path of Y into the juxtaposition of independent and identically distributed excursions from the maximum it is now clear that the event that there is no ruin corresponds to the event that there are N finite excursions which when pasted end to end have a right end point which is no higher than x followed by an infinite excursion. Here, as above, $N+1$ is geometrically distributed with parameter $1-\rho$. It follows immediately that

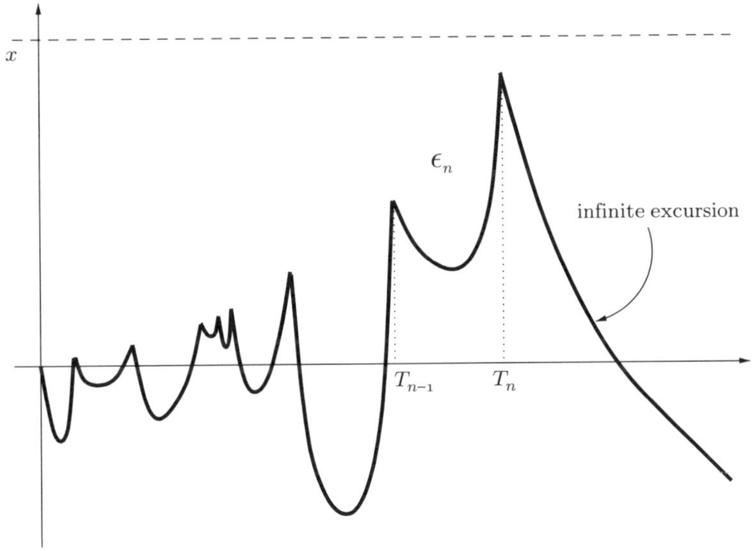

Fig. 4.1. A symbolic sketch of the decomposition of the path of Y when it fails to cross the level x.

$$\mathbb{P}(\sigma_x^+ = \infty) = \sum_{n\geq 0}(1-\rho)\rho^n \mathbb{P}(H_n \leq x | \epsilon_1, ..., \epsilon_n \text{ are finite}).$$

Recalling that the increments of H_n are independent and identically distributed with the same law as σ_0^+ under $\mathbb{P}$ it follows that the probabilities in the sum are each equal to $\mu^{*n}(x)$ where

$$\mu(\mathrm{d}x) = \mathbb{P}(H_1 \leq x | T_1 < \infty) = \mathbb{P}(-X_{\tau_0^-} \leq x | \tau_0^- < \infty)$$

thus explaining the form of the Pollaczek–Khintchine formula.

Note that in our reasoning above, we have not proved that $\mu(\mathrm{d}x) = (\mathrm{d}\rho)^{-1}\nu(x,\infty)\mathrm{d}x$. However, by comparing the conclusions of the previous discussion with the conclusion of Theorem 4.10, we obtain the following corollary.

Corollary 4.12. *Suppose that X is a spectrally negative Lévy process of bounded variation such that $\psi'(0+) > 0$. Then $\mathbb{P}(\tau_0^- < \infty) = \rho$ and*

$$\mathbb{P}(-X_{\tau_0^-} \leq x | \tau_0^- < \infty) = \frac{1}{\mathrm{d}\rho}\int_0^x \nu(y,\infty)\mathrm{d}y.$$

Exercises

4.1. Suppose that $X = \{X_t : t \geq 0\}$ is a spectrally negative process of bounded variation. Suppose that we define for each $t \geq 0$,

$$L_t^0 = \#\{0 < s \leq t : X_t = 0\}.$$

(i) Show that the process $\{L_t^0 : t \geq 0\}$ is almost surely integer valued with paths that are right continuous with left limits.

(ii) Suppose now that f is a function which is equal to a $C^1(\mathbb{R})$ function on $(-\infty, 0)$ and equal to another $C^1(\mathbb{R})$ function on $(0, \infty)$ but may have a discontinuity at 0. Show that for each $t \geq 0$,

$$
\begin{aligned}
f(X_t) = f(X_0) &+ d \int_0^t f'(X_s)\mathrm{d}s \\
&+ \int_{(0,t]} \int_{(-\infty,0)} (f(X_{s-} + x) - f(X_{s-}))N(\mathrm{d}s \times \mathrm{d}x) \\
&+ \int_0^t (f(X_s) - f(X_{s-}))\mathrm{d}L_s^0.
\end{aligned}
$$

4.2. Suppose that $X = \{X_t : t \geq 0\}$ is a spectrally negative Lévy process of bounded variation with drift d (see the discussion following Lemma 2.14). Show that for $f(s, x) \in C^{1,1}([0, \infty) \times \mathbb{R})$,

$$
\begin{aligned}
f(\overline{X}_t, X_t) = f(\overline{X}_0, X_0) &+ d \int_0^t \frac{\partial f}{\partial x}(\overline{X}_s, X_s)\mathrm{d}s + \int_0^t \frac{\partial f}{\partial s}(\overline{X}_s, X_s)\mathrm{d}\overline{X}_s \\
&+ \int_{[0,t]} \int_{(-\infty,0)} f(\overline{X}_s, X_{s-} + x) - f(\overline{X}_s, X_{s-})N(\mathrm{d}s \times \mathrm{d}x).
\end{aligned}
$$

4.3. Suppose that X is a Lévy process of bounded variation with Lévy measure Π and drift $d \in \mathbb{R}$. We exclude the case of a compound Poisson process where the jump distribution has lattice support. Suppose further that $f \in C^1(\mathbb{R})$ and $\lambda \geq 0$ are such that $\{\mathrm{e}^{-\lambda t} f(X_t) : t \geq 0\}$ is a martingale. Assume further that for each $y \in \mathcal{D}$,

$$
\int_{\mathbb{R}} |f(x + y) - f(y)|\Pi(\mathrm{d}x) < \infty,
$$

where $\mathcal{D} = [0, \infty)$ if X is a subordinator and $\mathcal{D} = \mathbb{R}$ otherwise.

(i) Let

$$
T_n = \inf\left\{t \geq 0 : \left|\int_{\mathbb{R}} (f(X_t + x) - f(X_t))\Pi(\mathrm{d}x)\right| > n\right\}.
$$

Deduce that T_n is a stopping time with respect to $\mathbb{F}$.

(ii) Show that $\{M_{t \wedge T_n} : t \geq 0\}$ is a martingale where

$$
\begin{aligned}
M_t = &\int_{[0,t]} \int_{\mathbb{R}} \mathrm{e}^{-\lambda s}(f(X_{s-} + x) - f(X_{s-}))N(\mathrm{d}s \times \mathrm{d}x) \\
&- \int_{[0,t]} \int_{\mathbb{R}} \mathrm{e}^{-\lambda s}(f(X_s + x) - f(X_s))\mathrm{d}s\Pi(\mathrm{d}x), \quad t \geq 0.
\end{aligned}
$$

(iii) Deduce with the help of the change of variable formula that, under the assumption that the support of the distribution of X_s is $\mathbb{R}$ for all $s > 0$,

$$d\frac{\partial f}{\partial y}(y) + \int_{\mathbb{R}} (f(y+x) - f(y))\Pi(\mathrm{d}x) = \lambda f(y)$$

for all $y \in \mathbb{R}$.

(iv) Assume now that for given $\beta \in \mathbb{R}$, $\int_{|x|>1} \mathrm{e}^{\beta x}\Pi(\mathrm{d}x) < \infty$. Show that if $f(x) = \mathrm{e}^{\beta x}$ then necessarily $\lambda = \psi(\beta)$.

4.4. Suppose that ϕ fulfils the conditions of Theorem 4.4 and that for each $t > 0$, $\mathbb{E}(\int_{[0,t]}\int_{\mathbb{R}} |\phi(s,x)|\mathrm{d}s\Pi(\mathrm{d}x)) < \infty$. If $M = \{M_t : t \geq 0\}$ is the martingale given in Corollary 4.6 and further, it is assumed that for all $t \geq 0$, $\mathbb{E}(\int_{[0,t]}\int_{\mathbb{R}} \phi(s,x)^2\mathrm{d}s\Pi(\mathrm{d}x)) < \infty$ show that

$$\mathbb{E}(M_t^2) = \mathbb{E}\left(\int_{[0,t]}\int_{\mathbb{R}} \phi(s,x)^2\mathrm{d}s\Pi(\mathrm{d}x)\right).$$

4.5. In this exercise, we use ideas coming from the proof of the Lévy–Itô decomposition to prove Itô's formula in Theorem 4.3 but for the case that $\sigma = 0$. Henceforth we will assume that X is a Lévy process with no Gaussian component and $f(s,x) \in C^{1,2}([0,\infty) \times \mathbb{R})$ which is uniformly bounded along with its first derivative in s and first two derivatives in x.

(i) Suppose that X has characteristic exponent

$$\Psi(\theta) = \mathrm{i}\theta a + \int_{\mathbb{R}} (1 - \mathrm{e}^{\mathrm{i}\theta x} + \mathrm{i}\theta x\mathbf{1}_{(|x|<1)})\Pi(\mathrm{d}x).$$

For each $1 > \varepsilon > 0$ let $X^{(\varepsilon)} = \{X_t^{(\varepsilon)} : t \geq 0\}$ be the Lévy process with characteristic exponent

$$\Psi^{(\varepsilon)}(\theta) = \mathrm{i}\theta a + \int_{\mathbb{R}\setminus(-\varepsilon,\varepsilon)} (1 - \mathrm{e}^{\mathrm{i}\theta x} + \mathrm{i}\theta x\mathbf{1}_{(|x|<1)})\Pi(\mathrm{d}x).$$

Show that

$$\begin{aligned} f(t, X_t^{(\varepsilon)}) &\\ = f(0, X_0) &+ \int_0^t \frac{\partial f}{\partial s}(s, X_s^{(\varepsilon)})\mathrm{d}s \\ + \int_{[0,t]}\int_{|x|>\varepsilon} &(f(s, X_{s-}^{(\varepsilon)} + x) - f(s, X_{s-}^{(\varepsilon)}) - x\frac{\partial f}{\partial x}(s, X_{s-}^{(\varepsilon)}))N(\mathrm{d}s \times \mathrm{d}x) \\ + \int_0^t &\frac{\partial f}{\partial x}(s, X_{s-}^{(\varepsilon)})\mathrm{d}X_s^{(2)} + M_t^{(\varepsilon)}, \end{aligned} \tag{4.20}$$

where $X^{(2)}$ is a Lévy process with characteristic exponent $a\mathrm{i}\theta + \int_{|x|\geq 1}(1 - \mathrm{e}^{\mathrm{i}\theta x})\Pi(\mathrm{d}x)$ and $M^{(\varepsilon)} = \{M_t^{(\varepsilon)} : t \geq 0\}$ is a right continuous, square integrable martingale with respect to the filtration $\{\mathcal{F}_t : t \geq 0\}$ of X which you should specify.

(ii) Fix $T > 0$. Show that $\{M^{(\varepsilon)} : 0 < \varepsilon < 1\}$ is a Cauchy family in the martingale space $\mathcal{M}_T^2$ (see Definition 2.11).

(iii) Denote the limiting martingale in part (ii) by M. By taking limits as $\varepsilon \downarrow 0$ along a suitable subsequence, show that the Itô formula holds where we may define

$$\int_0^t \frac{\partial f}{\partial x}(s, X_{s-})\mathrm{d}X_s = \int_0^t \frac{\partial f}{\partial x}(s, X_{s-})\mathrm{d}X^{(2)} + M_t.$$

Explain why the left-hand side above is a suitable choice of notation.

(iv) Show that if the restrictions of uniform boundedness of f and its derivatives using stopping times are removed then the same conclusion may be drawn as in (iii) except now there exists an increasing sequence of stopping times tending to infinity, say $\{T_n : n \geq 1\}$, such that for each $n \geq 1$ the process M is a martingale when stopped at time T_n. In other words, M is a local martingale and not necessarily a martingale.

4.6. Consider the workload process W of an $M/G/1$ queue as described in Sect. 1.3.2. Suppose that $W_0 = w = 0$ and the service distribution F has Laplace transform $\hat{F}(\beta) = \int_{(0,\infty)} \mathrm{e}^{-\beta x} F(\mathrm{d}x)$.

(i) Show that the first busy period (the time from the moment of first service to the first moment thereafter that the queue is again empty), denoted B, fulfils

$$\mathbb{E}(\mathrm{e}^{-\beta B}) = \hat{F}(\Phi(\beta))$$

where $\Phi(\beta)$ is the largest solution to the equation

$$\theta - \int_{(0,\infty)} (1 - \mathrm{e}^{-\theta x})\lambda F(\mathrm{d}x) = \beta.$$

(ii) When $\rho > 1$, show that there are a geometrically distributed number of busy periods. Hence give another proof of the first part of Theorem 4.1 when $w = 0$ by using this fact.

(iii) Suppose further that the service distribution F is that of an exponential random variable with parameter $\mu > \lambda$. This is the case of an $M/M/1$ queue. Show that the workload process has limiting distribution given by

$$\left(1 - \frac{\lambda}{\mu}\right)\left(\delta_0(\mathrm{d}x) + \mathbf{1}_{(x>0)}\lambda \mathrm{e}^{-(\mu-\lambda)x}\mathrm{d}x\right).$$

4.7. This exercise is only for the reader familiar with the general theory of stochastic calculus with respect to semi-martingales. Suppose that X is a general spectrally negative Lévy process. Recall the notation $\mathcal{E}_t(\alpha) = \exp\{\alpha X_t - \psi(\alpha)t\}$ for $t \geq 0$.

(i) If M is the Kella–Whitt martingale, show that

$$\mathrm{d}M_t = -\mathrm{e}^{-\overline{X}_t + \psi(\alpha)t}\mathrm{d}\mathcal{E}_t(\alpha)$$

and hence deduce that M is a local martingale.

(ii) Show that $\mathbb{E}(\overline{X}_t) < \infty$ for all $t > 0$.
(iii) Deduce that $\mathbb{E}(\sup_{s \le t} |M_s|) < \infty$ and hence that M is a martingale.

4.8. Suppose that X is a spectrally negative Lévy process of bounded variation with characteristic exponent Ψ.

(i) Show that for each $\alpha, \beta \in \mathbb{R}$,

$$M_t = -\Psi(\alpha) \int_0^t \mathrm{e}^{\mathrm{i}\alpha(X_s - \overline{X}_s) + \mathrm{i}\beta \overline{X}_s} \mathrm{d}s + 1 - \mathrm{e}^{\mathrm{i}\alpha(X_t - \overline{X}_t) + \mathrm{i}\beta \overline{X}_t}$$
$$-\mathrm{i}(\alpha - \beta) \int_0^t \mathrm{e}^{\mathrm{i}\alpha(X_s - \overline{X}_s) + \mathrm{i}\beta \overline{X}_s} d\overline{X}_s, \ t \ge 0$$

is a martingale. Note, for the reader familiar with general stochastic calculus for semi-martingales, one may equally prove that the latter is a martingale for a general spectrally negative Lévy process.
(ii) Use the fact that $\mathbb{E}(M_{\mathbf{e}_q}) = 0$, where $\mathbf{e}_q$ is an independent exponentially distributed random variable with parameter q, to show that

$$\mathbb{E}(\mathrm{e}^{\mathrm{i}\alpha(X_{\mathbf{e}_q} - \overline{X}_{\mathbf{e}_q}) + \mathrm{i}\beta \overline{X}_{\mathbf{e}_q}}) = \frac{q(\Phi(q) - \mathrm{i}\alpha)}{(\Psi(\alpha) + q)(\mathrm{i}\beta - \Phi(q))}, \tag{4.21}$$

where Φ is the right inverse of the Laplace exponent $\psi(\beta) = -\Psi(-\mathrm{i}\beta)$.
(iii) Deduce that $\overline{X}_{\mathbf{e}_q} - X_{\mathbf{e}_q}$ and $\overline{X}_{\mathbf{e}_q}$ are independent.

4.9. Suppose that X is *any* Lévy process of bounded variation with drift $d > 0$ (excluding subordinators).

(i) Show that

$$\lim_{t \downarrow 0} \frac{X_t}{t} = d$$

almost surely.
(ii) Define $\tau_0^- = \inf\{t > 0 : X_t < 0\}$. By reasoning along similar lines for the case of a spectrally negative process, show that $\mathbb{P}(\tau_0^- > 0) > 0$
(iii) Suppose now that $\lim_{t\uparrow\infty} X_t = \infty$. Let $\eta(\mathrm{d}x) = \mathbb{P}(-X_{\tau_0^-} \in \mathrm{d}x | \tau_0^- < \infty)$. Conclude that the Pollaczek–Khinchine formula

$$\mathbb{P}_x(\tau_0^- = \infty) = (1 - \rho) \sum_{k=0}^{\infty} \rho^k \eta^{*k}(x)$$

is still valid under these circumstances.

5

Subordinators at First Passage and Renewal Measures

In this chapter we look at subordinators; Lévy processes which have paths that are non-decreasing. In addition, we consider *killed subordinators*. That is subordinators which are sent to a "graveyard state" (in other words an additional point that is not on $[0, \infty)$) at an independent time that is exponentially distributed. Principally we are interested in first passage over a fixed level and some asymptotic features of the processes as this level tends to infinity. In particular, the (asymptotic) law of the overshoot and undershoot as well as the phenomena of crossing a level by hitting it. These three points of interest turn out to be very closely related to renewal measures. The results obtained in this chapter will be of significance later on when we consider first passage of a general Lévy process over a fixed level. As part of the presentation on asymptotic first passage, we will review some basic facts about regular variation. Regular variation will also be of use in later chapters.

5.1 Killed Subordinators and Renewal Measures

Recall that a subordinator is a Lévy process with paths which are non-decreasing almost surely. Equivalently, a subordinator is a Lévy process of bounded variation, positive drift $d > 0$ and jump measure concentrated on $(0, \infty)$. In this section we shall consider a slightly more general class of process, *killed subordinators*. Let Y be a subordinator and $\mathbf{e}_\eta$ an independent exponentially distributed random variable for some $\eta > 0$. Then a killed subordinator is the process

$$X_t = \begin{cases} Y_t \text{ if } t < \mathbf{e}_\eta \\ \partial \text{ if } t \geq \mathbf{e}_\eta, \end{cases}$$

where ∂ is a "graveyard state". We shall also refer to X as "Y killed at rate η". If we agree that $\mathbf{e}_\eta = \infty$ when $\eta = 0$ then the definition of a killed subordinator includes the class of regular subordinators.[1] This will prove to be useful for making general statements. The Laplace exponent of a killed subordinator X is defined for all $\theta \geq 0$ by the formula

$$\Phi(\eta) = -\log \mathbb{E}(\mathrm{e}^{-\theta X_1}) = -\log \mathbb{E}(\mathrm{e}^{-\theta Y_1}\mathbf{1}_{(1<\mathbf{e}_\eta)}) = \eta - \log \mathbb{E}(\mathrm{e}^{-\theta Y_1}) = \eta + \Psi(i\theta),$$

where Ψ is the Lévy–Khintchine exponent of Y. From the Lévy–Khintchine formula given in the form (2.21) we easily deduce that

$$\Phi(\theta) = \eta + d\theta + \int_{(0,\infty)} (1 - \mathrm{e}^{-\theta x})\Pi(\mathrm{d}x), \tag{5.1}$$

where $d \geq 0$ and $\int_{(0,\infty)}(1 \wedge x)\Pi(\mathrm{d}x) < \infty$; recall Exercise 2.11.

With each killed subordinator we associate a family of potential measures. Define for each $q \geq 0$ the *q-potential measure* by

$$U^{(q)}(\mathrm{d}x) = \mathbb{E}\left(\int_0^\infty \mathrm{e}^{-qt}\mathbf{1}_{(X_t \in \mathrm{d}x)}\mathrm{d}t\right) = \int_0^\infty \mathrm{e}^{-qt}\mathbb{P}(X_t \in \mathrm{d}x)\mathrm{d}t.$$

For notational ease we shall simply write $U^{(0)} = U$ and call it the *potential measure*. Note that the q-potential measure of a killed subordinator with killing at rate $\eta > 0$ is equal to the $(q+\eta)$-potential measure of a subordinator. Note also that for each $q > 0$, $(q+\eta)U^{(q)}$ is a probability measure on $[0,\infty)$ and also that for each $q \geq 0$ $U^{(q)}(x) := U^{(q)}[0,x]$ is right continuous. Roughly speaking, a q-potential measure is an expected discounted measure of how long the process X occupies different regions of space.[2]

These potential measures will play an important role in the study of how subordinators cross fixed levels. For this reason we will devote the remainder of this section to studying some of their analytical properties. One of the most important facts about q-potential measures is that they are closely related to renewal measures. Recall that the renewal measure associated with a distribution F concentrated on $[0,\infty)$ is defined by

$$V(\mathrm{d}x) = \sum_{k \geq 0} F^{*k}(\mathrm{d}x), \;\; x \geq 0,$$

where we understand $F^{*0}(\mathrm{d}x) := \delta_0(\mathrm{d}x)$. As with potential measures we work with the notation $V(x) := V[0,x]$. For future reference let us recall the two versions of the classical Renewal Theorem.

Theorem 5.1 (Renewal Theorem). *Suppose that V is the renewal function given above and assume that $\mu := \int_{(0,\infty)} xF(\mathrm{d}x) < \infty$.*

[1] A killed subordinator is only a Lévy process when $\eta = 0$ however, it is still a Markov process even when $\eta > 0$.

[2] From the general theory of Markov processes, $U^{(q)}$ also comes under name of resolvent measure or Green's measure.

(i) If F does not have lattice support then for all $y > 0$,

$$\lim_{x\uparrow\infty}\{V(x+y)-V(x)\}=\frac{y}{\mu}.$$

(ii) [3]If F does not have lattice support and $h:\mathbb{R}\to\mathbb{R}$ is directly Riemann integrable, then

$$\lim_{x\uparrow\infty}\int_0^x h(x-y)V(\mathrm{d}y)=\frac{1}{\mu}\int_0^\infty h(y)\mathrm{d}y.$$

(iii) Without restriction on the support of F,

$$\lim_{x\uparrow\infty}\frac{V(x)}{x}=\frac{1}{\mu}.$$

The precise relationship between q-potential measures of subordinators and renewal measures is given in the following lemma.

Lemma 5.2. *Suppose that X is a subordinator (no killing). Let $F = U^{(1)}$ and let V be the renewal measure associated with the distribution F. Then $V(\mathrm{d}x)$ is equal to the measure $\delta_0(\mathrm{d}x)+U(\mathrm{d}x)$ on $[0,\infty)$.*

Proof. First note that for all $\theta > 0$,

$$\begin{aligned}\int_{[0,\infty)}\mathrm{e}^{-\theta x}U^{(1)}(\mathrm{d}x)&=\int_0^\infty \mathrm{d}t\cdot\mathrm{e}^{-t}\int_{[0,\infty)}\mathrm{e}^{-\theta x}\mathbb{P}(X_t\in\mathrm{d}x)\\&=\int_0^\infty\mathrm{d}t\cdot\mathrm{e}^{-(1+\Phi(\theta))t}\\&=\frac{1}{1+\Phi(\theta)},\end{aligned}$$

where Φ is the Laplace exponent of the underlying subordinator. Note that in the final equality we have used the fact that $\Phi(\theta) > 0$.

Next compute the Laplace transform of V for all $\theta > 0$ as follows,

$$\begin{aligned}\int_{[0,\infty)}\mathrm{e}^{-\theta x}V(\mathrm{d}x)&=\sum_{k\geq 0}\left(\int_{[0,\infty)}\mathrm{e}^{-\theta x}U^{(1)}(\mathrm{d}x)\right)^k\\&=\sum_{k\geq 0}\left(\frac{1}{1+\Phi(\theta)}\right)^k\\&=\frac{1}{1-(1+\Phi(\theta))^{-1}}\\&=1+\frac{1}{\Phi(\theta)}.\end{aligned}\tag{5.2}$$

[3]This part of the theorem is also known on its own as the Key Renewal Theorem.

Note that in the third equality we have used the fact that $|1/(1+\Phi(\theta))| < 1$.

On the other hand, a similar computation to the one in the first paragraph of this proof shows that the Laplace transform of $\delta_0(\mathrm{d}x) + U(\mathrm{d}x)$ equals the right-hand side of (5.2). Since distinct measures have distinct Laplace transforms the proof is complete. □

The conclusion of the previous Lemma means that the Renewal Theorem can, and will, be employed at a later stage to understand the asymptotic behaviour of U. Specifically we have the following two asymptotics.

Corollary 5.3. *Suppose that X is a subordinator (no killing) such that $\mu := \mathbb{E}(X_1) < \infty$.*

(i) If U does not have lattice support, then for all $y > 0$,

$$\lim_{x\uparrow\infty} \{U(x+y) - U(x)\} = \frac{y}{\mu}.$$

(ii) Without restriction on the support of U,

$$\lim_{x\uparrow\infty} \frac{U(x)}{x} = \frac{1}{\mu}.$$

Proof. The proof is a direct consequence of Theorem 5.1 once one notes that

$$\mu = \int_{[0,\infty)} xU^{(1)}(\mathrm{d}x) = \int_0^\infty \mathrm{e}^{-t}\mathbb{E}(X_t)\mathrm{d}t = \int_0^\infty t\mathrm{e}^{-t}\mathbb{E}(X_1)\mathrm{d}t = \mathbb{E}(X_1)$$

and that $U^{(1)}$ has the same support as U. □

In the previous corollary, the requirement that U does not have a lattice support is not a serious restriction as there are analogues to Corollary 5.3 (i); see for example Feller (1971). The following theorem, based on standard results in Blumenthal and Getoor (1968), shows that in principle one can only find examples of potential measures with lattice support when X is a killed compound Poisson subordinator. The statement of the theorem gives in fact a stronger conclusion than the latter.

Theorem 5.4. *Suppose that X is a killed subordinator with Lévy measure Π.*

(i) If $\Pi(0,\infty) = \infty$ then for any $q \geq 0$, $U^{(q)}$ has no atoms.
(ii) If $\Pi(0,\infty) < \infty$ and Π has a non-lattice support then for all $q \geq 0$, $U^{(q)}$ does not have a lattice support.
(iii) If $\Pi(0,\infty) < \infty$ and Π has a lattice support, then for all $q \geq 0$, $U^{(q)}$ has the same lattice support in $(0,\infty)$.

Proof. (i) Recall the definition

$$U^{(q)}(\mathrm{d}x) = \mathbb{E}\left(\int_0^\infty \mathrm{e}^{-qt}\mathbf{1}_{(X_t\in \mathrm{d}x)}\mathrm{d}t\right)$$

and note that, on account of monotonicity of the paths of X, an atom at $x > 0$ occurs only if, with positive probability, the path of X remains at level x over some period of time (a, b) where $0 \leq a < b < \infty$. However, since $\Pi(0, \infty) = \infty$, we know the latter behaviour is impossible; see Exercise 2.7.

(ii) – (iii) Now suppose that X is equal in law to a compound Poisson subordinator with jump distribution F and arrival rate $\lambda > 0$ which is killed at rate $\eta \geq 0$. (Note $\lambda F = \Pi$). By conditioning on the number of jumps up to time $t > 0$ we have

$$\mathbb{P}(X_t \in \mathrm{d}x) = \mathrm{e}^{-\eta t} \sum_{k \geq 0} \mathrm{e}^{-\lambda t} \frac{(\lambda t)^k}{k!} F^{*k}(\mathrm{d}x),$$

where as usual we understand $F^{*0} = \delta_0(\mathrm{d}x)$. Using this representation of the transition measure we compute

$$\begin{aligned} U^{(q)}(\mathrm{d}x) &= \sum_{k \geq 0} \frac{1}{k!} F^{*k}(\mathrm{d}x) \int_0^\infty \mathrm{e}^{-(\lambda+q+\eta)t} (\lambda t)^k \mathrm{d}t \\ &= \frac{\rho}{\lambda} \sum_{k \geq 0} \rho^k F^{*k}(\mathrm{d}x), \end{aligned} \tag{5.3}$$

where $\rho = \lambda/(\lambda + \eta + q)$. The second and third statements of the theorem now follow from the last equality. In the case that F does not have a lattice support in $(0, \infty)$, then neither does F^{*k} for any $k \geq 1$ and hence neither does $U^{(q)}$. On the other hand, if F has a lattice support in $(0, \infty)$, then so does F^{*k} for any $k \geq 1$ (the sum of k independent and identically distributed lattice valued random variables is also lattice valued). □

Note that the above theorem shows that rescaling the Lévy measure of a subordinator, that is $\Pi \mapsto c\Pi$ for some $c > 0$, has no effect on the presence of atoms in the potential measure.

In addition to the close association of the potential measure U with classical renewal measures, the connection of a subordinator with renewal processes[4] can be seen in a pathwise sense when X is a compound Poisson subordinator with arrival rate $\lambda > 0$ and non-negative jumps with distribution F. In this case it is clear that the range of the process X, that is the projection of the graph of $\{X_t : t \geq 0\}$ onto the spatial axis, is nothing more

[4]We recall briefly that a renewal process $N = \{N_x : x \geq 0\}$ counts the points in $[0, x]$ for $x \geq 0$ of a point process (that is a random scattering of points) on $[0, \infty)$ in which points are laid down as follows. Let F be a distribution function on $(0, \infty)$ and suppose that $\{\xi_i : i = 1, 2, ...\}$ is a sequence of independent random variables with common distribution F. Points are positioned at $\{T_1, T_2, ...\}$ where for each $k \geq 1$, $T_k = \sum_{i=1}^k \xi_i$. In other words, the underlying point process is nothing more than the range of a random walk with jump distribution F. We may now identify for each $x \geq 0$, $N_x = \sup\{i : T_i \leq x\}$. Note that if F is an exponential distribution then N is nothing more than a Poisson process.

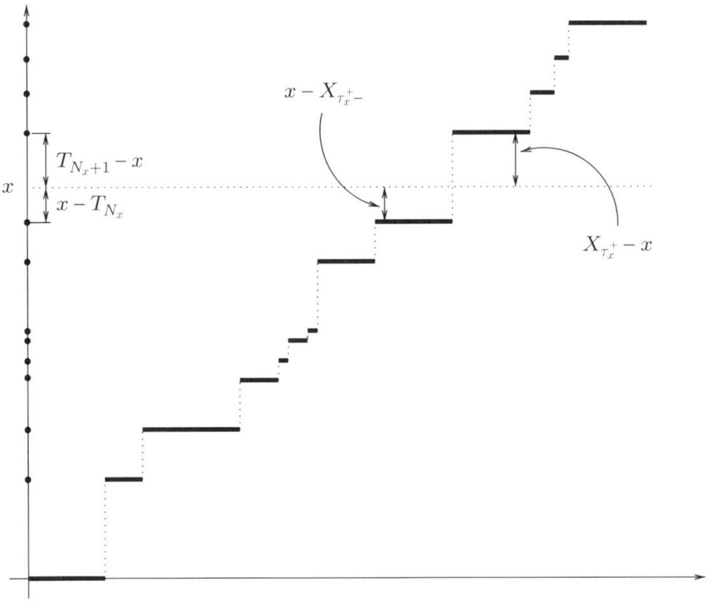

Fig. 5.1. A realisation of a compound Poisson subordinator. The range of the process, projected onto the vertical axis, forms a renewal process thus relating overshoot and undershoot to excess and current lifetimes.

than a renewal process. Note that in this renewal process the spatial domain of X plays the role of time and the inter-arrival times are precisely distributed according to F. See Fig. 5.1.

Denote this renewal process $N = \{N_x : t \geq 0\}$ and let $\{T_i : i \geq 0\}$ be the renewal epochs starting with $T_0 = 0$. Then the excess lifetime of N at time $x > 0$ is defined by $T_{N_x+1} - x$ and the current lifetime by $x - T_{N_x}$. On the other hand recall the stopping time (first passage time)

$$\tau_x^+ = \inf\{t > 0 : X_t > x\}.$$

Then the overshoot and undershoot at first passage of level x are given by $X_{\tau_x^+} - x$ and $x - X_{\tau_x^+-}$, respectively. Excess and current lifetimes and overshoots and undershoots are thus related by

$$X_{\tau_x^+} - x = T_{N_x+1} - x \text{ and } x - X_{\tau_x^+-} = x - T_{N_x}. \tag{5.4}$$

See Fig. 5.1.

Classical renewal theory presents the following result for the excess and current lifetime; see for example Feller (1971) or Dynkin (1961). We give the proof for the sake of later reference.

Lemma 5.5. *Suppose that N is a renewal process with F as the distribution for the spacings. Then the following hold.*

(i) For $u > 0$ and $y \in (0, x]$,

$$\mathbb{P}(T_{N_x+1} - x \in \mathrm{d}u, x - T_{N_x} \in \mathrm{d}y) = V(x - \mathrm{d}y)F(\mathrm{d}u + y), \qquad (5.5)$$

where V is the renewal measure constructed from F.

(ii) Suppose that F has mean $\mu < \infty$ and is non-lattice, then for $u > 0$ and $y > 0$,

$$\lim_{x\uparrow\infty} \mathbb{P}(T_{N_x+1} - x > u, x - T_{N_x} > y) = \frac{1}{\mu}\int_{u+y}^{\infty} \overline{F}(z)\mathrm{d}z,$$

where $\overline{F}(x) = 1 - F(x)$.

Proof. (i) The key to the proof of the first part is to condition on the number of renewal epochs at time x. We have for $k \geq 0$,

$$\mathbb{P}(T_{N_x+1} - x > u, x - T_{N_x} > y | N_x = k) = \int_{[0,x-y)} F^{*k}(\mathrm{d}v)\overline{F}(x - v + u)$$

as the probability on the left-hand side requires that the kth renewal epoch occurs sometime before $x-y$. Further, this epoch occurs in $\mathrm{d}v$ with probability $F^{*k}(\mathrm{d}v)$ and hence the probability that the excess exceeds u requires that the next inter-arrival time exceeds $x-v+u$ with probability $\overline{F}(x-v+u)$. Summing over k and changing variable in the integral via $z = x - v$ gives

$$\mathbb{P}(T_{N_x+1} - x > u, x - T_{N_x} > y) = \int_{(y,x]} V(x - \mathrm{d}z)\overline{F}(z + u).$$

In differential form this gives the distribution given in the statement of part (i).

(ii) From part (i) we may write for $u > 0$ and $y \in [0, x)$,

$$\begin{aligned}
&\mathbb{P}(T_{N_x+1} - x > u, x - T_{N_x} > y) \\
&= \int_{(u,\infty)} \int_{[0,x-y)} V(\mathrm{d}v)F(x - v + \mathrm{d}\theta) \\
&= \int_{(0,\infty)} F(\mathrm{d}t) \int_{[0,x)} V(\mathrm{d}v)\mathbf{1}_{(t>u+x-v)}\mathbf{1}_{(v\in[0,x-y))} \\
&= \int_{(0,\infty)} F(\mathrm{d}t) \int_{[0,x)} V(\mathrm{d}v)\mathbf{1}_{(v>u+x-t)}\mathbf{1}_{(v\in[0,x-y))},
\end{aligned}$$

where we have applied the change of variables $t = \theta + x - v$ in the second equality. Now note that the indicators and integral delimiters require that

$$v \geq (u + x - t) \vee 0 \text{ and } u + x - t \leq x - y.$$

Hence for $u > 0$ and $y \in [0, x)$,

$$
\begin{aligned}
&\mathbb{P}(T_{N_x+1} - x > u, x - T_{N_x} > y)\\
&= \int_{(0,\infty)} F(\mathrm{d}t)\{V(x-y) - V((u+x-t)\vee 0)\}\mathbf{1}_{(t\geq u+y)}\\
&= \int_{(u+y,\infty)} F(\mathrm{d}t)\{V(x-y) - V(u+x-t)\}\mathbf{1}_{(t<u+x)}\\
&\qquad + \int_{[u+x,\infty)} F(\mathrm{d}t)V(x-y).
\end{aligned}
\tag{5.6}
$$

To deal with the second term on the right-hand side of (5.6) we may use the Renewal Theorem 5.1 (iii) to show that for some $\varepsilon > 0$ and x sufficiently large

$$
\int_{(u+x,\infty)} F(\mathrm{d}t)V(x-y) \leq \frac{1+\varepsilon}{\mu}\int_{(u+x,\infty)} tF(\mathrm{d}t)
$$

which tends to zero as x tends to infinity as $\mu = \int_{(0,\infty)} tF(\mathrm{d}t) < \infty$.

For the first term on the right-hand side of (5.6) suppose that X is a compound Poisson subordinator whose jump distribution is F and arrival rate is 1, then for this subordinator

$$
\mathbb{E}(\tau_x^+) = \int_0^\infty \mathbb{P}(\tau_x^+ > t)\mathrm{d}t = \int_0^\infty \mathbb{P}(X_t \leq x)\mathrm{d}t = V(x),
$$

where the final equality follows from (5.3) with $q = \eta = 0$ and $\lambda = 1$. Now applying the strong Markov property we can establish that

$$
\begin{aligned}
V(x+y) &= \mathbb{E}(\tau_{x+y}^+)\\
&= \mathbb{E}(\tau_x^+ + \mathbb{E}_{X_{\tau_x^+}}(\tau_{x+y}^+))\\
&\leq \mathbb{E}(\tau_x^+) + \mathbb{E}(\tau_y^+)\\
&= V(x) + V(y).
\end{aligned}
$$

Using then the bound $V(x-y) - V(u+x-t) \leq V(t-u-y)$, the right continuity of V and the Renewal Theorem 5.1 (iii) we know that the integrand in first term on the hand side of (5.6) is bounded by a constant times t. Hence as $\int_{(0,\infty)} tF(\mathrm{d}t) < \infty$ dominated convergence applies together with Theorem 5.1 (i) to give

$$
\begin{aligned}
&\lim_{x\uparrow\infty} \int_{(u+y,\infty)} F(\mathrm{d}t)\{V(x-y) - V(u+x-t)\}\mathbf{1}_{(t<u+x)}\\
&= \frac{1}{\mu}\int_{(u+y,\infty)} (t-u-y)F(\mathrm{d}t)\\
&= \frac{1}{\mu}\int_{u+y}^\infty \overline{F}(t)\mathrm{d}t,
\end{aligned}
$$

where the final equality follows on integration by parts. □

In light of (5.4) we see that Lemma 5.5 gives the exact and asymptotic distribution of the overshoot and undershoot at first passage of a compound Poisson subordinator with jump distribution F (with finite mean and non-lattice support in the case of the asymptotic behaviour). In this spirit, the aim of the remainder of this chapter is to study the exact and asymptotic joint distributions of the overshoot and undershoot of a killed subordinator at first passage.

There are a number of differences in the range of a killed subordinator compared to the range of a compound Poisson subordinator which create additional difficulties to the situation described above. Firstly, in the case of a killed subordinator, the process may be killed before reaching a specified fixed level and hence one should expect an atom in the distribution of the overshoot at ∞. Secondly, the number of jumps over a finite time horizon may be infinite, which occurs if and only if $\Pi(0,\infty) = \infty$, and hence in this case the analysis in the proof of Lemma 5.5 (i) is no longer valid. Finally, in the case of a compound Poisson subordinator when F has no atoms, it is clear that the probability that there is first passage over a given level by hitting the level is zero. However, when moving to a killed subordinator for which either $\Pi(0,\infty) = \infty$ or there is a drift present, one should not exclude the possibility that first passage over a fixed level occurs by hitting the level with positive probability. The latter behaviour is called *creeping over a fixed level* and is equivalent to there being an atom at zero in the distribution of the overshoot at that level. As one might intuitively expect, creeping over a specified fixed level turns out to occur only in the presence of a drift in which case it is possible to creep over all fixed levels. These points will be dealt with in more detail in Sect. 5.3

5.2 Overshoots and Undershoots

We begin with the following theorem which gives the generalisation of Lemma 5.5 (i); indeed containing it as a corollary. Weaker versions of this theorem can be found in Kesten (1969) and Horowitz (1972). The format we give is from Bertoin (1996a).

Theorem 5.6. *Suppose that X is a killed subordinator. Then for $u > 0$ and $y \in [0, x]$,*

$$\mathbb{P}(X_{\tau_x^+} - x \in \mathrm{d}u, x - X_{\tau_x^+ -} \in \mathrm{d}y) = U(x - \mathrm{d}y)\Pi(y + \mathrm{d}u). \tag{5.7}$$

Proof. The proof principally makes use of the compensation formula. Suppose that f and g are two positive, bounded, continuous functions satisfying $f(0) = 0$. The latter ensures that the product $f(X_{\tau_x^+} - x)g(x - X_{\tau_x^+ -})$ is nonzero only if X jumps strictly above x when first crossing x. This means that we may write the expectation of the latter variable in terms of the Poisson random measure associated with the jumps of X whilst avoiding the issue of creeping.

To this end, let us assume that X is equal in law to a subordinator Y killed at rate $\eta \geq 0$. Then

$$\mathbb{E}(f(X_{\tau_x^+} - x)g(x - X_{\tau_x^+-})) = \mathbb{E}\left(\int_{[0,\infty)}\int_{(0,\infty)} \mathrm{e}^{-\eta t}\phi(t,\theta)N(\mathrm{d}t \times \mathrm{d}\theta)\right),$$

where

$$\phi(t,\theta) = \mathbf{1}_{(Y_{t-}\leq x)}\mathbf{1}_{(Y_{t-}+\theta > x)}f(Y_{t-} + \theta - x)g(x - Y_{t-}),$$

and N is the Point random measure associated with the jumps of Y. Note that it is straightforward to see that ϕ satisfies the conditions of Theorem 4.4; in particular that it is left continuous in t. Then with the help of the aforementioned theorem

$$\begin{aligned}
&\int_{[0,x]} g(y)\int_{(0,\infty)} f(u)\mathbb{P}(X_{\tau_x^+} - x \in \mathrm{d}u, x - X_{\tau_x^+-} \in \mathrm{d}y)\\
&= \mathbb{E}\left(\int_0^\infty \mathrm{d}t\cdot \mathrm{e}^{-\eta t}\mathbf{1}_{(Y_{t-}\leq x)}g(x - Y_{t-})\int_{(x-Y_{t-},\infty)} f(Y_{t-} + \theta - x)\Pi(\mathrm{d}\theta).\right)\\
&= \mathbb{E}\left(\int_0^\infty \mathrm{d}t\cdot \mathrm{e}^{-\eta t}\mathbf{1}_{(Y_t\leq x)}g(x - Y_t)\int_{(x-Y_t,\infty)} f(Y_t + \theta - x)\Pi(\mathrm{d}\theta)\right)\\
&= \int_{[0,x]} g(x-z)\int_{(x-z,\infty)} f(z+\theta - x)\Pi(\mathrm{d}\theta)\int_0^\infty \mathrm{d}t\cdot \mathrm{e}^{-\eta t}\mathbb{P}(Y_t \in \mathrm{d}z)\\
&= \int_{[0,x]} g(x-z)\int_{(x-z,\infty)} f(z+\theta - x)\Pi(\mathrm{d}\theta)U(\mathrm{d}z)\\
&= \int_{[0,x]} g(y)\int_{(0,\infty)} f(u)\Pi(\mathrm{d}u + y)U(x - \mathrm{d}y), \qquad (5.8)
\end{aligned}$$

where the final equality follows by changing variables with $y = x - z$ and then again with $u = \theta - y$. (Note also that U is the potential measure of X and not Y). As f and g are arbitrary within their prescribed classes, we read off from the left and right-hand sides of (5.8) the required distributional identity. □

Intutively speaking, the proof of Theorem 5.6 follows the logic of the proof of Lemma 5.5 (i). The compensation formula serves as a way of "decomposing" the event of first passage by a jump over level x according to the position of X prior to its first passage even when there are an unbounded number of jumps over finite time horizons.

To make the connection with the expression given for renewal processes in Lemma 5.5 (i), recall from (5.3) that $U(\mathrm{d}x) = \lambda^{-1}V(\mathrm{d}x)$ on $(0,\infty)$ where U is the potential measure associated with a compound Poisson subordinator with jump distribution F and arrival rate $\lambda > 0$ and V is the renewal measure associated with the distribution F. For this compound Poisson subordinator, we also know that $\Pi(\mathrm{d}x) = \lambda F(\mathrm{d}x)$ so that $U(x - \mathrm{d}y)\Pi(\mathrm{d}u + y) = V(x - \mathrm{d}y)$

$F(u + \mathrm{d}y)$. Setting $f(\cdot) = \mathbf{1}_{(\cdot > u)}$ and $g(\cdot) = \mathbf{1}_{(\cdot > y)}$, an easy calculation now brings one from the right hand side of (5.8) to the right-hand side of (5.5).

As (5.7) is the analogue of the statement in Lemma 5.5 (i) it is now natural to reconsider the proof of part (ii) of the same lemma in the more general context. The following result was presented in Bertoin et al. (1999).

Theorem 5.7. *Suppose that X is a subordinator (no killing) with finite mean $\mu := \mathbb{E}(X_1)$ and such that U does not have lattice support (cf. Theorem 5.4). Then for $u > 0$ and $y \geq 0$, in the sense of weak convergence*

$$\lim_{x \uparrow \infty} \mathbb{P}(X_{\tau_x^+} - x \in \mathrm{d}u, x - X_{\tau_x^+ -} \in \mathrm{d}y) = \frac{1}{\mu} \mathrm{d}y \Pi(y + \mathrm{d}u).$$

In particular it follows that the asymptotic probability of creeping satisfies

$$\lim_{x \uparrow \infty} \mathbb{P}(X_{\tau_x^+} = x) = \frac{d}{\mu}.$$

The proof of this result is a straightforward adaptation of the proof of Lemma 5.5 (ii) taking advantage of Corollary 5.3 and is left to the reader to verify in Exercise 5.1.

5.3 Creeping

Now let us turn to the issue of creeping. Although τ_x^+ is the first time that X strictly exceeds the level $x > 0$ it is possible that in fact $\mathbb{P}(X_{\tau_x^+} = x) > 0$; recall the statement and proof of Theorem 3.3. The following conclusion, found for example in Horowitz (1972), shows that crossing the level $x > 0$ by hitting it cannot occur by jumping onto it from a position strictly below x in the case where the jump measure is infinite. In other words, if our killed subordinator makes first passage above x with a jump then it must do so by jumping it clear, so $\{X_{\tau_x^+} = x\} = \{X_{\tau_x^+} - x = 0, x - X_{\tau_x^+ -} = 0\}$, which is of implicit relevance when computing the atom at zero in the overshoot distribution.

Lemma 5.8. *Let X be any killed subordinator with $\Pi(0, \infty) = \infty$. For all $x > 0$ we have*

$$\mathbb{P}(X_{\tau_x^+} - x = 0, x - X_{\tau_x^+ -} > 0) = 0. \tag{5.9}$$

Proof. Suppose for a given $x > 0$ that

$$\mathbb{P}(X_{\tau_x^+} - x = 0, x - X_{\tau_x^+ -} > 0) > 0.$$

Then this implies that there exists a $y < x$ such that

$$\mathbb{P}(X_{\tau_y^+} = x) > 0.$$

However the latter cannot happen because of the combined conclusions of Theorem 5.6 and Theorem 5.4 (i). Hence by contradiction (5.9) holds. □

Although one may write with the help of Theorem 5.6 and Lemma 5.8

$$\mathbb{P}(X_{\tau_x^+} = x) = 1 - \mathbb{P}(X_{\tau_x^+} > x) = 1 - \int_{(0,x]} U(x - \mathrm{d}y)\Pi(y, \infty)$$

this does not bring one closer to understanding when the probability on the left-hand side above is strictly positive. In fact, although the answer to this question is intuitively obvious, it turns out to be difficult to prove. It was resolved by Kesten (1969); see also Bretagnolle (1972). The result is given below.

Theorem 5.9. *For any killed subordinator with jump measure Π satisfying $\Pi(0,\infty) = \infty$, and drift coefficient d we have the following.*

(i) If $d = 0$ then $\mathbb{P}(X_{\tau_x^+} = x) = 0$ for all $x > 0$.
(ii) If $d > 0$ then U has a strictly positive and continuous density on $(0,\infty)$, say u, satisfying

$$\mathbb{P}(X_{\tau_x^+} = x) = du(x).$$

The version of the proof we give here for the above result follows the reasoning in Andrew (2005) (see also Section III.2 of Bertoin (1996a)) and first requires two auxiliary Lemmas given below. In the proof of both we shall make use of the following two key estimates for the probabilities $p_x := \mathbb{P}(X_{\tau_x^+} = x)$, $x > 0$. For all $0 < y < x$,

$$p_x \leq p_y p_{x-y} + (1 - p_{x-y}) \tag{5.10}$$

and

$$p_x \geq p_y p_{x-y}. \tag{5.11}$$

The upper bound is a direct consequence of the fact that

$$\begin{aligned}\mathbb{P}(X_{\tau_x^+} = x) &= \mathbb{P}(X_{\tau_{x-y}^+} = x - y,\ X_{\tau_x^+} = x) \\ &\quad + \mathbb{P}(X_{\tau_{x-y}^+} > x - y,\ X_{\tau_x^+} = x) \\ &\leq \mathbb{P}(X_{\tau_{x-y}^+} = x - y)\mathbb{P}(X_{\tau_x^+} = x | X_0 = x - y) \\ &\quad + \mathbb{P}(X_{\tau_{x-y}^+} > x - y),\end{aligned}$$

where in the last line the Strong Markov Property has been used. In a similar way the lower bound is a consequence of the fact that

$$\mathbb{P}(X_{\tau_x^+} = x) \geq \mathbb{P}(X_{\tau_{x-y}^+} = x - y)\mathbb{P}(X_{\tau_x^+} = x | X_0 = x - y).$$

Lemma 5.10.

(i) If for some $x > 0$ we have $p_x > 0$ then $\lim_{\varepsilon \downarrow 0} \sup_{\eta \in (0,\varepsilon)} p_\eta = 1$.
(ii) If for some $x > 0$ we have $p_x > 3/4$ then

$$p_y \geq 1/2 + \sqrt{p_x - 3/4}$$

for all $y \in (0, x]$.

Proof. (i) From Lemma 5.8 we know that X cannot jump onto x. In other words

$$\mathbb{P}(X_{\tau_x^+} = x > X_{\tau_x^+ -}) = 0.$$

This implies that

$$\{X_{\tau_x^+} = x\} \subseteq \bigcap_{n \geq 1} \{X \text{ visits } (x - 1/n, x)\}$$

almost surely. On the other hand, on the event $\bigcap_{n \geq 1}\{X \text{ visits } (x - 1/n, x)\}$ we also have by quasi-left-continuity (cf. Lemma 3.2) that $X_\sigma = x$ where $\sigma = \lim_{n \uparrow \infty} \tau^+_{x-1/n}$ (the limit exists because of monotonicity). Since

$$\{\sigma \leq t\} = \bigcap_{n \geq 1} \{\tau^+_{x-1/n} \leq t\}$$

almost surely it follows that σ is a stopping time with respect to $\mathbb{F}$. Applying the Strong Markov Property at time σ, since $X_\sigma = x$ and X is not a compound Poisson subordinator, we have that $X_t > x$ for all $t > \sigma$ showing that in fact $\sigma = \tau_x^+$. In conclusion

$$\{X_{\tau_x^+} = x\} = \bigcap_{n \geq 1} \{X \text{ visits } (x - 1/n, x)\}$$

almost surely.

We may now write

$$p_x = \lim_{n \uparrow \infty} \mathbb{P}(X \text{ visits } (x - 1/n, x)). \tag{5.12}$$

Also we may upper estimate

$$p_x \leq \mathbb{P}(X \text{ visits } (x - 1/n, x)) \sup_{z \in (0,1/n)} p_z.$$

Letting $n \uparrow \infty$ in the above inequality and taking (5.12) into account we see that $\lim_{\varepsilon \downarrow 0} \sup_{\eta \in (0,\varepsilon)} p_\eta = 1$.

(ii) Suppose that $0 < y < x$. We may assume without loss of generality that $p_y < p_x$ for otherwise it is clear that $p_y \geq p_x > 3/4 \geq 1/2 + \sqrt{p_x - 3/4}$.

From (5.10) it is a simple algebraic manipulation replacing y by $x - y$ to show that

$$p_y \leq \frac{1 - p_x}{1 - p_{x-y}}.$$

and again replacing y by $x - y$ in the above inequality to deduce that

$$1 - p_{x-y} \geq \frac{p_x - p_y}{1 - p_y}.$$

Combining the last two inequalities we therefore have

$$p_y \leq \frac{(1-p_x)(1-p_y)}{p_x - p_y}$$

and hence the quadratic inequality $p_y^2 - p_y + 1 - p_x \geq 0$. This in turn implies that

$$p_y \in [0, 1/2 - \sqrt{p_x - 3/4}] \cup [1/2 + \sqrt{p_x - 3/4}, 1]. \tag{5.13}$$

The remainder of the proof is thus dedicated to showing that inclusion of p_y in the first of the two intervals cannot happen.

Suppose then for contradiction that (5.13) holds for all $y \leq x$ and there exists a $y \in (0, x)$ such that $p_y \leq 1/2 - \sqrt{p_x - 3/4}$. Now define

$$g = \sup\{z \in [0, y); p_z \geq 1/2 + \sqrt{p_x - 3/4}\}$$

which is well defined since at least $p_0 = 1$. Note from this definition that it could be the case that $g = y$. Reconsidering the definition of g and (5.13), we see that either there exists an $\varepsilon > 0$ such that $p_z \leq 1/2 - \sqrt{p_x - 3/4}$ for all $z \in (g - \varepsilon, g)$ or for all $\varepsilon > 0$, there exists a sequence of $z \in (g - \varepsilon, g)$ such that $p_z \geq 1/2 + \sqrt{p_x - 3/4}$. In the former case it is clear by the definition of g that $p_g \geq 1/2 + \sqrt{p_x - 3/4}$. In the latter case, we see when referring back to (5.12) that

$$p_g = \lim_{z \uparrow g} \mathbb{P}(X \text{ visits } (z, g)) \geq \lim_{\varepsilon \downarrow 0} \sup_{\eta \in (0,\varepsilon)} p_{g-\eta}$$

and hence $p_g \geq 1/2 + \sqrt{p_x - 3/4}$. For both cases this implies in particular that $g < y$. On the other hand, using (5.11) and the conclusion of part (i) we see that

$$\lim_{\varepsilon \downarrow 0} \sup_{\eta \in (0,\varepsilon)} p_{g+\eta} \geq p_g \times \lim_{\varepsilon \downarrow 0} \sup_{\eta \in (0,\varepsilon)} p_\eta = p_g \geq 1/2 + \sqrt{p_x - 3/4}.$$

Since (5.13) is in force for all $y \leq x$ and $g < y$, this implies that there exists a $g' > g$ such that $p_{g'} \geq 1/2 + \sqrt{p_x - 3/4}$ which contradicts the definition of g. The consequence of this contradiction is that there does not exist a $y \in (0, x)$ for which $p_y < 1/2 + \sqrt{p_x - 3/4}$ and hence from (5.13) it necessarily follows that $p_y \geq 1/2 + \sqrt{p_x - 3/4}$ for all $y \in (0, x)$. □

Lemma 5.11. *Suppose there exists an $x > 0$ such that $p_x > 0$, then*

(i) $\lim_{\varepsilon \downarrow 0} p_\varepsilon = 1$ *and*
(ii) $x \mapsto p_x$ *is strictly positive and continuous on* $[0, \infty)$.

Proof. (i) The first part is a direct consequence of parts (i) and (ii) of Lemma 5.10.

(ii) Positivity follows from a repeated use of the lower estimate in (5.11) to obtain $p_x \geq (p_{x/n})^n$ and the conclusion of part (i).

To show continuity, note with the help of (5.10),

$$\limsup_{\varepsilon \downarrow 0} p_{x+\varepsilon} \leq \limsup_{\varepsilon \downarrow 0} \{p_\varepsilon p_x + 1 - p_\varepsilon\} = p_x$$

and from (5.11) and part (i),

$$\liminf_{\varepsilon \downarrow 0} p_{x+\varepsilon} \geq \liminf_{\varepsilon \downarrow 0} p_x p_\varepsilon = p_x.$$

Further, arguing in a similar manner,

$$\limsup_{\varepsilon \downarrow 0} p_{x-\varepsilon} \leq \limsup_{\varepsilon \downarrow 0} \frac{p_x}{p_\varepsilon} = p_x$$

and

$$\liminf_{\varepsilon \downarrow 0} p_{x-\varepsilon} \geq \liminf_{\varepsilon \downarrow 0} \frac{p_x + p_\varepsilon - 1}{p_\varepsilon} = p_x.$$

Thus continuity is confirmed. □

Finally we return to the proof of Theorem 5.9.

Proof (of Theorem 5.9). Consider the function

$$M(a) := \mathbb{E}\left(\int_0^a \mathbf{1}_{(X_{\tau_x^+} = x)} \mathrm{d}x\right) = \int_0^a p_x \mathrm{d}x$$

for all $a \geq 0$.

For convenience, suppose further that X is equal in law to a subordinator Y killed at rate η. Let N be the Poisson random measure associated with the jumps of X (or equivalently Y). Then we may write with the help of the Lévy–Itô decomposition for subordinators,

$$M(a) = \mathbb{E}\left(Y_{(\tau_a^+ \wedge \mathbf{e}_\eta)-} - \int_{[0,\tau_a^+ \wedge \mathbf{e}_\eta)} \int_{(0,\infty)} xN(\mathrm{d}s \times \mathrm{d}x)\right) = d\mathbb{E}(\tau_a^+ \wedge \mathbf{e}_\eta).$$

(i) If $d = 0$ then $p_x = 0$ for Lebesgue almost every x. Lemma 5.11 now implies that $p_x = 0$ for all $x > 0$.

(ii) If $d > 0$ then three exists an $x > 0$ such that $p_x > 0$. Hence from Lemma 5.11, $x \mapsto p_x$ is strictly positive and continuous. Further, we see that

$$M(a) = d \int_0^\infty \mathbb{P}(\tau_a^+ \wedge \mathbf{e}_\eta > t)\mathrm{d}t = d \int_0^\infty \mathbb{P}(X_t \leq a)\mathrm{d}t = dU(a).$$

The latter implies that U has a density which may be taken as equal to $d^{-1}p_x$ for all $x \geq 0$. □

We close this section by noting that the reason why the case of a killed compound Poisson subordinator ($\Pi(0,\infty) < \infty$) is excluded from Theorem 5.9 is because of possible atoms in Π. Take for example the case that Π consists of a single atom at $x_0 > 0$ then the same is true of the jump distribution. This means that $\mathbb{P}(X_{\tau^+_{nx_0}} = nx_0) > 0$ for all $n \geq 1$. None the less $\mathbb{P}(X_{\tau^+_x} = x) = 0$ for all other x.

5.4 Regular Variation and Tauberian Theorems

The inclusion of the forthcoming discussion on regular variation and Tauberian theorems is a prerequisite to Sect. 5.5 which gives the Dynkin–Lamperti asymptotics for the joint law of the overshoot and undershoot. However, the need for facts concerning regular variation will also appear in later sections and chapters.

Suppose that U is a measure supported on $[0,\infty)$ and with Laplace–transform

$$\Lambda(\theta) = \int_{[0,\infty)} \mathrm{e}^{-\theta x} U(\mathrm{d}x)$$

for $\theta \geq 0$ which may be infinite. Note that if there exists a θ_0 such that $\Lambda(\theta_0) < \infty$ then $\Lambda(\theta) < \infty$ for all $\theta \geq \theta_0$. The point of this chapter is to present some classic results which equivalently relate certain types of tail behaviour of the measure U to a similar type of behaviour of Λ. Our presentation will only offer the bare essentials based on the so called Karamata theory of regularly varying functions. Aside from their intrinsic analytic curiosity, regularly varying functions have proved to be of great practical value within probability theory, not least in the current context. The highly readable account given in Feller (1971) is an important bridging text embedding into probability theory the classic work of Karamata and his collaborators which dates back to the period between 1930 and the 1960s. For a complete account, the reader is referred to Bingham et al. (1987) or Embrechts et al. (1997). The presentation here is principally based on these books.

Definition 5.12. *A function $f : [0,\infty) \to (0,\infty)$ is said to be regularly varying at zero with index $\rho \in \mathbb{R}$ if for all $\lambda > 0$,*

$$\lim_{x\downarrow 0} \frac{f(\lambda x)}{f(x)} = \lambda^{\rho}.$$

If the above limit holds as x tends to infinity then f is said to be regularly varying at infinity with index ρ. The case that $\rho = 0$ is referred to as slow variation.

Note that any regularly varying function f may always be written in the form

$$f(x) = x^{\rho} L(x)$$

where L is a slowly varying function. Any function which has a strictly positive and finite limit at infinity is slowly varying at infinity so the class of slowly (and hence regularly) varying functions is clearly non-empty due to this trivial example. There are however many non-trival examples of slowly varying functions. Examples include $L(x) = \log x$, $L(x) = \log_k x$ (the kth iterate of $\log x$) and $L(x) = \exp\{(\log x)/\log\log x\}$. All of these examples have the property that they are functions which tend to infinity at infinity. The function

$$L(x) := \exp\{(\log x)^{\frac{1}{3}} \cos[(\log x)^{\frac{1}{3}}]\}$$

is an example of a regularly varying function at infinity which oscillates. That is to say $\liminf_{x\uparrow\infty} L(x) = 0$ and $\limsup_{x\uparrow\infty} L(x) = \infty$.

The main concern of this section are the following remarkable results.

Theorem 5.13. *Suppose that L is slowly varying at infinity, $\rho \in [0,\infty)$ and U is a measure supported on $[0,\infty)$. Then the following two statements are equivalent.*

(i) $\Lambda(\theta) \sim \theta^{-\rho} L(1/\theta)$, *as* $\theta \to 0$,
(ii)$U(x) \sim x^{\rho} L(x)/\Gamma(1+\rho)$ *as* $x \to \infty$.

In the above theorem we are using the notation $f \sim g$ for functions f and g to mean that $\lim f(x)/g(x) = 1$.

Theorem 5.14. *Suppose that L is slowly varying at infinity, $\rho \in (0,\infty)$ and U is a measure on $[0,\infty)$ which has an ultimately monotone density, u. Then the following two statements are equivalent.*

(i) $\Lambda(\theta) \sim \theta^{-\rho} L(1/\theta)$, *as* $\theta \to 0$,
(ii) $u(x) \sim x^{\rho-1} L(x)/\Gamma(\rho)$ *as* $x \to \infty$.

Recalling that $\Gamma(1+\rho) = \rho\Gamma(\rho)$, Theorem 5.14 is a natural statement next to Theorem 5.13. It says that, up to a slowly varying function, the asymptotic behaviour of the derivative of $U(x)$ behaves like the derivative of the polynomial that $U(x)$ asymptotically mimics; providing of course the density u exists and is ultimately monotone. The methods used to prove these results also produce the following corollary with virtually no change at all.

Corollary 5.15. *The statement of Theorems 5.13 and 5.14 are still valid when instead the limits in their parts (i) and (ii) are simultaneously changed to $\theta \to \infty$ and $x \to 0$.*

We now give the proof of Theorem 5.13 which, in addition to the assumed regular variation, uses little more than the Continuity Theorem for Laplace transforms of positive random variables.

Proof (of Theorem 5.13). It will be helpful for this proof to record the well known fact that for any $\infty > \rho \geq 0$ and $\lambda > 0$,

$$\int_0^\infty x^\rho \mathrm{e}^{-\lambda x}\mathrm{d}x = \frac{\Gamma(1+\rho)}{\lambda^{1+\rho}}. \tag{5.14}$$

In addition we will also use the fact that for all $\lambda > 0$ and $\theta > 0$,

$$\int_{[0,\infty)} \mathrm{e}^{-\lambda x} U(\mathrm{d}x/\theta) = \Lambda(\lambda\theta). \tag{5.15}$$

First we prove that (i) implies (ii). Fix $\lambda_0 > 0$. From (5.15) we have for $\theta > 0$ that $\mathrm{e}^{-\lambda_0 x}U(\mathrm{d}x/\theta)/\Lambda(\lambda_0\theta)$ is a probability distribution. Again from (5.15) we can compute its Laplace transform as $\Lambda((\lambda+\lambda_0)\theta)/\Lambda(\lambda_0\theta)$. The regular variation assumed in (i) together with (5.14) implies that

$$\lim_{\theta\downarrow 0}\int_{[0,\infty)} \mathrm{e}^{-(\lambda+\lambda_0)x}\frac{U(\mathrm{d}x/\theta)}{\Lambda(\lambda_0\theta)} = \frac{\lambda_0^\rho}{(\lambda_0+\lambda)^\rho} = \frac{\lambda_0^\rho}{\Gamma(\rho)}\int_0^\infty x^{\rho-1}\mathrm{e}^{-(\lambda+\lambda_0)x}\mathrm{d}x,$$

where the right-hand side is the Laplace transform of a gamma distribution. It follows from the Continuity Theorem for Laplace transforms of probability distributions that $\mathrm{e}^{-\lambda_0 x}U(\mathrm{d}x/\theta)/\Lambda(\lambda_0\theta)$ converges weakly to $\mathrm{e}^{-\lambda_0 x}\lambda_0^\rho x^{\rho-1}/\Gamma(\rho)\mathrm{d}x$ as θ tends to zero. Using the regular variation of Λ again, this implies that for all $x > 0$,

$$\lim_{\theta\downarrow 0}\frac{U(y/\theta)}{L(1/\theta)}\lambda_0^\rho\theta^\rho = \frac{\lambda_0^\rho y^\rho}{\rho\Gamma(\rho)}.$$

Now setting $y = 1$, rewriting $x = 1/\theta$ and recalling that $\Gamma(1+\rho) = \rho\Gamma(\rho)$, statement (ii) follows.

Now to prove that (ii) implies (i). The assumption in (ii) expressed in terms of weak convergence implies that for each $y \geq 0$,

$$\lim_{x\uparrow\infty}\frac{U(x\,\mathrm{d}y)}{U(x)} = \rho y^{\rho-1}\mathrm{d}y,$$

in particular for any $t > 0$ and $\lambda > 0$,

$$\lim_{x\uparrow\infty}\int_0^t \mathrm{e}^{-\lambda y}\frac{U(x\,\mathrm{d}y)}{U(x)} = \rho\int_0^t y^{(\rho-1)}\mathrm{e}^{-\lambda y}\mathrm{d}y. \tag{5.16}$$

In view of (5.15), the Laplace transform of the measure $U(x\,\mathrm{d}y)/U(x)$ is given by $\Lambda(\lambda/x)/U(x)$ for $\lambda > 0$. Now suppose that for some $0 < \lambda_0 < 1$ and $x_0 > 0$, the sequence $\{\Lambda(\lambda_0/x)/U(x) : x > x_0\}$ is uniformly bounded by some $C > 0$. With this further assumption in place, we may pick a sufficiently large $t > 0$ such that

$$\int_t^\infty \mathrm{e}^{-y}\frac{U(x\,\mathrm{d}y)}{U(x)} < \mathrm{e}^{-(1-\lambda_0)t}\int_t^\infty \mathrm{e}^{-\lambda_0 y}\frac{U(x\,\mathrm{d}y)}{U(x)} < C\mathrm{e}^{-(1-\lambda_0)t}.$$

Together with (5.16), the above estimate is sufficient to deduce that

$$\lim_{x\uparrow\infty}\frac{\Lambda(1/x)}{U(x)} = \lim_{x\uparrow\infty}\int_0^\infty \mathrm{e}^{-y}\frac{U(x\,\mathrm{d}y)}{U(x)} = \rho\int_0^\infty y^{(\rho-1)}\mathrm{e}^{-y}\mathrm{d}y = \Gamma(1+\rho).$$

Choosing $\lambda = 1$ and writing $\theta = 1/x$, the statement in (i) follows.

It remains then to show that for some $0 < \lambda_0 < 1$ and x_0, the sequence $\{\Lambda(\lambda_0/x)/U(x) : x > x_0\}$ is uniformly bounded by some $C > 0$. This is done by partitioning the domain of integration over the lattice $\{2^k x : k \geq 0\}$ for some $x > 0$. The assumed regular variation of U implies that for all x sufficiently large $U(2x) < 2^{\rho+1}U(x)$. This can be iterated to deduce for x sufficiently large, $U(2^n x) < 2^{n(1+\rho)}U(x)$ for each $n \geq 1$. With this inequality in hand we may quite coarsely estimate for all sufficiently large x,

$$\frac{\Lambda(\lambda_0/x)}{U(x)} \leq \sum_{n\geq 1} \mathrm{e}^{-\lambda_0 2^{n-1}} \frac{U(2^n x)}{U(x)} < \sum_{n\geq 1} 2^{n(1+\rho)} \mathrm{e}^{-\lambda_0 2^{n-1}} < \infty$$

and the proof is complete. □

Next we turn to the proof of Theorem 5.14 which implicitly uses the statement of Theorem 5.13.

Proof (of Theorem 5.14). First we prove that (ii) implies (i). It suffices to prove that (ii) implies Theorem 5.13 (ii). However this is a simple issue of weak convergence and regular variation, since for any $y > 0$,

$$\frac{\rho U(x\,\mathrm{d}y)}{xu(x)} = \frac{\rho u(xy)x}{xu(x)}\mathrm{d}y \to \rho y^{\rho-1}\mathrm{d}y$$

as x tends to infinity in the sense of weak convergence. This implies that

$$\frac{\rho U(xy)}{xu(x)} \sim y^{\rho}.$$

Now choosing $y = 1$ and taking account of the fact that $xu(x)/\rho \sim x^{\rho}L(x)/\Gamma(1+\rho)$ (here we use that $\Gamma(1+\rho) = \rho\Gamma(\rho)$), the result follows.

Next we prove that (i) implies (ii). From Theorem 5.13 we see that $U(x) \sim x^{\rho}L(x)/\Gamma(1+\rho)$ for some slowly varying function L. Let us assume that u is eventually monotone non-decreasing. For any $0 < a < b < \infty$ we have

$$U(bx) - U(ax) = \int_{ax}^{bx} u(y)\mathrm{d}y$$

and hence for x large enough,

$$\frac{(b-a)xu(ax)}{x^{\rho}L(x)/\Gamma(1+\rho)} \leq \frac{U(bx)-U(ax)}{x^{\rho}L(x)/\Gamma(1+\rho)} \leq \frac{(b-a)xu(bx)}{x^{\rho}L(x)/\Gamma(1+\rho)}. \tag{5.17}$$

Using the regular variation of U we also have that

$$\lim_{x\uparrow\infty} \frac{U(bx)-U(ax)}{x^{\rho}L(x)/\Gamma(1+\rho)} = (b^{\rho} - a^{\rho}).$$

Hence from the left inequality of (5.17) we have

$$\limsup_{x\uparrow\infty}\frac{u(ax)}{x^{\rho-1}L(x)/\Gamma(1+\rho)}\leq\frac{(b^{\rho}-a^{\rho})}{(b-a)}.$$

Now taking $a=1$ and letting $b\downarrow 1$ gives

$$\limsup_{x\uparrow\infty}\frac{u(x)}{x^{\rho-1}L(x)}\leq\frac{\rho}{\Gamma(1+\rho)}.$$

A similar treatment for the right inequality with $b=1$ and letting $a\uparrow 1$ shows that

$$\liminf_{x\uparrow\infty}\frac{u(x)}{x^{\rho-1}L(x)}\geq\frac{\rho}{\Gamma(1+\rho)}.$$

Recalling that $\Gamma(1+\rho)=\rho\Gamma(\rho)$, the statement of the theorem follows.

The proof when u is eventually non-increasing is essentially the same with minor adjustments. □

5.5 Dynkin–Lamperti Asymptotics

Let us return to the issue of the asymptotic behaviour of overshoots and undershoots of subordinators. The following theorem, due to Dynkin (1961) and Lamperti (1962), shows that obtaining an asymptotic bivariate limit of the overshoot and undershoot via rescaling by the level of the barrier is equivalent to an assumption of regular variation on the Laplace exponent of the subordinator.

Theorem 5.16. *Suppose that X is any subordinator with Laplace exponent Φ which is regularly varying at zero (resp. infinity) with some index $\alpha\in(0,1)$. Then, in the sense of weak convergence,*

$$\begin{aligned}\mathbb{P}&\left(\frac{X_{\tau_x^+}-x}{x}\in \mathrm{d}u,\frac{x-X_{\tau_x^+-}}{x}\in \mathrm{d}y\right)\\&\to\frac{\alpha\sin\pi\alpha}{\pi}(1-y)^{\alpha-1}(y+u)^{-\alpha-1}\mathrm{d}y\,\mathrm{d}u\end{aligned}\qquad(5.18)$$

for $u>0$ and $y\in[0,1)$ as x tends to infinity (resp. zero).

Note that the statement of the theorem is not empty as one sees that any stable subordinator fulfills the assumptions. Recall from Exercise 3.7 that a stable subordinator necessarily has Laplace exponent on $[0,\infty)$ given by $\Phi(\theta)=c\theta^{\alpha}$ for some $c>0$ and $\alpha\in(0,1)$.

It is possible to prove more than the above statement. For example, one may prove conversely that the pair

$$\left(\frac{X_{\tau_x^+}-x}{x},\frac{x-X_{\tau_x^+-}}{x}\right)$$

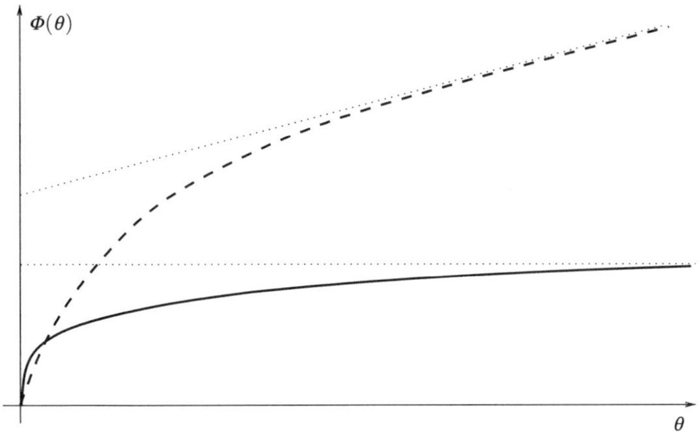

Fig. 5.2. Examples of the shape of the Laplace exponent $\Phi(\theta)$. The *solid concave curve* corresponds to the case of a compound Poisson process with infinite mean ($\Phi'(0+) = \infty$ and $\Phi(\infty) < \infty$). The *dashed concave curve* corresponds to the case of a finite mean subordinator with strictly positive linear drift ($\Phi'(0+) < \infty$ and $\lim_{\theta\uparrow\infty} \Phi(\theta)/\theta = d$).

has a non-degenerate limit in distribution as $x \uparrow \infty$ only if Φ is regularly varying with index $\alpha \in (0,1)$. See Exercise 5.7.

It is also possible to calculate the marginal laws of (5.18) as follows,

$$\mathbb{P}\left(\frac{X_{\tau_x^+} - x}{x} \in \mathrm{d}u\right) \to \frac{\sin \pi\alpha}{\pi} u^{-\alpha}(1+u)^{-1}\mathrm{d}u$$

and

$$\mathbb{P}\left(\frac{x - X_{\tau_x^+ -}}{x} \in \mathrm{d}y\right) \to \frac{\sin \pi\alpha}{\pi} y^{-\alpha}(1-y)^{\alpha-1}\mathrm{d}y$$

in the sense of weak convergence as $x \uparrow \infty$ or $x \downarrow 0$. The latter case is known as the generalised arcsine law; the arcsine law itself being a special case when $\alpha = 1/2$. In that case, the density $(\pi\sqrt{y(1-y)})^{-1}$ is related (via a linear transform) to the derivative of the arcsine function.

Before moving to the proof of Theorem 5.16 let us make some remarks about regular variation of the Laplace exponent Φ of a subordinator. It is easy to deduce with the help of dominated convergence that Φ is infinitely differentiable and strictly concave. In addition $\Phi'(0+) = \mathbb{E}(X_1) \in (0,\infty]$, $\Phi(0) = 0$ and $\Phi(\infty) = -\log \mathbb{P}(X_1 = 0)$ (which is only finite in the case that X is a compound Poisson subordinator). Finally recall again from Exercise 2.11 that $\lim_{\theta\uparrow\infty} \Phi(\theta)/\theta = d$. See Fig. 5.2 for a visualization of these facts.

Suppose now that Φ is regularly varying at the origin with index $\alpha \in \mathbb{R}$. As $\Phi(0) = 0$ this implies that necessarily $\alpha \geq 0$. If $\mathbb{E}(X_1) < \infty$ then clearly $\Phi(\theta)/\theta \sim \mathbb{E}(X_1)$ as $\theta \downarrow 0$ forcing $\alpha = 1$. On the other hand, if $\mathbb{E}(X_1) = \infty$ then $\Phi(\theta)/\theta$ explodes as $\theta \downarrow 0$ forcing $\alpha < 1$. In conclusion, regular variation at the origin of Φ with index $\alpha \in \mathbb{R}$ means necessarily that $\alpha \in [0, 1]$.

Now suppose that Φ is regularly varying at infinity with index $\alpha \in \mathbb{R}$. Since $\Phi(\infty) > 0$ (actually infinite in the case that X is not a compound Poisson subordinator) again implies that $\alpha \geq 0$. On the other hand, the fact that $\Phi(\theta)/\theta$ tends to the constant d at infinity also dictates that $\alpha \leq 1$. Hence regular variation at infinity of Φ with index $\alpha \in \mathbb{R}$ again necessarily implies that $\alpha \in [0, 1]$.

We now turn to the proof of Theorem 5.16, beginning with the following preparatory Lemma. Recall U is the potential measure of the given subordinator.

Lemma 5.17. *Suppose that the Laplace exponent of a subordinator, Φ, is regularly varying at zero (resp. infinity) with index $\alpha \in [0, 1]$. Then for all $\lambda > 0$*

(i) $U(\lambda x)\Phi(1/x) \to \lambda^\alpha/\Gamma(1+\alpha)$ as $x \uparrow \infty$ (resp. $x \downarrow 0$) and
(ii) when α is further restricted to $[0, 1)$, $\Pi(\lambda x, \infty)/\Phi(1/x) \to \lambda^{-\alpha}/\Gamma(1-\alpha)$ as $x \uparrow \infty$ (resp. $x \downarrow 0$).

Proof. (i) Recall that

$$\int_{[0,\infty)} \mathrm{e}^{-qx} U(\mathrm{d}x) = \frac{1}{\Phi(q)}.$$

The assumption on Φ means that $\Phi(\theta) \sim \theta^\alpha L(1/\theta)$ as θ tends to zero where L is slowly varying at infinity. That is to say $1/\Phi(1/x) \sim x^\alpha/L(x)$ as x tends to infinity. Noting that $1/L$ is still slowly varying at infinity, theorem 5.13 thus implies that $U(x) \sim x^\alpha/L(x)\Gamma(1+\alpha)$ as $x \uparrow \infty$. Regular variation now implies the statement in part (i). The same argument works when Φ is regularly varying at infinity rather than zero.

(ii) Now recall that

$$\frac{\Phi(\theta)}{\theta} = d + \int_0^\infty \mathrm{e}^{-\theta x} \Pi(x, \infty)\mathrm{d}x$$

showing that $\Phi(\theta)/\theta$ is the Laplace transform of the measure $d\delta_0(\mathrm{d}x) + \Pi(x, \infty)\mathrm{d}x$. The assumed regular variation on Φ implies that $\Phi(\theta)/\theta \sim \theta^{-(1-\alpha)}L(1/\theta)$, for some slowly varying L at infinity. Theorem 5.14 now implies that $\Pi(x, \infty) \sim x^{-\alpha}L(x)/\Gamma(1-\alpha)$. Regular variation now implies the statement in part (ii). As usual, the same argument works when instead it is assumed that Φ is regularly varying at infinity. Note also in this case the assumption that $\alpha \in [0, 1)$ implies that $d = 0$ as otherwise if $d > 0$ then necessarily $\alpha = 1$. □

Finally we are ready for the proof of the Dynkin–Lamperti Theorem.

Proof (of Theorem 5.16). We give the proof for the case that $x \uparrow \infty$. The proof for $x \downarrow 0$ requires minor modification.

Starting from the conclusion of Theorem 5.6 we have for $\theta \in (0,1]$ and $\phi > 0$,

$$\mathbb{P}\left(\frac{X_{\tau_x^+} - x}{x} \in \mathrm{d}\phi, \frac{x - X_{\tau_x^+ -}}{x} \in \mathrm{d}\theta\right) = U(x(1-\mathrm{d}\theta))\Pi(x(\theta + \mathrm{d}\phi))$$

and hence for $0 < a < b < 1$ and $c > 0$,

$$\begin{aligned}
&\mathbb{P}\left(\frac{X_{\tau_x^+} - x}{x} > c, \frac{x - X_{\tau_x^+ -}}{x} \in (a,b)\right) \\
&= \int_{(a,b)} \Pi(x(\theta + c), \infty)U(x(1\mathrm{d}\theta)) \\
&= \int_{(1-b,1-a)} \frac{\Pi(x(1-\eta+c),\infty)}{\Phi(1/x)} U(x\,\mathrm{d}\eta)\Phi(1/x), \qquad (5.19)
\end{aligned}$$

where in the last equality we have changed variables. From Lemma 5.17 (i) we see on the one hand that $U(x\,\mathrm{d}\eta)\Phi(1/x)$ converges weakly to $\eta^{\alpha-1}\mathrm{d}\eta/\Gamma(\alpha)$ (we have used that $\Gamma(1+\alpha) = \alpha\Gamma(\alpha)$). On the other hand we have seen from part (ii) of the same lemma that

$$\lim_{x\downarrow 0} \frac{\Pi(x(1-\eta+c),\infty)}{\Phi(1/x)} = \frac{(1-\eta+c)^{-\alpha}}{\Gamma(1-\alpha)}.$$

Since $\Pi(x(1-\eta+\phi),\infty)$ is monotone in η, it follows that the convergence above is uniform on $(1-b, 1-a)$; recall Exercise 2.4. It follows that the right-hand side of (5.19) converges to

$$\begin{aligned}
&\int_{(1-b,1-a)} \frac{(1-\eta+c)^{-\alpha}}{\Gamma(1-\alpha)} \frac{\eta^{\alpha-1}}{\Gamma(\alpha)}\mathrm{d}\eta \\
&= \frac{1}{\Gamma(\alpha)\Gamma(1-\alpha)} \int_{(a,b)} (\theta+c)^{-\alpha}(1-\theta)^{\alpha-1}\mathrm{d}\theta
\end{aligned}$$

as $x \uparrow \infty$ which is tantamount to

$$\begin{aligned}
&\lim_{x\uparrow\infty} \mathbb{P}\left(\frac{x - X_{\tau_x^+ -}}{x} \in \mathrm{d}y, \frac{X_{\tau_x^+ - x}}{x} \in \mathrm{d}u\right) \\
&= \frac{\alpha}{\Gamma(\alpha)\Gamma(1-\alpha)}(y+u)^{-\alpha-1}(1-y)^{\alpha-1}\mathrm{d}y\,\mathrm{d}u
\end{aligned}$$

in the sense of weak convergence. Finally, a special property of gamma functions gives $1/(\Gamma(\alpha)\Gamma(1-\alpha)) = (\sin\pi\alpha)/\pi$ and hence the proof is complete. □

Exercises

5.1. Prove Theorem 5.7.

5.2. Suppose that Y is a spectrally positive Lévy process of bounded variation drifting to $-\infty$ with Laplace exponent written in the usual form

$$\log \mathbb{E}(\mathrm{e}^{-\theta Y_1}) = \psi(\theta) = d\theta - \int_{(0,\infty)} (1 - \mathrm{e}^{-\theta x})\nu(\mathrm{d}x),$$

where necessarily $d > 0$, $\int_{(0,\infty)}(1 \wedge x)\nu(\mathrm{d}x) < \infty$ and $\psi'(0+) > 0$. Define $\sigma_x^+ = \inf\{t > 0 : Y_t > x\}$ and $\overline{Y}_t = \sup_{s\le t} Y_s$.

(i) Suppose that $X = \{X_t : t \ge 0\}$ is a compound Poisson subordinator with jump distribution $(d-\psi'(0+))^{-1}\nu(x,\infty)\mathrm{d}x$. By following similar reasoning to the explanation of the Pollaczeck–Khintchin formula in Chap. 4 show that

$$\mathbb{P}(Y_{\sigma_x^+}-x \in \mathrm{d}u, x-\overline{Y}_{\sigma_x^+-} \in \mathrm{d}y|\sigma_x^+ < \infty) = \mathbb{P}(X_{\tau_x^+}-x \in \mathrm{d}u, x-X_{\tau_x^+-} \in \mathrm{d}y).$$

(ii) Hence deduce that if $\int_0^\infty x\nu(x,\infty)\mathrm{d}x < \infty$ then for $u, y > 0$, in the sense of weak convergence

$$\lim_{x\uparrow\infty} \mathbb{P}(Y_{\sigma_x^+} - x \in \mathrm{d}u, x - \overline{Y}_{\sigma_x^+-} \in \mathrm{d}y|\sigma_x^+ < \infty) = \frac{1}{\int_0^\infty x\nu(x,\infty)\mathrm{d}x}\nu(u+y,\infty)\mathrm{d}u\,\mathrm{d}y$$

(iii) Give an interpretation of the result in (ii) in the context of modelling insurance claims.

5.3. Suppose that X is a finite mean subordinator and that its associated potential measure U does not have lattice support. Suppose that Z is a random variable whose distribution is equal to that of the limiting distribution of $X_{\tau_x^+} - X_{\tau_x^+-}$ as $x \uparrow \infty$. Suppose further that (V, W) is a bivariate random variable whose distribution is equal to the limiting distribution of $(X_{\tau_x^+}-x, x-X_{\tau_x^+-})$ as $x \uparrow \infty$ and U is independent of V, W, Z and uniformly distributed on $[0, 1]$. Show that (V, W) is equal in distribution to $((1 - U)Z, UZ)$.

5.4. Let X and Y be two (possibly correlated) subordinators killed independently at the rate $\eta \ge 0$. Denote their bivariate jump measure by $\mathbf{\Pi}(\cdot,\cdot)$. Define their bivariate renewal function

$$\mathcal{U}(\mathrm{d}x, \mathrm{d}y) = \int_0^\infty \mathrm{d}t \cdot \mathbb{P}(X_t \in \mathrm{d}x, Y_t \in \mathrm{d}y),$$

and suppose that as usual

$$\tau_x^+ = \inf\{t > 0 : X_t > x\}.$$

Use a generalised version of the compensation formula to establish the following quadruple law

$$P(\Delta X_{\tau_x^+} \in \mathrm{d}t, X_{\tau_x^+-} \in \mathrm{d}s, x - Y_{\tau_x^+-} \in \mathrm{d}y, Y_{\tau_x^+} - x \in \mathrm{d}u)$$
$$= \mathcal{U}(\mathrm{d}s, x - \mathrm{d}y)\mathbf{\Pi}(\mathrm{d}t, \mathrm{d}u + y)$$

for $u > 0$, $y \in [0, x]$ and $s, t \geq 0$. This formula will be of use later on while considering the first passage of a general Lévy process over a fixed level.

5.5. Let X be any subordinator with Laplace exponent Φ and recall that $\tau_x^+ = \inf\{t > 0 : X_t > x\}$. Let $\mathbf{e}_\alpha$ be an exponentially distributed random variable which is independent of X.

(i) By applying the Strong Markov Property at time τ_x^+ in the expectation $\mathbb{E}(\mathrm{e}^{-\beta X_{\mathbf{e}_\alpha}} \mathbf{1}_{(X_{\mathbf{e}_\alpha} > x)})$ show that for all $\alpha, \beta, x \geq 0$ we have

$$\mathbb{E}\left(\mathrm{e}^{-\alpha\tau_x^+ - \beta X_{\tau_x^+}}\right) = (\alpha + \Phi(\beta)) \int_{(x,\infty)} \mathrm{e}^{-\beta z} U^{(\alpha)}(\mathrm{d}z) \tag{5.20}$$

for all $x > 0$.

(ii) Show further with the help of the identity in (i) that when $q > \beta$,

$$\int_0^\infty \mathrm{e}^{-qx}\mathbb{E}\left(\mathrm{e}^{-\alpha\tau_x^+ - \beta(X_{\tau_x^+} - x)}\right) \mathrm{d}x = \frac{1}{q - \beta}\left(1 - \frac{\alpha + \Phi(\beta)}{\alpha + \Phi(q)}\right).$$

(iii) Deduce with the help of Theorem 5.9 that

$$\mathbb{E}(\mathrm{e}^{-\alpha\tau_x^+} \mathbf{1}_{(X_{\tau_x^+} = x)}) = du^{(\alpha)}(x)$$

where, if $d = 0$ the term $u^{(\alpha)}(x)$ may be taken as equal to zero and otherwise the potential measure $U^{(\alpha)}$ has a density and $u^{(\alpha)}$ is a continuous and strictly positive version thereof.

(iv) Show that for this version of the density, $u^{(\alpha)}(0+) = 1/d$ where d is the drift of X.

5.6. Suppose that X is a stable subordinator with parameter $\alpha \in (0, 1)$ thus having Laplace exponent $\Phi(\theta) = c\theta^\alpha$ for $\theta \geq 0$ and some $c > 0$ which in this exercise we will take equal to unity.

(i) Show from Exercise 1.4 that the precise expression for the jump measure is

$$\Pi(\mathrm{d}x) = \frac{x^{-(1+\alpha)}}{-\Gamma(-\alpha)}\mathrm{d}x.$$

(ii) By considering the Laplace transform of the potential measure U, show that

$$U(\mathrm{d}x) = \frac{x^{\alpha-1}}{\Gamma(\alpha)}\mathrm{d}x.$$

(iii) Hence deduce that

$$\mathbb{P}(X_{\tau_x^+} - x \in du, x - X_{\tau_x^+-} \in dy) = \frac{\alpha \sin \alpha\pi}{\pi}(x-y)^{\alpha-1}(y+u)^{-(\alpha+1)}du\, dy$$

for $u > 0$ and $y \in [0, x]$. Note further that the distribution of the pair

$$\left(\frac{x - X_{\tau_x^+-}}{x}, \frac{X_{\tau_x^+} - x}{x} \right) \tag{5.21}$$

is independent of x.

(iv) Show that stable subordinators do not creep.

5.7. Suppose that X is any subordinator.

(i) Use the joint law of the overshoot and undershoot of a subordinator to deduce that for $\beta, \gamma \geq 0$,

$$\int_0^\infty dx \cdot e^{-qx}\mathbb{E}(e^{-\beta X_{\tau_x^+-} - \gamma(X_{\tau_x^+} - x)} \mathbf{1}_{(X_{\tau_x^+} > x)}) = \frac{1}{q-\gamma}\left(\frac{\Phi(q) - \Phi(\gamma)}{\Phi(q+\beta)} \right) - \frac{d}{\Phi(q+\beta)}.$$

(ii) Taking account of creeping, use part (i) to deduce that

$$\int_0^\infty dx \cdot e^{-qx}\mathbb{E}(e^{-\beta(X_{\tau_{tx}^+-}/t) - \gamma(X_{\tau_{tx}^+} - tx)/t}) = \frac{1}{(q-\gamma)} \frac{\Phi(q/t) - \Phi(\gamma/t)}{\Phi((q+\beta)/t)}.$$

(iii) Show that if Φ is regularly varying at zero (resp. infinity) with index α equal to 0 or 1, then the limiting distribution of the pair in (5.21) is trivial as x tends to infinity (resp. zero).

(iv) It is possible to show that if a function $f : [0, \infty) \to (0, \infty)$ satisfies

$$\lim \frac{f(\lambda t)}{f(t)} = g(\lambda)$$

for all $\lambda > 0$ as t tends to zero (resp. infinity) then in fact f must be regularly varying. Roughly speaking, the reason for this is that for $\lambda, \mu > 0$,

$$g(\lambda\mu) = \lim \frac{f(\lambda\mu t)}{f(\lambda t)} \frac{f(\lambda t)}{f(t)} = g(\mu)g(\lambda)$$

showing that g is a multiplicative function. With a little measure theory, one can show that necessarily, it must follow that $g(\lambda) = \lambda^\rho$ for some $\rho \in \mathbb{R}$. See Theorem 1.4.1 of Bingham et al. (1987) for the full details. Use the above remarks to deduce that if (5.21) has a limiting distribution as x tends to infinity (resp. zero), then necessarily Φ is regularly varying at zero (resp. infinity) with index $\alpha \in [0, 1]$. Hence conclude that (5.21) has a non-trivial limiting distribution if and only if $\alpha \in (0, 1)$.

5.8. An important class of distributions, denoted $\mathcal{L}^{(\alpha)}$ where α is a parameter in $[0,\infty)$, are defined as follows. Suppose that F is a probability distribution function. Write $\overline{F}(x) = 1 - F(x)$. Then F belongs to $\mathcal{L}^{(\alpha)}$ if the support of F is in $[0,\infty)$ and non-lattice, $\overline{F}(x) > 0$ for all $x \geq 0$ and for all $y > 0$,

$$\lim_{x\uparrow\infty} \frac{\overline{F}(x+y)}{\overline{F}(x)} = \mathrm{e}^{-\alpha y}.$$

Note that the requirement that F is a probability measure can be weakened to a finite measure as one may always normalise by total mass to fulfill the conditions given earlier.

We are interested in establishing an asymptotic conditional distribution for the overshoot of a killed subordinator. To this end we assume that X is a killed subordinator with killing rate $\eta > 0$, Laplace exponent Φ, jump measure Π, drift $d \geq 0$ and potential measure U which is assumed to belong to class $\mathcal{L}^{(\alpha)}$ for some $\alpha \geq 0$ such that $\Phi(-\alpha) < \infty$.

(i) Show that

$$\mathbb{P}(\tau_x^+ < \infty) = \eta U(x,\infty)$$

where $\tau_x^+ = \inf\{t > 0 : X_t > x\}$.

(ii) Show that for all $\beta \geq 0$,

$$\mathbb{E}(\mathrm{e}^{-\beta(X_{\tau_x^+}-x)}|\tau_x^+ < \infty) = \frac{\Phi(\beta)}{\eta U(x,\infty)} \int_{(x,\infty)} \mathrm{e}^{-\beta(y-x)} U(\mathrm{d}y).$$

(iii) Applying integration by parts, deduce that

$$\lim_{x\uparrow\infty} \mathbb{E}(\mathrm{e}^{-\beta(X_{\tau_x^+}-x)}|\tau_x^+ < \infty) = \frac{\Phi(\beta)}{\eta}\left(\frac{\alpha}{\alpha+\beta}\right).$$

(iv) Now take the distribution G defined by its tail

$$G(x,\infty) = \frac{\mathrm{e}^{-\alpha x}}{\eta}\left\{\Phi(-\alpha) + \int_{(x,\infty)} (\mathrm{e}^{\alpha y} - \mathrm{e}^{\alpha x})\Pi(\mathrm{d}y)\right\}.$$

Show that G has an atom at zero and

$$\int_{(0,\infty)} \mathrm{e}^{-\beta y} G(\mathrm{d}y) = \frac{\Phi(\beta)}{\eta}\left(\frac{\alpha}{\alpha+\beta}\right) - \frac{d\alpha}{\eta},$$

where d is the drift coefficient of X.

(v) Deduce that

$$\lim_{x\uparrow\infty} \mathbb{P}(X_{\tau_x^+} - x > u|\tau_x^+ < \infty) = \overline{G}(u)$$

and

$$\lim_{x \uparrow \infty} \mathbb{P}(X_{\tau_x^+} = x | \tau_x^+ < \infty) = \frac{d\alpha}{\eta}.$$

5.9. Suppose that X is any Lévy process of bounded variation with zero drift which is not a compound Poisson process. By writing it as the difference of two independent subordinators prove that for any $x \in \mathbb{R}$,

$$\mathbb{P}(\inf\{t > 0 : X_t = x\} < \infty) = 0.$$

6

The Wiener–Hopf Factorisation

This chapter gives an account of the theory of excursions of a Lévy process from its maximum and the *Wiener–Hopf factorisation* that follows as a consequence from it.

In Sect. 4.6 the analytical form of the Pollaczek–Khintchine formula was explained in terms of the decomposition of the path of the underlying Lévy process into independent and identically distributed sections of path called excursions from the supremum. The decomposition made heavy use of the fact that for the particular class of Lévy processes considered, namely spectrally positive processes of bounded variation, the times of new maxima form a discrete set.

For a general Lévy process, it is still possible to decompose its path into "excursions from the running maximum". However conceptually this decomposition is a priori somewhat more tricky as, in principle, a general Lévy process may exhibit an infinite number of excursions from its maximum over any finite period of time. Nonetheless, when considered in the right mathematical framework, excursions from the maximum can be given a sensible definition and can be shown to form the support of a Poisson random measure. The theory of excursions presents one of the more mathematically challenging aspects of the theory of Lévy processes. This means that in order to keep to the level outlined in the preface of this text, there will be a number of proofs in the forthcoming sections which are excluded or only discussed at an intuitive level.

Amongst a very broad spectrum of probabilistic literature, *the Wiener–Hopf factorisation* may be thought of as a common reference to a multitude of statements concerning the distributional decomposition of the excursions of any Lévy process when sampled at an independent and exponentially distributed time. (We devote a little time later in this text to explain the origin of the name "Wiener–Hopf factorisation".) In the light of the excursion theory given in this chapter we explain in detail the Wiener–Hopf factorisation. The collection of conclusions which fall under the umbrella of the Wiener–Hopf factorisation turn out to provide a robust tool from which one may analyse a number of problems concerning the fluctuations of Lévy processes; in particular problems which have relevance to the applications we shall consider in later chapters. The chapter concludes with some special classes of Lévy processes for which the Wiener–Hopf factorisation may be exemplified in more detail.

6.1 Local Time at the Maximum

Unlike the Lévy processes presented in Sect. 4.6, a general Lévy process may have an infinite number of new maxima over any given finite period of time. As our ultimate goal is to show how to decompose events concerning the path of a Lévy process according to the behaviour of the path in individual excursions, we need a way of indexing them. To this end we introduce the notion of *local time at the maximum.*

To avoid trivialities we shall assume throughout this section that neither X nor $-X$ is a subordinator. Recall also the definition $\overline{X}_t = \sup_{s\leq t} X_s$. In the sequel we shall repeatedly refer to the process $\overline{X} - X = \{\overline{X}_t - X_t : t \geq 0\}$, which we also recall from Exercise 3.2, which can be shown to be Strong Markov.

Definition 6.1 (Local time at the maximum). *A continuous, non-decreasing, $[0,\infty)$-valued, $\mathbb{F}$-adapted process $L = \{L_t : t \geq 0\}$ is called a local time at the maximum (or just local time for short) if the following hold.*

(i) The support of the Stieltjes measure $\mathrm{d}L_t$ is the closure of the (random) set of times $\{t \geq 0 : \overline{X}_t = X_t\}$ and is finite for each $t \geq 0$.
(ii) For every $\mathbb{F}$-stopping time T such that $\overline{X}_T = X_T$ on $\{T < \infty\}$ almost surely, the shifted trivariate process

$$\{(X_{T+t} - X_T, \overline{X}_{T+t} - X_{T+t}, L_{T+t} - L_T) : t \geq 0\}$$

is independent of $\mathcal{F}_T$ on $\{T < \infty\}$ and has the same law as $\left(X, \overline{X} - X, L\right)$ under $\mathbb{P}$.

(The process which is identically zero is excluded).

Let us make some remarks about the above definition. Firstly if L is a local time then so is kL for any constant $k > 0$. Hence local times can at best be defined uniquely up to a multiplicative constant. On occasion we shall need to talk about both local time and the time scale on which the Lévy process itself is defined. In such cases we shall refer to the latter as *real time.* Finally, by applying this definition of local time to $-X$ it is clear that one may talk of a local time at the minimum. The latter shall always be referred to as $\widehat{L}$.

Local times as defined above do not always exist on account of the requirement of continuity. Nonetheless, in such cases, it turns out that one may construct right continuous processes which satisfy conditions (i) and (ii) of

Definition 6.1 and which serve their purpose equally well in the forthcoming analysis of the Wiener–Hopf factorisation. We provide more details shortly. We give next, however, some examples for which a continuous local time can be identified explicitly.

Example 6.2 (Spectrally negative processes). Recall that the formal definition of a spectrally negative Lévy process is simply one for which the Lévy measure satisfies $\Pi(0,\infty) = 0$ and whose paths are not monotone. As there are no positive jumps the process $\overline{X}$ must therefore be continuous. It is easy to check that $L = \overline{X}$ fulfils Definition 6.1.

Example 6.3 (Compound Poisson processes with drift $d \geq 0$). Just by considering the piece-wise linearity of the paths of these processes, one has obviously that over any finite time horizon, the time spent at the maximum has strictly positive Lebesgue measure with probability one. Hence the quantity

$$L_t := \int_0^t \mathbf{1}_{(\overline{X}_s = X_s)} \mathrm{d}s, \quad t \geq 0 \tag{6.1}$$

is almost surely positive and may be taken as a candidate for local time. Indeed it increases on $\{t : \overline{X}_t = X_t\}$, is continuous, non-decreasing and is an $\mathbb{F}$-adapted process. Taking T as in part (ii) of Definition 6.1 we also see that on $\{T < \infty\}$,

$$L_{T+t} - L_T = \int_T^{T+t} \mathbf{1}_{(\overline{X}_s = X_s)} \mathrm{d}s \tag{6.2}$$

which is independent of $\mathcal{F}_T$ because $\{\overline{X}_{T+t} - X_{T+t} : t \geq 0\}$ is and has the same law as L by the Strong Markov Property applied to the process $\overline{X} - X$ and the fact that $\overline{X}_T - X_T = 0$. The distributional requirement in part (ii) for the trivariate process $(X, \overline{X} - X, L)$ follows by similar reasoning.

If we only allow negative jumps and $d > 0$, then according to the previous example $\overline{X}$ also fulfils the definition of local time. However, as we have seen in the proof of Theorem 4.1,

$$\overline{X}_t = d \int_0^t \mathbf{1}_{(\overline{X}_s = X_s)} \mathrm{d}s$$

for all $t \geq 0$.

Next we would like to identify the class of Lévy processes for which continuous local time cannot be constructed and for which a right continuous alternative is introduced instead. In a nutshell, the aforementioned class are those Lévy processes whose moments of new maxima form a discrete set. The qualifying criterion for this turns out to be related to the behaviour of the Lévy process at arbitrarily small times; a sense of this has already been given in the discussion of Sect. 4.6. We spend a little time developing the relevant notions, namely path regularity, in order to complete the discussion on local time.

Definition 6.4. *For a Lévy process X, the point $x \in \mathbb{R}$ is said to be regular (resp. irregular) for an open or closed set B if*

$$\mathbb{P}_x\left(\tau^B = 0\right) = 1 \text{ (resp. 0)},$$

where the stopping time $\tau^B = \inf\{t > 0 : X_t \in B\}$. Intuitively speaking, x is regular for B if, when starting from x, the Lévy process hits B immediately.

Note that as τ^B is a stopping time, it follows that

$$\mathbf{1}_{(\tau^B=0)} = \mathbb{P}_x(\tau^B = 0|\mathcal{F}_0).$$

On the other hand, since $\mathcal{F}_0$ contains only null sets, Kolmogorov's definition of conditional expectation implies

$$\mathbb{P}_x(\tau^B = 0|\mathcal{F}_0) = \mathbb{P}_x(\tau^B = 0)$$

and hence $\mathbb{P}_x(\tau^B = 0)$ is either zero or one. In fact one may replace $\{\tau^B = 0\}$ by any event $A \in \mathcal{F}_0$ and reach the same conclusion about $\mathbb{P}(A)$. This is nothing but Blumenthal's zero–one law.

We know from the Lévy–Itô decomposition that the range of a Lévy process over any finite time horizon is almost surely bounded and thanks to right continuity, $\lim_{t\downarrow 0} \max\{-\underline{X}_t, \overline{X}_t\} = 0$. Hence for any given open or closed B, the points $x \in \mathbb{R}$ for which $\mathbb{P}_x(\tau^B = 0) = 1$ are necessarily those belonging to $B \cup \partial B$. However, $x \in \partial B$ is not a sufficient condition for $\mathbb{P}_x(\tau^B = 0) = 1$. To see why, consider the case that $B = (0,\infty)$ and X is any compound Poisson process.

Finally note that the notion of regularity can be established for any Markov process with analogous definition to the one given earlier. However, for the special case of a Lévy process, stationary independent increments allow us reduce the discussion of regularity of x for open and closed sets to simply the regularity of 0 for open and closed sets. Indeed, for any Lévy process, x is regular for B if and only if 0 is regular for $B - x$.

As we shall see shortly, it is regularity of 0 for $[0,\infty)$ which dictates whether one may find a continuous local time. The following result, due in parts to Rogozin (1968), Shtatland (1965) and Bertoin (1997), gives precise conditions for the slightly different issue of regularity of 0 for $(0,\infty)$.

Theorem 6.5. *For any Lévy process, X, other than a compound Poisson process, the point 0 is regular for $(0,\infty)$ if and only if*

$$\int_0^1 \frac{1}{t}\mathbb{P}\left(X_t > 0\right) \mathrm{d}t = \infty \tag{6.3}$$

and the latter holds if and only if one of the following three conditions hold,

(i) X is a process of unbounded variation,
(ii) X is a process of bounded variation and $d > 0$,

(iii) X is a process of bounded variation, d = 0 and

$$\int_{(0,1)} \frac{x\Pi(\mathrm{d}x)}{\int_0^x \Pi(-\infty,-y)\,\mathrm{d}y} = \infty.$$

Here d is the drift coefficient in the representation (2.21) of a Lévy process of bounded variation.

Recall that if N is the Poisson random measure associated with the jumps of X, then the time of arrival of a jump of size $\varepsilon > 0$ or greater, say, $T(\varepsilon)$, is exponentially distributed since

$$\mathbb{P}(T(\varepsilon) > t) = \mathbb{P}(N([0,t] \times \{\mathbb{R}\backslash(-\varepsilon,\varepsilon)\}) = 0) = \exp\{-t\Pi(\mathbb{R}\backslash(-\varepsilon,\varepsilon))\}.$$

This tells us that, if $\Pi(\mathbb{R}) = \infty$, then jumps of size greater than ε become less and less probable as $t \downarrow 0$. Hence that the jumps that have any influence over the initial behaviour of the path of X, if at all, will necessarily be arbitrarily small. With this in mind, one may intuitively see the conditions (i)–(iii) in Theorem 6.5 in the following way.

In case (i) when $\sigma > 0$ regularity follows as a consequence of the presence of Brownian motion whose behaviour on the small time scale always dominates the path of the Lévy process. If on the other hand $\sigma = 0$, the high intensity of small jumps causes behaviour on the small time scale close to that of Brownian motion even though the paths are not continuous. (We use the words "high intensity" here in the sense that $\int_{(-1,1)} |x|\Pi(\mathrm{d}x) = \infty$). Case (ii) says that when the Poisson point process of jumps fulfils the condition $\int_{(-1,1)} |x|\Pi(\mathrm{d}x) < \infty$, over small time scales, the sum of the jumps grows sub-linearly in time almost surely. Therefore if a drift is present, this dominates the initial motion of the path. In case (iii) when there is no dominant drift, the integral test may be thought of as a statement about what the "relative weight" of the small positive jumps compared to the small negative jumps needs to be in order for regularity to occur.

In the case of bounded variation, the integral $\int_0^x \Pi(-\infty,-y)\,\mathrm{d}y$ is finite for all $x > 0$. This can be deduced by noting that for any Lévy process of bounded variation $\int_{(-1,0)} |x|\Pi(\mathrm{d}x) < \infty$ and then integrating by parts. Note also that Theorem 6.5 implies that processes of unbounded variation are such that 0 is regular for both $(0,\infty)$ and $(-\infty,0)$ and that processes of bounded variation with $d > 0$ are irregular for $(-\infty,0]$. For processes of bounded variation with $d = 0$ it is possible to find examples where 0 is regular for both $(0,\infty)$ and $(-\infty,0)$. See Exercise 6.1.

We offer no proof of Theorem 6.5 here. However it is worth recalling that from Lemma 4.11 and the follow up Exercise 4.9 (i) we know that for any Lévy process, X, of bounded variation,

$$\lim_{t\downarrow 0} \frac{X_t}{t} = d$$

almost surely where d is the drift coefficient. This shows that if $d > 0$ (resp. $d < 0$) then for all $t > 0$ sufficiently small, X_t must be strictly positive (resp. negative). That is to say, 0 is regular for $(0,\infty)$ and irregular for $(-\infty,0]$ if $d>0$ (resp. regular for $(-\infty,0)$ and irregular for $[0,\infty)$ if $d<0$). For the case of a spectrally negative Lévy process of unbounded variation, Exercise 6.2 deduces regularity properties in agreement with Theorem 6.5. In addition, Exercise 6.8 shows how to establish the criterion (6.3).

There is a slight difference between regularity of 0 for $(0,\infty)$ and regularity of 0 for $[0,\infty)$. Consider for example the case of a compound Poisson Process. The latter is regular for $[0,\infty)$ but not for $(0,\infty)$ due to the initial exponentially distributed period of time during which the process remains at the origin. On the other hand, we also have the following conclusion.

Corollary 6.6. *The only Lévy process for which* 0 *is regular for* $[0,\infty)$ *but not for* $(0,\infty)$ *is a compound Poisson process.*

Proof. If a process is regular for $[0,\infty)$ but not for $(0,\infty)$ then, according to the cases given in Theorem 6.5 it must be a process of bounded variation with zero drift. If the underlying Lévy process is not a compound Poisson process then this would require that the process jumps into $(-\infty,0)$ and back on to 0 infinitely often for arbitrarily small times. This is impossible when one takes into account Exercise 5.9 which shows that a Lévy process of bounded variation with no drift which is not a compound Poisson process cannot hit a pre-specified point. □

By definition, when 0 is irregular for $[0,\infty)$ the Lévy process takes a strictly positive period of time to return to reach a new maximum above the origin. Hence, applying the Strong Markov Property at the time of first entry into $[0,\infty)$ we see that in a finite interval of time there are almost surely a finite number of new maxima, in other words, $\{0<s\leq t:\overline{X}_s=X_s\}$ is a discrete set. (Recall that this type of behaviour has been observed for spectrally positive Lévy process of bounded variation in Chap. 4). In this case we may then define the counting process $n=\{n_t:t\geq 0\}$ by

$$n_t=\#\{0<s\leq t:\overline{X}_s=X_s\}. \tag{6.4}$$

We are now ready to make the distinction between those processes which admit continuous local times in Definition 6.1 and those that do not.

Theorem 6.7. *Let* X *be any Lévy process.*

(i) *There exists a continuous version of* L *which is unique up to a multiplicative constant if and only if* 0 *is regular for* $[0,\infty)$.

(ii) If 0 is irregular for $[0,\infty)$ then we can take as our definition of local time

$$L_t = \sum_{i=0}^{n_t} \mathbf{e}_\lambda^{(i)}, \tag{6.5}$$

satisfying (i) and (ii) of Definition 6.1, where $\{\mathbf{e}_\lambda^{(i)} : i \geq 0\}$ are independent and exponentially distributed random variables with parameter $\lambda > 0$ (chosen arbitrarily) and $\{n_t : t \geq 0\}$ is the counting process defined in (6.4).

We offer no proof for case (i) which is a particular example of a classic result from potential theory of stochastic processes, a general account of which can be found in Blumenthal and Getoor (1968). The proof of part (ii) is quite accessible and we leave it as an Exercise. Note that one slight problem occurring in the definition of L in (6.5) is that it is not adapted to the filtration of X. However this is easily resolved by simply broadening out the filtration to include $\sigma(\mathbf{e}_\lambda^{(i)} : i \geq 0)$ before completing it with null sets. Also the choice of exponential distribution used in (6.5) is of no effective consequence. One could in principle always work with the definition

$$L'_t = \sum_{i-0}^{n_t} \mathbf{e}_1^{(i)}.$$

Scaling properties of exponential distributions would then allow us to construct the $\mathbf{e}_\lambda^{(i)}$s and $\mathbf{e}_1^{(i)}$s on the same probability space so that $\lambda\mathbf{e}_\lambda^{(i)} = \mathbf{e}_1^{(i)}$ for each $i = 0, 1, 2...$ and this would imply that $L' = \lambda L$ where L is local time constructed using exponential distributions with parameter λ. So in fact within the specified class of local times in part (ii) of the above theorem, the only effective difference is a multiplicative constant. The reason why we do not simply define $L_t = n_t$ has to do with the fact that we shall require some special properties of the inverse L^{-1}. This will be discussed in Sect. 6.2.

In accordance with the conclusion of Theorem 6.7, in the case that 0 is regular for $[0,\infty)$ we shall henceforth work with a continuous version of L and in the case that 0 is irregular for $[0,\infty)$ we shall work with the definition (6.5) for L.

In Example 6.3 we saw that we may use a multiple of the Lebesgue measure of the real time spent at the maximum to give a version of local time which is continuous. The fact that the latter is non-zero is a clear consequence of piecewise linearity of the process. Although compound Poisson processes (with drift) are the only Lévy processes which are piecewise linear, it is nonetheless natural to investigate to what extent one may work with the Lebesgue measure of the time spent at the maximum for local time in the case that 0 is regular for $[0,\infty)$. Rubinovitch (1971) supplies us with the following charactarisation of such processes.

Theorem 6.8. *Suppose that X is a Lévy process for which 0 is regular for $[0,\infty)$. Let L be some continuous version of local time. The constant $a \geq 0$ satisfying the relation*

$$\int_0^t \mathbf{1}_{(\overline{X}_s = X_s)} \mathrm{d}s = aL_t$$

is strictly positive if and only if X is a Lévy process of bounded variation and 0 is irregular for $(-\infty, 0)$.

Proof. Note that $\int_0^\infty \mathbf{1}_{(\overline{X}_t = X_t)} \mathrm{d}t > 0$ with positive probability if and only if

$$\mathbb{E}\left(\int_0^\infty \mathbf{1}_{(\overline{X}_t - X_t = 0)} \mathrm{d}t\right) > 0.$$

By Fubini's Theorem and the Duality Lemma the latter occurs if and only if $\int_0^\infty \mathbb{P}(\underline{X}_t = 0)\mathrm{d}t = \mathbb{E}(\int_0^\infty \mathbf{1}_{(\underline{X}_t = 0)} \mathrm{d}t) > 0$ (recall that $\underline{X}_t := \inf_{s \leq t} X_s$). Due to the fact that $\underline{X}$ has paths that are right continuous and monotone decreasing, the positivity of the last integral happens if and only if it takes an almost surely strictly positive time for X to visit $(-\infty, 0)$. In short we have that $\int_0^\infty \mathbf{1}_{(\overline{X}_t = X_t)} \mathrm{d}t > 0$ with positive probabiltiy if and only if 0 is irregular for $(-\infty, 0)$. By Theorem 6.5 the latter can only happen when X has bounded variation.

Following the same reasoning as used in Example 6.3 it is straightforward to deduce that

$$\int_0^t \mathbf{1}_{(\overline{X}_s = X_s)} \mathrm{d}s, \quad t \geq 0$$

may be used as a local time. Theorem 6.7 (i) now gives us the existence of a constant $a > 0$ so that for a given local time L,

$$aL_t = \int_0^t \mathbf{1}_{(\overline{X}_s = X_s)} \mathrm{d}s.$$

When 0 is regular for $(-\infty, 0)$ the reasoning above also shows that $\int_0^t \mathbf{1}_{(\overline{X}_s = X_s)} \mathrm{d}s = 0$ almost surely for all $t \geq 0$ and hence it is clear that the constant $a = 0$. □

We can now summarise the discussion on local times as follows. There are three types of Lévy processes which are associated with three types of local times.

1. Processes of bounded variation for which 0 is irregular for $[0, \infty)$. The set of maxima forms a discrete set and we take a right continuous version of local time in the form
$$L_t = \sum_{i=0}^{n_t} \mathbf{e}_1^{(i)},$$
where n_t is the count of the number of maxima up to time t and $\{\mathbf{e}_1^{(i)} : i = 0, 1, ...\}$ are independent and exponentially distributed random variables with parameter 1.

2. Processes of bounded variation for which 0 is irregular for $(-\infty, 0)$. There exists a continuous version of local time given by,

$$L_t = a \int_0^t \mathbf{1}_{(\overline{X}_s = X_s)} \mathrm{d}s$$

for some arbitrary $a > 0$. In the case that X is spectrally negative, we have that L is equal to a multiplicative constant times $\overline{X}$.
3. Processes of unbounded variation, in which case 0 is implicitly regular for $[0, \infty)$. We take a continuous version of local time exists but cannot be identified explicitly as a functional of the path of X in general. However, if X is a spectrally negative Lévy process, then this local time may be taken as $\overline{X}$.

6.2 The Ladder Process

Define the inverse local time process $L^{-1} = \{L_t^{-1} : t \geq 0\}$ where

$$L_t^{-1} := \begin{cases} \inf\{s > 0 : L_s > t\} & \text{if } t < L_\infty \\ \infty & \text{otherwise.} \end{cases}$$

Next define the process $H = \{H_t : t \geq 0\}$ where

$$H_t := \begin{cases} X_{L_t^{-1}} & \text{if } t < L_\infty \\ \infty & \text{otherwise.} \end{cases}$$

The range of inverse local time, L^{-1}, corresponds to the set of real times at which new maxima occur, called the *ascending ladder times.* The range of process H corresponds to the set of new maxima called the *ascending ladder heights.* The bivariate process $(L^{-1}, H) := \{(L_t^{-1}, H_t) : t \geq 0\}$, called the *ascending ladder process*, is the main object of study of this section. Note that it is implicit from their definition that L^{-1} and H are processes with paths that are right continuous with left limits.

The word "ascending" distinguishes the process (L^{-1}, H) from the analogous object $(\widehat{L}^{-1}, \widehat{H})$, constructed from $-X$, which is called the *descending ladder process* (note that local time at the maximum of $-X$ is the local time at the minimum of X which was previously referred to as $\widehat{L}$). When the context is obvious we shall drop the use of the words ascending or descending.

The ladder processes we have defined here are the continuous time analogue of the processes with the same name for random walks[1]. In the case of random walks, one defines L_n to be the number of times a maxima is reached after n steps, $T_n = \inf\{k \geq 1 : L_k = n\}$ as the number of steps required to achieve n new maxima and H_n as the n-th new maxima.

[1]Recall that $\{S_n : n = 0, 1, 2, ...\}$ is a random walk if $S_0 = 0$ and $S_n = \sum_{i=1}^n \xi_i$ where $\{\xi_i : i = 1, 2, ...\}$ are independent and identically distributed.

An additional subtlety for random walks is that the count L_n may be taken to include visits to previous maxima (consider for example a simple random walk which may visit an existing maximum several times before generating a strictly greater maximum). In that case the associated ascending ladder process is called *weak*. When $\{L_n : n \geq 0\}$ only counts the number of new maxima which exceed all previous maxima, the associated ascending ladder process is called *strict*.

The same subtlety appears in the definition of L for Lévy processes when 0 is irregular for $[0, \infty)$ and our definition of the process $\{n_t : t \geq 0\}$ is analogous to a count of weak ascending ladder heights. However, this is of no consequence in the forthcoming discussion as we shall see that, it turns out that in this setting with probability one no two maxima can be equal anyway. When 0 is regular for $[0, \infty)$ but X is not a compound Poisson process we shall again see in due course that ladder heights at different ladder times are distinct. Finally when X is a compound Poisson process, although one cannot count the times spent at the maximum, the distinction between weak and strict maxima will become an issue at some point in later discussion. Indeed the choice of local time

$$L_t = a \int_0^t \mathbf{1}_{(\overline{X}_s = X_s)} \mathrm{d}s, \quad t \geq 0$$

is analogous to the count of weak ascending ladder heights in a random walk. Consider for example the continuous time version of a simple random walk; that is a compound Poisson process with jumps in $\{-1, 1\}$.

Our interest in this section will be to characterise the ladder process. We start with the following lemma which will be used at several places later on.

Lemma 6.9. *For each $t \geq 0$, both L_t^{-1} and L_{t-}^{-1} are $\mathbb{F}$-stopping times.*

Proof. Recall from Sect. 3.1 that thanks to the assumed right continuity of $\mathbb{F}$ it suffices to prove that for each $s > 0$, $\{L_t^{-1} < s\} \in \mathcal{F}_s$ and a similar notion for L_{t-}^{-1}. For all $s, t \geq 0$, $\{L_t^{-1} < s\} = \{L_{s-} > t\}$ which belongs to $\mathcal{F}_s$ as the process L is $\mathbb{F}$-adapted. To prove that L_{t-}^{-1} is a stopping time note that

$$\{L_{t-}^{-1} < s\} = \bigcap_{n \geq 1} \{L_{t-1/n}^{-1} < s\} \in \mathcal{F}_s.$$

□

In the next theorem we shall link the process (L^{-1}, H) to a bivariate subordinator. Let us remind ourselves what the latter object is. With Exercise 2.10 in mind, recall that a bivariate subordinator is a two-dimensional $(0, \infty)^2$-valued stochastic processes $\mathbf{X} = \{\mathbf{X}_t : t \geq 0\}$ with paths that are right continuous with left limits as well as having stationary independent increments[2] and further, each component is non-decreasing. It is important to note that in general it is not correct to think of a bivariate subordinator simply as a vector process composed of two independent subordinators. Correlation between the subordinators in each of the co-ordinates may be represented as follows. Write $\mathbf{X}$ in the form

$$\mathbf{X}_t = \mathbf{d}t + \int_{[0,t]} \int_{(0,\infty)^2} \mathbf{x} N(\mathrm{d}s \times \mathrm{d}\mathbf{x}), \quad t \geq 0,$$

where $\mathbf{d} \in [0,\infty)^2$ and N is the Poisson random measure describing the jumps defined on $[0,\infty) \times (0,\infty)^2$ with intensity measure given by $\mathrm{d}t \times \Lambda(\mathrm{d}x, \mathrm{d}y)$ for a bivariate measure Λ on $(0,\infty)^2$ satisfying

$$\int_{(0,\infty)^2} (1 \wedge \sqrt{x^2 + y^2}) \Lambda(\mathrm{d}x, \mathrm{d}y) < \infty.$$

Independence of the two individual co-ordinate processes of $\mathbf{X}$ corresponds to the case that Λ is a product measure; say $\Lambda(\mathrm{d}x, \mathrm{d}y) = \Lambda_1(\mathrm{d}x)\Lambda_2(\mathrm{d}y)$.

For a general bivariate subordinator, positivity allows us to talk about their Laplace exponent $\phi(\alpha, \beta)$ given by

$$\mathbb{E}\left(\exp\left\{-\binom{\alpha}{\beta} \cdot \mathbf{X}_t\right\}\right) = \exp\{-\phi(\alpha,\beta)\, t\}.$$

Referring back to Chap. 2, it is a straightforward exercise to deduce that that

$$\phi(\alpha,\beta) = \mathbf{d} \cdot \binom{\alpha}{\beta} + \int_{(0,\infty)^2} (1 - \mathrm{e}^{-\binom{\alpha}{\beta} \cdot \binom{x}{y}}) \Lambda(\mathrm{d}x, \mathrm{d}y).$$

Theorem 6.10. *Let X be a Lévy process and $\mathbf{e}_q$ an independent and exponentially distributed random variable with parameter $q \geq 0$. Then*

$$\mathbb{P}\left(\limsup_{t \uparrow \infty} X_t < \infty\right) = 0 \text{ or } 1$$

and the ladder process (L^{-1}, H) satisfies the following properties.

(i) If $\mathbb{P}\left(\limsup_{t\uparrow\infty} X_t = \infty\right) = 1$ then (L^{-1}, H) has the law of a bivariate subordinator.

(ii) If $\mathbb{P}\left(\limsup_{t\uparrow\infty} X_t < \infty\right) = 1$ then for some $q > 0$ $L_\infty \stackrel{d}{=} \mathbf{e}_q$, the exponential distribution with parameter q, and $\{(L_t^{-1}, H_t) : t < L_\infty\}$ has the same law as $(\mathcal{L}^{-1}, \mathcal{H}) := \{(\mathcal{L}_t^{-1}, \mathcal{H}_t) : t < \mathbf{e}_q\}$ where $(\mathcal{L}^{-1}, \mathcal{H})$ is a bivariate subordinator independent of $\mathbf{e}_q$.

[2] Recall that right continuous with left limits and stationary independent increments for a multi-dimensional process means that the same properties hold component-wise. However, it does not necessarily mean that the individual component processes are independent.

Proof. Since

$$\{\limsup_{t\uparrow\infty} X_t < \infty\} = \{\limsup_{\mathbb{Q}\ni t\uparrow\infty} X_t < \infty\}$$

and the latter event is in the tail sigma algebra $\bigcap_{t\in\mathbb{Q}\cap[0,\infty)} \sigma\left(X_s : s \geq t\right)$, then Kolmogorov's zero–one law for tail events tells us that $\mathbb{P}\left(\limsup_{t\uparrow\infty} X_t < \infty\right) = 0$ or 1.

To deal with (i) and (ii) in the case that 0 is irregular for $[0,\infty)$, the analysis proceeds in the sprit of the discussion around the Pollaczek–Khintchine formula in Chap. 4. We give a brief outline of the arguments again. If we agree that a geometric distribution with parameter 1 is infinite with probability one then, the total number of excursions from the maximum n_∞ defined in (6.4) is geometrically distributed with parameter $1-\rho = \mathbb{P}(\tau_0^+ = \infty)$ where $\tau_0^+ = \inf\{t > 0 : X_t > 0\}$. Now define the sequence of times $T_0 = 0$,

$$\begin{aligned} T_{n+1} &= \inf\{t > T_n : X_t > X_{T_n}\} \\ &= \inf\{t > T_n : \Delta L_t > 0\} \\ &= \inf\{t > T_n : \Delta n_t = 1\} \end{aligned}$$

for $n = 0, 1, ..., n_\infty$ where $\Delta L_t = L_t - L_{t-}$, $\Delta n_t = n_t - n_{t-}$ and $\inf\emptyset = \infty$. It is easy to verify that the latter times form a sequence of stopping times with $T_n < \infty$ for all $n \geq 0$ and otherwise $T_{n_\infty} < \infty$ and $T_{n_\infty+1} = \infty$ when $n_\infty < \infty$. Further, by the Strong Markov Property for Lévy processes, if $n_\infty < \infty$ then the successive excursions of X from its maximum,

$$\epsilon_n := \{X_t : t \in (T_{n-1}, T_n]\}$$

for $n = 1, ..., n_\infty$ are equal in law to an independent sample of $n_\infty - 1$ copies of the first excursion from the maximum conditioned to be finite followed by a final independent copy conditioned to be infinite in length. If $n_\infty = \infty$ then the sequence $\{\epsilon_n : n = 1, 2, ...\}$ is equal in law to an independent sample of the first excursion from the maximum.

By considering Fig. 6.1 we see that L^{-1} (the reflection of L about the diagonal) is a step function and its successive jumps (the flat sections of L) correspond precisely to the sequence $\{T_{n+1} - T_n : n = 0, ..., n_\infty\}$. From the previous paragraph, it follows that L^{-1} has independent and identically distributed jumps and is sent independently to infinity (which we may consider as a "graveyard" state) on the n_∞-th jump accordingly with the arrival of the first infinite excursion. As the jumps of L are independent and exponentially distributed it also follows that the periods between jumps of L^{-1} are independent and exponentially distributed. According to Exercise 6.3 the process L^{-1} is now equal in law to a compound Poisson subordinator killed independently after an exponentially distributed time with parameter $\lambda(1-\rho)$. (Again we work with the notation that an exponential distribution with parameter 0 is infinite with probability one). It follows by construction that H is also a compound Poisson subordinator killed at the same rate.

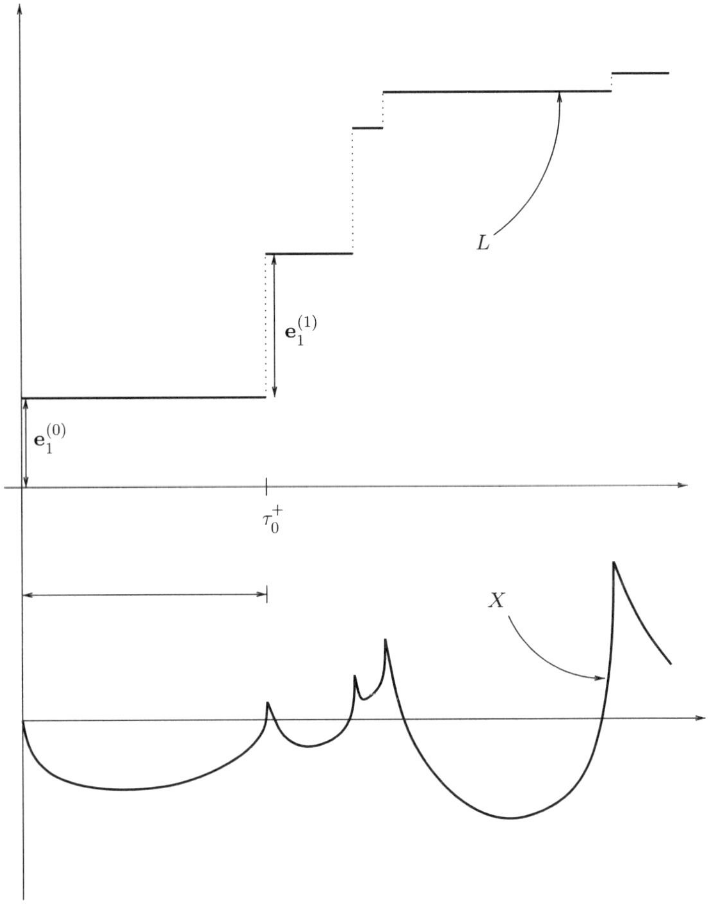

Fig. 6.1. A realisation of local time and inverse local time for a Lévy process for which 0 is irregular for $[0,\infty)$. The upper graph plots the paths of L and the lower graph plots the path of X symbolically in terms of the excursions from the maximum.

Next we prove (i) and (ii), for the case that 0 is regular for $[0,\infty)$ so that the version of local time we work with is assumed to have continuous paths. Fix $t > 0$. In Definition 6.1 let us work with the stopping time $T = L_t^{-1}$ for $t > 0$. Note that $L_{L_t^{-1}} = t$ on the event that $t < L_\infty$. Define the processes

$$\widetilde{X} = \{\widetilde{X}_t := X_{L_t^{-1}+s} - X_{L_t^{-1}} : s \geq 0\} \quad \text{and} \quad \widetilde{L} = \{\widetilde{L}_s := L_{L_t^{-1}+s} - t : s \geq 0\}.$$

From Lemma 6.9 we know that L_t^{-1} is a stopping time and hence according to Definition 6.1, on the event $L_t^{-1} < \infty$, that is $t < L_\infty$, the process $\widetilde{L}$ is the local time of $\widetilde{X}$ at its maximum. Note also from Theorem 3.1 and again

Definition 6.1 that $\widetilde{X}$ and $\widetilde{L}$ are independent of $\mathcal{F}_{L_t^{-1}}$. It is clear that

$$\widetilde{L}_s^{-1} = L_{t+s}^{-1} - L_t^{-1}. \tag{6.6}$$

Further

$$\widetilde{H}_s := \widetilde{X}_{\widetilde{L}_s^{-1}} = X_{L_{t+s}^{-1}} - X_{L_t^{-1}} = H_{t+s} - H_s. \tag{6.7}$$

In conclusion we have established that on $t < L_\infty$,

$$\{(L_{t+s}^{-1} - L_t^{-1}, H_{t+s} - H_t) : s \geq 0\}$$

is independent of $\mathcal{F}_{L_t^{-1}}$ and equal in law to (L^{-1}, H). With this in hand, note that for any $\alpha, \beta \geq 0$,

$$\begin{aligned}
&\mathbb{E}\left(\mathrm{e}^{-\alpha L_{t+s}^{-1} - \beta H_{t+s}} \mathbf{1}_{(t+s<L_\infty)}\right) \\
&= \mathbb{E}\left(\mathrm{e}^{-\alpha L_t^{-1} - \beta H_t} \mathbf{1}_{(t<L_\infty)} \mathbb{E}\left(\mathrm{e}^{-\alpha \widetilde{L}_s^{-1} - \beta \widetilde{H}_s} \mathbf{1}_{(s<\widetilde{L}_s)} \Big| \mathcal{F}_{L_t^{-1}}\right)\right) \\
&= \mathbb{E}\left(\mathrm{e}^{-\alpha L_t^{-1} - \beta H_t} \mathbf{1}_{(t<L_\infty)}\right) \\
&\quad \times \mathbb{E}\left(\mathrm{e}^{-\alpha L_s^{-1} - \beta H_s} \mathbf{1}_{(s<L_\infty)}\right).
\end{aligned}$$

As the expectation on the left-hand side above is also right continuous in t (on account of the same being true of L^{-1} and H) a standard argument shows that the multiplicative decomposition implies that

$$\mathbb{E}\left(\mathrm{e}^{-\alpha L_t^{-1} - \beta H_t} \mathbf{1}_{(t<L_\infty)}\right) = \mathrm{e}^{-\kappa(\alpha,\beta)t}, \tag{6.8}$$

where $\kappa(\alpha, \beta) = -\log \mathbb{E}(e^{-\alpha L_1^{-1} - \beta H_1} \mathbf{1}_{(1<L_\infty)}) \geq 0$. In particular we see that L_∞ must follow an exponential distribution with parameter $\kappa(0,0)$ if $\kappa(0,0) > 0$ and $L_\infty = \infty$ otherwise. Now writing for each α, β,

$$\kappa(\alpha, \beta) = \kappa(0,0) + \phi(\alpha, \beta), \tag{6.9}$$

Bayes formula applied to (6.8) shows that for all $t \geq 0$,

$$\begin{aligned}
\mathrm{e}^{-\phi(\alpha,\beta)} &= \mathbb{E}\left(\mathrm{e}^{-\alpha L_1^{-1} - \beta H_1} \Big| 1 < L_\infty\right) \\
&= \left\{\mathbb{E}\left(\mathrm{e}^{-\alpha L_t^{-1} - \beta H_t} \Big| t < L_\infty\right)\right\}^{1/t}
\end{aligned} \tag{6.10}$$

illustrating that $\phi(\alpha, \beta)$ is the Laplace exponent of a bivariate, component-wise positive, infinitely divisible distribution

$$\eta(\mathrm{d}x, \mathrm{d}y) = \mathbb{P}(L_1^{-1} \in \mathrm{d}x, H_1 \in \mathrm{d}y | 1 < L_\infty).$$

(Consider (6.10) for $t = 1/n$ where n is a positive integer). In the spirit of the Lévy–Itô decomposition, there exists a bivariate subordinator, say $(\mathcal{L}^{-1}, \mathcal{H})$

whose Laplace exponent is $\phi(\alpha, \beta)$. We now see from (6.8) and (6.9) that (L^{-1}, H) is equal in law to $(\mathcal{L}^{-1}, \mathcal{H})$ killed independently (with "graveyard" state ∞) after an exponentially distributed time with parameter $q = \kappa(0,0)$. In particular $\overline{X}_\infty = \infty$ almost surely if and only if $q = 0$ and otherwise $\overline{X}_\infty = \mathcal{H}_{\mathbf{e}_q} < \infty$ almost surely. □

Corollary 6.11. *In the previous Theorem, the subordinator associated with L^{-1} has drift a where a is the constant appearing in Theorem 6.8.*

Proof. Note that $\Delta L_t^{-1} := L_t^{-1} - L_{t-}^{-1}$ is greater than $\varepsilon > 0$ follows as a consequence of the path of X moving away from its maximum for a period of real time exceeding ε. That is to say, individual jumps of L^{-1} correspond to individual *excursions* of X from $\overline{X}$. Let us denote by $N_{L^{-1}}$ the Poisson random measure associated with the jumps of L^{-1}. Then the time it takes to accumulate $t < L_\infty$ units of local time is the sum of the periods of time that X has spent away from its maximum plus the real time that X has spent at its maximum (if at all). The latter qualification is only of significance when X is of bounded variation with 0 irregular for $(-\infty, 0)$ in which case the constant a in Theorem 6.8 is strictly positive and then the local time is taken as the Lebesgue measure of the time spent at the maximum. We have on $\{t < L_\infty\}$,

$$L_t^{-1} = \int_0^{L_t^{-1}} \mathbf{1}_{(\overline{X}_s = X_s)} \mathrm{d}s + \int_{[0,t]} \int_{(0,\infty)} x N_{L^{-1}}(\mathrm{d}s \times \mathrm{d}x).$$

From Theorem 6.8 we know that the integral is equal to $aL_{L_t^{-1}} = at$ and hence a is the drift of the subordinator L^{-1}. □

Finally note in the case of compound Poisson process for which, with positive probability, the same maxima may be visited over two intervals of time separated by at least one excursion, we have that $\Delta H_t = 0$ when $\Delta L_t^{-1} > 0$. In other words, the jump measure of H may have an atom at zero. This would be the case for the earlier given example of a compound Poisson process with jumps in $\{-1, 1\}$. Strictly speaking this violates our definition of a subordinator. However, this does not present a serious problem since H is necessarily a compound Poisson subordinator and hence its paths are well defined with the presence of this atom. Further, this does not affect the forthcoming analysis unless otherwise mentioned.

Example 6.12 (Spectrally negative processes). Suppose that X is a spectrally negative Lévy process with Laplace exponent ψ having right inverse Φ; recall the latter part of Sect. 3.3 for a reminder of what this means. As noted earlier, we may work with local time given by $L = \overline{X}$. It follows that L_x^{-1} is nothing more than the first passage time above $x > 0$. (Note that in general it is *not* true that L_x^{-1} is the first passage above x). As X is spectrally negative we have in particular that $H_x = X_{L_x^{-1}} = x$ on $\{x < L_\infty\}$. Recalling

Corollary 3.14 we already know that L^{-1} is a subordinator killed at rate $\Phi(0)$. Hence we may easily identify for $\alpha, \beta \geq 0$, $\kappa(\alpha,\beta) = \Phi(0) + \phi(\alpha,\beta)$ where

$$\phi(\alpha,\beta) = [\Phi(\alpha) - \Phi(0)] + \beta$$

is the Laplace exponent of a bivariate subordinator. Note in particular that $L_\infty < \infty$ if and only if $\Phi(0) > 0$ if and only if $\psi'(0+) = \mathbb{E}(X_1) \in [-\infty, 0)$ and, on account of the fact that L^{-1} is the first passage process, the latter occurs if and only if $\mathbb{P}(\limsup_{t\uparrow\infty} X_t < \infty) = 1$.

In the special case that X is a Brownian motion with drift ρ, we know explicitly that $\psi(\theta) = \rho\theta + \frac{1}{2}\theta^2$ and hence $\Phi(\alpha) = -\rho + \sqrt{\rho^2 + 2\alpha}$. Inverse local time can then be identified precisely as an inverse Gaussian process (killed at rate $2|\rho|$ if $\rho < 0$).

We close this section by making the important remark that the brief introduction to excursion theory offered here has not paid fair dues to their general setting. Excursion theory formally begins with Itô (1970) with further foundational contributions coming from Maisonneuve (1975) and can be applied to a much more general class of Markov processes than just Lévy processes. Recall that $\overline{X} - X$ is a Markov process and hence one may consider L as the local time at 0 of the latter process. In general it is possible to identify excursions of well defined Markov processes from individual points in their state space with the help of local time. The reader interested in a comprehensive account should refer to the detailed but nonetheless approachable account given in Chap. IV of Bertoin (1996) or Blumenthal (1992).

6.3 Excursions

Recall that in Sect. 4.6 we gave an explanation of the Pollaczek–Khintchine formula by decomposing the path of the Lévy processes considered there in terms of excursions from the maximum. Clearly this decomposition relied heavily on the fact that the number of new maxima over any finite time horizon is finite. That is to say, that 0 is irregular for $[0,\infty)$ and the local time at the maximum is a step function as in case 1 listed at the end of Sect. 6.1. Now that we have established the concept of local time at the maximum for any Lévy process we can give the general decomposition of the path of a Lévy process in terms of its excursions from the maximum.

Definition 6.13. *For each moment of local time $t > 0$ we define*

$$\epsilon_t = \begin{cases} \{X_{L^{-1}_{t-}+s} - X_{L^{-1}_{t-}} : 0 < s \leq L^{-1}_t - L^{-1}_{t-}\} & \textit{if } L^{-1}_{t-} < L^{-1}_t \\ \partial & \textit{if } L^{-1}_{t-} = L^{-1}_t \end{cases}$$

where we take $L^{-1}_{0-} = 0$ and ∂ is some "dummy" state. Note that when $L^{-1}_{t-} < L^{-1}_t$ the object ϵ_t is a process and hence is double indexed with

$\epsilon_t(s) = X_{L_{t-}^{-1}+s} - X_{L_{t-}^{-1}}$ for $0 < s \leq L_t^{-1} - L_{t-}^{-1}$. When $\epsilon_t \neq \partial$ we refer to it as the excursion (from the maximum) associated with local time t.

Note also that excursion paths are right continuous with left limits and, with the exception of its terminal value (if $L_t^{-1} < \infty$), is valued in $(-\infty, 0)$.

Definition 6.14. *Let $\mathcal{E}$ be the space of excursions of X from its running supremum. That is the space of mappings which are right continuous with left limits satisfying*

$$\begin{aligned} \epsilon &: (0, \zeta) \to (-\infty, 0) \quad \text{for some } \zeta \in (0, \infty] \\ \epsilon &: \{\zeta\} \to [0, \infty) \qquad\qquad \text{if } \zeta < \infty \end{aligned}$$

where $\zeta = \zeta(\epsilon)$ is the excursion length. Write also $h = h(\epsilon)$ for the terminal value of the excursion so that $h(\epsilon) = \epsilon(\zeta)$. Finally let $\overline{\epsilon} = -\inf_{s \in (0,\zeta)} \epsilon(s)$ for the excursion height.

We will shortly state the fundamental result of excursion theory which relates the process $\{(t, \epsilon_t) : t < L_\infty \text{ and } \epsilon_t \neq \partial\}$ to a Poisson point process on $[0, \infty) \times \mathcal{E}$. The latter process has not yet been discussed in this text however and so we devote a little time to its definition first. Recall that in Chap. 2 the existence of a Poisson random measure on an arbitrary sigma finite measure space $(S, \mathcal{S}, \eta)$ was proved in Theorem 2.4. If we reconsider the proof of Theorem 2.4 then in fact what was proved was the existence of a random set of points in S, each of which is assigned a unit mass thus defining the Poisson random measure N. Instead of referring to the Poisson random measure on $(S, \mathcal{S}, \eta)$ we can instead refer to its support. The latter is called a *Poisson point process* on $(S, \mathcal{S}, \eta)$ (or sometimes the Poisson point process on S with intensity η). In the case that $S = [0, \infty) \times \mathcal{E}$ we may think of the associated Poisson point process as a process of $\mathcal{E}$-valued points appearing in time.

Theorem 6.15. *There exists a σ-algebra Σ and σ-finite measure n such that $(\mathcal{E}, \Sigma, n)$ is a measure space and Σ is rich enough to contain sets of the form*

$$\{\epsilon \in \mathcal{E} : \zeta(\epsilon) \in A, \ \overline{\epsilon} \in B, \ h(\epsilon) \in C\}$$

where, for a given $\epsilon \in \mathcal{E}$, $\zeta(\epsilon)$, $\overline{\epsilon}$ and $h(\epsilon)$ were all given in Definition 6.14. Further, A, B and C are Borel sets of $[0, \infty]$.

(i) If $\mathbb{P}(\limsup_{t\uparrow\infty} X_t = \infty) = 1$ then $\{(t, \epsilon_t) : t \geq 0 \text{ and } \epsilon_t \neq \partial\}$ is a Poisson point process on $([0, \infty) \times \mathcal{E}, \mathcal{B}[0, \infty) \times \Sigma, \mathrm{d}t \times \mathrm{d}n)$.
(ii) If $\mathbb{P}(\limsup_{t\uparrow\infty} X_t < \infty) = 1$ then $\{(t, \epsilon_t) : t < L_\infty \text{ and } \epsilon_t \neq \partial\}$ is a Poisson point process on $([0, \infty) \times \mathcal{E}, \mathcal{B}[0, \infty) \times \Sigma, \mathrm{d}t \times \mathrm{d}n)$ stopped at the first arrival of an excursion which has infinite length.

We offer no proof for this result as it goes beyond the scope of this book. We refer instead to Bertoin (1996) who gives a rigorous treatment. However, the intuition behind this theorem lies with the observation that for each $t > 0$, by Lemma 6.9, L_{t-}^{-1} is a stopping time and hence by Theorem 3.1 the progression of $X_{L_{t-+s}^{-1}} - X_{L_{t-}^{-1}}$ in the time interval $(L_{t-}^{-1}, L_t^{-1}]$ is independent of $\mathcal{F}_{L_{t-}^{-1}}$. As alluded to earlier, this means that the paths of X may be decomposed into the juxtaposition of independent excursions from the maximum. The case that the drift coefficient a of L^{-1} is strictly positive is the case of a bounded variation Lévy process with 0 irregular for $(-\infty, 0)$ and hence local time which may be taken as proportional to the Lebesgue measure of the time that $X = \overline{X}$. In this case excursions from the maximum are interlaced by moments of real time where X can be described as *drifting at its maximum*. That is to say, moments of real time which contribute to a strict increase in the Lebesgue measure of the real time the process spends at its maximum. If there is a last maximum, then the process of excursions is stopped at the first arrival in $\mathcal{E}_\infty := \{\epsilon \in \mathcal{E} : \zeta(\epsilon) = \infty\}$.

Theorem 6.15 generalises the statement of Theorem 6.10. To see why, suppose that we write

$$\Lambda(\mathrm{d}x, \mathrm{d}y) = n(\zeta(\epsilon) \in \mathrm{d}x, h(\epsilon) \in \mathrm{d}y). \tag{6.11}$$

On $\{t < L_\infty\}$ the jumps of the ladder process (L^{-1}, H) form a Poisson process on $[0, \infty) \times (0, \infty)^2$ with intensity measure $\mathrm{d}t \times \Lambda(\mathrm{d}x, \mathrm{d}y)$. We can write L_t^{-1} as the sum of the Lebesgue measure of the time X spends drifting at the maximum (if at all) plus the jumps L^{-1} makes due to excursions from the maximum. Hence, if N is the counting measure associated with the Poisson point process of excursions, then on $\{L_\infty > t\}$,

$$\begin{aligned} L_t^{-1} &= \int_0^{L_t^{-1}} \mathbf{1}_{(\epsilon_s = \partial)} \mathrm{d}s + \int_{[0,t]} \int_{\mathcal{E}} \zeta(\epsilon) N(\mathrm{d}s \times \mathrm{d}\epsilon) \\ &= \int_0^{L_t^{-1}} \mathbf{1}_{(\overline{X}_s = X_s)} \mathrm{d}s + \int_{[0,t]} \int_{\mathcal{E}} \zeta(\epsilon) N(\mathrm{d}s \times \mathrm{d}\epsilon) \\ &= at + \int_{[0,t]} \int_{(0,\infty)} x N_{L^{-1}}(\mathrm{d}s \times \mathrm{d}x). \end{aligned} \tag{6.12}$$

We can also write the ladder height process H in terms of a drift, say $b \geq 0$, and its jumps which are given by the terminal values of excursions. Hence, on $\{t < L_\infty\}$,

$$H_t = bt + \int_{[0,t]} \int_{\mathcal{E}} h(\epsilon) N(\mathrm{d}s \times \mathrm{d}\epsilon). \tag{6.13}$$

Also we can see that $\mathbb{P}(L_\infty > t)$ is the probability that in the process of excursions the first arrival in $\mathcal{E}_\infty$ is after time t. Written in terms of the Poisson point process of excursions we see that

$$\mathbb{P}(L_\infty > t) = \mathbb{P}(N([0,t] \times \mathcal{E}_\infty) = 0) = \mathrm{e}^{-n(\mathcal{E}_\infty)t}.$$

This reinforces the earlier conclusion that L_∞ is exponentially distributed and we equate the parameters

$$\kappa(0,0) = n(\mathcal{E}_\infty). \tag{6.14}$$

6.4 The Wiener–Hopf Factorisation

A fundamental aspect of the theory of Lévy processes is a set of conclusions which in modern times are loosely referred to as *the Wiener–Hopf factorisation*. Historically the identities around which the Wiener–Hopf factorisation is centred are the culmination of a number of works initiated from within the theory of random walks. These include Spitzer, (1956, 1957, 1964), Feller (1971), Borovkov (1976), Pecherskii and Rogozin (1969), Gusak and Korolyuk (1969), Greenwood and Pitman (1980), Fristedt (1974) and many others; although the analytical roots of the so-called Wiener–Hopf method go much further back than these probabilistic references (see Sect. 6.6). The importance of the Wiener–Hopf factorisation is that it gives us information concerning the characteristics of the ascending and descending ladder processes. As indicated earlier, we shall use this knowledge in later chapters to consider a number of applications as well as to extract some generic results concerning course and fine path properties of Lévy process.

In this section we treat the Wiener–Hopf factorisation following closely the presentation of Greenwood and Pitman (1980) which relies heavily on the decomposition of the path of a Lévy process in terms of excursions from the maximum. Examples of the Wiener–Hopf factorisation will be treated in Sect. 6.5.

We begin by recalling that for $\alpha, \beta \geq 0$ the Laplace exponents $\kappa(\alpha,\beta)$ and $\widehat{\kappa}(\alpha,\beta)$ of the ascending ladder process (L^{-1}, H) and the descending ladder process $(\widehat{L}^{-1}, \widehat{H})$ are defined, respectively, by,

$$\mathbb{E}\left(\mathrm{e}^{-\alpha L_1^{-1} - \beta H_1}\mathbf{1}_{(1<L_\infty)}\right) = \mathrm{e}^{-\kappa(\alpha,\beta)} \text{ and } \mathbb{E}\left(\mathrm{e}^{-\alpha \widehat{L}_1^{-1} - \beta \widehat{H}_1}\mathbf{1}_{(1<\widehat{L}_\infty)}\right) = \mathrm{e}^{-\widehat{\kappa}(\alpha,\beta)}.$$

Further, on account of Theorems 6.10 and 6.15,

$$\kappa(\alpha,\beta) = q + \phi(\alpha,\beta) \tag{6.15}$$

where ϕ is the Laplace exponent of a bivariate subordinator and L_∞ is exponentially distributed with parameter $q = n(\mathcal{E}_\infty) \geq 0$. The exponent ϕ can be written in the form

$$\phi(\alpha,\beta) = \alpha a + \beta b + \int_{(0,\infty)^2} (1 - \mathrm{e}^{-\alpha x - \beta y}) \Lambda(\mathrm{d}x, \mathrm{d}y), \tag{6.16}$$

where the constant a was identified in Corollary 6.8, b is some non-negative constant representing the drift of H and $\Lambda(\mathrm{d}x, \mathrm{d}y)$ is given in terms of the excursion measure n in (6.11). It is also important to remark that both $\kappa(\alpha, \beta)$ and $\widehat{\kappa}(\alpha, \beta)$ can be analytically extended in β to $\mathbb{C}^+ = \{z \in \mathbb{C} : \Re z \geq 0\}$.

The next theorem gives the collection of statements which are known as the Wiener–Hopf factorisation. We need to introduce some additional notation first. As in earlier chapters, we shall understand $\mathbf{e}_p$ to be an independent random variable which is exponentially distributed with mean $1/p$. Further, we define

$$\overline{G}_t = \sup\{s < t : \overline{X}_s = X_s\}$$

and

$$\underline{G}_t = \sup\{s < t : \underline{X}_s = X_s\}.$$

An important fact concerning the definition of $\overline{G}_t$ which is responsible for the first sentence in the statement of the Wiener–Hopf factorisation (Theorem 6.16 below) is that, if X is not a compound Poisson process, then its maxima are obtained at unique times. To see this first suppose that 0 is regular for $[0, \infty)$. Since we have excluded Poisson processes, then by Corollary 6.6 this implies that 0 is regular for $(0, \infty)$. In this case, for any stopping time T such that $\overline{X}_T = X_T$ it follows by the Strong Markov Property and regularity that $\overline{X}_{T+u} > \overline{X}_T$ for all $u > 0$; in particular, we may consider the stopping times L_t^{-1} which run through all the times at which X visits its maximum. If the aforementioned regularity fails, then since X is assumed not to be a compound Poisson process, then 0 must be regular for $(-\infty, 0)$. In that case, the conclusions of the previous case apply to $-X$. However, $-X$ has the same law as X time reversed. Hence the path of X over any finite time horizon when time reversed has new maxima which are obtained at unique times. This implies that X itself cannot touch the same maxima at two different times when sampled over any finite time horizon.

As mentioned earlier however, if X is a compound Poisson process with an appropriate jump distribution, it is possible that X visits the same maxima at distinct ladder times.

Theorem 6.16 (The Wiener–Hopf factorisation). *Suppose that X is any Lévy process other than a compound Poisson process. As usual, denote by $\mathbf{e}_p$ an independent and exponentially distributed random variable.*

(i) The pairs

$$(\overline{G}_{\mathbf{e}_p}, \overline{X}_{\mathbf{e}_p}) \text{ and } (\mathbf{e}_p - \overline{G}_{\mathbf{e}_p}, \overline{X}_{\mathbf{e}_p} - X_{\mathbf{e}_p})$$

are independent and infinitely divisible, yielding the factorisation

$$\frac{p}{p - \mathrm{i}\vartheta + \Psi(\theta)} = \Psi_p^+(\vartheta, \theta) \cdot \Psi_p^-(\vartheta, \theta) \tag{6.17}$$

where $\theta, \vartheta \in \mathbb{R}$,

$$\Psi_p^+(\vartheta, \theta) = \mathbb{E}\left(\mathrm{e}^{\mathrm{i}\vartheta \overline{G}_{\mathbf{e}_p} + \mathrm{i}\theta \overline{X}_{\mathbf{e}_p}}\right) \text{ and } \Psi_p^-(\vartheta, \theta) = \mathbb{E}\left(\mathrm{e}^{\mathrm{i}\vartheta \underline{G}_{\mathbf{e}_p} + \mathrm{i}\theta \underline{X}_{\mathbf{e}_p}}\right).$$

The pair $\Psi_p^+(\vartheta, \theta)$ and $\Psi_p^-(\vartheta, \theta)$ are called the Wiener–Hopf factors.

(ii) *The Wiener–Hopf factors may themselves be identified in terms of the analytically extended Laplace exponents* $\kappa(\alpha,\beta)$ *and* $\widehat{\kappa}(\alpha,\beta)$ *via the Laplace transforms,*

$$\mathbb{E}\left(\mathrm{e}^{-\alpha\overline{G}_{\mathbf{e}_p}-\beta\overline{X}_{\mathbf{e}_p}}\right)=\frac{\kappa(p,0)}{\kappa(p+\alpha,\beta)}\text{ and }\mathbb{E}\left(\mathrm{e}^{-\alpha\underline{G}_{\mathbf{e}_p}+\beta\underline{X}_{\mathbf{e}_p}}\right)=\frac{\widehat{\kappa}(p,0)}{\widehat{\kappa}(p+\alpha,\beta)}\tag{6.18}$$

for $\alpha,\beta\in\mathbb{C}^+$.

(iii) *The Laplace exponents* $\kappa(\alpha,\beta)$ *and* $\widehat{\kappa}(\alpha,\beta)$ *may also be identified in terms of the law of* X *in the following way,*

$$\kappa(\alpha,\beta)=k\exp\left(\int_0^\infty\int_{(0,\infty)}(\mathrm{e}^{-t}-\mathrm{e}^{-\alpha t-\beta x})\,\frac{1}{t}\mathbb{P}(X_t\in\mathrm{d}x)\mathrm{d}t\right)\tag{6.19}$$

and

$$\widehat{\kappa}(\alpha,\beta)=\widehat{k}\exp\left(\int_0^\infty\int_{(-\infty,0)}(\mathrm{e}^{-t}-\mathrm{e}^{-\alpha t+\beta x})\,\frac{1}{t}\mathbb{P}(X_t\in\mathrm{d}x)\mathrm{d}t\right)\tag{6.20}$$

where $\alpha,\beta\in\mathbb{R}$ *and* k *and* $\widehat{k}$ *are strictly positive constants.*

(iv) *By setting* $\vartheta=0$ *and taking limits as* p *tends to zero in (6.17) one obtains for some constant* $k'>0$ *(which may be taken equal to unity by a suitable normalisation of local time),*

$$k'\Psi(\theta)=\kappa(0,-\mathrm{i}\theta)\widehat{\kappa}(0,\mathrm{i}\theta).\tag{6.21}$$

Let us now make some notes concerning the statement of the Wiener–Hopf factorisation.

Firstly, there are a number of unidentified constants in the expressions concerning the Wiener–Hopf factorisation. To some extent, these constants are meaningless since, as a little reflection reveals, they are dependent on the normalisation chosen in the definition of local time (cf. Definition 6.1). In this context local time is nothing other than an artificial clock to measure intrinsic time spent at the maximum. Naturally a different choice of local time will induce a different inverse local time and hence ladder height process. Nonetheless the range of the bivariate ladder process will be invariant to this choice as this will always correspond to the range of the real times and positions of the new maxima of the underlying Lévy process.

Secondly, the exclusion of the compound Poisson processes from the statement of the theorem is not to say that a Wiener–Hopf factorisation for this class of Lévy processes does not exist. The case of the compound Poisson process is essentially the case of the random walk and has some subtle differences which we shall come back to later on.

The proof of Theorem 6.16 we shall give makes use of a simple fact about infinitely divisible distributions as well as the fundamental properties of the Poisson point processes describing the excursions of X. We give these facts in the following two preparatory Lemmas. For the first, it may be useful to recall Exercise 2.10

Lemma 6.17. *Suppose that* $\mathbf{X} = \{\mathbf{X}_t : t \geq 0\}$ *is any d-dimensional Lévy process with characteristic exponent* $\Psi(\theta) = -\log \mathbb{E}(\mathrm{e}^{\mathrm{i}\theta\cdot\mathbf{X}_t})$ *for* $\theta \in \mathbb{R}^d$. *Then the pair* $(\mathbf{e}_p, \mathbf{X}_{\mathbf{e}_p})$ *has a bivariate infinitely divisible distribution with Lévy–Khintchine exponent given by*

$$\mathbb{E}\left(\mathrm{e}^{\mathrm{i}\vartheta\mathbf{e}_p+\mathrm{i}\theta\mathbf{X}_{\mathbf{e}_p}}\right) = \exp\left\{-\int_0^\infty\int_{\mathbb{R}^d}(1-\mathrm{e}^{\mathrm{i}\vartheta t+\mathrm{i}\theta\cdot x})\frac{1}{t}\mathrm{e}^{-pt}\mathbb{P}\left(\mathbf{X}_t\in\mathrm{d}x\right)\mathrm{d}t\right\}$$
$$= \frac{p}{p-\mathrm{i}\vartheta+\Psi(\theta)}.$$

Proof. Noting from the Lévy–Khintchine formula (cf. Exercise 2.10) that $\Re\Psi(\theta) = \theta\cdot\mathbf{A}\theta/2 + \int_{\mathbb{R}^d}(1-\cos\theta\cdot x)\Pi(\mathrm{d}x)$ where $\mathbf{A}$ is a $d\times d$ matrix which has non-negative eigenvalues, Π is the Lévy measure on $\mathbb{R}^d$ and $\theta\in\mathbb{R}^d$, it follows that $\Re\Psi(\theta)\geq 0$. Hence we have

$$\mathbb{E}\left(\mathrm{e}^{\mathrm{i}\vartheta\mathbf{e}_p+\mathrm{i}\theta X_{\mathbf{e}_p}}\right) = \int_0^\infty p\mathrm{e}^{-pt+\mathrm{i}\vartheta t-\Psi(\theta)t}\mathrm{d}t = \frac{p}{p-\mathrm{i}\vartheta+\Psi(\theta)}.$$

On the other hand using the Frullani integral in Lemma 1.7 we see that

$$\exp\left\{-\int_0^\infty\int_{\mathbb{R}}(1-\mathrm{e}^{\mathrm{i}\vartheta t+\mathrm{i}\theta\cdot x})\frac{1}{t}\mathrm{e}^{-pt}\mathbb{P}\left(\mathbf{X}_t\in\mathrm{d}x\right)\mathrm{d}t\right\}$$
$$= \exp\left\{-\int_0^\infty(1-\mathrm{e}^{-(\Psi(\theta)-\mathrm{i}\vartheta)t})\frac{1}{t}\mathrm{e}^{-pt}\mathrm{d}t\right\}$$
$$= \frac{p}{p-\mathrm{i}\vartheta+\Psi(\theta)}.$$

The result now follows. □

The next result concerns the Poisson point process of excursions $\{(t,\epsilon_t) : t\geq 0 \text{ and } \epsilon_t\neq\partial\}$. However, the conclusion depends only on the fact that points of the latter process belong to a product measure space $([0,\infty)\times\mathcal{E}, \mathcal{B}([0,\infty))\times\Sigma, \mathrm{d}t\times\mathrm{d}n)$ and not that $\mathcal{E}$ is the space of excursions.

Lemma 6.18. *Suppose that* $\{(t,\epsilon_t) : t\geq 0\}$ *is a Poisson point process on* $([0,\infty)\times\mathcal{E}, \mathcal{B}[0,\infty)\times\Sigma, \mathrm{d}t\times\mathrm{d}n)$. *Choose* $A\in\Sigma$ *such that* $n(A)<\infty$ *and define*

$$\sigma^A = \inf\{t>0 : \epsilon_t\in A\}.$$

(i) The random time σ^A *is exponentially distributed with parameter* $n(A)$.
(ii) The process $\{(t,\epsilon_t) : t<\sigma^A\}$ *is equal in law to a Poisson point process on* $[0,\infty)\times\mathcal{E}\backslash A$ *with intensity* $\mathrm{d}t\times\mathrm{d}n'$ *where* $n'(\mathrm{d}\epsilon) = n(\mathrm{d}\epsilon\cap\mathcal{E}\backslash A)$ *which is stopped at an independent exponential time with parameter* $n(A)$.
(iii) The process $\{(t,\epsilon_t) : t<\sigma^A\}$ *is independent of* ϵ_{σ^A}.

Proof. Let $S_1 = [0,\infty) \times A$ and $S_2 = [0,\infty) \times (\mathcal{E}\backslash A)$. Suppose that N is the Poisson random measure associated with the given Poisson point process. All three conclusions follow from Corollary 2.5 applied to the restriction of N to the disjoint sets S_1 and S_2, say $N^{(1)}$ and $N^{(2)}$, respectively.

Specifically, for (i) note that $\mathbb{P}(\sigma^A > t) = \mathbb{P}(N^{(1)}([0,t]\times A) = 0) = \mathrm{e}^{-n(A)t}$ as $N^{(1)}$ has intensity $\mathrm{d}t \times n(\mathrm{d}\epsilon \cap A)$. For (ii) and (iii) it suffices to note that $N^{(2)}$ has intensity $\mathrm{d}t \times n(\mathrm{d}\epsilon \cap \mathcal{E}\backslash A)$, that

$$\{(t,\epsilon_t) \in [0,\infty) \times \mathcal{E} : t < \sigma^A\} = \{(t,\epsilon_t) \in [0,\infty) \times (\mathcal{E}\backslash A) : t < \sigma^A\}$$

and that the first arrival in A is a point belonging to the process $N^{(1)}$ which is independent of $N^{(2)}$. □

Note in fact that, since $\{\sigma^A \leq t\} = \{N([0,t] \times A) \geq 1\}$, it is easily seen that σ^A is a stopping time with respect to the filtration $\mathbb{G} = \{\mathcal{G}_t : t \geq 0\}$ where

$$\mathcal{G}_t = \sigma(N(U \times V) : \ U \in \mathcal{B}[0,t] \text{ and } V \in \Sigma).$$

In the case that $\mathcal{E}$ is taken as the space of excursions, one may take $\mathbb{G} = \mathbb{F}$.

Now we are ready to give the proof of the Wiener–Hopf factorisation.

Proof (of Theorem 6.16 (i)). The crux of the first part of the Wiener–Hopf factorisation lies with the following important observation. Consider the Poisson point process of marked excursions on

$$([0,\infty) \times \mathcal{E} \times [0,\infty), \mathcal{B}[0,\infty) \times \Sigma \times \mathcal{B}[0,\infty), \mathrm{d}t \times \mathrm{d}n \times \mathrm{d}\eta)$$

where $\eta(\mathrm{d}x) = p\mathrm{e}^{-px}\mathrm{d}x$ for $x \geq 0$. That is to say, a Poisson point process whose points are described by $\{(t,\epsilon_t,\mathbf{e}_p^{(t)}) : t \geq 0 \text{ and } \epsilon_t \neq \partial\}$ where $\mathbf{e}_p^{(t)}$ is an independent copy of an exponentially distributed random variable if t is such that $\epsilon_t \neq \partial$ and otherwise $\mathbf{e}_p^{(t)} := \partial$. The Poisson point process of excursions is then just a projection of the latter on to $[0,\infty) \times \mathcal{E}$. Sampling the Lévy process X up to an independent exponentially distributed random time $\mathbf{e}_p$ corresponds to sampling the Poisson process of excursions up to time $L_{\mathbf{e}_p}$; that is $\{(t,\epsilon_t) : t < L_{\mathbf{e}_p} \text{ and } t \neq \partial\}$. In turn, we claim that the latter is equal in law to the projection on to $[0,\infty) \times \mathcal{E}$ of

$$\{(t,\epsilon_t,\mathbf{e}_p^{(t)}) : t < \sigma_1 \wedge \sigma_2 \text{ and } \epsilon_t \neq \partial\} \tag{6.22}$$

where

$$\sigma_1 := \inf\{t > 0 : \int_0^{L_t^{-1}} \mathbf{1}_{(\overline{X}_s = X_s)}\mathrm{d}s > \mathbf{e}_p\}$$

and

$$\sigma_2 := \inf\{t > 0 : \zeta(\epsilon_t) > \mathbf{e}_p^{(t)}\}$$

where we recall that $\zeta(\epsilon_t)$ is the duration of the excursion indexed by local time t. Note that in the case that the constant a in Theorem 6.8 is zero, in other words $\int_0^{\cdot} \mathbf{1}_{(\overline{X}_s = X_s)} \mathrm{d}s = 0$, we have simply that $\sigma_1 = \infty$. A formal proof of this claim would require the use of some additional mathematical tools. However, for the sake of brevity, we shall lean instead on an intuitive explanation as follows.

We recall that the path of the Lévy process up to time $\mathbf{e}_p$ is the independent juxtaposition of excursions and further, in the case that the constant a in Theorem 6.8 is strictly positive, excursions are interlaced with moments of real time when $X = \overline{X}$ and which accumulate positive Lebesgue measure. By the lack of memory property the event $\{t < L_{\mathbf{e}_p}\}$, that is the event that there are at least t units of local time for a given stretch of $\mathbf{e}_p$ units of real time, is equivalent to the event that the total amount of real time accumulated at the maximum by local time t has survived independent exponential killing at rate p as well as each of the excursion lengths up to local time t have survived independent exponential killing at rate p. This idea is easier to visualise when one considers the case that X is a compound Poisson process with strictly positive or strictly negative drift; see Figs. 6.2 and 6.3.

The times σ_1 and σ_2 are independent and further σ_2 is of the type of stopping time considered in Lemma 6.18 with $A = \{\zeta(\epsilon) > \mathbf{e}_p\}$ when applied to the Poisson point process (6.22). From each of the three statements given in Lemma 6.18 we deduce three facts concerning the Poisson point process (6.22).

(1) Since $\int_0^{L_t^{-1}} \mathbf{1}_{(\overline{X}_s = X_s)} \mathrm{d}s = at$ we have

$$\mathbb{P}(\sigma_1 > t) = \mathbb{P}\left(\int_0^{L_t^{-1}} \mathbf{1}_{(\overline{X}_s = X_s)} \mathrm{d}s < \mathbf{e}_p\right) = \mathrm{e}^{-apt}.$$

As mentioned earlier, if the constant $a = 0$ then we have that $\sigma_1 = \infty$. Further, with the help of Lemma 6.18 (i) we also have that

$$\begin{aligned}
&\mathbb{P}(\sigma_2 > t)\\
&= \exp\left\{-t \int_0^\infty p\mathrm{e}^{-px} \mathrm{d}x \cdot n(\zeta(\epsilon) > x)\right\}\\
&= \exp\left\{-t \int_0^\infty p\mathrm{e}^{-px} \mathrm{d}x \cdot [n(\infty > \zeta(\epsilon) > x) + n(\zeta(\epsilon) = \infty)]\right\}\\
&= \exp\left\{-n(\mathcal{E}_\infty)t - t \int_{(0,\infty)} (1 - \mathrm{e}^{-px}) n(\zeta(\epsilon) \in \mathrm{d}x)\right\}
\end{aligned}$$

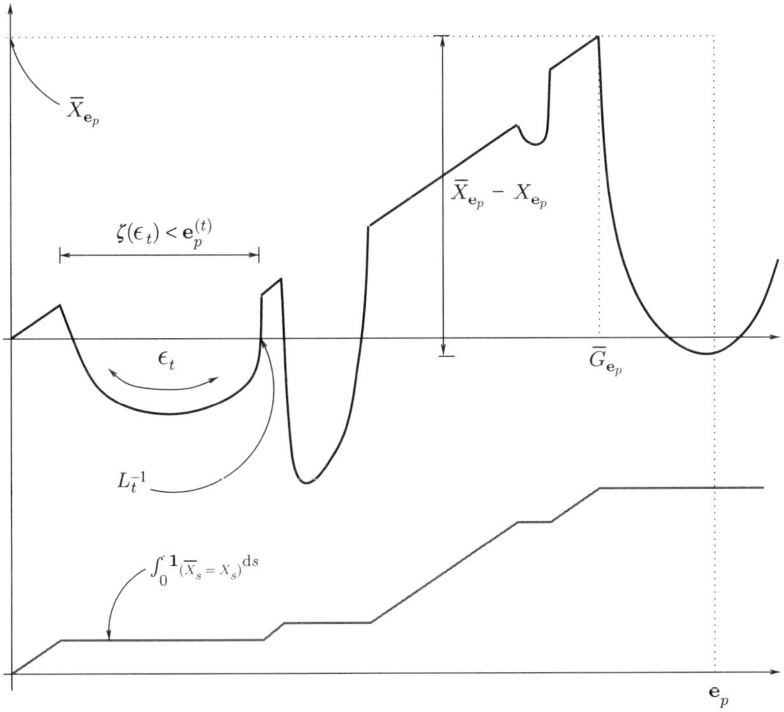

Fig. 6.2. A symbolic sketch of the decomposition of the path of a compound Poisson process with strictly positive drift over an independent and exponentially distributed period of time. The situation for bounded variation Lévy processes for which 0 is irregular for $(-\infty, 0)$ is analogous to the case in this sketch in the sense that the Lebesgue measure of the time spent at the maximum over any finite time horizon is strictly positive.

where we recall that $\mathcal{E}_\infty = \{\epsilon \in \mathcal{E} : \zeta(\epsilon) = \infty\}$. As σ_1 and σ_2 are independent and exponentially distributed it follows[3] that

$$\mathbb{P}(\sigma_1 \wedge \sigma_2 > t) = \exp\left\{ -t \left(n(\mathcal{E}_\infty) + ap + \int_{(0,\infty)} (1 - \mathrm{e}^{-px}) n(\zeta(\epsilon) \in \mathrm{d}x) \right) \right\}.$$

However, recall from (6.11) and (6.14) that $\kappa(0,0) = n(\mathcal{E}_\infty)$ and $\Lambda(\mathrm{d}x, [0,\infty)) = n(\zeta(\epsilon) \in \mathrm{d}x)$ and hence the exponent above is equal to $\kappa(p, 0)$ where κ is given by (6.15) and (6.16).

(2) From Lemma 6.18 (ii) and the observation (1) above, we see that the Poisson point process (6.22) is equal in law to a Poisson point process on

[3]Recall that the minimum of two independent exponential random variables is again exponentially distributed with the sum of their rates

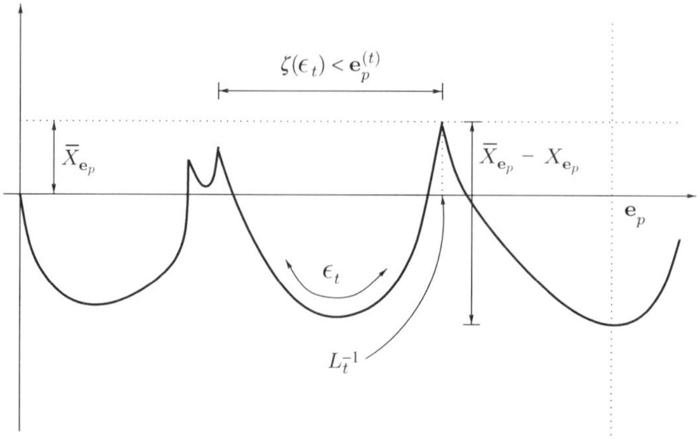

Fig. 6.3. A symbolic sketch of the decomposition of the path of a compound Poisson process with strictly negative drift over an independent and exponentially distributed period of time. The situation for a Lévy process of unbounded variation or a Lévy process for which 0 is irregular for $[0, \infty)$ is analogous to the case in this sketch in the sense that the Lebesgue measure of the time spent at the maximum is zero.

$[0, \infty) \times \mathcal{E} \times [0, \infty)$ with intensity

$$\mathrm{d}t \times n(\mathrm{d}\epsilon; \zeta(\epsilon) < x) \times \eta(\mathrm{d}x) \tag{6.23}$$

which is stopped at an independent time which is exponentially distributed with parameter $\kappa(p, 0)$.

(3) Lemma 6.18 (iii) tells us that on the event $\sigma_2 < \sigma_1$, the process (6.22) is independent of $\epsilon_{\sigma_2} = \epsilon_{\sigma_1 \wedge \sigma_2}$. On the other hand, when $\sigma_1 < \sigma_2$ since at the local time σ_1 we have $\partial = \epsilon_{\sigma_1} = \epsilon_{\sigma_1 \wedge \sigma_2}$ We conclude that $\epsilon_{\sigma_1 \wedge \sigma_2}$ is independent of (6.22).

Now note with the help of (6.12) and (6.13) that

$$\overline{G}_{\mathbf{e}_p} = L^{-1}_{(\sigma_1 \wedge \sigma_2)-} = a(\sigma_1 \wedge \sigma_2) + \int_{[0, \sigma_1 \wedge \sigma_2)} \int_{\mathcal{E}} \zeta(\epsilon_t) N(\mathrm{d}t \times \mathrm{d}\epsilon)$$

and

$$\overline{X}_{\mathbf{e}_p} = H_{(\sigma_1 \wedge \sigma_2)-} = b(\sigma_1 \wedge \sigma_2) + \int_{[0, \sigma_1 \wedge \sigma_2)} \int_{\mathcal{E}} h(\epsilon_t) N(\mathrm{d}t \times \mathrm{d}\epsilon).$$

From point (3) above, the previous two random variables are independent of the excursion $\epsilon_{\sigma_1 \wedge \sigma_2}$. This last excursion occupies the final $\mathbf{e}_p - \overline{G}_{\mathbf{e}_p}$ units of real time in the interval $[0, \mathbf{e}_p]$ and reaches a depth of $X_{\mathbf{e}_p} - \overline{X}_{\mathbf{e}_p}$ relative to its spatial point of issue (see Figs. 6.2 and 6.3). Note that the last two random variables are both zero if and only if $\epsilon_{\sigma_1 \wedge \sigma_2} = \partial$ if and only if $\sigma_1 < \sigma_2$. In conclusion $(\overline{G}_{\mathbf{e}_p}, \overline{X}_{\mathbf{e}_p})$ is independent of $(\mathbf{e}_p - \overline{G}_{\mathbf{e}_p}, \overline{X}_{\mathbf{e}_p} - X_{\mathbf{e}_p})$.

From point (2), the process $\{(L_t^{-1}, H_t) : t < \sigma_1 \wedge \sigma_2\}$ behaves like a subordinator with characteristic measure

$$\int_0^\infty p\mathrm{e}^{-pt}\mathrm{d}t \cdot n(\zeta(\epsilon) \in \mathrm{d}x, h(\epsilon) \in \mathrm{d}y, x < t) = \mathrm{e}^{-px}\Lambda(\mathrm{d}x, \mathrm{d}y)$$

and drift (a, b) which is stopped at an independent exponentially distributed time with parameter $\kappa(p, 0)$. Suppose that we denote this subordinator $(\mathbb{L}^{-1}, \mathbb{H}) = \{(\mathbb{L}_t^{-1}, \mathbb{H}_t) : t \geq 0\}$. Then

$$(\mathbb{L}_{\mathbf{e}_\chi}^{-1}, \mathbb{H}_{\mathbf{e}_\chi}) \stackrel{d}{=} (\overline{G}_{\mathbf{e}_p}, \overline{X}_{\mathbf{e}_p})$$

where $\mathbf{e}_\chi$ is an independent exponential random variable with parameter $\chi = \kappa(p, 0)$. From Lemma 6.17 we also see that $(\overline{G}_{\mathbf{e}_p}, \overline{X}_{\mathbf{e}_p})$ is infinitely divisible. Now note that by appealing to the Duality Lemma and the fact that maxima are attained at unique times (recall the discussion preceding the statement of Theorem 6.16), one sees that

$$(\mathbf{e}_p - \overline{G}_{\mathbf{e}_p}, \overline{X}_{\mathbf{e}_p} - X_{\mathbf{e}_p}) \stackrel{d}{=} (\underline{G}_{\mathbf{e}_p}, -\underline{X}_{\mathbf{e}_p}). \tag{6.24}$$

(This is seen for example in Figs. 6.2 and 6.3 by rotating them about 180°). For reasons similar to those given above, the pair $(\underline{G}_{\mathbf{e}_p}, -\underline{X}_{\mathbf{e}_p})$ must also be infinitely divisible. The factorisation (6.17) now follows. □

Proof (of Theorem 6.16 (ii)). From the proof of part (i), the bivariate subordinator $(\mathbb{L}^{-1}, \mathbb{H})$ has Laplace exponent equal to

$$a\alpha + b\beta + \int_{(0,\infty)^2} (1 - \mathrm{e}^{-\alpha x - \beta y})\mathrm{e}^{-px}\Lambda(\mathrm{d}x, \mathrm{d}y) = \kappa(\alpha + p, \beta) - \kappa(p, 0)$$

for $\alpha, \beta \geq 0$, where the second equality follows from (6.15) and (6.16). Hence from the second equality in the statement of Lemma 6.17,

$$\begin{aligned}\mathbb{E}\left(\mathrm{e}^{-\alpha\overline{G}_{\mathbf{e}_p} - \beta\overline{X}_{\mathbf{e}_p}}\right) &= \mathbb{E}\left(\mathrm{e}^{-\alpha\mathbb{L}_{\mathbf{e}_\chi}^{-1} - \beta\mathbb{H}_{\mathbf{e}_\chi}}\right) \\ &= \frac{\chi}{\kappa(\alpha + p, \beta) - \kappa(p, 0) + \chi} \\ &= \frac{\kappa(p, 0)}{\kappa(\alpha + p, \beta)}.\end{aligned} \tag{6.25}$$

Part (ii) follows from (6.25) by analytically extending the identity from $\alpha, \beta \geq 0$ to $\mathbb{C}^+$. □

Proof (of Theorem 6.16 (iii)). According to Lemma 6.17 the bivariate random variable $(\mathbf{e}_p, X_{\mathbf{e}_p})$ is infinitely divisible and has Lévy measure given by

$$\pi(\mathrm{d}t, \mathrm{d}x) = \frac{1}{t}\mathrm{e}^{-pt}\mathbb{P}(X_t \in \mathrm{d}x)\mathrm{d}t.$$

Since by part (i) we can write $(\mathbf{e}_p, X_{\mathbf{e}_p})$ as the independent sum

$$(\overline{G}_{\mathbf{e}_p}, \overline{X}_{\mathbf{e}_p}) + (\mathbf{e}_p - \overline{G}_{\mathbf{e}_p}, X_{\mathbf{e}_p} - \overline{X}_{\mathbf{e}_p})$$

it follows that $\pi = \pi^+ + \pi^-$ where π^+ and π^- are the Lévy measures of $(\overline{G}_{\mathbf{e}_p}, \overline{X}_{\mathbf{e}_p})$, and $(\mathbf{e}_p - \overline{G}_{\mathbf{e}_p}, X_{\mathbf{e}_p} - \overline{X}_{\mathbf{e}_p})$, respectively. Further, π^+ must be supported on $[0,\infty)\times[0,\infty)$ and π^- must be supported on $[0,\infty)\times(-\infty,0]$ since these are the supports of $(\overline{G}_{\mathbf{e}_p}, \overline{X}_{\mathbf{e}_p})$ and $(\mathbf{e}_p - \overline{G}_{\mathbf{e}_p}, X_{\mathbf{e}_p} - \overline{X}_{\mathbf{e}_p})$.

As X is not a compound Poisson process we have that $\mathbb{P}(X_t = 0) = 0$ for Lebesgue almost all $t > 0$.[4] We can now identify π^+ as simply the restriction of π to $[0,\infty)\times(0,\infty)$ and π^- as the restriction of π to $[0,\infty)\times(-\infty,0)$. Using the Lévy–Khintchine formula (2.28) from Exercise 2.10 for a bivariate pair of infinitely divisible random variables we can conclude that for some constants $k \geq 0$ and $\mathrm{k} \geq 0$ we can identify the Wiener–Hopf factors in the form

$$\Psi_p^+(\vartheta,\theta) = \exp\left\{ \mathrm{i}k\vartheta + \mathrm{ik}\theta + \int_0^\infty \int_{(0,\infty)} (\mathrm{e}^{\mathrm{i}\vartheta t + \mathrm{i}\theta x} - 1)\frac{1}{t}\mathrm{e}^{-pt}\mathbb{P}(X_t \in \mathrm{d}x)\mathrm{d}t \right\}$$

and

$$\Psi_p^-(\vartheta,\theta) = \exp\left\{ -\mathrm{i}k\vartheta - \mathrm{ik}\theta + \int_0^\infty \int_{(-\infty,0)} (\mathrm{e}^{\mathrm{i}\vartheta t + \mathrm{i}\theta x} - 1)\frac{1}{t}\mathrm{e}^{-pt}\mathbb{P}(X_t \in \mathrm{d}x)\mathrm{d}t \right\}.$$

Note in particular that the identification of Ψ^+ and Ψ^- should also take account of the fact that Ψ^+ extends analytically to the upper half of the complex plane in θ and Ψ^- extends to the lower half of the complex plane in θ. Since $\mathbf{e}_p$ can take arbitrarily small values, then so can $\overline{G}_{\mathbf{e}_p}$ and $\overline{X}_{\mathbf{e}_p}$. In which case the Lévy–Khintchine exponent of $(\overline{G}_{\mathbf{e}_p}, \overline{X}_{\mathbf{e}_p})$ should not contain the drift term $\mathrm{i}k\vartheta + \mathrm{ik}\theta$.

From (6.25) we can now identify $\kappa(\alpha,\beta)$ up to a constant and formula (6.19) follows. Similarly we may identify the formula given for $\widehat{\kappa}(\alpha,\beta)$. □

Proof (of Theorem 6.16 (iv)). Note from the expressions established in part (iii) and Lemma 1.7 for the Frullani integral,

$$\begin{aligned}
\kappa(p,0)\widehat{\kappa}(p,0) &= k' \exp\left\{ \int_0^\infty (\mathrm{e}^{-t} - \mathrm{e}^{-pt})\frac{1}{t}\mathrm{d}t \right\} \\
&= k' \exp\left\{ \int_0^\infty (1 - \mathrm{e}^{-pt})\mathrm{e}^{-t}\frac{1}{t}\mathrm{d}t - \int_0^\infty (1 - \mathrm{e}^{-t})\mathrm{e}^{-pt}\frac{1}{t}\mathrm{d}t \right\} \\
&= k'p
\end{aligned}$$

[4]This statement is intuitively appealing; however it requires a rigorous proof. We refrain from giving it here in order to avoid distraction from the proof at hand. The basic idea however is to prove, in the spirit of Theorem 5.4, that for each $q > 0$, the potential measure $U^{(q)}(\mathrm{d}x) := \mathbb{E}(\int_0^\infty \mathbf{1}_{(X_t \in \mathrm{d}x)}\mathrm{d}t)$ has no atoms. See for example Proposition I.15 of Bertoin (1996).

where $k' = k\widehat{k}$. Equation (6.17) now reads

$$\frac{1}{p - \mathrm{i}\vartheta + \Psi(\theta)} = \frac{k'}{\kappa(p - \mathrm{i}\vartheta, -\mathrm{i}\theta) \cdot \widehat{\kappa}(p - \mathrm{i}\vartheta, \mathrm{i}\theta)}.$$

Setting $\vartheta = p = 0$ delivers the required result. □

Corollary 6.19. *Suppose that X is a Lévy process other than a compound Poisson process and a factorisation (6.17) exists where Ψ_p^+ and Ψ_p^- are characteristic functions of infinitely divisible laws, then the factorisation is unique.*

Proof. The proof is a consequence of the argument given in the proof of part (iii) of the Wiener–Hopf factorisation. □

We conclude this section with some remarks about the case that X is a compound Poisson process. In this case, most of the proof of Theorem 6.16 goes through as stated. However, the following subtleties need to be taken account of.

In the proof of the part (i) of Theorem 6.16 it is no longer true that (6.24) holds. One needs to be more careful concerning the definition of $\overline{G}_t$ and $\underline{G}_t$. For compound Poisson processes, it is necessary to work with the new definitions

$$\overline{G}_t = \sup\{s < t : X_s = \overline{X}_t\} \text{ and } \underline{G}_t^* = \inf\{s < t : X_s = \underline{X}_t\}, \tag{6.26}$$

instead. It was shown in the case that X is not a compound Poisson process that maxima are obtained at distinct times. Hence the above definitions are consistent with the original definitions of $\overline{G}_t$ and $\underline{G}_t$ outside the class of compound Poisson processes.

Still appealing to duality the statement (6.24) should now be replaced by

$$(\mathbf{e}_p - \overline{G}_{\mathbf{e}_p}, \overline{X}_{\mathbf{e}_p} - X_{\mathbf{e}_p}) \stackrel{d}{=} (\underline{G}^*_{\mathbf{e}_p}, -\underline{X}_{\mathbf{e}_p}) \tag{6.27}$$

and the factorisation (6.17) requires redefining

$$\Psi_p^-(\vartheta, \theta) = \mathbb{E}(\mathrm{e}^{\mathrm{i}\vartheta \underline{G}^*_{\mathbf{e}_p} + \mathrm{i}\theta \underline{X}_{\mathbf{e}_p}}).$$

Further, in the proof of parts (ii) and (iii) of Theorem 6.16 an adjustment is required in the definitions of κ and $\widehat{\kappa}$. Recall that in the decomposition $\pi = \pi^+ + \pi^-$, the respective supports of π^+ and π^- are $[0,\infty) \times (0,\infty)$ and $[0,\infty) \times (-\infty, 0)$. Unlike earlier, we are now faced with the difficulty of assigning the mass given by the probabilities $\mathbb{P}(X_t = 0)$ for $t \geq 0$ to one or other of the integrals that, respectively, define κ and $\widehat{\kappa}$. The way to do this is to first consider the process $X^\varepsilon = \{X_t^\varepsilon : t \geq 0\}$ where

$$X_t^\varepsilon := X_t + \varepsilon t, \quad t \geq 0$$

and $\varepsilon \in \mathbb{R}$. A little thought reveals that for each fixed $t \geq 0$, $\lim_{\varepsilon \downarrow 0} \overline{G}_t^\varepsilon = \overline{G}_t$ where $\overline{G}_t^\varepsilon$ is given by (6.26) applied to X^ε and $\overline{G}_t$ is also given by (6.26).

Similarly, $\lim_{\varepsilon\downarrow 0}\overline{X}_t^{\varepsilon} = \overline{X}_t$ where $\overline{X}^{\varepsilon} = \sup_{s\leq t} X_s^{\varepsilon}$. Next note, in the sense of weak convergence

$$\lim_{\varepsilon\downarrow 0}\frac{1}{t}\mathrm{e}^{-pt}\mathbb{P}(X_t^{\varepsilon}\in \mathrm{d}x)\mathrm{d}t\mathbf{1}_{(x>0)} = \frac{1}{t}\mathrm{e}^{-pt}\mathbb{P}(X_t\in \mathrm{d}x)\mathrm{d}t\mathbf{1}_{(x\geq 0)}$$

whilst

$$\lim_{\varepsilon\downarrow 0}\frac{1}{t}\mathrm{e}^{-pt}\mathbb{P}(X_t^{\varepsilon}\in \mathrm{d}x)\mathrm{d}t\mathbf{1}_{(x<0)} = \frac{1}{t}\mathrm{e}^{-pt}\mathbb{P}(X_t\in \mathrm{d}x)\mathrm{d}t\mathbf{1}_{(x<0)}.$$

Hence applying Theorem 6.16 to X^{ε} and taking limits as $\varepsilon\downarrow 0$ in (6.18) and (6.19) one recovers statements (ii), (iii) and (iv) of Theorem 6.16 for compound Poisson processes but now with

$$\kappa(\alpha,\beta) = k\exp\left(\int_0^{\infty}\int_{[0,\infty)}\left(\mathrm{e}^{-t}-\mathrm{e}^{-\alpha t-\beta x}\right)\frac{1}{t}\mathbb{P}(X_t\in \mathrm{d}x)\mathrm{d}t\right)$$

(there is now closure of the interval at zero on the delimiter of the inner integral).

The reader may be curious about what would happen if we considered applying the conclusion of Theorem 6.16 to $X^{-\varepsilon}$ as $\varepsilon\downarrow 0$. In this case, using obvious notation for $\overline{G}_t^{-\varepsilon}$, it would follow that $\lim_{\varepsilon\downarrow 0}\overline{G}_t^{-\varepsilon} = \overline{G}_t^{*}$ where now

$$\overline{G}_t^{*} = \inf\{s<t : X_s = \overline{X}_t\}.$$

This pertains to another version of the Wiener–Hopf factorisation. which states that

$$(\mathbf{e}_p - \overline{G}_{\mathbf{e}_p}^{*}, \overline{X}_{\mathbf{e}_p} - X_{\mathbf{e}_p}) \stackrel{d}{=} (\underline{G}_{\mathbf{e}_p}, -\underline{X}_{\mathbf{e}_p})$$

with the new definition

$$\underline{G}_t = \sup\{s<t : X_s = \underline{X}_t\}.$$

Further, one would also have that κ satisfies (6.19) but $\widehat{\kappa}$ satisfies (6.20) but with the delimiter $(-\infty,0)$ replaced by $(-\infty,0]$.

6.5 Examples of the Wiener–Hopf Factorisation

We finish this chapter by describing some examples for which the Wiener–Hopf factorisation is explicit.

6.5.1 Brownian Motion

The simplest example of all, the Wiener–Hopf factorisation for a Brownian motion $B=\{B_t : t\geq 0\}$, verifies what has otherwise been established via different means earlier on in this text. In this case $\Psi(\theta) = \theta^2/2$ for $\theta\in\mathbb{R}$

and

$$\frac{p}{p - \mathrm{i}\vartheta + \theta^2/2} = \frac{\sqrt{2p}}{\sqrt{2p - 2\mathrm{i}\vartheta} - \mathrm{i}\theta} \cdot \frac{\sqrt{2p}}{\sqrt{2p - 2\mathrm{i}\vartheta} + \mathrm{i}\theta}.$$

From the factorisation (6.17) and the transforms given in (6.18) we know by analytic extension that

$$\Psi_p^+(\vartheta, \theta) = \frac{\kappa(p, 0)}{\kappa(p - \mathrm{i}\vartheta, -\mathrm{i}\theta)} \tag{6.28}$$

and

$$\Psi_p^-(\vartheta, \theta) = \frac{\widehat{\kappa}(p, 0)}{\widehat{\kappa}(p - \mathrm{i}\vartheta, \mathrm{i}\theta)}. \tag{6.29}$$

By inspection and the above expression for Ψ_p^-, we can identify

$$\kappa(\alpha, \beta) = \widehat{\kappa}(\alpha, \beta) = \sqrt{2\alpha} + \beta. \tag{6.30}$$

The fact that both κ and $\widehat{\kappa}$ have the same expression is obvious by symmetry. Further, (6.30) tells us that the ladder process (L^{-1}, H) is a one-sided stable-$\frac{1}{2}$ process in the case of L^{-1} and a linear drift in the case of H. This is of course to be expected when one reconsiders Example 6.12. In particular, it was shown there that L^{-1} has Laplace exponent $\Phi(\alpha) - \Phi(0)$ where Φ is the inverse of the Lévy–Khintchine exponent of B. For Brownian motion

$$\Phi(q) = \sqrt{2q} = \int_0^\infty \left(1 - \mathrm{e}^{-qx}\right)(2\pi)^{-1/2} x^{-3/2} \mathrm{d}x$$

where the second equality uses Exercise 1.4.

6.5.2 Spectrally Negative Lévy Processes

The previous example could also be seen as a consequence of the following more general analysis for spectrally negative Lévy processes. For such processes we know from Example 6.2 that we may work with the definition $L = \overline{X}$. We also know from Example 6.12 that

$$L_x^{-1} = \inf\{s > 0 : \overline{X}_s > x\} = \inf\{s > 0 : X_s > x\} = \tau_x^+$$

and

$$H_x = X_{L_x^{-1}} = x$$

on $\{x < L_\infty\}$. Hence

$$\mathbb{E}\left(\mathrm{e}^{-\alpha L_1^{-1} - \beta H_1} \mathbf{1}_{(1 < L_\infty)}\right) = \mathrm{e}^{-\Phi(\alpha) - \beta}$$

showing that we may take

$$\kappa(\alpha, \beta) = \Phi(\alpha) + \beta.$$

In that case, taking account of (6.28), one of the Wiener–Hopf factors must be

$$\frac{\Phi(p)}{\Phi(p-\mathrm{i}\vartheta)-\mathrm{i}\theta}$$

and hence the other factor must be

$$\frac{p}{\Phi(p)}\frac{\Phi(p-\mathrm{i}\vartheta)-\mathrm{i}\theta}{p-\mathrm{i}\vartheta+\Psi(\theta)}.$$

By inspection of (6.29) we see then that

$$\widehat{\kappa}(\alpha,\beta)=\frac{\alpha+\Psi(-\mathrm{i}\beta)}{\Phi(\alpha)-\beta}=\frac{\alpha-\psi(\beta)}{\Phi(\alpha)-\beta}$$

where in the second equality we have used the relation $\psi(\theta)=-\Psi(-\mathrm{i}\theta)$ between the Laplace exponent and the Lévy–Khintchine exponent. Given this expression for $\widehat{\kappa}$ however, there is little we can say in particular about the descending ladder process $(\widehat{L}^{-1},\widehat{H})$. Nonetheless, as we shall see in later chapters, the identification of the Wiener–Hopf factors does form the basis of a semi-explicit account of a number of fluctuation identities for spectrally negative processes.

6.5.3 Stable Processes

Suppose that X is an α-stable process so that for each $t>0$, X_t is equal in distribution to $t^{1/\alpha}X_1$. This has the immediate consequence that for all $t>0$,

$$\mathbb{P}(X_t\geq 0)=\rho$$

for some $\rho\in[0,1]$ known as the positivity parameter. It is possible to compute ρ in terms of the original parameters; see Zolotarev (1986) who showed that

$$\rho=\frac{1}{2}+\frac{1}{\pi\alpha}\arctan(\beta\tan\frac{\pi\alpha}{2})$$

for $\alpha\in(0,1)\cup(1,2)$ and $\beta\in[-1,1]$. For $\alpha=1$ and $\beta=0$ we clearly have $\rho=1/2$. We exclude the cases $\rho=1$ and $\rho=0$ in the subsequent discussion as these correspond to the cases that X and $-X$ are subordinators.

Note now from (6.19) that for $\lambda\geq 0$,

$$\begin{aligned}\kappa(\lambda,0)&=k\exp\left(\int_0^\infty\int_{[0,\infty)}\left(\mathrm{e}^{-t}-\mathrm{e}^{-\lambda t}\right)\frac{1}{t}\mathbb{P}(X_t\in\mathrm{d}x)\mathrm{d}t\right)\\&=k\exp\left(\int_0^\infty\left(\mathrm{e}^{-t}-\mathrm{e}^{-\lambda t}\right)\frac{\rho}{t}\mathrm{d}t\right)\\&=k\lambda^\rho\end{aligned}\qquad(6.31)$$

where in the final equality we have used the Frullani integral. This tells us directly that the process L^{-1} is itself stable. We can proceed further and calculate for $\lambda \geq 0$,

$$\begin{aligned}
\kappa(0,\lambda) &= k \exp\left(\int_0^\infty \int_{[0,\infty)} \left(\mathrm{e}^{-t} - \mathrm{e}^{-\lambda x}\right) \frac{1}{t} \mathbb{P}(X_t \in \mathrm{d}x)\mathrm{d}t\right) \\
&= k \exp\left(\int_0^\infty \frac{1}{t} \mathbb{E}\left(\left(\mathrm{e}^{-t} - \mathrm{e}^{-\lambda X_t}\right) \mathbf{1}_{(X_t \geq 0)}\right) \mathrm{d}t\right) \\
&= k \exp\left(\int_0^\infty \frac{1}{t} \mathbb{E}\left(\left(\mathrm{e}^{-t} - \mathrm{e}^{-\lambda t^{1/\alpha} X_1}\right) \mathbf{1}_{(X_1 \geq 0)}\right) \mathrm{d}t\right) \\
&= k \exp\left(\int_0^\infty \frac{1}{s} \mathbb{E}\left(\left(\mathrm{e}^{-s\lambda^{-\alpha}} - \mathrm{e}^{-s^{1/\alpha} X_1}\right) \mathbf{1}_{(X_1 \geq 0)}\right) \mathrm{d}s\right) \\
&= k \exp\left(\int_0^\infty \frac{1}{s} \mathbb{E}\left(\left(\mathrm{e}^{-s} - \mathrm{e}^{-X_s}\right) \mathbf{1}_{(X_s > 0)}\right) \mathrm{d}s\right) \\
&\quad \times \exp\left(-\int_0^\infty \frac{\rho}{s} \left(\mathrm{e}^{-s} - \mathrm{e}^{-s\lambda^{-\alpha}}\right) \mathrm{d}s\right) \\
&= \kappa(0,1) \times \exp\left(-\int_0^\infty \frac{1}{s} \left(\mathrm{e}^{-s} - \mathrm{e}^{-s\lambda^{-\alpha}}\right) \mathrm{d}s\right) \\
&= \kappa(0,1)\lambda^{\alpha\rho},
\end{aligned}$$

where in the third and fifth equality we have used the fact that $s^{1/\alpha} X_1$ is equal in distribution to X_s. The term $\kappa(0,1)$ is nothing more than a constant and hence we deduce that the ascending ladder height process is also a stable process of index $\alpha\rho$. It is now immediate from (6.21) that the descending ladder height process is a stable process of index $\alpha(1-\rho)$ which is consistent with the fact that $\mathbb{P}(X_t \leq 0) = 1 - \rho$. Note that this necessarily implies that $0 < \alpha\rho \leq 1$ and $0 \leq \alpha(1-\rho) \leq 1$ when $\rho \in (0,1)$. The extreme cases $\alpha\rho = 1$ and $\alpha(1-\rho) = 1$ correspond to spectrally negative and spectrally positive processes, respectively. For example, when $\beta = -1$ and $\alpha \in (1,2)$ we have a spectrally negative process of unbounded variation. It is easily checked that $\rho = 1/\alpha$ and hence from the calculation above $\kappa(0,\lambda) = \text{const.} \times \lambda$ consistently with earlier established facts for spectrally negative Lévy processes. Note that $\kappa(0,0) = \widehat{\kappa}(0,0) = 0$ showing that the killing rates in the ascending and descending ladder height processes are equal to zero and hence

$$\limsup_{t \uparrow \infty} X_t = -\liminf_{t \uparrow \infty} X_t = \infty.$$

Unfortunately it is not as easy to establish a closed expression for the bivariate exponent κ. The only known results in this direction are those of Doney (1987) who only deals with a set of parameter values of α and β which are dense in the full parameter range $(0,2)$ and $[-1,1]$. The expressions obtained by Doney are quite complicated however and we refrain from including them here.

6.5.4 Other Examples

Other than the stable processes, there are very few examples where one knows of explicit details of the Wiener–Hopf factorisation for processes which have both positive and negative jumps. One example, found in Feller (1971), is the case that X is the difference of a compound Poisson process with exponentially distributed jumps and another independent subordinator. The reason why the latter example is, up to a certain point, analytically tractable boils down to the fact that the ascending ladder height process must again be a (possibly killed) compound Poisson subordinator with exponential jumps. This is obvious when one considers that if a new maxima is incurred then it is achieved by an exponentially distributed jump and hence by the lack of memory property, the overshoot over the previous maximum must again be exponentially distributed. Knowing the nature of the ascending ladder height process allows one to compute the factor $\Psi_p^+(0,\theta)$ and hence establish the function $\kappa(\alpha,\beta)$ as well as the $\widehat{\kappa}(\alpha,\beta)$ from the factorisation (6.17).

There is another class of distributions which observe a property similar in nature to the lack of memory property and these are phase-type distributions. A distribution F on $(0,\infty)$ is *phase-type* if it is the distribution of the absorption time in a finite state continuous time Markov process $J=\{J_t : t\geq 0\}$ with one state Δ absorbing and the remaining ones $1,...,m$ transient. The parameters of this system are m, the restriction $\mathbf{T}$ of the full intensity matrix to the m transient states and the initial probability (row) vector $\mathbf{a}=(a_1,...,a_m)$ where $a_i=\mathbb{P}(J_0=i)$. For any $i=1,...,m$, let t_i be the intensity of the transition $i\to\Delta$ and write $\mathbf{t}$ for the column vector of intensities into Δ. It follows that

$$1-F(x)=\mathbf{a}e^{x\mathbf{T}}\mathbf{u}$$

where $\mathbf{u}$ is the $m\times 1$ vector with unit entries. An easy differentiation shows that the density is given by $f(x)=\mathbf{a}e^{\mathbf{T}x}\mathbf{t}$.

The "lack of memory"-type property which phase-type distributions enjoy can be expressed as follows. Suppose that $\{P_j : j=1,...,m\}$ are the respective probabilities of J when started from state $j=1,...,m$ and write $P_{\mathbf{a}}=\sum_{j=1}^m a_j P_j$ for the probability of J with randomised initial state having distribution $\mathbf{a}$. Let τ^Δ be the absorption time into state Δ of J. Then for $t,s\geq 0$,

$$P_{\mathbf{a}}(\tau^\Delta>t+s|\tau^\Delta>t)=P_{J_t}(\tau^\Delta>s)$$

which is again a phase-type distribution but with the vector $\mathbf{a}$ replaced by the vector which has a 1 in position J_t and otherwise zeros for the other entries.

Recently, Mordecki (2002) and Asmussen et al. (2004) have used this property to establish another example of a Lévy process for which elements of the Wiener–Hopf factorisation can be expressed in more mathematical detail. See also Pistorius (2006) and Mordecki (2005). The process they consider takes the form

$$X_t = X_t^{(+)} - \sum_{j=1}^{N(t)} U_j, \quad t \geq 0$$

where $\{X_t^{(+)} : t \geq 0\}$ is a Lévy process which has no negative jumps, $\{N_t : t \geq 0\}$ is an independent Poisson process with rate λ and $\{U_j : j \geq 1\}$ are i.i.d. random variables with a common phase-type distribution F. The idea exploited by the aforementioned authors is that when X obtains a new maximum, it either does so continuously on account of the process $X^{(+)}$ or with a jump which is phase-type distributed. Suppose this jump is of length $s + t$ where t is the undershoot and s is the overshoot. If one thinks of this jump as the time it takes for absorption of an auxiliary Markov chain, J, from a randomised initial position with distribution $\mathbf{a}$, then s corresponds to the time it takes to absorption from a randomised initial position J_t. In conclusion, the overshoot must be again phase-type distributed but with a different parameter set. The above phase-type processes thus offer sufficient mathematical structure for the Laplace exponent $\widehat{\kappa}(\alpha, \beta)$ to be computed in more detail. We give a brief indication of the structure of $\widehat{\kappa}(\alpha, \beta)$ below without going into details however for the sake of brevity.

The characteristic exponent thus takes the form

$$\Psi(\theta) = a\mathrm{i}\theta + \frac{1}{2}\sigma^2\theta^2 + \int_{(0,\infty)} (1 - \mathrm{e}^{\mathrm{i}\theta x} - \mathrm{i}\theta x \mathbf{1}_{(x<1)})\Pi(\mathrm{d}x) + \lambda(\mathbf{E}(\mathrm{e}^{\mathrm{i}\theta U_1}) - 1)$$

for $\theta \in \mathbb{R}$. Since $\mathbf{E}(\mathrm{e}^{\mathrm{i}\theta U_1}) = -\mathbf{a}(\mathrm{i}\theta\mathbf{I} + \mathbf{T})^{-1}\mathbf{t}$, where $\mathbf{I}$ is the identity matrix, and the latter expression is effectively the ratio of two complex polynomials, it can be shown that Ψ, can be analytically extended to $\{z \in \mathbb{C} : \Re z \leq 0\}$ with the exception of a finite number of poles (the eigenvalues of $\mathbf{T}$). Define for each $\alpha > 0$, the finite set of roots with negative real part

$$\mathcal{I}_\alpha = \{\rho_i : \Psi(\rho_i) + \alpha = 0 \text{ and } \Re\rho_i < 0\},$$

where multiple roots are counted individually. Next, define for each $\alpha > 0$, a second set of roots with negative real part

$$\mathcal{J}_\alpha = \left\{\eta_i : \frac{1}{\alpha + \Psi(\eta_i)} = 0 \text{ and } \Re\eta_i < 0\right\}$$

again taking multiplicity into account. Asmussen et al. (2004) show that

$$\widehat{\kappa}(\alpha, \beta) = \frac{\prod_{i\in\mathcal{I}_\alpha}(\beta - \rho_i)}{\prod_{i\subset\mathcal{J}_\alpha}(\beta - \eta_i)}$$

thus giving a semi-explicit expression for the Wiener–Hopf factor,

$$\Psi_p^-(\vartheta, \theta) = \frac{\prod_{i\in\mathcal{I}_p}(-\rho_i)\prod_{i\in\mathcal{I}_{p-\mathrm{i}\vartheta}}(-\mathrm{i}\theta - \rho_i)}{\prod_{i\in\mathcal{J}_p}(-\eta_i)\prod_{i\in\mathcal{J}_{p-\mathrm{i}\vartheta}}(-\mathrm{i}\theta - \eta_i)}$$

for $\Re s \geq 0$.

Earlier examples of Wiener–Hopf factorisations for which both factors take a rational form as above were investigated by Borovkov (1976) and Feller (1971).

6.6 Brief Remarks on the Term "Wiener–Hopf"

Having now completed our exposition of the Wiener–Hopf factorisation, the reader may feel somewhat confused as to the association of the name "Wiener–Hopf" with Theorem 6.16. Indeed, in our presentation we have made no reference to works of Wiener nor Hopf. The connection between Theorem 6.16 and these two scientists lies with their analytic study of the solutions to integral equations of the form

$$Q(x) = \int_0^\infty Q(y)f(x-y)\mathrm{d}y, \qquad x > 0 \tag{6.32}$$

where $f : \mathbb{R} \to [0,\infty)$ is a pre-specified kernel; see Payley and Wiener (1934) and Hopf (1934). If one considers a compound Poisson process X which has the property that $\limsup_{t\uparrow\infty} X_t < \infty$ then the Strong Markov Property implies that $\overline{X}_\infty$ is equal in distribution to $(\xi + \overline{X}_\infty) \vee 0$ where ξ is independent of $\overline{X}_\infty$ and has the same distribution as the jumps of X. If the latter jump distribution has density f then one shows easily that $H(x) = \mathbb{P}(\overline{X}_\infty \leq x)$ satisfies

$$H(x) = \int_{-\infty}^x H(x-y)f(y)\mathrm{d}y = \int_0^\infty f(x-y)H(y)\mathrm{d}y$$

and hence one obtains immediately the existence of a solution to (6.32) for the given f. This observation dates back to the work of Spitzer (1957).

Embedded in the complex analytic techniques used to analyse (6.32) and generalisations thereof by Wiener, Hopf and many others that followed, are factorisations of operators (which can take the form of Fourier transforms). In the probabilistic setting here, this is manifested in the form of the independence seen in Theorem 6.16 (i) and how this is used to identify the factors Ψ^+ and Ψ^- in conjunction with analytic extension in the proof of part (iii) of the same theorem. The full extent of the analytic Wiener–Hopf factorisation techniques go far beyond the current setting and we make no attempt to expose them here. The name "Wiener–Hopf" factorisation thus appears as a rather obscure feature of what may otherwise be considered as a sophisticated path decomposition of a Lévy process.

Exercises

6.1. Give an example of a Lévy process which has bounded variation with zero drift for which 0 is regular for both $(0,\infty)$ and $(-\infty,0)$. Give an example of a Lévy process of bounded variation and zero drift for which 0 is only regular for $(0,\infty)$.

6.2. Suppose that X is a spectrally negative Lévy process of unbounded variation with Laplace exponent ψ and recall the definition $\tau_x^+ = \inf\{t > 0 : X_t > x\}$. Recall also that the process $\tau^+ := \{\tau_x^+ : x \geq 0\}$ is a (possibly killed) subordinator (see Corollary 3.14) with Laplace exponent Φ, the right inverse of ψ.

(i) Suppose that d is the drift of the process τ^+. Show that necessarily $d = 0$.
(ii) Deduce that

$$\lim_{x\downarrow 0} \frac{\tau_x^+}{x} = 0$$

and hence that

$$\limsup_{t\downarrow 0} \frac{X_t}{t} = \infty.$$

Conclude from the latter that 0 is regular for $(0,\infty)$ and hence that the jump measure of τ^+ cannot be finite.
(iii) From the Wiener–Hopf factorisation of X show that

$$\lim_{\theta\uparrow\infty} \mathbb{E}(\mathrm{e}^{\theta \underline{X}_{\mathbf{e}_q}}) = 0$$

and hence use this to give an alternative proof that 0 is regular for $(0,\infty)$.

6.3. Fix $\rho \in (0,1]$. Show that a compound Poisson subordinator with jump rate $\lambda\rho$ killed at an independent and exponentially distributed time with parameter $\lambda(1-\rho)$ is equal in law to a compound Poisson subordinator killed after an independent number of jumps which is distributed geometrically with parameter $1-\rho$.

6.4. Show that the only processes for which

$$\int_0^\infty \mathbf{1}_{(\overline{X}_t = X_t)}\mathrm{d}t > 0 \text{ and } \int_0^\infty \mathbf{1}_{(X_t = \underline{X}_t)}\mathrm{d}t > 0$$

almost surely are compound Poisson processes.

6.5. Suppose that X is spectrally negative with Lévy measure Π and Gaussian coefficient σ and suppose that $\mathbb{E}(X_t) > 0$. (Recall that in general $\mathbb{E}(X_t) \in [-\infty,\infty)$.)

(i) Show that

$$\int_{-\infty}^{-1} \Pi(-\infty,x)\mathrm{d}x < \infty.$$

(ii) Using Theorem 6.16 (iv) deduce that, up to a constant,

$$\widehat{\kappa}(0, \mathrm{i}\theta) = \left(-a + \int_{(-\infty,-1)} x\Pi(\mathrm{d}x)\right)$$
$$-\frac{1}{2}\mathrm{i}\theta\sigma^2 + \int_{(-\infty,0)} (1 - \mathrm{e}^{\mathrm{i}\theta x})\Pi(-\infty, x)\mathrm{d}x.$$

Hence deduce that there exists a choice of local time at the maximum for which the descending ladder height process has jump measure given by $\Pi(-\infty, -x)\mathrm{d}x$ on $(0, \infty)$, drift $\sigma^2/2$ and is killed at rate $\mathbb{E}(X_1)$.

6.6. Suppose that X is a spectrally negative stable process of index $1 < \alpha < 2$.

(i) Deduce with the help of Theorem 3.12 that up to a multiplicative constant

$$\kappa(\theta, 0) = \theta^{1/\alpha}$$

and hence that $\mathbb{P}(X_t \geq 0) = 1/\alpha$ for all $t \geq 0$.

(ii) By reconsidering the Wiener–Hopf factorisation, show that for each $t \geq 0$ and $\theta \geq 0$,

$$\mathbb{E}(\mathrm{e}^{-\theta \overline{X}_t}) = \sum_{n=0}^{\infty} \frac{(-\theta t^{1/\alpha})^n}{\Gamma(1 + n/\alpha)}.$$

This identity is taken from Bingham (1971, 1972).

6.7 (The second factorisation identity). In this exercise we derive what is commonly called the *second factorisation identity* which is due to Percheskii and Rogozin (1969). It uses the Laplace exponents κ and $\widehat{\kappa}$ to give an identity concerning the problem of first passage above a fixed level $x \in \mathbb{R}$. The derivation we use here makes use of calculations in Darling et al. (1972) and Alili and Kyprianou (2004). We shall use the derivation of this identity later to solve some optimal stopping problems.

Define as usual

$$\tau_x^+ = \inf\{t > 0 : X_t > x\}$$

where X is any Lévy process.

(i) Using the same technique as in Exercise 5.5, prove that for all $\alpha > 0$, $\beta \geq 0$ and $x \in \mathbb{R}$ we have

$$\mathbb{E}\left(\mathrm{e}^{-\alpha\tau_x^+ - \beta X_{\tau_x^+}} \mathbf{1}_{(\tau_x^+ < \infty)}\right) = \frac{\mathbb{E}\left(\mathrm{e}^{-\beta \overline{X}_{\mathbf{e}_\alpha}} \mathbf{1}_{(\overline{X}_{\mathbf{e}_\alpha} > x)}\right)}{\mathbb{E}\left(\mathrm{e}^{-\beta \overline{X}_{\mathbf{e}_\alpha}}\right)}. \tag{6.33}$$

(ii) Establish the second factorisation identity as follows. If X is not a subordinator then

$$\int_0^\infty \mathrm{e}^{-qx}\mathbb{E}\left(\mathrm{e}^{-\alpha\tau_x^+ - \beta(X_{\tau_x^+} - x)} \mathbf{1}_{(\tau_x^+ < \infty)}\right) \mathrm{d}x = \frac{\kappa(\alpha, q) - \kappa(\alpha, \beta)}{(q - \beta)\kappa(\alpha, q)}.$$

6.8. Suppose that X is any Lévy process which is not a subordinator and $\mathbf{e}_p$ is an independent random variable which is exponentially distributed with parameter $p > 0$. Note that 0 is regular for $(0, \infty)$ if and only if $P(\overline{X}_{\mathbf{e}_p} = 0) = 0$ where $\mathbf{e}_p$ is an independent exponential random variable with parameter p. Use the Wiener–Hopf factorisation to show that 0 is regular for $(0, \infty)$ if and only if

$$\int_0^1 \frac{1}{t}\mathbb{P}(X_t > 0)\mathrm{d}t = \infty.$$

Now noting that 0 is irregular for $[0, \infty)$ if and only if $\mathbb{P}(\overline{G}_{\mathbf{e}_p} = 0) > 0$. Show that 0 is regular for $[0, \infty)$ if and only if

$$\int_0^1 \frac{1}{t}\mathbb{P}(X_t \geq 0)\mathrm{d}t = \infty.$$

6.9. This exercise gives the random walk analogue of the Wiener–Hopf factorisation. In fact this is the original setting of the Wiener–Hopf factorisation. We give the formulation in Greenwood and Pitman (1980a). However, one may also consult Feller (1971) and Borovkov (1976) for other accounts.

Suppose that under P, $S = \{S_n : n \geq 0\}$ is a random walk with $S_0 = 0$ and increment distribution F. We assume that S can jump both upwards and downwards, in other words $\min\{F(-\infty, 0), F(0, \infty)\} > 0$ and that F has no atoms. Denote by $\mathbf{\Gamma}_p$ an independent random variable which has a geometric distribution with parameter $p \in (0, 1)$ and let

$$G = \min\{k = 0, 1, ..., \mathbf{\Gamma}_p : S_k = \max_{j=1,\ldots,\mathbf{\Gamma}_p} S_j\}.$$

Note that S_G is the last maximum over times $\{0, 1, ..., \mathbf{\Gamma}_p\}$. Define $N = \inf\{n > 0 : S_n > 0\}$ the first passage time into $(0, \infty)$ or equivalently the first strict ladder time. Our aim is to characterise the joint laws (G, S_G) and (N, S_N) in terms of F, the basic data of the random walk.

(i) Show that (even without the restriction that $\min\{F(0, \infty), F(-\infty, 0)\} > 0$),

$$E(s^{\mathbf{\Gamma}_p}\mathrm{e}^{\mathrm{i}\theta S_{\mathbf{\Gamma}_p}}) = \exp\left\{-\int_{\mathbb{R}}\sum_{n=1}^{\infty}(1 - s^n\mathrm{e}^{\mathrm{i}\theta x})q^n\frac{1}{n}F^{*n}(\mathrm{d}x)\right\}$$

where $0 < s \leq 1$, $\theta \in \mathbb{R}$, $q = 1 - p$ and E is expectation under P. Deduce that the pair $(\mathbf{\Gamma}_p, S_{\mathbf{\Gamma}_p})$ is infinitely divisible.

(ii) Let ν be an independent random variable which is geometrically distributed on $\{0, 1, 2, ...\}$ with parameter $P(N > \mathbf{\Gamma}_p)$. Using a path decomposition in terms of excursions from the maximum, show that the pair (G, S_G) is equal in distribution to the component-wise sum of ν independent copies of (N, S_N) conditioned on the event $\{N \leq \mathbf{\Gamma}_p\}$ and hence form an infinitely divisible 2D distribution.

(iii) Show that (G, S_G) and $(\mathbf{\Gamma}_p - G, S_{\mathbf{\Gamma}_p} - S_G)$ are independent. Further, that the latter pair is equal in distribution to (D, S_D) where

$$D = \max\{k = 0, 1, ..., \mathbf{\Gamma}_p : S_k = \min_{j=1,\ldots,\mathbf{\Gamma}_p} S_j\}.$$

(iv) Deduce that

$$E(s^G \mathrm{e}^{\mathrm{i}\theta S_G}) = \exp\left\{ - \int_{(0,\infty)} \sum_{n=1}^{\infty} (1 - s^n \mathrm{e}^{\mathrm{i}\theta x}) q^n \frac{1}{n} F^{*n}(\mathrm{d}x) \right\}$$

for $0 < s \leq 1$ and $\theta \in \mathbb{R}$. Note when $s = 1$, this identity was established by Spitzer (1956).

(v) Show that

$$E(s^G \mathrm{e}^{\mathrm{i}\theta S_G}) = \frac{P(\mathbf{\Gamma}_p < N)}{1 - E((qs)^N \mathrm{e}^{\mathrm{i}\theta S_N})}$$

and hence deduce that

$$\frac{1}{1 - E(s^N \mathrm{e}^{\mathrm{i}\theta S_N})} = \exp\left\{ \int_{(0,\infty)} \sum_{n=1}^{\infty} s^n \mathrm{e}^{\mathrm{i}\theta x} \frac{1}{n} F^{*n}(\mathrm{d}x) \right\}.$$

According to Feller (1971), this is known as Baxter's identity.

7

Lévy Processes at First Passage and Insurance Risk

This chapter is devoted to studying how the Wiener–Hopf factorisation can be used to characterise the behaviour of any Lévy process at first passage over a fixed level. The case of a subordinator will be excluded throughout this chapter as this has been dealt with in Chap. 5. Nonetheless, the understanding of how subordinators make first passage will play a crucial role in understanding the case of a general Lévy process.

Recall from Sects. 1.3.1 and 2.7.1 that a natural extension of the classical Cramér–Lundberg model for risk insurance is that of a spectrally negative Lévy process. In that case, if we take X to model the negative of the capital of an insurance firm, then due to spatial homogeneity a characterisation of first passage of X at $x > 0$ corresponds precisely to a characterisation of ruin when there is an initial capital at time zero of x units.

To some extent, the results we present on the first passage problem suffer from a lack of analytical explicitness which is due to the same symptoms being present in the Wiener–Hopf factorisation. None the less there is sufficient mathematical structure to establish qualitative statements concerning the characterisation of first passage, or equivalently ruinous behaviour of insurance risk processes. This becomes more apparent when looking at asymptotic properties of the established characterisations.

We start by looking at the general Lévy risk insurance model, when the probability of ruin is strictly less than one and further we consider a version of Cramér's estimate of ruin for a reasonably general class of Lévy processes. Then we move to a full characterisation of the ruin problem and conclude with some results concerning the asymptotic behaviour of ruin for a special class of insurance risk models.

7.1 Drifting and Oscillating

Suppose that X is any Lévy process, then define as usual

$$\tau_x^+ = \inf\{t > 0 : X_t > x\}$$

for $x \in \mathbb{R}$. If we now define $-X$ as the capital of an insurance firm, then the probability of ruin when starting from an initial capital of $x > 0$ is given by $\mathbb{P}(\tau_x^+ < \infty)$. In this section we shall establish precisely when the probability of ruin is strictly less than one. Further, in the latter case, we shall give sufficient conditions under which the ruin probability decays exponentially as $x \uparrow \infty$; that is to say we handle the case of Cramér's estimate of ruin.

Suppose now that $H = \{H_t : t \geq 0\}$ is the ascending ladder height process of X. If

$$T_x^+ = \inf\{t > 0 : H_t > x\},$$

then quite clearly

$$\mathbb{P}(\tau_x^+ < \infty) = \mathbb{P}(T_x^+ < \infty). \tag{7.1}$$

Recall from Theorem 6.10 that the process H behaves like a subordinator, possibly killed at an independent and exponentially distributed time. The criterion for killing is that $\mathbb{P}(\limsup_{t\uparrow\infty} X_t < \infty) = 1$. Suppose the latter fails. Then the probability on the right-hand side of (7.1) is equal to 1. If on the other hand there is killing, then since killing can occur at arbitrarily small times with positive probability, then $\mathbb{P}(T_x^+ < \infty) < 1$. In conclusion we know that for any $x > 0$,

$$\mathbb{P}(\text{ruin from initial reserve } x) < 1 \Leftrightarrow \mathbb{P}(\limsup_{t\uparrow\infty} X_t < \infty) = 1. \tag{7.2}$$

We devote the remainder of this section then to establishing conditions under which $\mathbb{P}(\lim\ \sup_{t\uparrow\infty} X_t < \infty) = 1$.

Theorem 7.1. *Suppose that X is a Lévy process.*

(i) If $\int_1^\infty t^{-1}\mathbb{P}(X_t \geq 0)\mathrm{d}t < \infty$ then

$$\lim_{t\uparrow\infty} X_t = -\infty$$

almost surely and X is said to drift to $-\infty$.

(ii) If $\int_1^\infty t^{-1}\mathbb{P}(X_t \leq 0)\mathrm{d}t < \infty$ then

$$\lim_{t\uparrow\infty} X_t = \infty$$

almost surely and X is said to drift to ∞.

(iii) If both the integral tests in (i) and (ii) fail, [1] *then*

$$\limsup_{t\uparrow\infty} X_t = -\liminf_{t\uparrow\infty} X_t = \infty$$

almost surely and X is said to oscillate.

---

[1]Note that $\int_1^\infty t^{-1}\mathbb{P}(X_t \geq 0)dt + \int_1^\infty t^{-1}\mathbb{P}(X_t \leq 0)dt \geq \int_1^\infty t^{-1}\mathrm{d}t = \infty$ and hence at least one of the integral tests in (i) or (ii) fails.

Proof. We follow a similar proof to the one given in Bertoin (1996).

(i) From Theorem 6.16 (see also the discussion at the end of Sect. 6.4 concerning the adjusted definitions of $\overline{G}_\infty$ and $\underline{G}_\infty$ for the case of compound Poisson processes) we have for all $\alpha \geq 0$,

$$\mathbb{E}\left(\mathrm{e}^{-\alpha \overline{G}_{\mathbf{e}_p}}\right) = \exp\left\{-\int_0^\infty (1-\mathrm{e}^{-\alpha t})\frac{1}{t}\mathrm{e}^{-pt}\mathbb{P}(X_t \geq 0)\mathrm{d}t\right\}. \qquad (7.3)$$

Letting p tend to zero in (7.3) and applying the Dominated Convergence Theorem on the left-hand side and the Monotone Convergence Theorem on the right-hand side we see that

$$\mathbb{E}\left(\mathrm{e}^{-\alpha \overline{G}_\infty}\right) = \exp\left\{-\int_0^\infty (1-\mathrm{e}^{-\alpha t})\frac{1}{t}\mathbb{P}(X_t \geq 0)\mathrm{d}t\right\}. \qquad (7.4)$$

If $\int_1^\infty t^{-1}\mathbb{P}(X_t \geq 0)\mathrm{d}t < \infty$ then since $0 \leq (1-\mathrm{e}^{-\alpha t}) \leq 1 \wedge t$ for all sufficiently small α, we see that

$$\begin{aligned}\int_0^\infty (1-\mathrm{e}^{-\alpha t})\frac{1}{t}\mathbb{P}(X_t \geq 0)\mathrm{d}t &\leq \int_0^\infty (1 \wedge t)\,\frac{1}{t}\mathbb{P}(X_t \geq 0)\mathrm{d}t \\ &\leq \int_1^\infty \frac{1}{t}\mathbb{P}(X_t \geq 0)\mathrm{d}t + \int_0^1 \mathbb{P}(X_t \geq 0)\mathrm{d}t < \infty.\end{aligned}$$

Hence once again appealing to the Dominated Convergence Theorem, taking α to zero in (7.4), it follows that

$$\lim_{\alpha \downarrow 0}\int_0^\infty (1-\mathrm{e}^{-\alpha t})\frac{1}{t}\mathbb{P}(X_t \geq 0)\mathrm{d}t = 0$$

and therefore $\mathbb{P}\left(\overline{G}_\infty < \infty\right) = 1$. This implies that $\mathbb{P}\left(\overline{X}_\infty < \infty\right) = 1$.

Now noting that $\int_1^\infty t^{-1}\mathbb{P}(X_t \geq 0)\mathrm{d}t = \int_1^\infty t^{-1}(1-\mathbb{P}(X_t < 0))\mathrm{d}t < \infty$, since $\int_1^\infty t^{-1}\mathrm{d}t = \infty$, we are forced to conclude that

$$\int_1^\infty \frac{1}{t}\mathbb{P}(X_t < 0)\mathrm{d}t = \infty.$$

The Wiener–Hopf factorisation also gives us

$$\mathbb{E}\left(\mathrm{e}^{-\alpha \underline{G}_{\mathbf{e}_p}}\right) = \exp\left\{-\int_0^\infty (1-\mathrm{e}^{-\alpha t})\frac{1}{t}e^{-pt}\mathbb{P}(X_t \leq 0)\mathrm{d}t\right\}.$$

Taking limits as $p \downarrow 0$ and noting that

$$\int_0^\infty (1-\mathrm{e}^{-\alpha t})\frac{1}{t}\mathbb{P}(X_t \leq 0)\mathrm{d}t \geq k\int_1^\infty \frac{1}{t}\mathbb{P}(X_t \leq 0)\mathrm{d}t = \infty$$

for some appropriate constant $k > 0$, we get $\mathbb{P}\left(\underline{G}_\infty = \infty\right) = 1$. Equivalently $\mathbb{P}\left(\underline{X}_\infty = -\infty\right) = 1$.

We have proved that $\limsup_{t\uparrow\infty} X_t < \infty$ and $\liminf_{t\uparrow\infty} X_t = -\infty$ almost surely. This means that

$$\tau^-_{-x} := \inf\{t > 0 : X_t < -x\}$$

is almost surely finite for each $x > 0$. Note that

$$\{X_t > x/2 \text{ for some } t > 0\} = \{\overline{X}_\infty > x/2\}$$

and hence, since $\mathbb{P}\left(\overline{X}_\infty < \infty\right) = 1$, for each $1 > \varepsilon > 0$, there exists an $x_\varepsilon > 0$ such that for all $x > x_\varepsilon$,

$$\mathbb{P}\left(X_t > x/2 \text{ for some } t > 0\right) < \varepsilon.$$

Since τ^-_{-x} is a stopping time which is almost surely finite, we can use the previous estimate together with the Strong Markov Property and conclude that for all $x > x_\varepsilon$,

$$\begin{aligned}&\mathbb{P}\left(X_t > -x/2 \text{ for some } t > \tau^-_{-x}\right)\\ &\leq \mathbb{P}\left(X_t > x/2 \text{ for some } t > 0\right) < \varepsilon.\end{aligned}$$

This gives us the uniform estimate for $x > x_\varepsilon$,

$$\begin{aligned}\mathbb{P}\left(\limsup_{t\uparrow\infty} X_t \leq -x/2\right) &\geq \mathbb{P}\left(X_t \leq -x/2 \text{ for all } t > \tau^-_{-x}\right)\\ &\geq 1 - \varepsilon.\end{aligned}$$

Since both x may be taken arbitrarily large and ε may be taken arbitrarily close to 0 the proof of part (i) is complete.

(ii) The second part follows from the first part applied to $-X$.

(iii) The argument in (i) shows that in fact when $\int_1^\infty t^{-1}\mathbb{P}(X_t \leq 0)\mathrm{d}t = \int_1^\infty t^{-1}\mathbb{P}(X_t \geq 0)\mathrm{d}t = \infty$, then $-\underline{X}_\infty = \overline{X}_\infty = \infty$ almost surely and the assertion follows. □

Whilst the last theorem shows that there are only three types of asymptotic behaviour, the integral tests which help to distinguish between the three cases are not particularly user friendly. What would be more appropriate is a criterion in terms of the triple (a, σ, Π). The latter is provided by Chung and Fuchs (1951) and Erickson (1973) for random walks; see also Bertoin (1997). To state their criteria, recall from Theorem 3.8 and Exericse 3.3 that the mean of X_1 is well defined if and only if

$$\int_1^\infty x\Pi(\mathrm{d}x) < \infty \text{ or } \int_{-\infty}^{-1} |x|\Pi(\mathrm{d}x) < \infty$$

in which case $\mathbb{E}(X_1) \in [-\infty, \infty]$. When both the above integrals are infinite the mean $\mathbb{E}(X_1)$ is undefined.

Theorem 7.2. *Suppose that X is a Lévy process with characteristic measure Π.*

(i) If $\mathbb{E}(X_1)$ is defined and valued in $[-\infty, 0)$ or if $\mathbb{E}(X_1)$ is undefined and

$$\int_1^\infty \frac{x\Pi(\mathrm{d}x)}{\int_0^x \Pi(-\infty, -y)\mathrm{d}y} < \infty,$$

then

$$\lim_{t\uparrow\infty} X_t = -\infty.$$

(ii) If $\mathbb{E}(X_1)$ is defined and valued in $(0, \infty]$ or if $\mathbb{E}(X_1)$ is undefined and

$$\int_{-\infty}^{-1} \frac{|x|\Pi(\mathrm{d}x)}{\int_0^{|x|} \Pi(y, \infty)\mathrm{d}y} < \infty,$$

then

$$\lim_{t\uparrow\infty} X_t = \infty.$$

(iii) If $\mathbb{E}(X_1)$ is defined and equal to zero or if $\mathbb{E}(X_1)$ is undefined and both of the integral tests in part (i) and (ii) fail, then

$$\limsup_{t\uparrow\infty} X_t = -\liminf_{t\uparrow\infty} X_t = \infty.$$

We give no proof here of this important result, although one may consult Exercise 7.2 for related results which lean on the classical Strong Law of Large Numbers.

It is interesting to compare the integral tests in the above theorem with those of Theorem 6.5. It would seem that the issue of regularity 0 for the half line may be seen as the "small time" analogue of drifting or oscillating, however there is no known formal path-wise connection.

In the case that X is spectrally negative, thanks to the finiteness and convexity of its Laplace exponent $\psi(\theta) = \log \mathbb{E}(\mathrm{e}^{\theta X_1})$ on $\theta \geq 0$ (see Exercise 3.5), one always has that $\mathbb{E}(X_1) \in [-\infty, \infty)$. That is to say, the asymptotic behaviour of a spectrally negative Lévy process can always be determined from its mean or equivalently $\psi'(0+)$. See Exercise 7.3 which shows how to derive this conclusion from Theorem 7.1 and the Wiener–Hopf factorisation.

On account of the trichotomy of drifting and oscillating, we may now revise the statement (7.2) as

$$\mathbb{P}(\text{ruin from initial capital } x) < 1 \Leftrightarrow \mathbb{P}(\lim_{t\uparrow\infty} X_t = -\infty) = 1.$$

We close this section by making some brief remarks on the link between drifting and oscillating and another closely related dichotomy known as transience and recurrence which is often discussed within the more general context of Markov Processes.

Definition 7.3. *A Lévy process X is said to be transient if for all $a > 0$,*

$$\mathbb{P}\left(\int_0^\infty \mathbf{1}_{(|X_t|<a)}\mathrm{d}t < \infty\right) = 1$$

and recurrent if for all $a > 0$,

$$\mathbb{P}\left(\int_0^\infty \mathbf{1}_{(|X_t|<a)}\mathrm{d}t = \infty\right) = 1$$

In the previous definition, the requirements for transience and recurrence may appear quite strong as in principle, the given probabilities could be less than one. The events in the definition however belong to the tail sigma algebra $\bigcap_{t\in\mathbb{Q}\cap[0,\infty)} \sigma(X_s : s \geq t)$ and hence according to Kolmogorov's zero-one can only have probabilities equal to zero or one. Nonetheless, one could argue that it could be the case for example that for small a it is the case that $\mathbb{P}(\int_0^\infty \mathbf{1}_{(|X_t|<a)}\mathrm{d}t = \infty) = 0$ and for large values of a the same probability is 1. It turns out that Lévy processes always adhere to one of the two cases given in the definition as the following classic analytic dichotomy due to Port and Stone (1971)[2] confirms.

Theorem 7.4. *Suppose that X is a Lévy process with characteristic exponent Ψ, then it is transient if and only if for some sufficiently small $\varepsilon > 0$,*

$$\int_{(-\varepsilon,\varepsilon)} \Re\left(\frac{1}{\Psi(\theta)}\right) \mathrm{d}\theta < \infty$$

and otherwise it is recurrent.

Probabilistic reasoning also leads to the following interpretation of the dichotomy.

Theorem 7.5. *Suppose that X is any Lévy process.*

(i) Then, X is transient if and only if

$$\lim_{t\uparrow\infty} |X_t| = \infty$$

almost surely.

(ii) Suppose that X is not a compound Poisson process, then X is recurrent if and only if for all $x \in \mathbb{R}$,

$$\liminf_{t\uparrow\infty} |X_t - x| = 0 \tag{7.5}$$

almost surely.

[2] Theorem 7.4 is built on the strength of the original, but weaker result of Chung and Fuchs (1951). See also Kingman (1964).

The reason for the exclusion of compound Poisson processes in part (ii) can be seen when one considers the case of a compound Poisson process where the jump distribution is supported on a lattice, say $\delta\mathbb{Z}$ for some $\delta > 0$. In that case it is clear that the set of points visited will be a subset of $\delta\mathbb{Z}$ and (7.5) no longer makes sense. Otherwise part (ii) says that recurrence is equivalent to the path of X approaching any given point arbitrarily closely over an infinite time horizon. Note that this behaviour does not imply that for a given $x \in \mathbb{R}$, $\mathbb{P}(\tau^{\{x\}} < \infty) > 0$ where $\tau^{\{x\}} = \inf\{t > 0 : X_t = x\}$. Precisely when the latter happens will be discussed later in Sect. 7.5.

By definition a process which is recurrent cannot drift to ∞ or $-\infty$ and therefore must oscillate. Whilst it is clear that a process drifting to ∞ or $-\infty$ is transient, an oscillating process may not necessarily be recurrent. Indeed it is possible to construct an example of a transient process which oscillates. Inspired by similar remarks for random walks in Feller (1971) one finds such an example in the form of a symmetric stable process ($\beta = 0$) of index $0 < \alpha < 1$. Up to a multiplicative constant, the characteristic exponent for this process is simply $\Psi(\theta) = |\theta|^\alpha$. According to the integral test in Theorem 7.4, the latter class of processes are transient. Nonetheless, since by symmetry $\mathbb{P}(X_t \geq 0) = 1/2 = \mathbb{P}(X_t \leq 0)$, it is clear from the Theorem 7.1 that X oscillates. In contrast, note however that for a linear Brownian motion, the definitions oscillation and recurrence coincide as do the definitions of transience and drifting to $\pm\infty$.

7.2 Cramér's Estimate of Ruin

In this section we extend the classical result of Cramér presented in Theorem 1.10 to the case of a general Lévy process; in particular, a Lévy process which is not necessarily spectrally negative (which was noted to be the natural generalisation of the Cramér–Lundberg model). We follow the treatment of Bertoin and Doney (1994). Recall from the previous section that the probability of ruin may be written in the form $\mathbb{P}(\tau_x^+ < \infty)$. Roughly speaking, our aim is to show that under suitable circumstances, there exists a constant $\nu > 0$ so that $e^{\nu x}\mathbb{P}(\tau_x^+ < \infty)$ has a precise asymptotic behaviour as $x \uparrow \infty$. That is to say, the probability of ruin decays exponentially when the initial capital becomes larger. The precise result is formulated as follows.

Theorem 7.6. *Assume that X is a Lévy process which does not have monotone paths, for which*

(i) $\lim_{t\uparrow\infty} X_t = -\infty$,
(ii) there exists a $\nu \in (0,\infty)$ *such that* $\psi(\nu) = 0$ *where* $\psi(\theta) = \log \mathbb{E}(\exp\{\theta X_1\})$ *is the Laplace exponent of* X,
(iii) the support of Π *is not lattice if* $\Pi(\mathbb{R}) < \infty$.

Then

$$\lim_{x\uparrow\infty} e^{\nu x}\mathbb{P}\left(\tau_x^+ < \infty\right) = \kappa(0,0)\left(\nu \left.\frac{\partial\kappa(0,\beta)}{\partial\beta}\right|_{\beta=-\nu}\right)^{-1}, \tag{7.6}$$

where the limit is interpreted to be zero if the derivative on the right-hand side is infinite.

Note that condition (ii) implies the existence of $\mathbb{E}(X_1)$ and because of the conclusion in Theorem 7.2, condition (i) implies further that $\mathbb{E}(X_1) < 0$. We know that if the moment generating function of X_1 exists to the right of the origin, then it must be convex (this may be shown using arguments similar to those in Exercise 3.5 or alternatively note the remarks in the proof of Theorem 3.9). Conditions (i) and (ii) therefore also imply that for $\theta \in (0,\nu)$, the function $\psi(\theta)$ is below the origin, eventually climbing above it at $\theta = \nu$. Condition (ii) is known as the *Cramér condition.* Essentially Cramér's estimate of ruin says that the existence of exponential moments of a Lévy process which drifts to $-\infty$ implies an exponentially decaying tail of its all time maximum. Indeed note that $\mathbb{P}(\tau_x^+ < \infty) = \mathbb{P}(\overline{X}_\infty > x)$. Since renewal theory will play a predominant role in the proof, the third condition of Theorem 7.6 is simply for convenience allowing the use of the Renewal Theorem without running into the special case of lattice supports. Nonetheless, it is possible to remove condition (iii). See Bertoin and Doney (1994) for further details.

Proof (of Theorem 7.6). The proof is long and we break it into steps.

Step 1. Define the potential measure for the ascending ladder height process,

$$U(A) = \mathbb{E}\left(\int_0^\infty \mathbf{1}_{(H_t\in A)}\mathrm{d}t\right)$$

for Borel sets A in $[0,\infty)$ where L and H are the local time at the supremum and the ascending ladder height process, respectively. Let $T_x^+ = \inf\{t > 0 : H_t > x\}$. Applying the Strong Markov Property at this stopping time for H we get

$$\begin{aligned}
U(x,\infty) &= \mathbb{E}\left(\int_0^\infty \mathbf{1}_{(H_t>x)}\mathrm{d}t; T_x^+ < L_\infty\right)\\
&= \mathbb{P}\left(T_x^+ < L_\infty\right)\mathbb{E}\left(\left.\int_{T_x^+}^{L_\infty} \mathbf{1}_{(H_t>x)}\mathrm{d}t\right| H_s : s \le T_x^+\right)\\
&= \mathbb{P}\left(T_x^+ < L_\infty\right)\mathbb{E}\left(\int_0^{L_\infty} \mathbf{1}_{(H_t\ge 0)}\mathrm{d}t\right)\\
&= \mathbb{P}\left(T_x^+ < L_\infty\right)\mathbb{E}\left(L_\infty\right).
\end{aligned} \tag{7.7}$$

Since $\lim_{t\uparrow\infty} X_t = -\infty$ we know that L_∞ is exponentially distributed with some parameter which is recovered from the joint Laplace exponent $\kappa(\alpha,\beta)$ by setting $\alpha = \beta = 0$. Note also that $\mathbb{P}\left(T_x^+ < L_\infty\right) = \mathbb{P}(\overline{X}_\infty > x) = \mathbb{P}(\tau_x^+ < \infty)$. Hence (7.7) now takes the form

$$\kappa(0,0)U(x,\infty) = \mathbb{P}(\tau_x^+ < \infty). \tag{7.8}$$

Step 2. In order to complete the proof, we henceforth need to establish a precise asymptotic for $\mathrm{e}^{\nu x}U(x,\infty)$. To this end, we shall show via a change of measure that in fact $U_\nu(\mathrm{d}x) := \mathrm{e}^{\nu x}U(\mathrm{d}x)$ on $(0,\infty)$ is a renewal measure in which case the Key Renewal Theorem 5.1 (ii) will help us clarify the required asymptotic.

Since $\psi(\nu) = 0$, we know (cf. Chap. 3) that $\{\exp\{\nu X_t\} : t \geq 0\}$ is a martingale with unit mean and hence it can be used to define a change of measure via

$$\left.\frac{\mathrm{d}\mathbb{P}^\nu}{\mathrm{d}\mathbb{P}}\right|_{\mathcal{F}_t} = \mathrm{e}^{\nu X_t}$$

which by Theorem 3.9 keeps the process X within the class of Lévy processes. From Theorem 6.9 we know that L_t^{-1} is a stopping time and hence we have with the help of the Strong Markov Property,

$$\begin{aligned}
\mathbb{P}^\nu\left(H_t \in \mathrm{d}x, L_t^{-1} < s\right) &= \mathbb{E}\left(\mathrm{e}^{\nu X_s}; H_t \in \mathrm{d}x, L_t^{-1} < s\right) \\
&= \mathbb{E}\left(\mathrm{e}^{\nu X_{L_t^{-1}}}; H_t \in \mathrm{d}x, L_t^{-1} < s\right) \\
&= \mathrm{e}^{\nu x}\mathbb{P}\left(H_t \in \mathrm{d}x, L_t^{-1} < s\right).
\end{aligned}$$

Appealing to monotone convergence and taking $s \uparrow \infty$,

$$\mathbb{P}^\nu\left(H_t \in \mathrm{d}x\right) = \mathrm{e}^{\nu x}\mathbb{P}\left(H_t \in \mathrm{d}x\right). \tag{7.9}$$

Now note that on $(0,\infty)$,

$$U_\nu(\mathrm{d}x) = \mathrm{e}^{\nu x}U(\mathrm{d}x) = \int_0^\infty \mathbb{P}^\nu(H_t \in \mathrm{d}x)\mathrm{d}t.$$

The final equality shows that $U_\nu(\mathrm{d}x)$ is equal to the potential measure of the ascending ladder height process H under $\mathbb{P}^\nu$ on $(0,\infty)$. According to Lemma 5.2, the latter is equal to a renewal measure providing that H is a subordinator under $\mathbb{P}^\nu$ (as opposed to a killed subordinator). This is proved in the next step.

Step 3. A similar argument to the one above yields

$$\mathbb{P}^\nu(\widehat{H}_t \in \mathrm{d}x) = \mathrm{e}^{-\nu x}\mathbb{P}(\widehat{H}_t \in \mathrm{d}x),$$

where now $\widehat{H}$ is the descending ladder height process. From the last two equalities we can easily deduce that the Laplace exponents $\widehat{\kappa}^\nu$ of the descending ladder processes under the measure $\mathbb{P}^\nu$ satisfy

$$\widehat{\kappa}^\nu(0,\beta) = \widehat{\kappa}(0,\beta+\nu)$$

showing in particular $\widehat{\kappa}^{\nu}(0,0) = \widehat{\kappa}(0,\nu) > 0$. The latter is the rate at which the local time $\widehat{L}$ is stopped under $\mathbb{P}^{\nu}$ and hence $\mathbb{P}^{\nu}(\liminf_{t\uparrow\infty} X_t > -\infty)$ almost surely. By the trichotomy given in Theorem 7.1 this is equivalent to $\mathbb{P}^{\nu}(\lim_{t\uparrow\infty} X_t = \infty)$. We now have, as required in the previous step, that H is a subordinator without killing.

Step 4. We should like to use the Renewal Theorem in conjunction with $U_{\nu}(\mathrm{d}x)$. Note from Lemma 5.2 that the underlying distribution of the latter renewal measure is given by $F(\mathrm{d}x) = U_{\nu}^{(1)}(\mathrm{d}x)$ on $[0,\infty)$. In order to calculate its mean, we need to reconsider briefly some properties of κ^{ν}.

From (7.9) one deduces that $\kappa(0,\beta) < \infty$ for $\beta \geq -\nu$. Convexity of ψ on $[0,\nu]$ (see the proof of Theorem 3.9) implies that it is also finite along this interval. We may now appeal to analytic extension to deduce from the Theorem 6.16 (iv) that

$$\Psi(\theta - \mathrm{i}\beta) = -\psi(\beta + \mathrm{i}\theta) = k'\kappa(0, -\beta - \mathrm{i}\theta)\widehat{\kappa}(0, \beta + \mathrm{i}\theta)$$

for some $k' > 0$, $\beta \in [0,\nu]$ and $\theta \in \mathbb{R}$. Now setting $\beta = \nu$ and $\theta = 0$ we further deduce that

$$-\psi(\nu) = 0 = k'\kappa(0, -\nu)\widehat{\kappa}(0,\nu).$$

Since $k'\widehat{\kappa}(0,\nu) > 0$, we conclude that $\kappa(0,-\nu) = 0$.

We may now compute the mean of the distribution F,

$$\begin{aligned}
\mu &= \int_{[0,\infty)} x U_{\nu}^{(1)}(\mathrm{d}x) \\
&= \int_0^{\infty} \mathrm{d}t \cdot \mathrm{e}^{-t} \int_{[0,\infty)} x\mathbb{P}^{\nu}(H_t \in \mathrm{d}x) \\
&= \int_0^{\infty} \mathrm{d}t \cdot \mathrm{e}^{-t}\mathbb{E}(H_t \mathrm{e}^{\nu H_t}) \\
&= \int_0^{\infty} \mathrm{d}t \cdot \mathrm{e}^{-t-\kappa(0,-\nu)t} \left.\frac{\partial \kappa(0,\beta)}{\partial \beta}\right|_{\beta=-\nu} \\
&= \left.\frac{\partial \kappa(0,\beta)}{\partial \beta}\right|_{\beta=-\nu}
\end{aligned}$$

which is possibly infinite in value.

Finally, appealing to the Key Renewal Theorem 5.1 (ii) we have that $U_{\nu}(\mathrm{d}x)$ converges weakly as a measure to $\mu^{-1}\mathrm{d}x$. Hence it now follows from (7.8) that

$$\begin{aligned}
\lim_{x\uparrow\infty} \mathrm{e}^{\nu x}\mathbb{P}(\tau_x^+ < \infty) &= \kappa(0,0) \lim_{x\uparrow\infty} \int_x^\infty \mathrm{e}^{-\nu(y-x)} U_\nu(\mathrm{d}y) \\
&= \kappa(0,0) \lim_{x\uparrow\infty} \int_0^\infty \mathrm{e}^{-\nu z} U_\nu(x+\mathrm{d}z) \\
&= \frac{\kappa(0,0)}{\mu} \int_0^\infty \mathrm{e}^{-\nu z}\mathrm{d}z \\
&= \frac{\kappa(0,0)}{\nu\mu},
\end{aligned}$$

where the limit is understood to be zero if $\mu = \infty$. □

Let us close this section by making a couple of remarks.

Firstly, in the case where X is spectrally negative, the Laplace exponent $\psi(\theta)$ is finite on $\theta \geq 0$. When $\psi'(0+) < 0$ condition (i) of Theorem 7.6 holds. In that case we know already from Theorem 3.12 that

$$\mathbb{E}(\mathrm{e}^{\Phi(q)x - q\tau_x^+}\mathbf{1}_{(\tau_x^+<\infty)}) = 1,$$

where Φ is the right inverse of ψ. Taking $q \downarrow 0$ we recover

$$\mathrm{e}^{\Phi(0)x}\mathbb{P}(\tau_x^+ < \infty) = 1$$

for all $x \geq 0$ which is a stronger statement than that of the above theorem. Taking account of the fact that the Wiener–Hopf factorisation shows that $\kappa(\alpha,\beta) = \beta + \Phi(\alpha)$ for $\alpha, \beta \geq 0$ one may also check that the constant on the right-hand side of (7.6) is consistently equal to 1. Of course, when X is spectrally negative, first passage at level x always occurs by hitting this level and hence this case is of less interest as far as ruin problems are concerned.

Secondly, when X is such that we are considering the classical Cramér–Lundberg model (the assumptions of the above theorem are not necessary), it is a straightforward exercise to show that formula (7.8) can be re-written to give the Pollaczek–Khintchine formula in (1.11). The point now being that the ascending ladder height process H whose potential measure is U is equal in law to a killed compound Poisson subordinator whose jumps have the integrated tail distribution given in (1.12).

7.3 A Quintuple Law at First Passage

In this section we shall give a quantitative account of how a general Lévy process undergoes first passage over a fixed barrier when it jumps clear over it. There will be a number of parallels between the analysis here and the analysis in Chap. 5 concerning first passage of a subordinator. Since subordinators have already been dealt with, they are excluded from the following discussion.

Recall the notation from Chap. 6

$$\overline{G}_t = \sup\{s < t : \overline{X}_s = X_s\}$$

and our standard notation already used in this chapter,

$$\tau_x^+ = \inf\{t > 0 : X_t > x\}.$$

The centre-piece of this section will concern a quintuple law at first passage of the following variables,

$\overline{G}_{\tau_x^+-}$: the time of the last maximum prior to first passage,
$\tau_x^+ - \overline{G}_{\tau_x^+-}$: the length of the excursion making first passage,
$X_{\tau_x^+} - x$: the overshoot at first passage,
$x - X_{\tau_x^+-}$: the undershoot at first passage,
$x - \overline{X}_{\tau_x^+-}$: the undershoot of the last maximum at first passage.

In order to state the main result of this section let us introduce some more notation. Recall from Chap. 6 that for $\alpha, \beta \geq 0$, $\kappa(\alpha, \beta)$ is the Laplace exponent of the ascending ladder process (L^{-1}, H); see (6.8). Associated with the latter is the bivariate potential measure

$$\mathcal{U}(\mathrm{d}s, \mathrm{d}x) = \int_0^\infty \mathrm{d}t \cdot \mathbb{P}(L_t^{-1} \in \mathrm{d}s, H_t \in \mathrm{d}x)$$

supported on $[0, \infty)^2$. On taking a bivariate Laplace transform we find with the help of Fubini's Theorem that

$$\int_{[0,\infty)^2} \mathrm{e}^{-\alpha s - \beta x} \mathcal{U}(\mathrm{d}s, \mathrm{d}x) = \int_0^\infty \mathrm{d}t \cdot \mathbb{E}(\mathrm{e}^{-\alpha L_t^{-1} - \beta H_t}) = \frac{1}{\kappa(\alpha, \beta)} \quad (7.10)$$

for $\alpha, \beta \geq 0$. An important fact to note here is that since L can only be defined up to a multiplicative constant, this in turn affects the exponent κ which in turn affects the measure $\mathcal{U}$. To see precisely how, suppose that $\mathcal{L} = cL$ where L is some choice of local time at the maximum (and hence so is $\mathcal{L}$). It is easily checked that $\mathcal{L}_t^{-1} = L_{t/c}^{-1}$ and if $\mathcal{H}$ is the ladder height process associated with $\mathcal{L}$ then $\mathcal{H}_t = X_{\mathcal{L}_t^{-1}} = X_{L_{t/c}^{-1}} = H_{t/c}$. If $\mathcal{U}^*$ is the measure associated with $\mathcal{L}$ instead of L then we see that

$$\mathcal{U}^*(\mathrm{d}s, \mathrm{d}x) = \int_0^\infty \mathrm{d}t \cdot \mathbb{P}(L_{t/c}^{-1} \in \mathrm{d}s, H_{t/c} \in \mathrm{d}x) = c\mathcal{U}(\mathrm{d}s, \mathrm{d}x)$$

where the final equality follows by the substitution $u = t/c$ in the integral.

We shall define the bivariate measure $\widehat{\mathcal{U}}$ on $[0, \infty)^2$ in the obvious way from the descending ladder process $(\widehat{L}^{-1}, \widehat{H})$.

The following main result is due to Doney and Kyprianou (2005), although similar ideas to those used in the proof can be found in Spitzer (1964), Borovkov (1976) and Bertoin (1996).

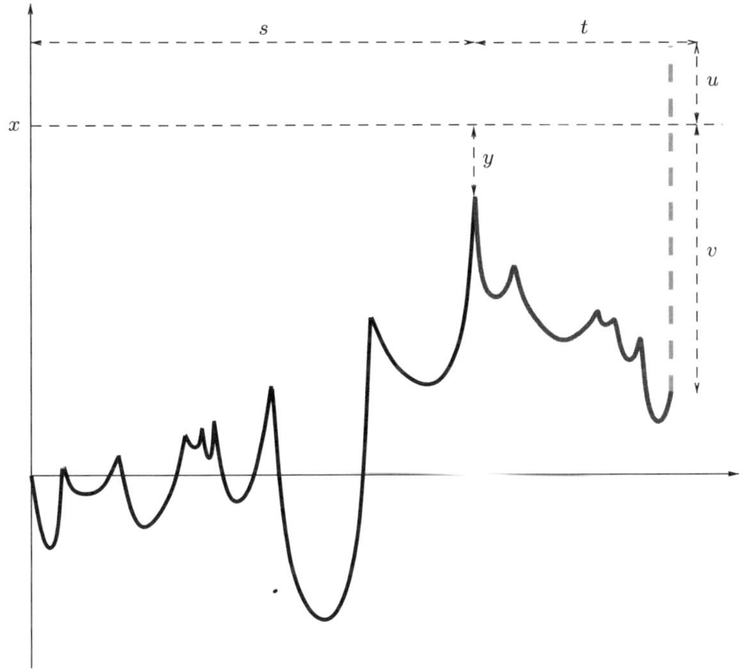

Fig. 7.1. A symbolic description of the quantities involved in the quintuple law.

Theorem 7.7. *Suppose that X is not a compound Poisson process. Then there exists a normalisation of local time at the maximum such that for each $x > 0$ we have on $u > 0$, $v \geq y$, $y \in [0, x]$, $s, t \geq 0$,*

$$\mathbb{P}(\tau_x^+ - \overline{G}_{\tau_x^+ -} \in \mathrm{d}t, \overline{G}_{\tau_x^+ -} \in \mathrm{d}s, X_{\tau_x^+} - x \in \mathrm{d}u, x - X_{\tau_x^+ -} \in \mathrm{d}v, x - \overline{X}_{\tau_x^+ -} \in \mathrm{d}y)$$
$$= \mathcal{U}(\mathrm{d}s, x - \mathrm{d}y)\widehat{\mathcal{U}}(\mathrm{d}t, \mathrm{d}v - y)\Pi(\mathrm{d}u + v)$$

where Π is the Lévy measure of X.

Before going to the proof, let us give some intuition behind the statement of this result with the help of Fig. 7.1. Roughly speaking the event on the left-hand side of the quintuple law requires that there is an ascending ladder height at the time-space point $(s, x - y)$ before going into the final excursion which crosses the level x. Recalling that excursions when indexed by local time at the maximum form a Poisson point process. This means that the behaviour of the last excursion is independent of the preceding ones and hence the quintuple law factorises into the laws of the last excursion and the preceding excursions. The first factor, $\mathcal{U}(\mathrm{d}s, x - \mathrm{d}y)$ thus measures the aforementioned event of an ascending ladder height at $(s, x - y)$. To measure the behaviour of the final excursion, one should look at it rotated about 180°. In the rotated excursion, one starts with a jump of size $u + v$ which is measured by $\Pi(\mathrm{d}u + v)$. The remaining path of the rotated excursion must meet the last ascending ladder height with one of its own descending ladder points. By the Duality Lemma 3.4, rotation of a finite segment of path of a Lévy process produces a path with the same law as the original process. Hence in the rotated excursion, independently of the initial jump of size $u+v$, the quintuple law requires the path descends to a ladder point at a time-space point $(t, v-y)$, and this has measure $\widehat{\mathcal{U}}(\mathrm{d}t, \mathrm{d}v - y)$.

The proof of Theorem 7.7 is somewhat more demanding however as in general there are a countably infinite number of excursions to deal with unlike the case of a random walk where there are only a finite number of excursions.

Proof (of Theorem 7.7). We prove the result in three steps.

Step 1. Let us suppose that m, k, f, g and h are all positive, continuous functions with compact support satisfying $f(0) = g(0) = h(0) = 0$. We prove in this step that

$$\mathbb{E}(m(\tau_x^+ - \overline{G}_{\tau_x^+-})k(\overline{G}_{\tau_x^+-})f(X_{\tau_x^+} - x)g(x - X_{\tau_x^+-})h(x - \overline{X}_{\tau_x^+-}))$$
$$= \widehat{\mathbb{E}}_x\left(\int_0^{\tau_0^-} m(t - \underline{G}_t)k(\underline{G}_t)h(\underline{X}_t)w(X_t)\mathrm{d}t\right), \tag{7.11}$$

where $w(z) = g(z)\int_{(z,\infty)} \Pi(\mathrm{d}u)f(u-z)$ and $\widehat{\mathbb{E}}_x$ is expectation under the law, $\widehat{\mathbb{P}}_x$, of $-X$ initiated from position $-X_0 = x$.

The proof of (7.11) follows by an application of the compensation formula applied to the Poisson random measure, N, with intensity measure $\mathrm{d}t\Pi(\mathrm{d}x)$ associated to the jumps of X. We have

$$\mathbb{E}(m(\tau_x^+ - \overline{G}_{\tau_x^+-})k(\overline{G}_{\tau_x^+-})f(X_{\tau_x^+} - x)g(x - X_{\tau_x^+-})h(x - \overline{X}_{\tau_x^+-}))$$
$$= \mathbb{E}\left(\int_{[0,\infty)}\int_{\mathbb{R}} m(t - \overline{G}_{t-})k(\overline{G}_{t-})g(x - X_{t-})h(x - \overline{X}_{t-}) \right.$$
$$\left. \times \mathbf{1}_{(x-\overline{X}_{t-}>0)}f(X_{t-} + z - x)\mathbf{1}_{(z>x-X_{t-})}N(\mathrm{d}t\times\mathrm{d}z)\right)$$
$$= \mathbb{E}\left(\int_0^\infty \mathrm{d}t\cdot m(t - \overline{G}_{t-})k(\overline{G}_{t-})g(x - X_{t-})h(x - \overline{X}_{t-}) \right.$$
$$\left. \times \mathbf{1}_{(x-\overline{X}_{t-}>0)}\int_{(x-X_{t-},\infty)} \Pi(\mathrm{d}\phi)f(X_{t-} + \phi - x)\right)$$
$$= \mathbb{E}\left(\int_0^\infty \mathrm{d}t\cdot m(t - \overline{G}_{t-})k(\overline{G}_{t-})h(x - \overline{X}_{t-})\mathbf{1}_{(x-\overline{X}_{t-}>0)}w(x - X_{t-})\right)$$
$$= \widehat{\mathbb{E}}_x\left(\int_0^\infty \mathrm{d}t\cdot\mathbf{1}_{(t<\tau_0^-)}m(t - \underline{G}_t)k(\underline{G}_t)h(\underline{X}_t)w(X_t)\right)$$

which is equal to the right-hand side of (7.11). In the last equality we have rewritten the previous equality in terms of the path of $-X$. Note that the condition $f(0) = g(0) = h(0) = 0$ has been used implicitly to exclude from the calculation the event $\{X_{\tau_x^+} = x\}$.

Step 2. Next we prove that

$$\mathbb{E}_x\left(\int_0^{\tau_0^-} m(t - \underline{G}_t)k(\underline{G}_t)h(\underline{X}_t)w(X_t)\mathrm{d}t\right)$$

$$= \int_{[0,\infty)}\int_{[0,\infty)} \mathcal{U}(\mathrm{d}t, \mathrm{d}\phi)$$

$$\cdot \int_{[0,\infty)}\int_{[0,x]} \widehat{\mathcal{U}}(\mathrm{d}s, \mathrm{d}\theta)m(t)k(s)h(x-\theta)w(x+\phi-\theta). \quad (7.12)$$

(In fact we need the same identity as above but with the expectation $\mathbb{E}$ replaced by $\widehat{\mathbb{E}}$, however for convenience we establish the identity as given).

For $q > 0$,

$$\mathbb{E}_x\left(\int_0^{\tau_0^-} \mathrm{d}t \cdot m(t - \underline{G}_t)k(\underline{G}_t)h(\underline{X}_t)w(X_t)\mathrm{e}^{-qt}\right)$$

$$= q^{-1}\mathbb{E}_x\left(m(\mathbf{e}_q - \underline{G}_{\mathbf{e}_q})k(\underline{G}_{\mathbf{e}_q})h(\underline{X}_{\mathbf{e}_q})w(X_{\mathbf{e}_q} - \underline{X}_{\mathbf{e}_q} + \underline{X}_{\mathbf{e}_q}); \mathbf{e}_q < \tau_0^-\right)$$

$$= q^{-1}\int_{[0,\infty)}\int_{[0,x]} \mathbb{P}(\underline{G}_{\mathbf{e}_q} \in \mathrm{d}s, -\underline{X}_{\mathbf{e}_q} \in \mathrm{d}\theta)k(s)$$

$$\cdot \int_{[0,\infty)}\int_{[0,\infty)} \mathbb{P}(\mathbf{e}_q - \underline{G}_{\mathbf{e}_q} \in \mathrm{d}t, X_{\mathbf{e}_q} - \underline{X}_{\mathbf{e}_q} \in \mathrm{d}\phi)m(t)h(x-\theta)w(x+\phi-\theta)$$

$$= q^{-1}\int_{[0,\infty)}\int_{[0,x]} \mathbb{P}(\underline{G}_{\mathbf{e}_q} \in \mathrm{d}s, -\underline{X}_{\mathbf{e}_q} \in \mathrm{d}\theta)k(s)$$

$$\cdot \int_{[0,\infty)}\int_{[0,\infty)} \mathbb{P}(\overline{G}_{\mathbf{e}_q} \in \mathrm{d}t, \overline{X}_{\mathbf{e}_q} \in \mathrm{d}\phi)m(t)h(x-\theta)w(x+\phi-\theta), \quad (7.13)$$

where the Wiener–Hopf factorisation[3] and duality have been used in the second and third equalities, respectively. Further it is also known from the Wiener–Hopf factorisation, Theorem 6.16, that

$$\frac{1}{\kappa(q,0)}\mathbb{E}\left(\mathrm{e}^{-\alpha\overline{G}_{\mathbf{e}_q} - \beta\overline{X}_{\mathbf{e}_q}}\right) = \frac{1}{\kappa(\alpha+q,\beta)}$$

and hence recalling (7.10) it follows that

$$\lim_{q\downarrow 0}\frac{1}{\kappa(q,0)}\mathbb{P}(\overline{G}_{\mathbf{e}_q} \in \mathrm{d}t, \overline{X}_{\mathbf{e}_q} \in \mathrm{d}\phi) = \mathcal{U}(\mathrm{d}t, \mathrm{d}\phi)$$

[3]Specifically we use the independence of the pairs $(\underline{G}_{\mathbf{e}_q}, \underline{X}_{\mathbf{e}_q})$ and $(\mathbf{e}_q - \underline{G}_{\mathbf{e}_q}, X_{\mathbf{e}_q} - \underline{X}_{\mathbf{e}_q})$.

in the sense of weak convergence. A similar convergence holds for

$$\mathbb{P}(\underline{G}_{\mathbf{e}_q} \in \mathrm{d}s, -\underline{X}_{\mathbf{e}_q} \in \mathrm{d}\theta)/\widehat{\kappa}(q,0).$$

Equality (7.12) thus follows by splitting the divisor q into the product $\kappa(q,0)\times\widehat{\kappa}(q,0)$ (this factorisation was observed in the proof of Theorem 6.16 (iv)) and taking limits in (7.13). Note that in general $q = k\kappa(q,0)\widehat{\kappa}(q,0)$ for some $k>0$ which depends on the normalisation of local time (at the maximum). It is thus here that we require a suitable normalisation of local time at the maximum in order to have $k=1$.

Step 3. We combine the conclusions of steps 1 and 2 (where step 2 is applied to $-X$) to conclude that

$$\begin{aligned}
&\mathbb{E}(m(\tau_x^+ - \overline{G}_{\tau_x^+-})k(\overline{G}_{\tau_x^+-})f(X_{\tau_x^+} - x)g(x - X_{\tau_x^+-})h(x - \overline{X}_{\tau_x^+-}))\\
&= \int_{u>0,y\in[0,x],0<y\le v,s\ge 0,t\ge 0} m(t)k(s)f(u)g(v)h(y)\\
&\mathbb{P}(\tau_x^+ - \overline{G}_{\tau_x^+-} \in \mathrm{d}t, \overline{G}_{\tau_x^+-} \in \mathrm{d}s, X_{\tau_x^+} - x \in \mathrm{d}u, x - X_{\tau_x^+-} \in \mathrm{d}v, x - \overline{X}_{\tau_x^+-} \in \mathrm{d}y)\\
&= \int_{[0,\infty)}\int_{[0,\infty)} \widehat{\mathcal{U}}(\mathrm{d}t,\mathrm{d}\phi)\int_{[0,\infty)}\int_{[0,x]} \mathcal{U}(\mathrm{d}s,\mathrm{d}\theta)m(t)k(s)\\
&\qquad\cdot h(x-\theta)g(x+\phi-\theta)\int_{(x+\phi-\theta,\infty)} \Pi(\mathrm{d}\eta)f(\eta-(x+\phi-\theta)).
\end{aligned}$$

Substituting $y = x-\theta$, then $y+\phi = v$ and finally $\eta = v+u$ in the right-hand side above yields

$$\begin{aligned}
&\mathbb{E}(m(\tau_x^+ - \overline{G}_{\tau_x^+-})k(\overline{G}_{\tau_x^+-})f(X_{\tau_x^+} - x)g(x - X_{\tau_x^+-})h(x - \overline{X}_{\tau_x^+-}))\\
&= \int_{[0,\infty)}\int_{[0,x]} \mathcal{U}(\mathrm{d}s, x-\mathrm{d}y)\int_{[0,\infty)}\int_{[y,\infty)} \widehat{\mathcal{U}}(\mathrm{d}t,\mathrm{d}v-y)\\
&\qquad\cdot\int_{(0,\infty)} \Pi(\mathrm{d}u+v)m(t)k(s)f(u)g(v)h(y)
\end{aligned}$$

and the statement of the theorem follows. □

The case of a compound Poisson process has been excluded from the statement of the theorem on account of the additional subtleties that occur in connection with the ascending and descending ladder height processes and their definitions in the weak or strict sense. (Recall the discussion of weak and strict ladder processes in Sect. 6.1). Nonetheless the result is still valid provided one takes the bivariate renewal measure $\mathcal{U}$ as that of the *weak (resp. strict)* ascending ladder process and $\widehat{\mathcal{U}}$ is taken as the bivariate renewal measure of the *strict (resp. weak)* descending ladder process.

To be realistic, the quintuple law in general does not necessarily bring one closer to explicit formulae for special examples of Lévy processes on account of the indirect involvement of the quantities κ and $\widehat{\kappa}$ which themselves are embedded into the Wiener–Hopf factorisation. There are two examples where one may make reasonable progress into making these formulae more explicit. These are the cases of stable processes, dealt with in Exercise 7.4 and spectrally positive processes which we shall now treat.

For any spectrally positive process X, let $U(\mathrm{d}x) = \int_{[0,\infty)} \mathcal{U}(\mathrm{d}s, \mathrm{d}x)$ and note from the Wiener–Hopf factorisation in Sect. 6.5.2, which gives an expression for $\kappa(\alpha, \beta)$, and the Laplace transform (7.10) that

$$\int_{[0,\infty)} \mathrm{e}^{-\beta x} U(\mathrm{d}x) = \frac{\beta - \Phi(0)}{\psi(\beta)}, \tag{7.14}$$

where Φ is the right inverse of the Laplace exponent ψ of $-X$. Using obvious notation, it is also clear from (7.10) that since we may take $\widehat{\kappa}(0,\beta) = \Phi(0) + \beta$ then $\widehat{U}(\mathrm{d}x) = \mathrm{e}^{-\Phi(0)x}\mathrm{d}x$.

The quintuple law for spectrally positive Lévy processes marginalises to the triple law

$$\begin{aligned} &\mathbb{P}(X_{\tau_x^+} - x \in \mathrm{d}u, x - X_{\tau_x^+ -} \in \mathrm{d}v, x - \overline{X}_{\tau_x^+ -} \in \mathrm{d}y) \\ &= \mathrm{e}^{-\Phi(0)(v-y)} U(x - \mathrm{d}y) \Pi(\mathrm{d}u + v)\mathrm{d}v \end{aligned} \tag{7.15}$$

for $y \in [0, x]$, $v \geq y$ and $u > 0$. If we assume further that $\liminf_{t\uparrow\infty} X_t = -\infty$ then we know that $\Phi(0) = 0$ and the right-hand side of (7.15) is equal to the inverse Laplace transform of $\beta/\psi(\beta)$.

7.4 The Jump Measure of the Ascending Ladder Height Process

Recall that the basic information on the ladder height process H is captured in its Laplace exponent $\kappa(0, \beta)$ which itself is embedded in the Wiener–Hopf factorisation. In this section we shall show that the quintuple law allows us to gain some additional insight into the analytical form of the jump measure of the ascending ladder height. Since the Wiener–Hopf factorisation played a major role in establishing the quintuple law, one may consider the main result below ultimately as extracting information concerning the ascending ladder height process out of the Wiener–Hopf factorisation. The result is due to Vigon (2002).

Theorem 7.8. *Suppose that Π_H is the jump measure associated with the ascending ladder height process of a Lévy process X, other than a compound Poisson process, with jump measure Π. Then for all $y > 0$ and a suitable normalisation of local time at the maximum*

$$\Pi_H(y, \infty) = \int_{[0,\infty)} \widehat{U}(\mathrm{d}z)\Pi(z + y, \infty),$$

where $\widehat{U}(\mathrm{d}z) = \int_{[0,\infty)} \widehat{\mathcal{U}}(\mathrm{d}s, \mathrm{d}z) = \mathbb{E}(\int_0^\infty \mathbf{1}_{(\widehat{H}_t \in \mathrm{d}z)}\mathrm{d}t)$.

Proof. The result follows from the joint law of the overshoot and undershoot of the maximum of X at first passage of some $x > 0$ as given by the quintuple law by comparing it against the overshoot and undershoot of the process H at the same level.

Define $T_x^+ = \inf\{t > 0 : H_t > x\}$ and use again the definition $U(\mathrm{d}x) = \int_{[0,\infty)} \mathcal{U}(\mathrm{d}x, \mathrm{d}s)$. Note that since the range of $\overline{X}$ is the same as the range of H it follows that $H_{T_x^+-} = \overline{X}_{\tau_x^+-}$. Hence from Theorem 5.6 we have for $u > 0$ and $y \in [0, x]$,

$$\begin{aligned}
&\mathbb{P}(X_{\tau_x^+} - x \in \mathrm{d}u, x - \overline{X}_{\tau_x^+-} \in \mathrm{d}y) \\
&= \mathbb{P}(H_{T_x^+} - x \in \mathrm{d}u, x - H_{T_x^+-} \in \mathrm{d}y) \\
&= U(x - \mathrm{d}y)\Pi_H(\mathrm{d}u + y).
\end{aligned} \tag{7.16}$$

On the other hand, the quintuple law gives for $u > 0$ and $y \in [0, x]$,

$$\begin{aligned}
&\mathbb{P}(X_{\tau_x^+} - x \in \mathrm{d}u, x - \overline{X}_{\tau_x^+-} \in \mathrm{d}y) \\
&= U(x - \mathrm{d}y) \int_{[y,\infty)} \widehat{U}(\mathrm{d}v - y)\Pi(\mathrm{d}u + v).
\end{aligned} \tag{7.17}$$

Equating the right-hand sides of (7.16) and (7.17) implies that

$$\Pi_H(\mathrm{d}u + y) = \int_{[y,\infty)} \widehat{U}(\mathrm{d}v - y)\Pi(\mathrm{d}u + v).$$

Integrating over $u > 0$ the statement of the theorem easily follows. □

Similar techniques allow one to make a more general statement concerning the bivariate jump measure of the ascending ladder process (L^{-1}, H). This is done in Exercise 7.5. As in the previous theorem however, the expression for jump measure still suffers from a lack of explicitness due to the involvement of the quantity $\widehat{U}$. However, if one considers the case of a spectrally positive Lévy process then the situation becomes somewhat more favourable for Π_H.

Corollary 7.9. *Under the conditions of Theorem 7.8, if X is spectrally positive then*

$$\Pi_H(y, \infty) = \int_{[0,\infty)} \mathrm{e}^{-\Phi(0)z}\Pi(z + y, \infty)\mathrm{d}z,$$

where Φ is the right inverse of the Laplace exponent ψ of $-X$.

Proof. Taking into account the remarks in the final paragraph of Sect. 7.3 the result follows easily. □

Note in particular that if the spectrally positive process in the above corollary has the property that $\liminf_{t\uparrow\infty} X_t = -\infty$ then $\Phi(0) = 0$ and hence for $x > 0$,

$$\Pi_H(\mathrm{d}x) = \Pi(x, \infty)\mathrm{d}x. \tag{7.18}$$

The same conclusion was drawn in Exercise 6.5 using the Wiener–Hopf factorisation. If we take further the case that $-X$ is a Cramér–Lundberg process then the previous statement is consistent with Theorem 1.9. In the notation of the latter theorem, on account of irregularity of 0 for $(-\infty, 0)$ for the process $-X$, the ascending ladder height process of X must be a compound Poisson subordinator and therefore the jump measure Π_H is proportional to η which is the jump distribution of the descending ladder process of the Cramér–Lundberg process. The constant of proportionality is the arrival rate of the compound Poisson subordinator. (The latter depends on the normalization of local time).

7.5 Creeping

As with the case of a subordinator, one may talk of a Lévy process *creeping* over a fixed level $x > 0$. To be precise, a Lévy process creeps over the level x when

$$\mathbb{P}(X_{\tau_x^+} = x) > 0. \tag{7.19}$$

The class of Lévy processes which creep upwards atleast one point can easily be seen to be non-empty by simply considering any spectrally negative Lévy process. By definition, any spectrally negative Lévy process has the property that for all $x \geq 0$

$$\mathbb{P}(X_{\tau_x^+} = x | \tau_x^+ < \infty) = 1.$$

From the latter, (7.19) easily follows when we recall from Theorem 3.12 that $\mathbb{P}(\tau_x^+ < \infty) = \mathrm{e}^{-\Phi(0)x} > 0$ where Φ is the right inverse of the Laplace exponent of X.

Lemma 7.10. *Suppose that X is not a compound Poisson process. Then X creeps upwards at some (and then all) $x > 0$ if and only if*

$$\lim_{\beta \uparrow \infty} \frac{\kappa(0, \beta)}{\beta} > 0. \tag{7.20}$$

Proof. The key to understanding when an arbitrary Lévy process creeps upwards is embedded within the problem of whether a subordinator creeps upwards. Indeed we know that X creeps across $x > 0$ if and only if with positive probability the point x lies in the range of $\{\overline{X}_t : t \geq 0\}$. The latter range is almost surely identical to the range of the ascending ladder process $\{H_t : t \geq 0\}$. Hence we see that X creeps across $x > 0$ if and only if H creeps across x. For this reason it follows that if a Lévy process creeps over some $x > 0$, then it will creep over all $x > 0$ provided H has the same behaviour. We know from Sect. 5.3 that the only subordinators which may creep over some levels but not others are compound Poisson subordinators whose jumps are distributed with lattice support and hence we need to rule this case out. Let us now split the discussion into two cases, according to the regularity of 0 for $(0, \infty)$.

Suppose that 0 is regular for $(0,\infty)$. Recall from Theorem 6.10 that H has the law of a (possibly killed) subordinator. This subordinator cannot be a compound Poisson process by the assumption of regularity. We are then within the scope of Theorem 5.9 which tells us that there is creeping if and only if the underlying subordinator has a strictly positive drift. By definition, however, the Laplace exponent $\kappa(0,\beta)$ of H, up to an additive constant (the killing rate), is the Laplace exponent of the underlying subordinator. Hence the presence of a strictly positive drift coefficient is exposed by taking the limit given in the statement of the lemma (recall Exercise 2.11). In other words, there is creeping if and only if (7.20) holds.

Suppose now that 0 is irregular for $(0,\infty)$ which has the consequence that the ascending ladder height must be a compound Poisson process subordinator. We may not appeal to Theorem 5.9 as in the previous paragraph. Since it has been assumed that X is not a compound Poisson process, it also means that 0 must be regular for $(-\infty,0)$ and hence the descending ladder height process cannot be a compound Poisson subordinator. According to Corollary 7.8 we know that $\Pi_H(\mathrm{d}x) = \int_{[0,\infty)} \widehat{U}(\mathrm{d}v)\Pi(\mathrm{d}x+v)$. Theorem 5.4 (i) shows that $\widehat{U}$ has no atoms on $(0,\infty)$ as $\widehat{H}$ is not a compound Poisson process. Hence Π_H has no atoms. In conclusion, whilst H is a compound Poisson process, its Lévy measure, which is proportional to the jump distribution of H, has no atoms and therefore the compound Poisson subordinator H cannot hit specified points. As X will creep over $x>0$ if and only if H can hit x with positive probability, then we conclude that X cannot creep over x and hence (7.20) fails. □

Note that, as in the case of subordinators, the exclusion of compound Poisson processes from the previous theorem is due to the possibility of examples in which the jump measure has atoms allowing the process X to jump on to specific points with positive probability.

The criterion given in the above corollary is not particularly useful in general for determining whether a process can creep upwards or not. Ideally one would like to establish a criterion in terms of the components of the Lévy–Khintchine exponent which is equivalent to upward creeping. The following result does precisely this.

Theorem 7.11. *Suppose that X is a Lévy process which is not a compound Poisson process. Then X creeps upwards if and only if one of the following three situations occur,*

(i) X has bounded variation with Lévy–Khintchine exponent

$$\Psi(\theta) = -\mathrm{i}\theta d + \int_{\mathbb{R}\setminus\{0\}} (1-\mathrm{e}^{\mathrm{i}\theta x})\Pi(\mathrm{d}x)$$

and $d>0$,

(ii) X has a Gaussian component,

(iii) X has unbounded variation, has no Gaussian component and its characteristic measure Π satisfies

$$\int_0^1 \frac{x\Pi(x,\infty)}{\int_0^x \Pi(-1,-u)\mathrm{d}u}\mathrm{d}x < \infty.$$

The proof of parts (i) and (ii) appear in Exercise 7.6. The precise formulation and proof of part (iii) remained a challenging open problem until recently when it was resolved by Vigon (2002). We do not give details of the proof which requires a deep analytical understanding of the Wiener–Hopf factorisation and goes far beyond the scope of this text. A recent, more probabilistic proof is given in Doney (2006).

We close this section by making some remarks on the difference between a Lévy process X creeping over x and *hitting the point* x. Formally speaking, we say that X hits the point x if $\mathbb{P}(\tau^{\{x\}} < \infty) > 0$ where

$$\tau^{\{x\}} = \inf\{t > 0 : X_t = x\}$$

with the usual convention that $\inf \emptyset = \infty$. Clearly if X creeps over x (either upwards or downwards) then it must hit x. When X is a subordinator, the converse is also obviously true. However, if X is not a subordinator, then it can be shown that the converse is not necessarily true. The following result due to Kesten (1969) and Bretagnolle (1971) gives a complete characterisation of the range of a Lévy process.

Theorem 7.12. *Suppose that X is not a compound Poisson process. Let*

$$C := \{x \in \mathbb{R} : \mathbb{P}(\tau^{\{x\}} < \infty) > 0\}$$

be the set of points that a Lévy process can hit. Then $C \neq \emptyset$ if and only if

$$\int_{\mathbb{R}} \Re\left(\frac{1}{1+\Psi(u)}\right) \mathrm{d}u < \infty. \tag{7.21}$$

Moreover,

(i) If $\sigma > 0$, then (7.21) is satisfied and $C = \mathbb{R}$.
(ii) If $\sigma = 0$ and X is of unbounded variation and (7.21) is satisfied then $C = \mathbb{R}$.
(iii) If X is of bounded variation, then (7.21) is satisfied if and only if $\mathrm{d} \neq 0$ where d is the drift in the representation (2.22) of its Lévy–Khintchine exponent Ψ. In that case $C = \mathbb{R}$ unless X or $-X$ is a subordinator and then $C = (0,\infty)$ or $C = (-\infty, 0)$, respectively.

From this characterisation one may deduce that, for example, a symmetric α-stable process where $\alpha \in (1,2)$ cannot creep and yet $C = \mathbb{R}$. See Exercise 7.6 for details.

7.6 Regular Variation and Infinite Divisibility

It has been pointed out at several points earlier in this chapter that the quintuple law lacks to some extent a degree of explicitness which would otherwise give it far greater practical value for the study of ruin problems. In Sect. 7.7 we shall give some indication of how the quintuple law gives some analytical advantage however when studying the asymptotic of ruin problems as the ruinous barrier tends to infinity. We need to make a short digression first however into the behaviour of infinitely divisible random variables whose Lévy measures have regularly varying tails.

Recall from Definition 5.12 that a function $f:[0,\infty)\to(0,\infty)$ is regularly varying at infinity with index $\rho\in\mathbb{R}$ if for all $\lambda>0$,

$$\lim_{x\uparrow\infty}\frac{f(\lambda x)}{f(x)}=\lambda^{\rho}.$$

Let us suppose that H is a random variable valued on $[0,\infty)$ which is infinitely divisible with Lévy measure Π_H. Throughout this section we shall suppose that $\Pi_H(\cdot,\infty)$ is regularly varying with index $-\alpha$ for some $\alpha>0$. Our interest here is to understand how this assumed tail behaviour of Π_H reflects on the tail behaviour of the distribution of the random variable H. We do this with a sequence of lemmas. The reader may skip their proofs at no cost to the understanding of their application in Sect. 7.7. The first of these lemmas is taken from Feller (1971).

Lemma 7.13. *Define the probability measure*

$$\nu(\mathrm{d}x)=\frac{\Pi_H(\mathrm{d}x)}{\Pi_H(1,\infty)}\mathbf{1}_{(x>1)}.$$

Then using the usual notation ν^n for the n-fold convolution of ν with itself, we have that

$$\nu^{*n}(x,\infty)\sim n\nu(x,\infty) \qquad (7.22)$$

as $x\uparrow\infty$ for each $n=2,3,\ldots$.

Proof. The result follows by proving a slightly more general result. Suppose that F_1 and F_2 are distribution functions on $[0,\infty)$ such that $F_i(x,\infty)\sim x^{-\alpha}L_i(x)$ for $i=1,2$ as $x\uparrow\infty$ where L_1 and L_2 are slowly varying at infinity. Then

$$(F_1*F_2)(x,\infty)\sim x^{-\alpha}(L_1(x)+L_2(x)) \qquad (7.23)$$

as $x\uparrow\infty$. Indeed from the latter one may argue that (7.22) clearly holds for $n=2$ and hence by induction it holds for all integers $n\geq 2$.

To prove (7.23), let Y_1 and Y_2 be independent random variables with distributions F_1 and F_2. Fix $\delta>0$ and write $x'=x(1+\delta)$. The event $\{Y_1+Y_2>x\}$

contains the event $\{Y_1 > x'\} \cup \{Y_2 > x'\}$ and hence

$$F_1 * F_2(x, \infty) \geq F_1(x', \infty) + F_2(x', \infty).$$

On the other hand set $1/2 > \delta > 0$, suppose $x'' = (1-\delta)x$ then the event $\{Y_1 + Y_2 > x\}$ is a subset of the event $\{Y_1 > x''\} \cup \{Y_2 > x''\} \cup \{\min(Y_1, Y_2) > \delta x\}$. On account of the assumptions made on F_1 and F_2, it is clear that as $x \uparrow \infty$ $\mathbb{P}(\min(Y_1, Y_2) > \delta x) = \mathbb{P}(Y_1 > \delta x)^2$ is of considerably smaller order than $\mathbb{P}(Y_i > x'')$ for each $i = 1, 2$. It follows that as $x \uparrow \infty$

$$F_1 * F_2(x, \infty) \leq (1+\varepsilon)(F_1(x'', \infty) + F_2(x'', \infty))$$

for all small $\varepsilon > 0$. The two inequalities for $F_1 * F_2$ together with the assumed regular variation imply that

$$\begin{aligned}(1+\delta)^{-\alpha} &\leq \liminf_{x\uparrow\infty} \frac{F_1 * F_2(x, \infty)}{x^{-\alpha}(L_1(x) + L_2(x))} \\ &\leq \limsup_{x\uparrow\infty} \frac{F_1 * F_2(x, \infty)}{x^{-\alpha}(L_1(x) + L_2(x))} \leq (1+\varepsilon)(1-\delta)^{-\alpha}.\end{aligned}$$

Since δ and ε may be made arbitrarily small, the required result follows. □

Note that any distribution on $[0, \infty)$ which fulfils the condition (7.22) in fact belongs to a larger class of distributions known as subexponential.[4] The latter class was introduced by Christyakov (1964) within the context of branching processes. The following Lemma, due to Kesten, thus gives a general property of all subexponential distributions.

Lemma 7.14. *Suppose that Y is any random variable whose distribution G, satisfying $G(x) > 0$ for all $x > 0$, has the same asymptotic convolution properties as (7.22). Then given any $\varepsilon > 0$ there exists a constant $C = C(\varepsilon) > 0$ such that*

$$\frac{G^{*n}(x, \infty)}{G(x, \infty)} \leq C(1+\varepsilon)^n$$

for all $n \in \{1, 2, ...\}$ and $x > 0$.

Proof. The proof is inductive. Suppose that for each $n = 1, 2, ...,$

$$\xi_n := \sup_{x \geq 0} \frac{G^{*n}(x, \infty)}{G(x, \infty)}.$$

It is clear that $\xi_1 \leq 1$. Next note that $1 - G^{*(n+1)} = 1 - G * (1 - G^{*n})$. Then for any $0 < T < \infty$,

[4] Any distribution F is thus subexponential if, when $X_1, ..., X_n$ are independent random variables with distribution F, $\mathbb{P}(X_1 + \cdots + X_n > x) \sim n\mathbb{P}(X_1 > x)$ as $x \uparrow \infty$.

$$
\begin{aligned}
\xi_{n+1} &\leq 1 + \sup_{0\leq x\leq T} \int_0^x \frac{1-G^{*n}(x-y)}{1-G(x)} G(\mathrm{d}x) \\
&\quad + \sup_{x>T} \int_0^x \frac{1-G^{*n}(x-y)}{1-G(y)} \frac{1-G(x-y)}{1-G(x)} G(\mathrm{d}x) \\
&\leq 1 + \frac{1}{1-G(T)} + \xi_n \sup_{t>T} \frac{G(x)-G^{*2}(x)}{1-G(x)}.
\end{aligned}
$$

Since G satisfies (7.22), given any $\varepsilon > 0$, we can choose $T > 0$ such that

$$
\xi_{n+1} \leq 1 + \frac{1}{1-G(T)} + \xi_n(1+\varepsilon).
$$

Hence, iterating we find

$$
\xi_{n+1} \leq \left(\frac{2-G(T)}{1-G(T)}\right) \frac{1}{\varepsilon}(1+\varepsilon)^{n+1}
$$

which establishes the claim with the obvious choice of C. □

In the next lemma, we use the asymptotic behaviour in Lemma 7.13 and the uniform bounds in Lemma 7.14 to show that the distribution of H must also have regularly varying tails. The result is due to Embrechts et al. (1979). Recall that we are assuming throughout that $\Pi_H(\cdot,\infty)$ is slowly varying with index $-\alpha$ for some $\alpha > 0$.

Lemma 7.15. *As $x \uparrow \infty$,*

$$
\mathbb{P}(H > x) \sim \Pi_H(x,\infty)
$$

implying that the distribution of H has a regularly varying tail at infinity with index $-\alpha$.

Proof. The relationship between H and Π_H is expressed via the Lévy–Khintchine formula. In this case, since H is $[0,\infty)$-valued, we may consider instead its Laplace exponent $\Phi(\theta) := -\log \mathbb{E}(\mathrm{e}^{-\theta H})$ which from the Lévy–Khintchine formula satisfies

$$
\begin{aligned}
\Phi(\theta) &= d\theta + \int_{(0,\infty)} (1-\mathrm{e}^{-\theta x}) \Pi_H(\mathrm{d}x) \\
&= d\theta + \int_{(0,1]} (1-\mathrm{e}^{-\theta x}) \Pi_H(\mathrm{d}x) && (7.24) \\
&\quad + \int_{(1,\infty)} (1-\mathrm{e}^{-\theta x}) \Pi_H(\mathrm{d}x). && (7.25)
\end{aligned}
$$

The second equality above allows the random variable H to be seen as equal in distribution to the independent sum of two infinitely divisible random variables, say H_1 and H_2, where H_1 has Laplace exponent given by (7.24) and H_2 has Laplace exponent given by (7.25). According to Theorem 3.6, $\mathbb{E}(e^{\lambda H_1}) < \infty$ for any $\lambda > 0$ because trivially $\int_{x\geq 1} e^{\lambda x} \Pi_{H_1}(dx) < \infty$ where $\Pi_{H_1}(dx) = \Pi_H(dx)\mathbf{1}_{(x\in(0,1])}$. It follows that one may upper estimate the tail of H_1 by any exponentially decaying function. Specifically, with the help of the Markov inequality, $\mathbb{P}(H_1 > x) \leq \mathbb{E}(e^{\lambda H_1})e^{-\lambda x}$ for any $\lambda > 0$.

On the other hand, by assumption, the tail of the measure $\Pi_{H_2}(dx) = \Pi_H(dx)\mathbf{1}_{(x\geq 1)}$ is regularly varying with index $-\alpha$. Since Π_{H_2} necessarily has finite total mass, we may consider H_2 as the distribution at time 1 of a compound Poisson process with rate $\eta := \Pi_H(1,\infty)$ and jump distribution ν (defined in Lemma 7.13). We know then that

$$\mathbb{P}(H_2 > x) = e^{-\eta} \sum_{k\geq 0} \frac{\eta^k}{k!} \nu^{*k}(x,\infty),$$

where as usual we interpret $\nu^{*0}(dx) = \delta_0(dx)$ (so in fact the first term of the above sum is equal to zero). Next use the conclusion of Lemma 7.14 with dominated convergence to establish that

$$\lim_{x\uparrow\infty} \frac{\mathbb{P}(H_2 > x)}{\Pi_H(x,\infty)/\eta} = \lim_{x\uparrow\infty} e^{-\eta} \sum_{k\geq 1} \frac{\eta^k}{k!} \frac{\nu^{*k}(x,\infty)}{\nu(x,\infty)}.$$

The conclusion of Lemma 7.13 allows the computation of the limiting sum explicitly. That is to say $\sum_{k\geq 1} e^{\eta}\eta^k/(k-1)! = \eta$. In conclusion, we have

$$\lim_{x\uparrow\infty} \frac{\mathbb{P}(H_2 > x)}{\Pi_H(x,\infty)} = 1.$$

The proof of this lemma is thus completed once we show that

$$\lim_{x\uparrow\infty} \frac{\mathbb{P}(H_1 + H_2 > x)}{\mathbb{P}(H_2 > x)} = 1. \tag{7.26}$$

However, this fact follows by reconsidering the proof of Lemma 7.13. If in this proof one takes F_i as the distribution of H_i for $i = 1, 2$, then with the slight difference that F_1 has exponentially decaying tails, one may follow the proof step by step to deduce that the above limit holds. Intuitively speaking, the tails of H_1 are considerably lighter than those of H_2 and hence for large x, the event in the nominator of (7.26) essentially occurs due to a large observation of H_2. The details are left as an exercise to the reader. □

7.7 Asymptotic Ruinous Behaviour with Regular Variation

In this section, we give the promised example of how to use the quintuple law to obtain precise analytic statements concerning asymptotic ruinous behaviour under assumptions of regular variation. The following theorem, due to

Asmussen and Klüppelberg (1996) and Klüppelberg and Kyprianou (2005), is our main objective.

Theorem 7.16. *Suppose that X is any spectrally positive Lévy process with mean $\mathbb{E}(X_1) < 0$. Suppose that $\Pi(\cdot, \infty)$ is regularly varying at infinity with index $-(\alpha + 1)$ for some $\alpha \in (0, \infty)$. Then we have the following ruinous behaviour:*

(i) As $x \uparrow \infty$ we have

$$\mathbb{P}(\tau_x^+ < \infty) \sim \frac{1}{|\mathbb{E}(X_1)|} \int_x^\infty \Pi(y, \infty)\mathrm{d}y$$

and consequently, the ruin probability is regularly varying at infinity with index $-\alpha$. (Note that convexity of the Laplace exponent of $-X$ dictates that $|\mathbb{E}(X_1)| < \infty$ when $\mathbb{E}(X_1) < 0$).

(ii) For all $u, v > 0$,

$$\lim_{x\uparrow\infty} \mathbb{P}\left(\frac{X_{\tau_x^+} - x}{x/\alpha} > u, \frac{-X_{\tau_x^+ -}}{x/\alpha} > v \middle| \tau_x^+ < \infty \right) = \left(1 + \frac{v+u}{\alpha}\right)^{-\alpha}. \tag{7.27}$$

Part (i) of the above theorem shows that when the so called *Cramér condition* appearing in Theorem 7.6 fails, conditions may exist where one may still gain information about the asymptotic behaviour of the ruin probability. Part (ii) shows that with re-scaling, the overshoot and undershoot converges to a non-trivial distribution. In fact the limiting distribution takes the form of a bivariate generalised Pareto distribution (cf. Definition 3.4.9 in Embrechts et al. (1997)). The result in part (ii) is also reminiscent of the following extraction from extreme value theory. It is known that a distribution, F, is in the domain of attraction of a generalised Pareto distribution if $F(\cdot, \infty)$ is regularly varying at infinity with index $-\alpha$ for some $\alpha > 0$; in which case

$$\lim_{x\uparrow\infty} \frac{F(x + xu/\alpha, \infty)}{F(x)} = \left(1 + \frac{u}{\alpha}\right)^{-\alpha}$$

for $\alpha > 0$ and $u > 0$.

Generalised Pareto distributions have heavy tails in the sense that their moment generating functions do not exist on the positive half of the real axis. Roughly speaking this means that there is a good chance to observe relatively large values when sampling from this distribution. This fact is important in the conclusion of the theorem as it confirms the folk-law that when modelling insurance risk where claims occur with heavy tails, when ruin occurs, it can occur when the insurance company has quite healthy reserves and the deficit at ruin will be quite devastating.

Proof (of Theorem 7.16). (i) Following the logic that leads to (7.8) we have that

$$\mathbb{P}(\tau_x^+ < \infty) = qU(x, \infty) = q \int_0^\infty \mathbb{P}(H_t > x)\mathrm{d}x,$$

where $q = \kappa(0,0)$ is the killing rate of the ascending ladder process. Writing $[t]$ for the integer part of t and note with the help of Lemma 7.14 that for $x > 0$,

$$\frac{\mathbb{P}(H_t > x)}{\mathbb{P}(H_1 > x)} \leq \frac{\mathbb{P}(H_{[t]+1} > x)}{\mathbb{P}(H_1 > x)} \leq C(1+\varepsilon)^{[t]+1}\mathrm{e}^{-q[t]}.$$

(To see where the exponential term on the right-hand side comes from, recall that H is equal in law to a subordinator killed independently at rate q). Now appealing to the Dominated Convergence Theorem, we have

$$\lim_{x\uparrow\infty} \frac{\mathbb{P}(\tau_x^+ < \infty)}{\mathbb{P}(H_1 > x)} = q\int_0^\infty \mathrm{d}t \cdot \lim_{x\uparrow\infty} \frac{\mathbb{P}(H_t > x)}{\mathbb{P}(H_1 > x)}. \tag{7.28}$$

In order to deal with the limit on the right-hand side above, we shall use the fact that $\mathbb{P}(H_t > x) = \mathrm{e}^{-qt}\mathbb{P}(\mathcal{H}_t > x)$ where $\mathcal{H}_t$ is an infinitely divisible random variable. To be more specific, one may think of $\{\mathcal{H}_t : t \geq 0\}$ as a subordinator which when killed at an independent and exponentially distributed time with parameter q, has the same law as $\{H_t : t \geq 0\}$. Associated to the random variable $\mathcal{H}_t$ is its Lévy measure, which necessarily takes the form $t\Pi_H$. By Lemma 7.15 it follows that

$$\lim_{x\uparrow\infty} \frac{\mathbb{P}(H_t > x)}{\mathbb{P}(H_1 > x)} = t\mathrm{e}^{-q(t-1)}.$$

Hence referring back to (7.28) we have that

$$\lim_{x\uparrow\infty} \frac{\mathbb{P}(\tau_x^+ < \infty)}{\Pi_H(x,\infty)} = q\int_0^\infty t\mathrm{e}^{-qt}\mathrm{d}t = \frac{1}{q}.$$

On the other hand, taking account of exponential killing, one easily computes

$$U(\infty) = \int_0^\infty \mathbb{P}(H_t < \infty)\mathrm{d}t = \int_0^\infty \mathrm{e}^{-qt}\mathrm{d}t = \frac{1}{q}.$$

The conclusion of the first part now follows once we note from (7.14) that $U(\infty) = \lim_{\beta\downarrow 0} \beta/\psi(\beta) = 1/\psi'(0+)$ where $\psi(\beta) = \log \mathbb{E}(\mathrm{e}^{-\beta X_1})$ and which in particular implies $q = |\mathbb{E}(X_1)|$.

(ii) Applying the quintuple law in marginalised form, we have

$$\begin{aligned}&\mathbb{P}\left(X_{\tau_x^+} - x > u^*, x - X_{\tau_x^+-} > v^*\right)\\&= \int_0^x U(x - \mathrm{d}y)\int_{[v^*\vee y,\infty)} \mathrm{d}z\Pi(u^* + z, \infty)\end{aligned}$$

for $u^*, v^* > 0$. For any spectrally positive Lévy process which drifts to $-\infty$, Vigon's identity in Theorem 7.8 reads

$$\Pi_H(u,\infty) = \int_u^\infty \Pi(z,\infty)\mathrm{d}z.$$

Choosing $u^* = ux/\alpha$ and $v^* = x + vx/\alpha$ we find that

$$\mathbb{P}\left(\frac{X_{\tau_x^+} - x}{x/\alpha} > u, \frac{-X_{\tau_x^+ -}}{x/\alpha} > v\right) = U(x)\Pi(x + x(v+u)/\alpha, \infty). \tag{7.29}$$

From part (i), if the limit exists then it holds that

$$\begin{aligned}\lim_{x\uparrow\infty} \mathbb{P}&\left(\frac{X_{\tau_x^+} - x}{x/\alpha} > u, \frac{-X_{\tau_x^+ -}}{x/\alpha} > v \,\middle|\, \tau_x < \infty\right)\\ &= \lim_{x\uparrow\infty} \frac{U(x)}{U(\infty)} \frac{\Pi(x + x(v+u)/\alpha, \infty)}{\Pi_H(x, \infty)}.\end{aligned} \tag{7.30}$$

Since $\Pi(\cdot, \infty)$ is regularly varying with index $-(\alpha+1)$ by assumption, the Monotone Density Theorem 5.14 implies that $\Pi_H(\cdot, \infty)$ is regularly varying with index $-\alpha$. Hence the limit in (7.30) exists and in particular (7.27) holds thus concluding the proof. □

Exercises

7.1 (Moments of the supremum). Fix $n = 1, 2, ...$ and suppose that

$$\int_{(1,\infty)} x^n \Pi(\mathrm{d}x) < \infty \tag{7.31}$$

(or equivalently $\mathbb{E}((\max\{X_1, 0\})^n) < \infty$ by Exercise 3.3).

(i) Suppose that X^K is the Lévy process with the same characteristics as X except that the measure Π is replaced by Π^K where

$$\Pi^K(\mathrm{d}x) = \Pi(\mathrm{d}x)\mathbf{1}_{(x>-K)} + \delta_{-K}(\mathrm{d}x)\Pi(-\infty, -K).$$

In other words, the paths of X^K are an adjustment of the paths of X in that all negative jumps of magnitude K or greater are replaced by a negative jump of precisely magnitude K.
Deduce that $\mathbb{E}(|X_t^K|^n) < \infty$ for all $t \geq 0$ and that the descending ladder height process of X^K has moments of all orders.

(ii) Use the Wiener–Hopf factorisation together with Maclaurin expansions up to order n of the terms therein to deduce that

$$\mathbb{E}(\overline{X}_{\mathbf{e}_q}^n) < \infty$$

holds for any $q > 0$.

(iii) Now suppose that $q = 0$ and $\limsup_{t\uparrow\infty} X_t < \infty$ and that (7.31) holds for $n = 2, 3, ...$ By adapting the arguments above, show that

$$\mathbb{E}(\overline{X}_\infty^{n-1}) < \infty.$$

7.2 (The Strong Law of Large Numbers for Lévy Processes). Suppose that X is a Lévy process such that $\mathbb{E}|X_1| < \infty$. For $n \geq 0$ let $Y_n = \sup_{t\in[n,n+1]} |X_t - X_n|$. Clearly this is a sequence of independent and identically distributed random variables.

(i) Use the previous exercise to show that $\mathbb{E}(Y_n) < \infty$.
(ii) Use the classical Strong Law of Large Numbers to deduce that $\lim_{n\uparrow\infty} n^{-1} Y_n = 0$.
(iii) Now deduce that

$$\lim_{t\uparrow\infty} \frac{X_t}{t} = \mathbb{E}(X_1).$$

(iv) Now suppose that $\mathbb{E}(X_1) = \infty$. Show that

$$\lim_{t\uparrow\infty} \frac{X_t}{t} = \infty.$$

(Hint: consider truncating the Lévy measure on $(0, \infty)$).

7.3. The idea of this exercise is to recover the conclusion of Theorem 7.2 for spectrally negative Lévy processes, X, using Theorem 7.1 and the Wiener–Hopf factorisation. As usual the Laplace exponent of X is denoted ψ and its right inverse is Φ.

(i) Using one of the Wiener–Hopf factors show that

$$\mathbb{E}\left(\mathrm{e}^{\beta \underline{X}_\infty} \mathbf{1}_{(-\underline{X}_\infty < \infty)}\right) = \begin{cases} 0 & \text{if } \psi'(0) < 0 \\ \psi'(0)\beta/\psi(\beta) & \text{if } \psi'(0) \geq 0. \end{cases}$$

(ii) Using the other Wiener–Hopf factor show that

$$\mathbb{E}\left(\mathrm{e}^{-\beta \overline{X}_\infty} \mathbf{1}_{(\overline{X}_\infty < \infty)}\right) = \begin{cases} \Phi(0)/(\beta + \Phi(0)) & \text{if } \psi'(0) < 0 \\ 0 & \text{if } \psi'(0) \geq 0. \end{cases}$$

(iii) Deduce from Theorem 7.1 that $\lim_{t\uparrow\infty} X_t = \infty$ when $\mathbb{E}(X_1) > 0$, $\lim_{t\uparrow\infty} X_t = -\infty$ when $\mathbb{E}(X_1) < 0$, and $\limsup_{t\uparrow\infty} X_t = -\liminf_{t\uparrow\infty} X_t = \infty$ when $\mathbb{E}(X_1) = 0$.
(iv) Deduce that a spectrally negative stable process of index $\alpha \in (1, 2)$ necessarily oscillates.

7.4. Let X be a stable process with index $\alpha \in (0, 2)$ which is not (the negative of) a subordinator and has positive jumps. Let $\rho = \mathbb{P}(X_t \geq 0)$.

(i) Explain why such processes cannot creep upwards. If further it experiences negative jumps, explain why it cannot creep downwards either.
(ii) Suppose that $U(dx) = \int_{[0,\infty)} \mathcal{U}(\mathrm{d}x, \mathrm{d}s)$ for $x \geq 0$. Show that (up to a multiplicative constant)

$$U(\mathrm{d}x) = \frac{x^{\alpha\rho - 1}}{\Gamma(\alpha\rho)} \mathrm{d}x$$

for $x \geq 0$. [Hint: reconsider Exercise 5.6].

(iii) Show that for $y \in [0, x]$, $v \geq y$ and $u > 0$,

$$\mathbb{P}(X_{\tau_x^+} - x \in \mathrm{d}u, x - X_{\tau_x^+ -} \in \mathrm{d}v, x - \overline{X}_{\tau_x^+ -} \in \mathrm{d}y) = c \cdot \frac{(x-y)^{\alpha\rho - 1}(v-y)^{\alpha(1-\rho)-1}}{(v+u)^{1+\alpha}} dy dv du$$

where c is a strictly positive constant.

(iv) Explain why the constant c must normalise the above triple law to a probability distribution. Show that

$$c = \frac{\sin \alpha\rho\pi}{\pi} \frac{\Gamma(\alpha+1)}{\Gamma(\alpha\rho)\Gamma(\alpha(1-\rho))}.$$

7.5. Suppose that X is a Lévy process but not a compound Poisson process with jump measure Π and ascending ladder process (L^{-1}, H) having jump measure $\mathbf{\Pi}(\mathrm{d}t, \mathrm{d}h)$. Using the conclusion of Exercise 5.4 show that

$$\mathbf{\Pi}(\mathrm{d}t, \mathrm{d}h) = \int_{[0,\infty)} \widehat{\mathcal{U}}(\mathrm{d}t, \mathrm{d}\theta)\Pi(\mathrm{d}h + \theta).$$

Show further that if X is spectrally positive then

$$\mathbf{\Pi}(\mathrm{d}t, \mathrm{d}h) = \int_0^\infty \mathrm{d}\theta \cdot \mathbb{P}(L_\theta^{-1} \in \mathrm{d}t)\Pi(\mathrm{d}h + \theta).$$

7.6. Here we deduce some statements about creeping and hitting points.

(i) Show that

$$\lim_{|\theta| \uparrow \infty} \frac{\Psi(\theta)}{\theta^2} = \frac{\sigma^2}{2}$$

where σ is the Gaussian coefficient of Ψ. Show with the help of the Wiener–Hopf factorisation, $k'\Psi(\theta) = \kappa(0, -\mathrm{i}\theta)\widehat{\kappa}(0, \mathrm{i}\theta)$, that a Lévy process creeps both upwards and downwards if and only if it has a Gaussian component (that is if and only if $\sigma > 0$).

(ii) Show that a Lévy process of bounded variation with Lévy–Khintchine exponent

$$\Psi(\theta) = -i\theta d + \int_{\mathbb{R}\setminus\{0\}} (1 - \mathrm{e}^{i\theta x})\Pi(\mathrm{d}x)$$

creeps upwards if and only if $d > 0$.

(iii) Show that any Lévy process for which 0 is irregular for $(0, \infty)$ cannot creep upwards.

(iv) Show that a spectrally negative Lévy process with no Gaussian component cannot creep downwards.

(v) Use part (i) to show that a symmetric α-stable process with $\alpha \in (1, 2)$ cannot creep. Use the integral test (7.21) to deduce that the latter Lévy process can hit all points.

7.7. Characterising the first passage over a fixed level for a general Lévy process is clearly non-trivial. Characterizing the first passage outside of a fixed interval for a general Lévy process (the so-called two-sided exit problem) has many more difficulties. This exercise concerns one of the few examples of Lévy processes for which an explicit characterisation can be obtained. The result is due to Rogozin (1972).

Suppose that X is a stable process with both positive and negative jumps and index $\alpha \in (0, 2)$ with symmetry parameter $\beta \in (-1, 1)$. The case that X or $-X$ is a subordinator or spectrally negative is thus excluded.[5] From the discussion in Sect. 6.5.3 we know that the positivity parameter $\rho \in (0, 1)$ and that $\alpha\rho \in (0, 1)$ and $\alpha(1 - \rho) \in (0, 1)$.

(i) With the help of the conclusion of Exercise 5.6 (ii) show that

$$\mathbb{P}_x(X_{\tau_1^+} \leq 1 + y) = \Phi_{\alpha\rho}\left(\frac{y}{1-x}\right)$$

for $x \leq 1$ and

$$\mathbb{P}_x(-X_{\tau_0^-} \leq y) = \Phi_{\alpha(1-\rho)}\left(\frac{y}{x}\right)$$

for $x \geq 0$ where

$$\Phi_q(u) = \begin{cases} \frac{\sin \pi q}{\pi} \int_0^u t^{-q}(1+t)^{-1}\mathrm{d}t & \text{for } u \geq 0 \\ 0 & \text{for } u < 0. \end{cases}$$

[Hint: it will be helpful to prove that

$$\int_0^{1/(1+\theta)} u^{\alpha-1}(1-u)^{-(\alpha+1)}\mathrm{d}v = \frac{\theta^{-\alpha}}{\alpha}$$

for any $\theta > 0$].

(ii) Define

$$r(x,y) = \mathbb{P}_x(X_{\tau_1^+} \leq 1+y; \tau_1^+ < \tau_0^-) \text{ and } l(x,y) = \mathbb{P}_x(X_{\tau_0^-} \geq -y; \tau_1^+ > \tau_0^-).$$

Show that the following system of equations hold

$$r(x,y) = \Phi_{\alpha\rho}\left(\frac{y}{1-x}\right) - \int_{(0,\infty)} \Phi_{\alpha\rho}\left(\frac{y}{1+z}\right) l(x,\mathrm{d}z)$$

and

$$l(x,y) = \Phi_{\alpha(1-\rho)}\left(\frac{y}{x}\right) - \int_{(0,\infty)} \Phi_{\alpha(1-\rho)}\left(\frac{y}{1+z}\right) r(x,\mathrm{d}z).$$

(iii) Assuming the above system of equations has a unique solution show that

$$r(x,y) = \frac{\sin \pi\alpha\rho}{\pi}(1-x)^{\alpha\rho}x^{\alpha(1-\rho)} \int_0^y t^{-\alpha\rho}(t+1)^{-\alpha(1-\rho)}(t+1-x)^{-1}\mathrm{d}t$$

and write down a similar expression for $l(x,y)$.

[5] The case that X or $-X$ is spectrally negative is dealt with later in Exercise 8.12.

8

Exit Problems for Spectrally Negative Processes

In this chapter we consider in more detail the special case of spectrally negative Lévy processes. As we have already seen in a number of examples in previous chapters, Lévy processes which have jumps in only one direction turn out to offer a significant advantage for many calculations. We devote our time in this chapter, initially, to gathering facts about spectrally negative processes from earlier chapters, and then to an ensemble of fluctuation identities which are semi-explicit in terms of a class of functions known as scale functions whose properties we shall also explore.

8.1 Basic Properties Reviewed

Let us gather what we have already established in previous chapters together with other easily derived facts.

The Laplace exponent. Rather than working with the Lévy–Khintchine exponent, it is preferable to work with the Laplace exponent

$$\psi(\lambda) := \frac{1}{t}\log\ \mathbb{E}\left(\mathrm{e}^{\lambda X_t}\right) = -\Psi(-\mathrm{i}\lambda),$$

which is finite at least for all $\lambda \geq 0$. The function $\psi : [0,\infty) \to \mathbb{R}$ is zero at zero and tends to infinity at infinity. Further, it is infinitely differentiable and strictly convex. In particular $\psi'(0+) = \mathbb{E}(X_1) \in [-\infty,\infty)$. Define the right inverse

$$\Phi(q) = \sup\{\lambda \geq 0 : \psi(\lambda) = q\}$$

for each $q \geq 0$. If $\psi'(0+) \geq 0$ then $\lambda = 0$ is the unique solution to $\psi(\lambda) = 0$ and otherwise there are two solutions to the latter with $\lambda = \Phi(0) > 0$ being the larger of the two, the other is $\lambda = 0$ (see Fig. 3.3).

First passage upwards. The first passage time above a level $x > 0$ has been defined by $\tau_x^+ = \inf\{t > 0 : X_t > x\}$. From Theorem 3.12 we know that for each $q \geq 0$,

$$\mathbb{E}(\mathrm{e}^{-q\tau_x^+}\mathbf{1}_{(\tau_x^+<\infty)}) = \mathrm{e}^{-\Phi(q)x}.$$

Further, the process $\{\tau_x^+ : x \geq 0\}$ is a subordinator with Laplace exponent $\Phi(q) - \Phi(0)$ killed at rate $\Phi(0)$.

Path variation. A spectrally negative Lévy process has paths of bounded or unbounded variation. Given the triple (a, σ, Π) as in Theorem 1.6 where necessarily $\mathrm{supp}\Pi \subseteq (-\infty, 0)$ we may always write

$$\psi(\lambda) = -a\lambda + \frac{1}{2}\sigma^2\lambda^2 + \int_{(-\infty,0)} (\mathrm{e}^{\lambda x} - 1 - \lambda x\mathbf{1}_{(x>-1)})\Pi(\mathrm{d}x),$$

where (a, σ, Π) are the same as those given in the Lévy–Khintchine formula. When X has bounded variation we may always write

$$\psi(\lambda) = d\lambda - \int_{(-\infty,0)} \left(1 - \mathrm{e}^{\lambda x}\right)\Pi(\mathrm{d}x), \tag{8.1}$$

where necessarily

$$d = -a - \int_{-1}^{0} x\Pi(\mathrm{d}x)$$

is strictly positive. Hence a spectrally negative Lévy process of bounded variation must always take the form of a positive drift minus a pure jump subordinator. Note that if $d \leq 0$ then we would see the Laplace exponent of a decreasing subordinator which is excluded from the definition of a spectrally negative process.

Regularity. From Theorem 6.5 (i) and (ii) one sees immediately then that 0 is regular for $(0, \infty)$ for X irrespective of path variation. Further, by considering the process $-X$, we can see from the same theorem that 0 is regular for $(-\infty, 0)$ if and only if X has unbounded variation. Said another way, 0 is regular for both $(0, \infty)$ and $(-\infty, 0)$ if and only if it has unbounded variation.

Creeping. We know from Corollary 3.13 and the fact that there are no positive jumps that

$$\mathbb{P}(X_{\tau_x^+} = x|\tau_x^+ < \infty) = 1.$$

Hence spectrally negative Lévy processes necessarily creep upwards. It was shown however in Exercise 7.6 that they creep downward if and only if $\sigma > 0$.

Wiener–Hopf factorisation. In Chap. 6 we identified up to a multiplicative constant

$$\kappa(\alpha, \beta) = \Phi(\alpha) + \beta \text{ and } \widehat{\kappa}(\alpha, \beta) = \frac{\alpha - \psi(\beta)}{\Phi(\alpha) - \beta}$$

for $\alpha, \beta \geq 0$. Appropriate choices of local time at the maximum and minimum allow the multiplicative constants to be taken as equal to unity. From Theorem 6.16 (ii) this leads to

$$\mathbb{E}\left(\mathrm{e}^{-\beta \overline{X}_{\mathbf{e}_p}}\right) = \frac{\Phi(p)}{\Phi(p) + \beta} \text{ and } \mathbb{E}\left(\mathrm{e}^{\beta \underline{X}_{\mathbf{e}_p}}\right) = \frac{p}{\Phi(p)} \frac{\Phi(p) - \beta}{p - \psi(\beta)}, \tag{8.2}$$

where $\mathbf{e}_p$ is an independent and exponentially distributed random variable with parameter $p \geq 0$. The first of these two expressions shows that $\overline{X}_{\mathbf{e}_p}$ is exponentially distributed with parameter $\Phi(p)$. Note that when $p = 0$ in the last statement, we employ our usual notation that an exponential variable with parameter zero is infinite with probability one.

Drifting and oscillating. From Theorem 7.2 or Exercise 7.3 we have the following asymptotic behaviour for X. The process drifts to infinity if and only if $\psi'(0+) > 0$, oscillates if and only if $\psi'(0+) = 0$ and drifts to minus infinity if and only if $\psi'(0+) < 0$.

Exponential change of measure. From Exercise 1.5 we know that

$$\{\mathrm{e}^{cX_t - \psi(c)t} : t \geq 0\}$$

is a martingale for each $c \geq 0$. Define for each $c \geq 0$ the change of measure

$$\left.\frac{d\mathbb{P}^c}{d\mathbb{P}}\right|_{\mathcal{F}_t} = \mathrm{e}^{cX_t - \psi(c)t}. \tag{8.3}$$

(Recall that $\mathbb{F} = \{\mathcal{F}_t : t \geq 0\}$ is the filtration generated by X which has been completed by null sets of $\mathbb{P}$, which in the case of a Lévy process means that it is right continuous). When X is a Brownian motion this is the same change of measure that appears in the most elementary form of the Cameron–Martin–Girsanov Theorem. In that case we know that the effect of the change of measure is that the process X under $\mathbb{P}^c$ has the same law as a Brownian motion with drift c. In Sect. 3.3 we showed that when $(X, \mathbb{P})$ is a spectrally negative Lévy process then $(X, \mathbb{P}^c)$ is also a spectrally negative Lévy process and a straightforward calculation shows that its Laplace exponent $\psi_c(\lambda)$ is given by

$$\begin{aligned}\psi_c(\lambda) &= \psi(\lambda + c) - \psi(c) \\ &= \left(\sigma^2 c - a + \int_{(-\infty,0)} x(\mathrm{e}^{cx} - 1)\mathbf{1}_{(x>-1)}\Pi(\mathrm{d}x)\right)\lambda \\ &\quad + \frac{1}{2}\sigma^2\lambda^2 + \int_{(-\infty,0)} (\mathrm{e}^{\lambda x} - 1 - \lambda x \mathbf{1}_{(x>-1)})\mathrm{e}^{cx}\Pi(\mathrm{d}x),\end{aligned} \tag{8.4}$$

for $\lambda \geq -c$.

When we set $c = \Phi(p)$ for $p \geq 0$ we discover that $\psi_{\Phi(p)}(\lambda) = \psi(\lambda + \Phi(p)) - p$ and hence $\psi'_{\Phi(p)}(0) = \psi'(\Phi(p)) \geq 0$ on account of the strict convexity of ψ. In particular, $(X, \mathbb{P}^{\Phi(p)})$ always drifts to infinity for $p > 0$. Roughly speaking, the effect of the change of measure has been to change the characteristics of X to those of a spectrally negative Lévy process with exponentially tilted Lévy measure. Note also that $(X, \mathbb{P})$ is of bounded variation if and only if $(X, \mathbb{P}^c)$ is of bounded variation. This statement is clear when $\sigma > 0$. When $\sigma = 0$ it is justified by noting that $\int_{(-1,0)} |x|\Pi(\mathrm{d}x) < \infty$ if and only if $\int_{(-1,0)} |x|\mathrm{e}^{cx}\Pi(\mathrm{d}x) < \infty$. In the case that X is of bounded variation and we write the Laplace exponent in the form (8.1) we also see from the second equality of (8.4) that

$$\psi_c(\theta) = d\theta - \int_{(-\infty,0)} (1 - e^{\theta x})e^{cx}\Pi(\mathrm{d}x).$$

Thus under $\mathbb{P}^c$ the process retains the same drift and only the Lévy measure is exponentially tilted.

8.2 The One-Sided and Two-Sided Exit Problems

In this section we shall develop semi-explicit identities concerning exiting from a half line and a strip. Recall that $\mathbb{P}_x$ and $\mathbb{E}_x$ are shorthand for $\mathbb{P}(\cdot|X_0 = x)$ and $\mathbb{E}(\cdot|X_0 = x)$ and for the special case that $x = 0$ we keep with our old notation, so that $\mathbb{P}_0 = \mathbb{P}$ and $\mathbb{E}_0 = \mathbb{E}$, unless we wish to emphasise the fact that $X_0 = 0$. Recall also

$$\tau_x^+ = \inf\{t > 0 : X_t > x\} \text{ and } \tau_x^- = \inf\{t > 0 : X_t < x\}$$

for all $x \in \mathbb{R}$. The main results of this section are the following.

Theorem 8.1 (One- and two-sided exit formulae). *There exist a family of functions $W^{(q)} : \mathbb{R} \to [0, \infty)$ and*

$$Z^{(q)}(x) = 1 + q\int_0^x W^{(q)}(y)\mathrm{d}y, \text{ for } x \in \mathbb{R}$$

defined for each $q \geq 0$ such that the following hold (for short we shall write $W^{(0)} = W$).

(i) For any $q \geq 0$, we have $W^{(q)}(x) = 0$ for $x < 0$ and $W^{(q)}$ is characterised on $[0, \infty)$ as a strictly increasing and continuous function whose Laplace transform satisfies

$$\int_0^\infty \mathrm{e}^{-\beta x}W^{(q)}(x)\mathrm{d}x = \frac{1}{\psi(\beta) - q} \text{ for } \beta > \Phi(q). \tag{8.5}$$

(ii) For any $x \in \mathbb{R}$ and $q \geq 0$,

$$\mathbb{E}_x\left(\mathrm{e}^{-q\tau_0^-}\mathbf{1}_{(\tau_0^- < \infty)}\right) = Z^{(q)}(x) - \frac{q}{\Phi(q)}W^{(q)}(x), \tag{8.6}$$

where we understand $q/\Phi(q)$ in the limiting sense for $q = 0$, so that

$$\mathbb{P}_x(\tau_0^- < \infty) = \begin{cases} 1 - \psi'(0+)W(x) & \text{if } \psi'(0+) > 0 \\ 1 & \text{if } \psi'(0+) \leq 0 \end{cases}. \tag{8.7}$$

(iii) For any $x \leq a$ and $q \geq 0$,

$$\mathbb{E}_x\left(\mathrm{e}^{-q\tau_a^+}\mathbf{1}_{(\tau_0^- > \tau_a^+)}\right) = \frac{W^{(q)}(x)}{W^{(q)}(a)}, \tag{8.8}$$

and

$$\mathbb{E}_x\left(e^{-q\tau_0^-}\mathbf{1}_{(\tau_0^- < \tau_a^+)}\right) = Z^{(q)}(x) - Z^{(q)}(a)\frac{W^{(q)}(x)}{W^{(q)}(a)}. \tag{8.9}$$

The function W has been called the scale function because of the analogous role it plays in (8.8) to scale functions for diffusions in the sense of equation (8.8). In keeping with existing literature we will refer to the functions $W^{(q)}$ and $Z^{(q)}$ as the q-scale functions.[1]

Note also that (8.7) should give the Pollaczek–Khintchine formula (1.11) when X is taken as the Cramér–Lundberg risk process. Exercise 8.3 shows how.

Let us make some remarks on the historical appearance of these formulae. Identity (8.6) appears in the form of its Fourier transform in Emery (1973) and for the case that Π is finite and $\sigma = 0$ in Korolyuk (1975a). Identity (8.8) first appeared for the case $q = 0$ in Zolotarev (1964) followed by Takács (1966) and then with a short proof in Rogers (1990). The case $q > 0$ is found in Korolyuk (1975a) for the case that Π is finite and $\sigma = 0$, in Bertoin (1996a) for the case of a purely asymmetric stable process and then again for a general spectrally negative Lévy process in Bertoin (1997a) (who refered to a method used for the case $q = 0$ in Bertoin (1996a)). See also Doney (2006) for further remarks on this identity. Finally (8.9) was proved for the case that Π is finite and $\sigma = 0$ by Korolyuk (1974, 1975a); see Bertoin (1997a) for the general case.

Proof (of Theorem 8.1 (8.8)). We prove (8.8) for the case that $\psi'(0+) > 0$ and $q = 0$, then for the case that $q > 0$ (no restriction on $\psi'(0+)$). Finally the case that $\psi'(0+) \leq 0$ and $q = 0$ is achieved by passing to the limit as q tends to zero.

Assume that $\psi'(0+) > 0$ so that $-\underline{X}_\infty$ is $\mathbb{P}$-almost surely finite. Now define the non-decreasing function

$$W(x) = \mathbb{P}_x(\underline{X}_\infty \geq 0).$$

[1] One may also argue that the terminology "scale function" is also inappropriate as the mentioned analogy breaks down in a number of other respects.

A simple argument using the law of total probability and the Strong Markov Property now yields for $x \in [0,a)$

$$\begin{aligned}\mathbb{P}_x(\underline{X}_\infty \geq 0) &= \mathbb{E}_x\left(\mathbb{P}_x(\underline{X}_\infty \geq 0|\mathcal{F}_{\tau_a^+})\right)\\ &= \mathbb{E}_x\left(\mathbf{1}_{(\tau_a^+<\tau_0^-)}\mathbb{P}_a(\underline{X}_\infty \geq 0)\right) + \mathbb{E}_x\left(\mathbf{1}_{(\tau_a^+>\tau_0^-)}\mathbb{P}_{X_{\tau_0^-}}(\underline{X}_\infty \geq 0)\right)\\ &= \mathbb{P}_a(\underline{X}_\infty \geq 0)\mathbb{P}_x(\tau_a^+ < \tau_0^-).\end{aligned}$$

To justify that the second term in the second equality disappears note the following. If X has no Gaussian component then it cannot creep downwards implying that $X_{\tau_0^-} < 0$ and then we use that $\mathbb{P}_x(\underline{X}_\infty \geq 0) = 0$ for $x < 0$. If X has a Gaussian component then $X_{\tau_0^-} \leq 0$ and we need to know that $W(0) = 0$. However since 0 is regular for $(-\infty, 0)$ and $(0,\infty)$ it follows that $\underline{X}_\infty < 0$ $\mathbb{P}$-almost surely which is the same as $W(0) = 0$.

We now have

$$\mathbb{P}_x(\tau_a^+ < \tau_0^-) = \frac{W(x)}{W(a)}. \tag{8.10}$$

It is trivial, but nonetheless useful for later use, to note that the same equality holds even when $x < 0$ since both sides are equal to zero.

Now assume that $q > 0$ or $q = 0$ and $\psi'(0+) < 0$. In this case, by the convexity of ψ, we know that $\Phi(q) > 0$ and hence $\psi'_{\Phi(q)}(0) = \psi'(\Phi(q)) > 0$ (again by convexity) which implies that under $\mathbb{P}^{\Phi(q)}$, the process X drifts to infinity. For $(X, \mathbb{P}^{\Phi(q)})$ we have already established the existence of a 0-scale function $W_{\Phi(q)}(x) = \mathbb{P}_x^{\Phi(q)}(\underline{X}_\infty \geq 0)$ which fulfils the relation

$$\mathbb{P}_x^{\Phi(q)}(\tau_a^+ < \tau_0^-) = \frac{W_{\Phi(q)}(x)}{W_{\Phi(q)}(a)}. \tag{8.11}$$

However by definition of $\mathbb{P}^{\Phi(q)}$, we also have that

$$\begin{aligned}\mathbb{P}_x^{\Phi(q)}(\tau_a^+ < \tau_0^-) &= \mathbb{E}_x(\mathrm{e}^{\Phi(q)(X_{\tau_a^+}-x)-q\tau_a^+}\mathbf{1}_{(\tau_a^+<\tau_0^-)})\\ &= \mathrm{e}^{\Phi(q)(a-x)}\mathbb{E}_x(\mathrm{e}^{-q\tau_a^+}\mathbf{1}_{(\tau_a^+<\tau_0^-)}).\end{aligned} \tag{8.12}$$

Combining (8.11) and (8.12) gives

$$\mathbb{E}_x\left(\mathrm{e}^{-q\tau_a^+}\mathbf{1}_{(\tau_a^+<\tau_0^-)}\right) = \mathrm{e}^{-\Phi(q)(a-x)}\frac{W_{\Phi(q)}(x)}{W_{\Phi(q)}(a)} = \frac{W^{(q)}(x)}{W^{(q)}(a)}, \tag{8.13}$$

where $W^{(q)}(x) = \mathrm{e}^{\Phi(q)x}W_{\Phi(q)}(x)$. Clearly $W^{(q)}$ is identically zero on $(-\infty, 0)$ and non-decreasing.

Suppose now the final case that $\psi'(0+) = 0$ and $q = 0$. Since the limit as $q \downarrow 0$ on the left-hand side of (8.13) exists then the same is true of the right-hand side-By choosing an arbitrary $b > a$ we can thus define, $W(x) = \lim_{q\downarrow 0} W^{(q)}(x)/W^{(q)}(b)$ for each $x \leq a$. Consequently,

$$\begin{aligned} W(x) &= \lim_{q\downarrow 0} \frac{W^{(q)}(x)}{W^{(q)}(b)} \\ &= \lim_{q\downarrow 0} \mathbb{E}_x \left(\mathrm{e}^{-q\tau_a^+} \mathbf{1}_{(\tau_a^+ < \tau_0^-)} \right) \frac{W^{(q)}(a)}{W^{(q)}(b)} \\ &= \mathbb{P}_x(\tau_a^+ < \tau_0^-) W(a). \end{aligned} \tag{8.14}$$

Again it is clear that W is identically zero on $(-\infty, 0)$ and non-decreasing.

It is important to note for the remaining parts of the proof that the definition of $W^{(q)}$ we have given above may be taken up to any multiplicative constant without affecting the validity of the arguments. □

Proof (of Theorem 8.1 (i)). Suppose again that X is assumed to drift to infinity so that $\psi'(0+) > 0$. First consider the case that $q = 0$. Recalling that the definition of W in (8.8) may be taken up to a multiplicative constant, let us work with

$$W(x) = \frac{1}{\psi'(0+)} \mathbb{P}_x(\underline{X}_\infty \geq 0). \tag{8.15}$$

We may take limits in the second Wiener–Hopf factor given in (8.2) to deduce that

$$\mathbb{E}\left(\mathrm{e}^{\beta \underline{X}_\infty}\right) = \psi'(0+) \frac{\beta}{\psi(\beta)}$$

for $\beta > 0$. Integrating by parts, we also see that

$$\begin{aligned} \mathbb{E}\left(\mathrm{e}^{\beta \underline{X}_\infty}\right) &= \int_{[0,\infty)} \mathrm{e}^{-\beta x} \mathbb{P}\left(-\underline{X}_\infty \in \mathrm{d}x\right) \\ &= \mathbb{P}\left(-\underline{X}_\infty = 0\right) + \int_{(0,\infty)} \mathrm{e}^{-\beta x} \, \mathrm{d}\mathbb{P}\left(-\underline{X}_\infty \in (0, x]\right) \\ &= \int_0^\infty \mathbb{P}\left(-\underline{X}_\infty = 0\right) \beta \, \mathrm{e}^{-\beta x} \, \mathrm{d}x + \beta \int_0^\infty \mathrm{e}^{-\beta x} \mathbb{P}\left(-\underline{X}_\infty \in (0, x]\right) \mathrm{d}x \\ &= \beta \int_0^\infty \mathrm{e}^{-\beta x} \mathbb{P}\left(-\underline{X}_\infty \leq x\right) \mathrm{d}x \\ &= \beta \int_0^\infty \mathrm{e}^{-\beta x} \mathbb{P}_x\left(\underline{X}_\infty \geq 0\right) \mathrm{d}x, \end{aligned}$$

and hence

$$\int_0^\infty \mathrm{e}^{-\beta x} W(x) \, \mathrm{d}x = \frac{1}{\psi(\beta)} \tag{8.16}$$

for all $\beta > 0 = \Phi(0)$.

Now for the case that $q > 0$ or $q = 0$ and $\psi'(0+) < 0$ take as before $W^{(q)}(x) = \mathrm{e}^{\Phi(q)x} W_{\Phi(q)}(x)$. As remarked earlier, X under $\mathbb{P}^{\Phi(q)}$ drifts to infinity and hence using the conclusion from the previous paragraph together with

(8.4) we have

$$\int_0^\infty \mathrm{e}^{-\beta x} W^{(q)}(x)\mathrm{d}x = \int_0^\infty \mathrm{e}^{-(\beta-\Phi(q))x} W_{\Phi(q)}(x)\mathrm{d}x$$
$$= \frac{1}{\psi_{\Phi(q)}(\beta - \Phi(q))}$$
$$= \frac{1}{\psi(\beta) - q}$$

provided $\beta - \Phi(q) > 0$. Since $W^{(q)}$ is an increasing function we can also talk of the measure $W^{(q)}(\mathrm{d}x)$ associated with the distribution $W^{(q)}(0,x]$. Integration by parts gives a characterisation of the measure $W^{(q)}$,

$$\int_{[0,\infty)} \mathrm{e}^{-\beta x}\ W^{(q)}(\mathrm{d}x) = W^{(q)}(0) + \int_{(0,\infty)} \mathrm{e}^{-\beta x}\ \mathrm{d}W^{(q)}(0,x]$$
$$= \int_0^\infty \beta\ \mathrm{e}^{-\beta x}\ W^{(q)}(0)\mathrm{d}x + \int_0^\infty \beta\ \mathrm{e}^{-\beta x}\ W^{(q)}(0,x]\mathrm{d}x$$
$$= \frac{\beta}{\psi(\beta) - q} \tag{8.17}$$

for $\beta > \Phi(q)$.

For the case that $q = 0$ and $\psi'(0+) = 0$ one may appeal to the Extended Continuity Theorem for Laplace Transforms (see Feller (1971), Theorem XIII.1.2a) to deduce that since

$$\lim_{q\downarrow 0} \int_{[0,\infty)} \mathrm{e}^{-\beta x}\ W^{(q)}(\mathrm{d}x) = \lim_{q\downarrow 0} \frac{\beta}{\psi(\beta) - q} = \frac{\beta}{\psi(\beta)},$$

then there exists a measure W^* such that $W^*(\mathrm{d}x) = \lim_{q\downarrow 0} W^{(q)}(\mathrm{d}x)$

$$\int_{[0,\infty)} \mathrm{e}^{-\beta x}\, W^*(\mathrm{d}x) = \frac{\beta}{\psi(\beta)}.$$

Clearly $W^*(x) := W^*[0,x]$ is a multiple of W given in (8.14) so we may take as definition $W = W^*$. Hence integration by parts shows that (8.16) holds again.

Next we turn to continuity and strict monotonicity of $W^{(q)}$. The argument is taken from Bertoin (1996a,b). Recall that $\{(t,\epsilon_t) : t \geq 0 \text{ and } \epsilon_t \neq \partial\}$ is the Poisson point process of excursions on $[0,\infty) \times \mathcal{E}$ with intensity $dt \times dn$ decomposing the path of X. Write $\overline{\epsilon}$ for the height of each excursion $\epsilon \in \mathcal{E}$; see Definition 6.14. Choose $a > x \geq 0$. For spectrally negative Lévy processes we work with the definition of local time $L = \overline{X}$ and hence $L_{\tau^+_{a-x}} = \overline{X}_{\tau^+_{a-x}} = a - x$. Therefore it holds that

$$\{\underline{X}_{\tau^+_{a-x}} > -x\} = \{\forall t \leq a-x \text{ and } \epsilon_t \neq \partial, \overline{\epsilon}_t < t + x\}.$$

It follows with the help of (8.10) that

$$\begin{aligned}\frac{W(x)}{W(a)} &= \mathbb{P}_x\left(\underline{X}_{\tau_a^+} > 0\right) \\ &= \mathbb{P}\left(\underline{X}_{\tau_{a-x}^+} > -x\right) \\ &= \mathbb{P}(\forall t \leq a - x \text{ and } \epsilon_t \neq \partial, \overline{\epsilon}_t < t + x) \\ &= \mathbb{P}(N(A) = 0),\end{aligned}$$

where N is the Poisson random measure associated with the process of excursions and $A = \{(t, \epsilon_t) : t \leq a-x \text{ and } \overline{\epsilon}_t \geq t+x\}$. Since $N(A)$ is Poisson distributed with parameter $\int \mathbf{1}_A \ \mathrm{d}t\ n(\mathrm{d}\epsilon) = \int_0^{a-x} n\left(\overline{\epsilon} \geq t + x\right) \mathrm{d}t = \int_x^a n\left(\overline{\epsilon} \geq t\right) \mathrm{d}t$ we have that

$$\frac{W(x)}{W(a)} = \exp\left\{-\int_x^a n\left(\overline{\epsilon} \geq t\right) \mathrm{d}t\right\}. \tag{8.18}$$

Since a may be chosen arbitrarily large, continuity and strict monotonicity follow from (8.18). Continuity of W also guarantees that it is uniquely defined via its Laplace transform on $[0, \infty)$. From the definition

$$W^{(q)}(x) = \mathrm{e}^{\Phi(q)x} W_{\Phi(q)}(x), \tag{8.19}$$

the properties of continuity, uniquecness and strict monotonicity carry over to the case $q > 0$. □

Proof (of Theorem 8.1 (ii)). Using the Laplace transform of $W^{(q)}(x)$ (given in (8.5)) as well as the Laplace–Stieltjes transform (8.17), we can interpret the second Wiener–Hopf factor in (8.2) as saying that for $x \geq 0$,

$$\mathbb{P}(-\underline{X}_{\mathbf{e}_q} \in \mathrm{d}x) = \frac{q}{\Phi(q)} W^{(q)}(\mathrm{d}x) - qW^{(q)}(x)\mathrm{d}x, \tag{8.20}$$

and hence for $x \geq 0$,

$$\begin{aligned}\mathbb{E}_x\left(\mathrm{e}^{-q\tau_0^-}\mathbf{1}_{(\tau_0^- < \infty)}\right) &= \mathbb{P}_x(\mathbf{e}_q > \tau_0^-) \\ &= \mathbb{P}_x(\underline{X}_{\mathbf{e}_q} < 0) \\ &= \mathbb{P}(-\underline{X}_{\mathbf{e}_q} > x) \\ &= 1 - \mathbb{P}(-\underline{X}_{\mathbf{e}_q} \leq x) \\ &= 1 + q\int_0^x W^{(q)}(y)\mathrm{d}y - \frac{q}{\Phi(q)} W^{(q)}(x) \\ &= Z^{(q)}(x) - \frac{q}{\Phi(q)} W^{(q)}(x)\end{aligned} \tag{8.21}$$

Note that since $Z^{(q)}(x) = 1$ and $W^{(q)}(x) = 0$ for all $x \in (-\infty, 0)$, the statement is valid for all $x \in \mathbb{R}$. The proof is now complete for the case that $q > 0$.

Finally we have that $\lim_{q\downarrow 0} q/\Phi(q) = \lim_{q\downarrow\infty} \psi(\Phi(q))/\Phi(q)$ which is either $\psi'(0+)$ if the process drifts to infinity or oscillates so that $\Phi(0) = 0$, or zero otherwise when $\Phi(0) > 0$. The proof is thus completed by taking the limit in q in (8.6). □

Proof (of Theorem 8.1 (8.9)). Fix $q > 0$. We have for $x \geq 0$,

$$\mathbb{E}_x(\mathrm{e}^{-q\tau_0^-}\mathbf{1}_{(\tau_0^- < \tau_a^+)}) = \mathbb{E}_x(\mathrm{e}^{-q\tau_0^-}\mathbf{1}_{(\tau_0^- < \infty)}) - \mathbb{E}_x(\mathrm{e}^{-q\tau_0^-}\mathbf{1}_{(\tau_a^+ < \tau_0^-)}).$$

Applying the Strong Markov Property at τ_a^+ and using the fact that X creeps upwards, we also have that

$$\mathbb{E}_x(\mathrm{e}^{-q\tau_0^-}\mathbf{1}_{(\tau_a^+ < \tau_0^-)}) = \mathbb{E}_x(\mathrm{e}^{-q\tau_a^+}\mathbf{1}_{(\tau_a^+ < \tau_0^-)})\mathbb{E}_a(\mathrm{e}^{-q\tau_0^-}\mathbf{1}_{(\tau_0^- < \infty)}).$$

Appealing to (8.6) and (8.8) we now have that

$$\begin{aligned}\mathbb{E}_x(\mathrm{e}^{-q\tau_0^-}\mathbf{1}_{(\tau_0^- < \tau_a^+)}) &= Z^{(q)}(x) - \frac{q}{\Phi(q)}W^{(q)}(x) \\ &\quad - \frac{W^{(q)}(x)}{W^{(q)}(a)}\left(Z^{(q)}(a) - \frac{q}{\Phi(q)}W^{(q)}(a)\right)\end{aligned}$$

and the required result follows. □

8.3 The Scale Functions $W^{(q)}$ and $Z^{(q)}$

Let us explore a little further the analytical properties of the functions $W^{(q)}$ and $Z^{(q)}$.

Lemma 8.2. *For all $q \geq 0$, the function $W^{(q)}$ has left and right derivatives on $(0,\infty)$ which agree if and only if the measure $n(\bar{\epsilon} \in \mathrm{d}x)$ has no atoms. In that case, $W^{(q)} \in C^1(0,\infty)$.*

Proof. First suppose that $q > 0$. Since $W^{(q)}(x) := \mathrm{e}^{\Phi(q)x}W_{\Phi(q)}(x)$, it suffices to prove the result for $q = 0$. However, in this case, we identified

$$W(x) = W(a)\exp\left\{-\int_x^a n(\bar{\epsilon} \geq t)\mathrm{d}t\right\}$$

for any arbitrary $a > x$. It follows then that the left and right first derivatives exist and are given by

$$W'_-(x) = n(\bar{\epsilon} \geq x)W(x) \text{ and } W'_+(x) = n(\bar{\epsilon} > x)W(x).$$

Since W is continuous, W' exists if and only if $n(\bar{\epsilon} \in \mathrm{d}x)$ has no atoms as claimed. In that case it is clear that it also belongs to the class $C^1(0,\infty)$. □

Although the proof is a little technical, it can be shown that $n(\overline{\epsilon} \in \mathrm{d}x)$ has no atoms if X is a process of unbounded variation. If X has bounded variation then it is very easy to construct an example where $n(\overline{\epsilon} \in \mathrm{d}x)$ has at least one atom. Consider for example the case of a compound Poisson process with positive drift and negative jumps whose distribution has an atom at unity. An excursion may therefore begin with a jump of size one. Since thereafter the process may fail to jump again before reaching its previous maximum, we see the excursion measure of heights must have at least an atom at 1; $n(\overline{\epsilon} = 1) > 0$. In Exercise 8.3 however, for each $k = 1, 2, ...$, conditions are established under which W belongs to the class of $C^k(0, \infty)$ when the underlying process has bounded variation.

Next we look at how $W^{(q)}$ and $Z^{(q)}$ extend analytically in the parameter q. This will turn out to be important in some of the exercises at the end of this chapter.

Lemma 8.3. *For each $x \geq 0$, the function $q \mapsto W^{(q)}(x)$ may be analytically extended to $q \in \mathbb{C}$.*

Proof. For a fixed choice of $q > 0$,

$$\begin{aligned}\int_0^\infty \mathrm{e}^{-\beta x} W^{(q)}(x)\mathrm{d}x &= \frac{1}{\psi(\beta) - q} \\ &= \frac{1}{\psi(\beta)} \frac{1}{1 - q/\psi(\beta)} \\ &= \frac{1}{\psi(\beta)} \sum_{k\geq 0} q^k \frac{1}{\psi(\beta)^k} \qquad (8.22)\end{aligned}$$

for $\beta > \Phi(q)$. The latter inequality implies that $0 < q/\psi(\beta) < 1$. Next note that

$$\sum_{k\geq 0} q^k W^{*(k+1)}(x)$$

converges for each $x \geq 0$ where W^{*k} is the k th convolution of W with itself. This is easily deduced once one has the estimates

$$W^{*(k+1)}(x) \leq \frac{x^k}{k!} W(x)^{(k+1)}, \qquad (8.23)$$

which itself can easily be proved by induction. Indeed note that if (8.23) holds for $k \geq 1$ then by monotonicity of W,

$$\begin{aligned}W^{*(k+1)}(x) &\leq \int_0^x \frac{y^{k-1}}{(k-1)!} W(y)^k W(x-y)\,\mathrm{d}y \\ &\leq \frac{1}{(k-1)!} W(x)^{k+1} \int_0^x y^{k-1}\,\mathrm{d}y \\ &= \frac{x^k}{k!} W(x)^{k+1}.\end{aligned}$$

Returning to (8.22) we may now apply Fubini's Theorem (justified by the assumption that $\beta > \Phi(q)$) and deduce that

$$\begin{aligned}\int_0^\infty \mathrm{e}^{-\beta x} W^{(q)}(x)\mathrm{d}x &= \sum_{k\geq 0} q^k \frac{1}{\psi(\beta)^{k+1}} \\ &= \sum_{k\geq 0} q^k \int_0^\infty \mathrm{e}^{-\beta x} W^{*(k+1)}(x)\,\mathrm{d}x \\ &= \int_0^\infty \mathrm{e}^{-\beta x} \sum_{k\geq 0} q^k W^{*(k+1)}(x)\,\mathrm{d}x.\end{aligned}$$

Thanks to continuity of W and $W^{(q)}$ we have that

$$W^{(q)}(x) = \sum_{k\geq 0} q^k W^{*(k+1)}(x). \tag{8.24}$$

Now noting that $\sum_{k\geq 0} q^k W^{*(k+1)}(x)$ converges for all $q \in \mathbb{C}$ we may extend the definition of $W^{(q)}$ for each fixed $x \geq 0$ by the equality given in (8.24). □

Suppose that for each $c \geq 0$ we call $W_c^{(q)}$ the function fulfilling the definitions given in Theorem 8.1 but with respect to the measure $\mathbb{P}^c$. The previous Lemma allows us to establish the following relationship for $W_c^{(q)}$ with different values of q and c.

Lemma 8.4. *For any $q \in \mathbb{C}$ and $c \in \mathbb{R}$ such that $\psi(c) < \infty$ we have*

$$W^{(q)}(x) = \mathrm{e}^{cx} W_c^{(q-\psi(c))}(x) \tag{8.25}$$

for all $x \geq 0$.

Proof. For a given $c \in \mathbb{R}$ such that $\psi(c) < \infty$ the identity (8.25) holds for $q - \psi(c) \geq 0$ on account of both left and right-hand side being continuous functions with the same Laplace transform. By Lemma 8.3 both left- and right-hand side of (8.25) are analytic in q for each fixed $x \geq 0$. The Identity Theorem for analytic functions thus implies that they are equal for all $q \in \mathbb{C}$. □

Unfortunately a convenient relation such as (8.25) cannot be given for $Z^{(q)}$. Nonetheless we do have the following obvious corollary.

Corollary 8.5. *For each $x > 0$ the function $q \mapsto Z^{(q)}(x)$ may be analytically extended to $q \in \mathbb{C}$.*

The final Lemma of this section shows that a discontinuity of $W^{(q)}$ at zero may occur even when $W^{(q)}$ belongs to $C^1(0, \infty)$.

Lemma 8.6. *For all $q \geq 0$, $W^{(q)}(0) = 0$ if and only if X has unbounded variation. Otherwise, when X has bounded variation, it is equal to $1/d$ where $d > 0$ is the drift.*

Proof. Note that for all $q > 0$,

$$W^{(q)}(0) = \lim_{\beta\uparrow\infty} \int_0^\infty \beta\, \mathrm{e}^{-\beta x} W^{(q)}(x)\mathrm{d}x$$
$$= \lim_{\beta\uparrow\infty} \frac{\beta}{\psi(\beta) - q}$$
$$= \lim_{\beta\uparrow\infty} \frac{\beta - \Phi(q)}{\psi(\beta) - q}$$
$$= \frac{\Phi(q)}{q} \lim_{\beta\uparrow\infty} \mathbb{E}\left(\mathrm{e}^{\beta \underline{X}_{\mathbf{e}_q}}\right)$$
$$= \frac{\Phi(q)}{q} \mathbb{P}(\underline{X}_{\mathbf{e}_q} = 0).$$

Now recall that $\mathbb{P}(\underline{X}_{\mathbf{e}_q} = 0) > 0$ if and only if 0 is irregular for $(-\infty, 0)$ which was shown earlier to be for the case of processes having bounded variation. The above calculation also shows that

$$W^{(q)}(0) = \lim_{\beta\uparrow\infty} \frac{\beta}{\psi(\beta) - q} = \lim_{\beta\uparrow\infty} \frac{\beta}{\psi(\beta)}$$

which in turn is equal to $1/d$ by Exercise 2.11.

To deal with the case that $q = 0$ note from (8.24) that for any $p > 0$, $W^{(p)}(0) = W(0)$. □

Returning to (8.8) we see that the conclusion of the previous lemma indicates that, precisely when X has bounded variation,

$$\mathbb{P}_0(\tau_a^+ < \tau_0^-) = \frac{W(0)}{W(a)} > 0. \tag{8.26}$$

Note that the stopping time τ_0^- is defined by strict first passage. Hence when X has the property that 0 is irregular for $(-\infty, 0)$, it takes an almost surely positive amount of time to exit the half line $[0, \infty)$. Since the aforementioned irregularity is equivalent to bounded variation for this class of Lévy processes, we see that (8.26) makes sense.

8.4 Potential Measures

In this section we give an example of how scale functions may be used to describe potential measures associated with the one- and two-sided exit problems. This gives the opportunity to study the overshoot distributions at first passage below a level. Many of the calculations in this section concerning potential measures are reproduced from Bertoin (1997a).

To introduce the idea of potential measures and their relevance in this context, fix $a > 0$ and suppose that

$$\tau = \tau_a^+ \wedge \tau_0^- .$$

A computation in the spirit of Theorem 5.6 and Lemma 5.8 with the help of the Compensation Formula (Theorem 4.4) gives for $x \in [0, a]$, A any Borel set in $[0, a)$ and B any Borel set in $(-\infty, 0)$,

$$\begin{aligned}
&\mathbb{P}_x(X_\tau \in B, X_{\tau-} \in A) \\
&= \mathbb{E}_x\left(\int_{[0,\infty)}\int_{(-\infty,0)} \mathbf{1}_{(\overline{X}_{t-} \leq a, \underline{X}_{t-} \geq 0, X_{t-} \in A)} \mathbf{1}_{(y \in B - X_{t-})} N(\mathrm{d}t \times \mathrm{d}y)\right) \\
&= \mathbb{E}_x\left(\int_0^\infty \mathbf{1}_{(t<\tau)} \Pi(B - X_t) \mathbf{1}_{(X_t \in A)} \mathrm{d}t\right) \\
&= \int_A \Pi(B - y) U(x, \mathrm{d}y),
\end{aligned} \tag{8.27}$$

where N is the Poisson random measure associated with the jumps of X and

$$U(x, \mathrm{d}y) := \int_0^\infty \mathbb{P}_x(X_t \in \mathrm{d}y, \tau > t)\mathrm{d}t.$$

The latter is called the *potential measure of X killed on exiting* $[0, a]$ when initiated from x. It is also known as the resolvent measure. More generally we can work with the q-potential measure where

$$U^{(q)}(x, \mathrm{d}y) := \int_0^\infty \mathrm{e}^{-qt}\, \mathbb{P}_x(X_t \in \mathrm{d}y, \tau > t)\mathrm{d}t$$

for $q \geq 0$ with the agreement that $U^{(0)} = U$. If a density of $U^{(q)}(x, \mathrm{d}y)$ exists with respect to Lebesgue measure for each $x \in [0, a]$ then we call it the potential density and label it $u^{(q)}(x, y)$ (with $u^{(0)} = u$). It turns out that for a spectrally negative process, not only does a potential density exist, but we can write it in semi-explicit terms. This is given in the next theorem due to Suprun (1976) and later Bertoin (1997a). Note, in the statement of the result, it is implicitly understood that $W^{(q)}(z)$ is identically zero for $z < 0$.

Theorem 8.7. *Suppose, for $q \geq 0$, that $U^{(q)}(x, \mathrm{d}y)$ is the q-potential measure of a spectrally negative Lévy process killed on exiting $[0, a]$ where $x, y \in [0, a]$. Then it has a density $u^{(q)}(x, y)$ given by*

$$u^{(q)}(x, y) = \frac{W^{(q)}(x) W^{(q)}(a - y)}{W^{(q)}(a)} - W^{(q)}(x - y). \tag{8.28}$$

Proof. We start by noting that for all $x, y \geq 0$ and $q > 0$,

$$R^{(q)}(x, \mathrm{d}y) := \int_0^\infty \mathrm{e}^{-qt}\, \mathbb{P}_x(X_t \in \mathrm{d}y, \tau_0^- > t)\mathrm{d}t = \frac{1}{q}\mathbb{P}_x(X_{\mathbf{e}_q} \in \mathrm{d}y, \underline{X}_{\mathbf{e}_q} \geq 0),$$

where $\mathbf{e}_q$ is an independent, exponentially distributed random variable with parameter $q > 0$. Note that one may think of $R^{(q)}$ as the q-potential measure of the process X when killed on exiting $[0, \infty)$.

Appealing to the Wiener–Hopf factorisation, specifically that $X_{\mathbf{e}_q} - \underline{X}_{\mathbf{e}_q}$ is independent of $\underline{X}_{\mathbf{e}_q}$, we have that

$$\begin{aligned} R^{(q)}(x, \mathrm{d}y) &= \frac{1}{q}\mathbb{P}((X_{\mathbf{e}_q} - \underline{X}_{\mathbf{e}_q}) + \underline{X}_{\mathbf{e}_q} \in \mathrm{d}y - x, -\underline{X}_{\mathbf{e}_q} \leq x) \\ &= \frac{1}{q}\int_{[x-y,x]} \mathbb{P}(-\underline{X}_{\mathbf{e}_q} \in \mathrm{d}z) \int_{[0,\infty)} \mathbb{P}(X_{\mathbf{e}_q} - \underline{X}_{\mathbf{e}_q} \in \mathrm{d}y - x + z). \end{aligned}$$

Recall however, that by duality $X_{\mathbf{e}_q} - \underline{X}_{\mathbf{e}_q}$ is equal in distribution to $\overline{X}_{\mathbf{e}_q}$ which itself is exponentially distributed with parameter $\Phi(q)$. In addition, the law of $-\underline{X}_{\mathbf{e}_q}$ has been identified in (8.20). We may therefore develop the expression for $R^{(q)}(x, \mathrm{d}y)$ as follows:

$$R^{(q)}(x, \mathrm{d}y) = \left\{ \int_{[x-y,x]} \left(\frac{1}{\Phi(q)} W^{(q)}(\mathrm{d}z) - W^{(q)}(z)\mathrm{d}z \right) \Phi(q)\mathrm{e}^{-\Phi(q)(y-x+z)} \right\} \mathrm{d}y.$$

This shows that there exists a density, $r^{(q)}(x, y)$, for the measure $R^{(q)}(x, \mathrm{d}y)$. Now applying integration by parts to the integral in the last equality, we have that

$$r^{(q)}(x, y) = \mathrm{e}^{-\Phi(q)y} W^{(q)}(x) - W^{(q)}(x - y).$$

Finally we may use the above established facts to compute the potential density $u^{(q)}$ as follows. First note that with the help of the Strong Markov Property,

$$\begin{aligned} qU^{(q)}(x, \mathrm{d}y) &= \mathbb{P}_x(X_{\mathbf{e}_q} \in \mathrm{d}y, \underline{X}_{\mathbf{e}_q} \geq 0, \overline{X}_{\mathbf{e}_q} \leq a) \\ &= \mathbb{P}_x(X_{\mathbf{e}_q} \in \mathrm{d}y, \underline{X}_{\mathbf{e}_q} \geq 0) \\ &\quad -\mathbb{P}_x(X_{\mathbf{e}_q} \in \mathrm{d}y, \underline{X}_{\mathbf{e}_q} \geq 0, \overline{X}_{\mathbf{e}_q} > a) \\ &= \mathbb{P}_x(X_{\mathbf{e}_q} \in \mathrm{d}y, \underline{X}_{\mathbf{e}_q} \geq 0) \\ &\quad -\mathbb{P}_x(X_\tau = a, \tau < \mathbf{e}_q)\mathbb{P}_a(X_{\mathbf{e}_q} \in \mathrm{d}y, \underline{X}_{\mathbf{e}_q} \geq 0). \end{aligned}$$

The first and third of the three probabilities on the right-hand side above have been computed in the previous paragraph, the second probability may be written

$$\mathbb{P}_x(\mathrm{e}^{-q\tau_a^+}; \tau_a^+ < \tau_0^-) = \frac{W^{(q)}(x)}{W^{(q)}(a)}.$$

In conclusion, we have then that $U^{(q)}(x, \mathrm{d}y)$ has a density

$$r^{(q)}(x, y) - \frac{W^{(q)}(x)}{W^{(q)}(a)} r^{(q)}(a, y),$$

which after a short amount of algebra is equal to the right-hand side of (8.28).

To complete the proof when $q = 0$, one may take limits in (8.28) noting that the right-hand side is analytic and hence continuous in q for fixed values x, a, y. The right-hand side of (8.28) tends to $u(x, y)$ by monotone convergence of $U^{(q)}$ as $q \downarrow 0$. □

Note that in fact the above proof contains the following corollary.

Corollary 8.8. *For $q \geq 0$, the q-potential measure of a spectrally negative Lévy process killed on exiting $[0, \infty)$ has density given by*

$$r^{(q)}(x, y) = \mathrm{e}^{-\Phi(q)y} W^{(q)}(x) - W^{(q)}(x - y)$$

for $x, y \geq 0$.

Define further the q-potential measure of X without killing by

$$\Theta^{(q)}(x, \mathrm{d}y) = \int_0^\infty \mathrm{e}^{-qt} \mathbb{P}_x(X_t \in \mathrm{d}y)\mathrm{d}t$$

for $x, y \in \mathbb{R}$. Note by spatial homogeneity $\Theta^{(q)}(x, \mathrm{d}y) = \Theta^{(q)}(0, \mathrm{d}y - x)$. If $\Theta^{(q)}(x, \mathrm{d}y)$ has a density, then we may always write it in the form $\theta^{(q)}(x - y)$ for some function $\theta^{(q)}$. The following corollary was established in Bingham (1975).

Corollary 8.9. *For $q > 0$, the q-potential density of a spectrally negative Lévy process is given by*

$$\theta^{(q)}(z) = \Phi'(q)\mathrm{e}^{-\Phi(q)z} - W^{(q)}(-z)$$

for all $z \in \mathbb{R}$.

Proof. The result is obtained from Corollary 8.8 by considering the effect of moving the killing barrier to an arbitrary large distance from the initial point. Formally

$$\theta^{(q)}(z) = \lim_{x \uparrow \infty} r^{(q)}(x, x + z) = \lim_{x \uparrow \infty} \mathrm{e}^{-\Phi(q)(x+z)} W^{(q)}(x + z) - W^{(q)}(-z).$$

Note however that from the proof of Theorem 8.1 (iii) we identified $W^{(q)}(x) = \mathrm{e}^{\Phi(q)x} W_{\Phi(q)}(x)$ where

$$\int_0^\infty \mathrm{e}^{-\theta x} W_{\Phi(q)}(x)\mathrm{d}x = \frac{1}{\psi_{\Phi(q)}(\theta)}.$$

It follows then that

$$\theta^{(q)}(z) = \mathrm{e}^{-\Phi(q)z} W_{\Phi(q)}(\infty) - W^{(q)}(-z).$$

Note that $(X, \mathbb{P}^{\Phi(q)})$ drifts to infinity and hence $W_{\Phi(q)}(\infty) < \infty$. Since $W_{\Phi(q)}$ is a continuous function, we have that

$$W_{\Phi(q)}(\infty) = \lim_{\theta \downarrow 0} \int_0^\infty \theta \ \mathrm{e}{-}\theta x W_{\Phi(q)}(x)\mathrm{d}x = \lim_{\theta \downarrow 0} \frac{\theta}{\psi_{\Phi(q)}(\theta)} = \frac{1}{\psi'_{\Phi(q)}(0+)}.$$

As $\psi(\Phi(q)) = q$, differentiation of the latter equality implies that the left-hand side above is equal to $\Phi'(q)$ and the proof is complete. □

To conclude this section, let us now return to (8.27). The above results now show that for $z \in (-\infty, 0)$ and $y \in (0, a]$,

$$\begin{aligned}&\mathbb{P}_x(X_\tau \in \mathrm{d}z, X_{\tau-} \in \mathrm{d}y)\\ &= \Pi(\mathrm{d}z - y)\left\{\frac{W(x)W(a-y) - W(a)W(x-y)}{W(a)}\right\}\mathrm{d}y. \qquad (8.29)\end{aligned}$$

(recall that $\tau = \tau_a^+ \wedge \tau_0^-$). Similarly, in the limiting case when a tends to infinity,

$$\mathbb{P}_x(X_{\tau_0^-} \in \mathrm{d}z, X_{\tau_0^- -} \in \mathrm{d}y) = \Pi(\mathrm{d}z - y)\left\{\mathrm{e}^{-\Phi(0)y}W(x) - W(x-y)\right\}\mathrm{d}y.$$

8.5 Identities for Reflected Processes

In this final section we give further support to the idea that the functions $W^{(q)}$ and $Z^{(q)}$ play a central role in many fluctuation identities concerning spectrally negative Lévy processes. We give a brief account of their appearance in a number of identities for spectrally negative Lévy processes reflected either at their supremum or their infimum.

We begin by reiterating what we mean by a Lévy process reflected in its supremum or reflected in its infimum. Fix $x \geq 0$. Then the process

$$Y_t^x := (x \vee \overline{X}_t) - X_t \ t \geq 0$$

is called the process reflected at its supremum (with initial value x) and the process

$$Z_t^x := X_t - (\underline{X}_t \wedge (-x))$$

is called the process reflected at its infimum (with inital value x).

For such processes we may consider the exit times

$$\overline{\sigma}_a^x = \inf\{t > 0 : Y_t^x > a\} \text{ and } \underline{\sigma}_a^x = \inf\{t > 0 : Z_t^x > a\}$$

for levels $a > 0$. In the spirit of Theorem 8.1 we have the following result.

Theorem 8.10. *Let X be a spectrally negative Lévy process with Lévy measure Π. Fix $a > 0$. We have*

(i) for $x \in [0,a]$ and $\theta \in \mathbb{R}$ such that $\psi(\theta) < \infty$,

$$\mathbb{E}(\mathrm{e}^{-q\overline{\sigma}_a^x - \theta Y_{\overline{\sigma}_a^x}^x}) = \mathrm{e}^{-\theta x}\left(Z_\theta^{(p)}(a-x) - W_\theta^{(p)}(a-x)\frac{pW^{(p)}(a) + \theta Z_\theta^{(p)}(a)}{W_\theta^{(p)\prime}(a) + \theta W_\theta^{(p)}(a)}\right),$$

where $p = q - \psi(\theta)$ and $W_\theta^{(q)\prime}(a)$ is the right derivative at a. Further, (ii) for $x \in [0,a]$,

$$\mathbb{E}(\mathrm{e}^{-q\underline{\sigma}_a^x}) = \frac{Z^{(q)}(x)}{Z^{(q)}(a)}.$$

Part (i) was proved[2] in Avram et al. (2004) and part (ii) in Pistorius (2004). The proofs for a general spectrally negative Lévy process however turn out to be quite complicated requiring the need for a theory which is beyond the scope of this text; that is to say Itô's excursion theory. Doney (2005, 2006) gives another proof of the above theorem, again based on excursion theory. If it is assumed that the underlying Lévy process has bounded variation, then a proof can be given using notions that have been presented earlier in this text. Part (ii) for processes of bounded variation is proved in Exercise 8.11. Below we give the proof of part (i) for bounded variation processes and $\theta = 0$. The proof, based on Korolyuk (1974), requires the slightly stronger assumption that $W^{(q)}$ is a $C^1(0,\infty)$ function, or equivalently from Exercise 8.3 that Π has no atoms.

Proof (of Theorem 8.10 (i) with bounded variation, $\theta = 0$ and Π has no atoms). Fix $a > 0$. First note that it suffices to prove the result for $x = 0$ since by the Strong Markov Property

$$\mathbb{E}(\mathrm{e}^{-q\overline{\sigma}_a^x}) = \mathbb{E}_{a-x}(\mathrm{e}^{-q\tau_a^+}\mathbf{1}_{(\tau_a^+ < \tau_0^-)})\mathbb{E}(\mathrm{e}^{-q\overline{\sigma}_a^0}) + \mathbb{E}_{a-x}(\mathrm{e}^{-q\tau_0^-}\mathbf{1}_{(\tau_0^- < \tau_a^+)}).$$

The required expression for $x = 0$ is

$$\mathbb{E}(\mathrm{e}^{-q\overline{\sigma}_a^0}) = Z^{(q)}(a) - \frac{qW^{(q)}(a)^2}{W^{(q)\prime}(a)}. \tag{8.30}$$

Using Theorem 8.1 (8.8) we have from the Strong Markov Property that for all $x \in \mathbb{R}$ and $t \geq 0$

$$\mathbb{E}_x(\mathrm{e}^{-q\tau_a^+}\mathbf{1}_{(\tau_0^- > \tau_a^+)}|\mathcal{F}_{t\wedge\tau_0^-\wedge\tau_a^+}) = \mathrm{e}^{-q(t\wedge\tau_0^-\wedge\tau_a^+)}\frac{W^{(q)}(X_{t\wedge\tau_0^-\wedge\tau_a^+})}{W^{(q)}(a)}.$$

It is important to note in this calculation that we have used the fact that $W^{(q)}(X_{\tau_a^+})/W^{(q)}(a) = 1$ and $W^{(q)}(X_{\tau_0^-})/W^{(q)}(a) = 0$. Note in particular in the last equality it is important to note that $\mathbb{P}(W^{(q)}(X_{\tau_0^-}) = W^{(q)}(0+)) = 0$

[2]See also the note at the end of this chapter.

as X cannot creep downwards. With similar reasoning one also deduces that for all $t \geq 0$,

$$\begin{aligned}&\mathbb{E}_x(\mathrm{e}^{-q\tau_0^-}\mathbf{1}_{(\tau_0^-<\infty)}|\mathcal{F}_{t\wedge\tau_0^-\wedge\tau_a^+})\\&= \mathrm{e}^{-q(t\wedge\tau_0^-\wedge\tau_a^+)}\left(Z^{(q)}(X_{t\wedge\tau_0^-\wedge\tau_a^+}) - \frac{q}{\Phi(q)}W^{(q)}(X_{t\wedge\tau_0^-\wedge\tau_a^+})\right).\end{aligned}$$

The last two computations together with linearity imply that

$$\{\mathrm{e}^{-q(t\wedge\tau_0^-\wedge\tau_a^+)}W^{(q)}(X_{t\wedge\tau_0^-\wedge\tau_a^+}) : t \geq 0\}$$

and

$$\{\mathrm{e}^{-q(t\wedge\tau_0^-\wedge\tau_a^+)}Z^{(q)}(X_{t\wedge\tau_0^-\wedge\tau_a^+}) : t \geq 0\}$$

are martingales.

Since the functions $W^{(q)}$ and $Z^{(q)}$ belong to the class $C^1(0,\infty)$ we may apply (a slighly adapted version of) the result in Exercise 4.3 (iii) to deduce that $(\mathcal{L} - q)W^{(q)}(x) = (\mathcal{L} - q)Z^{(q)}(x) = 0$ for $x \in (0,a)$ where $\mathcal{L}$ is the non-local operator satisfying

$$\mathcal{L}g(x) = dg'(x) + \int_{(-\infty,0)}(g(x+y) - g(x))\Pi(\mathrm{d}y)$$

for $g \in C^1(\mathbb{R})$ and $d > 0$ is the drift of X.

Next define

$$f(x) = Z^{(q)}(x) - W^{(q)}(x)\frac{qW^{(q)}(a)}{W^{(q)\prime}(a)}$$

and note that an easy calculation leads to $f'(a) = 0$. Note also that by linearity $(\mathcal{L} - q)f(x) = 0$ for $x \in (0,a)$.

Applying the version of the change of variable formula in Exericse 4.2 to the stochastic process we have on $\{t < \overline{\sigma}_a^0\}$,

$$\begin{aligned}&\mathrm{e}^{-qt}f(a - \overline{X}_t + X_t)\\&= f(a) - q\int_0^t \mathrm{e}^{-qs}f(a - \overline{X}_s + X_s)\mathrm{d}s + d\int_0^t \mathrm{e}^{-qs}f'(a - \overline{X}_s + X_s)\mathrm{d}s\\&+ \int_{[0,t]}\int_{(-\infty,0)} \mathrm{e}^{-qs}(f(a-\overline{X}_{s-}+X_{s-} + x)) - f(a - \overline{X}_{s-}+X_{s-})N(\mathrm{d}s \times \mathrm{d}x)\\&+ \int_0^\infty f'(a-)\mathrm{d}\overline{X}_s,\end{aligned} \tag{8.31}$$

where N is the Poisson random measure associated with the jumps of X. Now consider the process

$$\begin{aligned}&M_t =\\&\int_{[0,t]}\int_{(-\infty,0)} \mathrm{e}^{-qs}(f(a - \overline{X}_{s-} + X_{s-} + x) - f(a - \overline{X}_{s-} + X_{s-})N(\mathrm{d}s \times \mathrm{d}x)\\&\quad - \int_0^t \mathrm{e}^{-qs}\int_{(-\infty,0)}(f(a - \overline{X}_{s-} + X_{s-} + x) - f(a - \overline{X}_{s-} + X_{s-})\mathrm{d}s\Pi(\mathrm{d}x)\end{aligned}$$

for $t \geq 0$. As the function f is continuous and bounded on $(0, a]$ and $\{\overline{X}_{t-} - X_{t-} : t \geq 0\}$ is left continuous, we may use the compensation formula to prove that $\{M_t : t \geq 0\}$ is a martingale. The proof of this fact is similar in nature to calculations that appeared in the proof of Theorem 4.7 and we leave it as an exercise for the reader. Returning then to (8.31), we have that on $\{t < \overline{\sigma}_a^0\}$,

$$\begin{aligned}
&\mathrm{e}^{-qt} f(a - \overline{X}_t + X_t) \\
&= f(a) + \int_0^t \mathrm{e}^{-qs} (\mathcal{L} - q) f(a - \overline{X}_{s-} + X_{s-}) \mathrm{d}s \\
&+ M_t + \int_0^t f'(a-) \mathrm{d}\overline{X}_s.
\end{aligned}$$

Since $(\mathcal{L} - q)f(x) = 0$ for $x \in (0, a)$ and $f'(a-) = 0$ it follows that

$$\{\mathrm{e}^{-qt} f(a - \overline{X}_t + X_t) : t < \overline{\sigma}_a^0\},$$

is a martingale. In particular using the fact that f is bounded, taking its expectation and letting $t \uparrow \infty$ we discover that

$$f(a) = \mathbb{E}(\mathrm{e}^{-q\overline{\sigma}_a^0} f(a - Y^0_{\overline{\sigma}_a^0})) = \mathbb{E}(\mathrm{e}^{-q\overline{\sigma}_a^0}),$$

where in the final equality we have used the fact that $Z^{(q)}(a - Y^0_{\overline{\sigma}_a^0})) = 1$ and $W^{(q)}(a - Y^0_{\overline{\sigma}_a^0}) = 0$. Note again that in the last equality we have used the fact that X cannot creep downwards and hence Y^0 cannot hit a at first passage above a so that $\mathbb{P}(W^{(q)}(a - Y^0_{\overline{\sigma}_a^0}) = W^{(q)}(0+)) = 0$. Reconsidering the expression for $f(a)$ we see that the equality (8.30) and hence the theorem has been proved. □

It turns out that it is also possible to say something about the q-potential functions of Y^x and Z^x with killing at first passage over a specified level $a > 0$. The latter two are defined, respectively, by

$$\overline{U}^{(q)}(x, \mathrm{d}y) = \int_0^\infty \mathrm{e}^{-qt} \mathbb{P}(Y_t^x \in \mathrm{d}y, \overline{\sigma}_a^x > t) \mathrm{d}t$$

for $x, y \in [0, a)$ and

$$\underline{U}^{(q)}(x, \mathrm{d}y) = \int_0^\infty \mathrm{e}^{-qt} \mathbb{P}(Z_t^x \in \mathrm{d}y, \underline{\sigma}_a^x > t) \mathrm{d}t$$

for $x, y \in [0, a)$. The following results are due to Pistorius (2004). Alternative proofs are also given in Doney (2005, 2006). Once again, we offer no proofs here on account of their difficulty.

Theorem 8.11. *Fix $a > 0$. For each $q \geq 0$,*

(i) for $x, y \in [0, a)$

$$\overline{U}^{(q)}(x, \mathrm{d}y) = \left(W^{(q)}(a-x)\frac{W^{(q)}(0)}{W^{(q)\prime}(a)}\right)\delta_0(\mathrm{d}x) + \left(W^{(q)}(a-x)\frac{W^{(q)\prime}(y)}{W^{(q)\prime}(a)} - W^{(q)}(y-x)\right)\mathrm{d}y.$$

(ii) for $x, y \in [0, a)$ the measure $\underline{U}^{(q)}(x, \mathrm{d}y)$ has a density given by

$$\underline{u}^{(q)}(x, y) = W^{(q)}(a-y)\frac{Z^{(q)}(x)}{Z^{(q)}(a)} - W^{(q)}(x-y).$$

As in Theorem 8.10 we take $W^{(q)\prime}$ to mean the right derivative. Note in particular that when the underlying Lévy process is of unbounded variation, the q-potential for Z^x killed on first passage above a is absolutely continuous with respect to Lebesgue measure which otherwise has an atom at zero.

On a final note, we emphasise that there exists an additional body of literature written in Russian by members of the Kiev school of probability which considers the type of boundary problems described above for spectrally one-sided Lévy processes using a so-called "potential method" developed in Korolyuk (1974). For example Theorem 8.10 (i) can be found for the case that Π has finite total mass and $\sigma = 0$ in Korolyuk (1975a,b) and Bratiychuk and Gusak (1991). The reader is also referred to Korolyuk et al. (1976) and Korolyuk and Borovskich (1981) and references therein.[3]

8.6 Brief Remarks on Spectrally Negative GOUs

Very closely related to Brownian motion is the Ornstein–Uhlenbeck process. The latter processes are defined as the almost sure path-wise solution $Y = \{Y_t : t \geq 0\}$ to the equation

$$Y_t = Y_0 - \lambda \int_0^t Y_s \, \mathrm{d}s + B_t, \; t \geq 0, \tag{8.32}$$

where $\lambda > 0$ is called the *mean reverting drift* and $\{B_t : t \geq 0\}$ is a standard Brownian motion. It turns out that Y may be uniquely identified by

$$Y_t = \mathrm{e}^{-\lambda t}\left(Y_0 + \int_0^t \mathrm{e}^{\lambda s} \, \mathrm{d}B_s\right), \; t \geq 0, \tag{8.33}$$

[3] I am greatful to Professors V.S. Korolyuk and M.S. Bratiychuk for bringing this literature to my attention.

where the integral on the right-hand side is understood as a classical stochastic Itô-integral (see the discussion around (4.5)).

Suppose now that in (8.32) we replace the Brownian motion B by a spectrally negative Lévy processes, say $X = \{X_t : t \geq 0\}$. The resulting object, known as a *generalised Ornstein–Uhlenbeck process* (GOU for short), also satisfies (8.33) where B is replaced by X. Such processes were studied by Hadjiev (1985). In particular Hadjiev gave an expression for the first passage time above a specified level of the process which exhibits some similarities to the analogous expression for spectrally negative Lévy processes as follows.

Theorem 8.12. *Suppose that* $Y = \{Y_t : t \geq 0\}$ *is a GOU with mean reverting drift* $\lambda > 0$ *driven by a spectrally negative Lévy process,* $X = \{X_t : t \geq 0\}$ *with Laplace exponent* ψ*, Lévy measure* Π*, and Gaussian coefficient* σ*. Assume that* $\int_{x<-1} \log |x| \Pi(\mathrm{d}x) < \infty$*. Fix* $a > 0$ *and assume that either* X *has unbounded variation or that is has bounded variation and* $d > \lambda a$ *where* d *is the drift in the decomposition (8.1).*

Then for $y \leq a$*, the Laplace transform of* $\tau_a^+ := \inf\{t > 0 : Y_t > a\}$ *is given by*

$$\mathbb{E}_y(\mathrm{e}^{-\theta \tau_a^+}) = \frac{H^{(\theta/\lambda)}(y)}{H^{(\theta/\lambda)}(a)}$$

for all $\theta \geq 0$ *where for* $\nu > 0$ *and* $x \in \mathbb{R}$

$$H^{(\nu)}(x) = \int_0^\infty \exp\left(xu - \frac{1}{\lambda} \int_0^u \psi(v) \frac{\mathrm{d}v}{v} \right) u^{\nu - 1} \mathrm{d}u.$$

Further work in this area is found in Novikov (2004) and Patie (2004, 2005). In general, compared to the case of Lévy processes, there is relatively little known about the fluctuations of generalised Ornstein–Uhlenbeck processes.

Other work on generalised Ornstein–Uhlenbeck processes concern their long term behaviour. Specifically in the classical case that X is a Brownian motion, it is well known that Y converges in distribution. In the case that the underlying source of randomness is a spectrally negative Lévy process, the stationary distribution is known to exist with Laplace transform given in explicit form provided the Lévy measure associated with X satisfies a certain integrability condition; see for example the discussion in Patie (2005).

Lindner and Maller (2005) also consider stationary distributions of generalised Ornstein–Uhlenbeck processes to yet a further degree of abstraction. In their case, they define

$$Y_t = \mathrm{e}^{-\xi_t} \left(Y_0 + \int_0^t \mathrm{e}^{\xi_{s-}} \mathrm{d}\eta_s \right),$$

where $\{(\xi_t, \eta_t) : t \geq 0\}$ is an arbitrary bivarate Lévy process. This in turn motivates the study of distribution of the complete integral

$$\int_0^\infty \mathrm{e}^{\xi_{s-}} \, \mathrm{d}\eta_s.$$

It turns out that such integrals also appear in the characterisation of so-called self-simliar Markov processes, risk theory, the theory of Brownian diffusions in random environments, mathematical finance and the theory of self-similar fragmentation; see the review paper of Bertoin and Yor (2005) and references therein. Particular issues of concern are the almost sure convergence of the above integral as well as its moments and the tail behaviour of its distribution. See Erickson and Maller (2004), Bertoin and Yor (2005) and Maulik and Zwart (2006).

Exercises

8.1. Suppose that X is a spectrally negative Lévy process with Laplace exponent ψ such that $\psi'(0+) < 0$. Show that for $t \geq 0$ and any A in $\mathcal{F}_t$,

$$\lim_{x\uparrow\infty} \mathbb{P}(A|\tau_x^+ < \infty) = \mathbb{P}^{\Phi(0)}(A),$$

where, as usual, Φ is the right inverse of ψ.

8.2. Suppose that X is a spectrally negative stable process with index $\alpha \in (1, 2)$ and assume without loss of generality that its Laplace exponent is given by $\psi(\theta) = \theta^\alpha$ for $\theta \geq 0$ (cf. Exercise 3.7).

(i) Show that for $q > 0$ and $\beta > q^{1/\alpha}$,

$$\int_0^\infty e^{-\beta x}\overline{W}^{(q)}(x)dx = \frac{1}{\beta(\beta^\alpha - q)} = \sum_{n\geq 1} q^{n-1}\beta^{-\alpha n-1},$$

where $\overline{W}^{(q)}(x) = \int_0^x W^{(q)}(y)dy$.

(ii) Deduce that for $x \geq 0$

$$Z^{(q)}(x) = \sum_{n\geq 0} q^n \frac{x^{\alpha n}}{\Gamma(1+\alpha n)}.$$

Note that the right-hand side above is also equal to $E_\alpha(qx^\alpha)$ where $E_\alpha(\cdot)$ is the Mittag–Leffler function of parameter α (a generalisation of the exponential function with parameter α).

(iii) Deduce that for $q \geq 0$,

$$W^{(q)}(x) = \alpha x^{\alpha-1}E'_\alpha(qx^\alpha)$$

for $x \geq 0$.

(iv) Show that for standard Brownian motion that

$$W^{(q)}(x) = \sqrt{\frac{2}{q}}\sinh(\sqrt{2q}x) \text{ and } Z^{(q)}(x) = \cosh(\sqrt{2q}x)$$

for $x \geq 0$ and $q \geq 0$.

8.3. Suppose that X is a spectrally negative Lévy process of bounded variation such that $\lim_{t\uparrow\infty} X_t = \infty$. For convenience, write $X_t = dt - S_t$ where $S = \{S_t : t \geq 0\}$ is a subordinator with jump measure Π.

(i) Show that necessarily $d^{-1}\int_0^\infty \Pi(y,\infty)\mathrm{d}y < 1$.

(ii) Show that the scale function W satisfies

$$\int_{[0,\infty)} \mathrm{e}^{-\beta x} W(\mathrm{d}x) = \frac{1}{d - \int_0^\infty \mathrm{e}^{-\beta y}\Pi(y,\infty)\mathrm{d}y}$$

and deduce that

$$W(\mathrm{d}x) = \frac{1}{d}\sum_{n\geq 0} \nu^{*n}(\mathrm{d}x),$$

where $\nu(\mathrm{d}x) = d^{-1}\Pi(x,\infty)\mathrm{d}x$ and as usual we understand $\nu^{*0}(\mathrm{d}x) = \delta_0(\mathrm{d}x)$.

(iii) Suppose that S is a compound Poisson process with rate $\lambda > 0$ and jump distribution which is exponential with parameter $\mu > 0$. Show that

$$W(x) = \frac{1}{d}\left(1 + \frac{\lambda}{d\mu - \lambda}(1 - \mathrm{e}^{-(\mu - d^{-1}\lambda)x})\right).$$

8.4. This exercise is based on Chan and Kyprianou (2005). Suppose that X is a spectrally negative Lévy process of bounded variation and adopt the notation of the previous question.

(i) Suppose that $\lim_{t\uparrow\infty} X_t = \infty$. By considering the term for $n = 1$ in the sum in part (ii) of Exercise 8.3, conclude that W has a continuous derivative if and only if Π has no atoms.

(ii) Deduce further that for each $n \geq 2$, W is n times differentiable (with continuous derivatives) if and only if $\Pi(x,\infty)$ is $n-1$ times differentiable (with continuous derivatives).

(iii) Now remove the restriction that $\lim_{t\uparrow} X_t = \infty$. Show using change of measure that for $q > 0$ or $q = 0$ and $\lim_{t\uparrow\infty} X_t = -\infty$, $W^{(q)}$ has the same smoothness criteria as above. That is to say $W^{(q)}$ has a continuous derivative if and only if Π has no atoms and for $n \geq 2$, $W^{(q)}$ is n times differentiable with continuous derivatives if and only if $\Pi(x,\infty)$ is $n-1$ times differentiable with continuous derivatives.

(iv) For the final case that $q = 0$ and X oscillates, use (8.24) to deduce that the same smoothness criteria apply as in the previous part of the question.

8.5. Let X be any spectrally negative Lévy process with Laplace exponent ψ.

(i) Use (8.9) and (8.6) to establish that for each $q \geq 0$,

$$\lim_{x\uparrow\infty} \frac{Z^{(q)}(x)}{W^{(q)}(x)} = \frac{q}{\Phi(q)},$$

where the right-hand side is understood in the limiting sense when $q = 0$. In addition, show that

$$\lim_{a\uparrow\infty} \frac{W^{(q)}(a-x)}{W^{(q)}(a)} = \mathrm{e}^{-\Phi(q)x}.$$

(ii) Taking account of a possible atom at the origin, write down the Laplace transform of $W(q)(\mathrm{d}x)$ on $[0,\infty)$ and show that if X has unbounded variation then $W^{(q)\prime}(0) = 2/\sigma^2$ where σ is the Gaussian coefficient in the Lévy–Itô decomposition and it is understood that $1/0 = \infty$. If however, X has bounded variation then

$$W^{(q)\prime}(0) = \frac{\Pi(-\infty,0)+q}{d^2}.$$

where it is understood that the right hand side is infinite if $\Pi(-\infty,0)=\infty$.

8.6. Suppose that X is a spectrally negative Lévy process. Using the results of Chap. 5, show with the help of the Wiener–Hopf factorisation and scale functions that

$$\mathbb{P}(X_{\tau_x^-} = x) = \frac{\sigma^2}{2}[W'(-x) - \Phi(0)W(-x)]$$

for all $x \leq 0$. As usual W is the scale function, Φ is the inverse of the Laplace exponent ψ of X and σ is the Gaussian coefficient.

8.7. This exercise deals with first hitting of points below zero of spectrally negative Lévy processes following the work of Doney (1991). For each $x > 0$ define

$$T(-x) = \inf\{t > 0 : X_t = -x\},$$

where X is a spectrally negative Lévy process with Laplace exponent ψ and right inverse Φ.

(i) Show that for all $c \geq 0$ and $q \geq 0$,

$$\Phi_c(q) = \Phi(q + \psi(c)) - c.$$

(ii) Show for $x > 0$, $c \geq 0$ and $p \geq \psi(c) \vee 0$,

$$\mathbb{E}(\mathrm{e}^{-p\tau_{-x}^- + c(X_{\tau_{-x}^-}+x)}\mathbf{1}_{(\tau_{-x}^-<\infty)}) = \mathrm{e}^{cx}\left(Z_c^{(q)}(x) - \frac{q}{\Phi_c(q)}W_c^{(q)}(x)\right),$$

where $q = p - \psi(c)$. Use analytic extension to justify that the above identity is in fact valid for all $x > 0, c \geq 0$ and $p \geq 0$.

(iii) By noting that $T(-x) \geq \tau_{-x}^-$, condition on $\mathcal{F}_{\tau_{-x}^-}$ to deduce that for p, $u \geq 0$,

$$\mathbb{E}(\mathrm{e}^{-pT(-x)-u(T(-x)-\tau_{-x}^-)}\mathbf{1}_{(T(-x)<\infty)}) = \mathbb{E}(\mathrm{e}^{-p\tau_{-x}^- + \Phi(p+u)(X_{\tau_{-x}^-}+x)}\mathbf{1}_{(\tau_{-x}^-<\infty)}).$$

(iv) By taking a limit as $u \downarrow 0$ in part (iii) and making use of the identity in part (ii) deduce that

$$\mathbb{E}(\mathrm{e}^{-pT(-x)}\mathbf{1}_{(T(-x)<\infty)}) = \mathrm{e}^{\Phi(p)x} - \psi'(\Phi(p))W^{(p)}(x)$$

and hence by taking limits again as $x \downarrow 0$,

$$\mathbb{E}\left(\mathrm{e}^{-pT(0)}\mathbf{1}_{(T(0)<\infty)}\right) = \begin{cases} 1 - \psi'(\Phi(p))\frac{1}{d} & \text{if} X \text{has bounded variation} \\ 1 & \text{if } X \text{ has unbounded variation,} \end{cases}$$

where d is the drift term in the Laplace exponent if X has bounded variation.

8.8. Again relying on Doney (1991) we shall make the following application of part (iii) of the previous exercise. Suppose that $B = \{B_t : t \geq 0\}$ is a Brownian motion. Denote

$$\sigma = \inf\{t > 0 : B_t = \overline{B}_t = t\}.$$

(i) Suppose that X is a descending stable-$\frac{1}{2}$ subordinator with upward unit drift. Show that

$$\mathbb{P}(\sigma < \infty) = \mathbb{P}(T(0) < \infty),$$

where $T(0)$ is defined in Exercise 8.7.

(ii) Deduce from part (i) that $\mathbb{P}(\sigma < \infty) = \frac{1}{2}$.

8.9. The following exercise is based on results found in Huzak et al. (2004a). Suppose that we consider a generalisation of the Cramér–Lundberg process, X, in the form of a spectrally negative Lévy process which drifts to ∞. In particular, it will be the case that $X = \sum_{i=1}^{n} X^{(i)}$ where each of the $X^{(i)}$s are independent spectrally negative Lévy processes with Lévy measures $\Pi^{(i)}$ concentrated on $(-\infty, 0)$. One may think of them as competing risk processes.

(i) With the help of the compensation formula, show that for $x, y \geq 0, z < 0$ and $i = 1, ..., n$,

$$\begin{aligned} &\mathbb{P}_x(X_{\tau_0^- -} \in \mathrm{d}y, X_{\tau_0^-} \in \mathrm{d}z, \Delta X_{\tau_0^-} = \Delta X^{(i)}_{\tau_0^-}) \\ &= r(x, y)\Pi^{(i)}(-y + \mathrm{d}z)\mathrm{d}y, \end{aligned}$$

where $r(x, y)$ is the potential density of the process killed on first passage into $(-\infty, 0)$.

(ii) Suppose now that $x = 0$ and each of the processes $X^{(i)}$ are of bounded variation. Recall that any such spectrally negative Lévy process is the difference of a linear drift and a subordinator. Let d be the drift of X. Show that for $y > 0$,

$$\begin{aligned} &\mathbb{P}(X_{\tau_0^- -} \in \mathrm{d}y, X_{\tau_0^-} \in \mathrm{d}z, \Delta X_{\tau_0^-} = \Delta X^{(i)}_{\tau_0^-}) \\ &= \frac{1}{d}\Pi^{(i)}(-y + \mathrm{d}z)\mathrm{d}y. \end{aligned}$$

(iii) For each $i = 1, ..., n$ let d_i be the drift of $X^{(i)}$. Note that necessarily $d = \sum_{i=1}^n d_i$. Suppose further that for each $i = 1..., n$, $\mu_i := \mathbb{E}(X_1^{(i)}) - d_i < \infty$. Show that the probability that ruin[4] occurs as a result of a claim from the ith process when $x = 0$ is equal to μ_i/d.

8.10. This exercise is based on the results of Chiu and Yin (2005). Suppose that X is any spectrally negative Lévy process with Laplace exponent ψ, satisfying $\lim_{t\uparrow\infty} X_t = \infty$. Recall that this necessarily implies that $\psi'(0+) > 0$. Define for each $x \in \mathbb{R}$,

$$\Lambda_0 = \sup\{t > 0 : X_t < 0\}.$$

Here we work with the definition $\sup \emptyset = 0$ so that the event $\{\Lambda_0 = 0\}$ corresponds to the event that X never enters $(-\infty, 0)$.

(i) Using the equivalent events $\{\Lambda_0 < t\} = \{X_t \geq 0, \inf_{s\geq t} X_s \geq 0\}$ and the Markov Property, show that for each $q > 0$ and $y \in \mathbb{R}$

$$\mathbb{E}_y(\mathrm{e}^{-q\Lambda_0}) = q \int_0^\infty \theta^{(q)}(x - y)\mathbb{P}_x(\underline{X}_\infty \geq 0)\mathrm{d}x,$$

where $\theta^{(q)}$ is the q-potential density of X.

(ii) Hence show that for $y \leq 0$,

$$\mathbb{E}_y(\mathrm{e}^{-q\Lambda_0}) = \psi'(0+)\Phi'(q)\mathrm{e}^{\Phi(q)y},$$

where Φ is the right inverse of ψ and in particular

$$\mathbb{P}(\Lambda_0 = 0) = \begin{cases} \psi'(0+)/d & \text{if } X \text{ has bounded variation with drift } d \\ 0 & \text{if } X \text{ has unbounded variation.} \end{cases}$$

(iii) Suppose now that $y > 0$. Use again the Strong Markov Property to deduce that for $q > 0$,

$$\mathbb{E}_y(\mathrm{e}^{-q\Lambda_0}\mathbf{1}_{(\Lambda_0>0)}) = \psi'(0+)\Phi'(q)\mathbb{E}_y(\mathrm{e}^{-q\tau_0^- + \Phi(q)X_{\tau_0^-}}\mathbf{1}_{(\tau_0^-<\infty)}).$$

(iv) Deduce that for $y > 0$ and $q > 0$,

$$\mathbb{E}_y(\mathrm{e}^{-q\Lambda_0}\mathbf{1}_{(\Lambda_0>0)}) = \psi'(0+)\Phi'(q)\mathrm{e}^{\Phi(q)y} - \psi'(0+)W^{(q)}(y).$$

8.11 (Proof of Theorem 8.10 (ii) with Bounded Variation). Adopt the setting of Theorem 8.10 (ii). It may be assumed that $\underline{\sigma}_a^x$ is a stopping time with respect to the filtration $\mathbb{F}$ (recall in our standard notation this is the filtration generated by the underlying Lévy process X which satisfies the usual conditions of completion and right continuity).

---

[4]Recall that the event "ruin" means $\{\tau_0^- < \infty\}$.

(i) Show that for any $x \in (0, a]$,

$$\mathbb{E}(\mathrm{e}^{-q\underline{\sigma}_a^x}) = \mathbb{E}_x(\mathrm{e}^{-q\tau_0^-}\mathbf{1}_{(\tau_0^- < \tau_a^+)})\mathbb{E}(\mathrm{e}^{-q\underline{\sigma}_a^0}) + \mathbb{E}_x(\mathrm{e}^{-q\tau_a^+}\mathbf{1}_{(\tau_a^+ < \tau_0^-)}).$$

(ii) By taking limits as x tends to zero in part (i) deduce that

$$\mathbb{E}(\mathrm{e}^{-q\underline{\sigma}_a^x}) = \frac{Z^{(q)}(x)}{Z^{(q)}(a)}$$

for all $x \in [0, a]$. [Hint: recall that $W^{(q)}(0) > 0$ if X has paths of bounded variation].

(iii) The following application comes from Dube et al. (2004). Let W be a general storage process as described at the beginning of Chap. 4. Now suppose that this storage process has a limited capacity say $c > 0$. This means that when the work load exceeds c units, the excess of work is removed and dumped. Prove that the Laplace transform (with parameter $q > 0$) of the first time for the workload of this storage process to become zero when started from $0 < x < c$ is given by $Z^{(q)}(c - x)/Z^{(q)}(c)$ where $Z^{(q)}$ is the scale function associated with the underlying Lévy process driving W.

8.12. Suppose that X is a spectrally negative α-stable process for $\alpha \in (1, 2)$. We are interested in establishing the distribution of the overshoot below the origin when the process, starting from $x \in (1, 2)$ first exists in the latter interval below. In principle one could attempt to invert the formula given in Exercise 8.7 (ii). However, the following technique from Rogozin (1972) offers a more straightforward method. It will be helpful to first review Exercise 7.7.

(i) Show that

$$\mathbb{P}_x(-X_{\tau_0^-} \leq y; \tau_0^- < \tau_1^+) = \Phi_{\alpha-1}\left(\frac{y}{x}\right) - \mathbb{P}_x(\tau_1^+ < \tau_0^-)\Phi_{\alpha-1}(y),$$

where $\Phi_{\alpha-1}$ was defined in Exercise 7.7.

(ii) Hence deduce that

$$\begin{aligned}&\mathbb{P}_x(-X_{\tau_0^-} \leq y; \tau_0^- < \tau_1^+)\\ &= \frac{\sin\pi(\alpha-1)}{\pi} x^{\alpha-1}(1-x)\int_0^y t^{-(\alpha-1)}(t+1)^{-1}(t+x)^{-1}\mathrm{d}t.\end{aligned}$$

9

Applications to Optimal Stopping Problems

The aim of this chapter is to show how some of the established fluctuation identities for Lévy processes and reflected Lévy processes can be used to solve quite specific, but nonetheless exemplary optimal stopping problems. To some extent this will be done in an unsatisfactory way as we shall do so *without* first giving a thorough account of the general theory of optimal stopping. However we shall give rigorous proofs relying on the method of "guess and verify." That is to say, our proofs will simply start with a candidate solution, the choice of which being inspired by intuition and then we shall prove that this candidate verifies sufficient conditions in order to confirm its status as the actual solution. For a more complete overview of the theory of optimal stopping the reader is referred to the main three texts Chow et al. (1971), Shiryaev (1978) and Peskir and Shiryaev (2006); see also Chap. 10 of Øksendal (2003) and Chap. 2 of Øksendal and Sulem (2004). See also the foundational work of Snell (1952) and Dynkin (1963).

The optimal stopping problems we consider in this chapter will be of the form

$$v(x) = \sup_{\tau \in \mathcal{T}} \mathbb{E}_x(\mathrm{e}^{-q\tau} G(X_\tau)) \tag{9.1}$$

(or variants thereof), where $X = \{X_t : t \geq 0\}$ a Lévy process. Further, G is a non-negative measurable function, $q \geq 0$ and $\mathcal{T}$ is a family of stopping times with respect to the filtration[1] $\mathbb{F}$. Note that when talking of a solution to (9.1) it is meant that the function v is finite for all $x \in \mathbb{R}$.

In the final section, we give an example of a stochastic game driven by a Lévy process. Stochastic games are generalised versions of optimal stopping problems in which there are two players whose objectives are to maximise and minimise, respectively, an expected value based on the path of an underlying Lévy process.

[1]Recall that $\mathbb{F} = \{\mathcal{F}_t : t \geq 0\}$ where $\mathcal{F}_t = \sigma(\mathcal{F}_t^0, \mathcal{O})$ such that $\mathcal{F}_t^0 = \sigma(X_s : s \leq t)$ and $\mathcal{O}$ is the set of $\mathbb{P}$-null sets.

9.1 Sufficient Conditions for Optimality

Here we give sufficient conditions with which one may verify that a candidate solution solves the optimal stopping problem (9.1).

Lemma 9.1. *Consider the optimal stopping problem (9.1) for $q \geq 0$ under the assumption that for all $x \in \mathbb{R}$,*

$$\mathbb{P}_x(\textit{there exists } \lim_{t\uparrow\infty} \mathrm{e}^{-qt}G(X_t) < \infty) = 1. \tag{9.2}$$

Suppose that $\tau^* \in \mathcal{T}$ is a candidate optimal strategy for the optimal stopping problem (9.1) and let $v^*(x) = \mathbb{E}_x(\mathrm{e}^{-q\tau^*}G(X_{\tau^*}))$. Then the pair (v^*, τ^*) is a solution if

(i) $v^*(x) \geq G(x)$ for all $x \in \mathbb{R}$,
(ii) the process $\{\mathrm{e}^{-qt}v^*(X_t) : t \geq 0\}$ is a right continuous supermartingale.

Proof. The definition of v^* implies that

$$\sup_{\tau\in\mathcal{T}} \mathbb{E}_x(\mathrm{e}^{-q\tau}G(X_\tau)) \geq v^*(x)$$

for all $x \in \mathbb{R}$. On the other hand, property (ii) together with Doob's Optional Stopping Theorem[2] imply that for all $t \geq 0$, $x \in \mathbb{R}$ and $\sigma \in \mathcal{T}$,

$$v^*(x) \geq \mathbb{E}_x(\mathrm{e}^{-q(t\wedge\sigma)}v^*(X_{t\wedge\sigma}))$$

and hence by property (i), Fatou's Lemma, the non-negativity of G and assumption (9.2)

$$\begin{aligned} v^*(x) &\geq \liminf_{t\uparrow\infty} \mathbb{E}_x(\mathrm{e}^{-q(t\wedge\sigma)}G(X_{t\wedge\sigma})) \\ &\geq \mathbb{E}_x(\liminf_{t\uparrow\infty} \mathrm{e}^{-q(t\wedge\sigma)}G(X_{t\wedge\sigma})) \\ &= \mathbb{E}_x(\mathrm{e}^{-q\sigma}G(X_\sigma)). \end{aligned}$$

As $\sigma \in \mathcal{T}$ is arbitrary, it follows that for all $x \in \mathbb{R}$

$$v^*(x) \geq \sup_{\tau\in\mathcal{T}} \mathbb{E}_x(\mathrm{e}^{-q\tau}G(X_\tau)).$$

In conclusion it must hold that

$$v^*(x) = \sup_{\tau\in\mathcal{T}} \mathbb{E}_x(\mathrm{e}^{-q\tau}G(X_\tau))$$

for all $x \in \mathbb{R}$. □

[2]Right continuity of paths is used implicitly here.

When G is a monotone function and $q > 0$, a reasonable class of candidate solutions that one may consider in conjunction with the previous lemma are those based on first passage times over a specified threshold. That is, either first passage above a given constant in the case that G is monotone increasing or first passage below a given constant in the case that G is monotone decreasing. An intuitive justification may be given as follows.

Suppose that G is monotone increasing. In order to optimise the value $G(X_\tau)$ one should stop at some time τ for which X_τ is large. On the other hand, this should not happen after too much time on account of the exponential discounting. This suggests that there is a threshold, which may depend on time, over which one should stop X in order to maximise the expected discounted gain. Suppose however, that by time $t > 0$ one has not reached this threshold. Then, by the Markov property, given $X_t = x$, any stopping time τ which depends only on the continuation of the path of X from the space-time point (x, t) would yield an expected gain $\mathrm{e}^{-qt}\mathbb{E}_x(\mathrm{e}^{-q\tau}G(X_\tau))$. The optimisation of this expression over the latter class of stopping times is essentially the same procedure as in the original problem (9.1). Note that since X is a Markov process, there is nothing to be gained by considering stopping times which take account of the history of the process $\{X_s : s < t\}$. These arguments suggest that threshold should not vary with time and hence a candidate for the optimal strategy takes the form

$$\tau_y^+ = \inf\{t > 0 : X_t \in A\},$$

where $A = [y, \infty)$ or (y, ∞) for some $y \in \mathbb{R}$. Similar reasoning applies when G is monotone decreasing.

When $q = 0$ and G is monotone increasing then it may be optimal never to stop. To avoid this case, one may impose the added assumption that $\limsup_{t\uparrow\infty} X_t < \infty$ almost surely. One may then again expect to describe the optimal stopping strategy as first passage above a threshold. The reason being that one may not predict when the Lévy process is close to its all time maximum. Again the threshold should be time invariant due to the Markov property. If $q = 0$ and G is monotone decreasing then in light of the aforementioned, one may impose the condition that $\liminf_{t\uparrow\infty} X_t > -\infty$ almost surely and expect to see an optimal strategy consisting of first passage below a time invariant threshold.

9.2 The McKean Optimal Stopping Problem

This optimal stopping problem is given by

$$v(x) = \sup_{\tau\in\mathcal{T}} \mathbb{E}_x(\mathrm{e}^{-q\tau}(K - \mathrm{e}^{X_\tau})^+), \tag{9.3}$$

where $\mathcal{T}$ is all $\mathbb{F}$-stopping times and in the current context we consider the cases that either

$$q > 0 \textit{ or } q = 0 \text{ and } \lim_{t\uparrow\infty} X_t = \infty \text{ a.s.}$$

The solution to this optimal stopping problem was first considered by McKean (1965) for the case that X is linear Brownian motion in the context of the optimal time to sell a risky asset for a fixed price K and in the presence of discounting, where the value of the risky asset follows the dynamics of an exponential Brownian motion.

In Darling et al. (1972) a solution to a discrete-time analogue of (9.3) was obtained. In that case, the process X is replaced by a random walk. Some years later and again within the context of the the optimal time to sell a risky asset (the pricing of an American put), a number of authors dealt with the solution to (9.3) for a variety of special classes of Lévy processes.[3] Below we give the solution to (9.3) as presented in Mordecki (2002). The proof we shall give here comes however from Alili and Kyprianou (2005) and remains close in nature to the random walk proofs of Darling et al. (1972).

Theorem 9.2. *The solution to (9.3) under the stated assumption is given by*

$$v(x) = \frac{\mathbb{E}\left(\left(K\mathbb{E}\left(\mathrm{e}^{\underline{X}_{\mathbf{e}_q}}\right) - \mathrm{e}^{x+\underline{X}_{\mathbf{e}_q}}\right)^+\right)}{\mathbb{E}\left(\mathrm{e}^{\underline{X}_{\mathbf{e}_q}}\right)}$$

and the optimal stopping time is given by

$$\tau^* = \inf\{t > 0 : X_t < x^*\}$$

where

$$x^* = \log K\mathbb{E}\left(\mathrm{e}^{\underline{X}_{\mathbf{e}_q}}\right).$$

Here, as usual, $\mathbf{e}_q$ denotes an independent random variable which is independent of X and exponentially distributed with the understanding that when $q = 0$ this variable is infinite valued with probability one. Further, $\underline{X}_t = \inf_{s\leq t} X_s$. Note then in the case that $q = 0$, as we have assumed that $\lim_{t\uparrow\infty} X_t = \infty$, by Theorem 7.1 we know that $|\underline{X}_\infty| < \infty$ almost surely.

[3] Gerber and Shiu (1994) dealt with the case of bounded variation spectrally positive Lévy processes; Boyarchenko and Levendorskiĭ (2002) handled a class of tempered stable processes; Chan (2004) covers the case of spectrally negative processes; Avram et al. (2002, 2002a) deal with spectrally negative Lévy processes again; Asmussen et al. (2004) look at Lévy processes which have phase-type jumps and Chesney and Jeanblanc (2004) again for the spectrally negative case.

Proof (of Theorem 9.2). First note that the assumption (9.2) is trivially satisfied. In view of the remarks following Lemma 9.1 let us define the bounded functions

$$v_y(x) = \mathbb{E}_x\left(\mathrm{e}^{-q\tau_y^-}(K - \mathrm{e}^{X_{\tau_y^-}})^+\right). \tag{9.4}$$

We shall show that the solution to (9.3) is of the form (9.4) for a suitable choice of $y \le \log K$ by using Lemma 9.1.

According to the conclusion of Exercise 6.7 (i) we have that

$$\mathbb{E}_x\left(\mathrm{e}^{-\alpha\tau_y^- + \beta X_{\tau_y^-}}\mathbf{1}_{(\tau_y^- < \infty)}\right) = \mathrm{e}^{\beta x}\frac{\mathbb{E}(\mathrm{e}^{\beta \underline{X}_{\mathbf{e}_\alpha}}\mathbf{1}_{(-\underline{X}_{\mathbf{e}_\alpha} > x-y)})}{\mathbb{E}(\mathrm{e}^{\beta\underline{X}_{\mathbf{e}_\alpha}})} \tag{9.5}$$

for $\alpha, \beta \ge 0$ and $x - y \ge 0$ and hence it follows that

$$v_y(x) = \frac{\mathbb{E}\left((K\mathbb{E}(\mathrm{e}^{\underline{X}_{\mathbf{e}_q}}) - \mathrm{e}^{x+\underline{X}_{\mathbf{e}_q}})\mathbf{1}_{(-\underline{X}_{\mathbf{e}_q} > x-y)}\right)}{\mathbb{E}(\mathrm{e}^{\underline{X}_{\mathbf{e}_q}})}. \tag{9.6}$$

Lower bound (i). The lower bound $v_y(x) \ge (K - \mathrm{e}^x)^+$ is respected if and only if $v_y(x) \ge 0$ and $v_y(x) \ge (K - \mathrm{e}^x)$. From (9.4) we see that $v_y(x) \ge 0$ always holds. On the other hand, a straightforward manipulation shows that

$$v_y(x) = (K - \mathrm{e}^x) + \frac{\mathbb{E}\left((\mathrm{e}^{x+\underline{X}_{\mathbf{e}_q}} - K\mathbb{E}(\mathrm{e}^{\underline{X}_{\mathbf{e}_q}}))\mathbf{1}_{(-\underline{X}_{\mathbf{e}_q} \le x-y)}\right)}{\mathbb{E}(\mathrm{e}^{\underline{X}_{\mathbf{e}_q}})}. \tag{9.7}$$

From (9.7) we see that a sufficient condition that $v_y(x) \ge (K - \mathrm{e}^x)$ is that

$$\mathrm{e}^y \ge K\mathbb{E}(\mathrm{e}^{\underline{X}_{\mathbf{e}_q}}). \tag{9.8}$$

Supermartingale property (ii). On the event $\{t < \mathbf{e}_q\}$ the identity $\underline{X}_{\mathbf{e}_q} = \underline{X}_t \wedge (X_t + I)$ holds where conditionally on $\mathcal{F}_t$, I has the same distribution as $\underline{X}_{\mathbf{e}_q}$. In particular it follows that on $\{t < \mathbf{e}_q\}$, $\underline{X}_{\mathbf{e}_q} \le X_t + I$. If

$$\mathrm{e}^y \le K\mathbb{E}(\mathrm{e}^{\underline{X}_{\mathbf{e}_q}}) \tag{9.9}$$

then for $x \in \mathbb{R}$

$$\begin{aligned} v_y(x) &\ge \frac{\mathbb{E}\left(\mathbf{1}_{(t<\mathbf{e}_q)}\mathbb{E}\left((K\mathbb{E}(\mathrm{e}^{\underline{X}_{\mathbf{e}_q}}) - \mathrm{e}^{x+X_t+I})\mathbf{1}_{(-(X_t+I)>x-y)}\middle|\mathcal{F}_t\right)\right)}{\mathbb{E}(\mathrm{e}^{\underline{X}_{\mathbf{e}_q}})} \\ &\ge \mathbb{E}\left(\mathrm{e}^{-qt}v_y(x + X_t)\right) \\ &= \mathbb{E}_x\left(\mathrm{e}^{-qt}v_y(X_t)\right). \end{aligned}$$

Stationary independent increments now imply that for $0 \le s \le t < \infty$,

$$\mathbb{E}(\mathrm{e}^{-rt}v_y(X_t)|\mathcal{F}_s) = \mathrm{e}^{-rs}\mathbb{E}_{X_s}(\mathrm{e}^{-r(t-s)}v_y(X_{t-s})) \le \mathrm{e}^{-rs}v_y(X_s) \tag{9.10}$$

showing that $\{e^{-qt}v_y(X_t) : t \geq 0\}$ is a $\mathbb{P}_x$-supermartingale. Right continuity of its paths follow from the right continuity of the paths of X and right continuity of v_y which can be seen from (9.7).

To conclude, we see then that it would be sufficient to take $y = \log K\mathbb{E}(e^{\underline{X}_{\mathbf{e}_q}})$ in order to satisfy conditions (i) and (ii) of Lemma 9.1 and establish a solution to (9.3). □

In the case that X is spectrally negative, the solution may be expressed in terms of the scale functions. This was shown by Avram et al. (2002a) and Chan (2004).

Corollary 9.3. *Suppose that X is spectrally negative. Then*

$$v(x) = KZ^{(q)}(x - x^*) - e^x Z_1^{(p)}(x - x^*),$$

where $p = q - \psi(1)$ and

$$x^* = \log\left(K \frac{q}{\Phi(q)} \frac{\Phi(q) - 1}{q - \psi(1)}\right).$$

Recall that Φ_1 is the right inverse of ψ_1 which in turn is the Laplace exponent of X under the measure $\mathbb{P}^1$. Note that we have

$$\psi_1(\lambda) = \psi(\lambda + 1) - \psi(1)$$

for all $\lambda \geq -1$. Hence as $\Phi(q) - 1 > -1$,

$$\psi_1(\Phi(q) - 1) = q - \psi(1) = p$$

and this implies $\Phi_1(p) = \Phi(q) - 1$, where for negative values of p we understand

$$\Phi_1(p) = \sup\{\lambda \geq -1 : \psi_1(\lambda) = p\}.$$

The subscript on the functions $W_1^{(p)}$ and $Z_1^{(p)}$ indicate that they are the scale functions associated with the measure $\mathbb{P}^1$.

Proof (of Corollary 9.3). We know from Theorem 9.2 that $v = v_y$ for $y = x^*$. Hence from (9.4) and the conclusion of Exercise 8.7 (ii) we may write down the given expression for v immediately as

$$v(x) = K\left(Z^{(q)}(x - x^*) - W^{(q)}(x - x^*)\frac{q}{\Phi(q)}\right)$$
$$-e^x\left(Z_1^{(p)}(x - x^*) - W_1^{(p)}(x - x^*)\frac{p}{\Phi_1(p)}\right).$$

Next note that the general form of x^* given in Theorem 9.2 together with the expression for one of the Wiener–Hopf factors in (8.2) shows directly that

$$e^{x^*} = K\frac{q}{\Phi(q)}\frac{\Phi(q) - 1}{q - \psi(1)}.$$

From (8.25) we have that $e^x W_1^{(p)}(x) = W^{(q)}(x)$. Hence taking into account the definition of $\Phi_1(p)$, two of the terms in the expression for v given above cancel to give the identity in the statement of the corollary. □

9.3 Smooth Fit versus Continuous Fit

It is clear that the solution to (9.3) is lower bounded by the gain function G and further is equal to the gain function on the domain on which the distribution of X_{τ^*} is concentrated. It turns out that there are different ways in which the function v "fits" on to the gain function G according to certain path properties of the underlying Lévy process. The McKean optimal stopping problem provides a good example of where a dichotomy appears in this respect. We say that there is *continuous fit* at x^* if the left and right limit points of v at x^* exist and are equal. In addition, if the left and right derivatives of v exist at the boundary x^* and are equal then we say that there is *smooth fit* at x^*. The remainder of this section is devoted to explaining the dichotomy of smooth and continuous fit in (9.3).

Consider again the McKean optimal stopping problem. The following Theorem is again taken from Alili and Kyprianou (2005).

Theorem 9.4. *The function $v(\log y)$ is convex in $y > 0$ and in particular there is continuous fit of v at x^*. The right derivative at x^* is given by $v'(x^*+) = -\mathrm{e}^{x^*} + K\mathbb{P}(\underline{X}_{\mathbf{e}_q} = 0)$. Thus, the optimal stopping problem (9.3) exhibits smooth fit at x^* if and only if 0 is regular for $(-\infty, 0)$.*

Proof. Note that for a fixed stopping time $\tau \in \mathcal{T}$ the expression $\mathbb{E}(\mathrm{e}^{-q\tau}(K - \mathrm{e}^{x+X_\tau})^+)$ is convex in e^x as the same is true of the function $(K - c\mathrm{e}^x)^+$ where $c > 0$ is a constant. Further, since taking the supremum is a subadditive operation, it can easily be deduced that $v(\log y)$ is a convex function in y. In particular v is continuous.

Next we establish necessary and sufficient conditions for smooth fit. Since $v(x) = K - \mathrm{e}^x$ for all $x < x^*$, and hence $v'(x^*-) = -\mathrm{e}^{x^*}$, we are required to show that $v'(x^*+) = -\mathrm{e}^{x^*}$ for smooth fit. Starting from (9.6) and recalling that $\mathrm{e}^{x^*} = K\mathbb{E}(\mathrm{e}^{\underline{X}_{\mathbf{e}_q}})$, we have

$$\begin{aligned} v(x) &= -K\mathbb{E}\left((\mathrm{e}^{x-x^*+\underline{X}_{\mathbf{e}_q}} - 1)\mathbf{1}_{(-\underline{X}_{\mathbf{e}_q} > x-x^*)}\right) \\ &= -K(\mathrm{e}^{x-x^*} - 1)\mathbb{E}\left(\mathrm{e}^{\underline{X}_{\mathbf{e}_q}}\mathbf{1}_{(-\underline{X}_{\mathbf{e}_q} > x-x^*)}\right) \\ &\quad -K\mathbb{E}\left((\mathrm{e}^{\underline{X}_{\mathbf{e}_q}} - 1)\mathbf{1}_{(-\underline{X}_{\mathbf{e}_q} > x-x^*)}\right). \end{aligned}$$

From the last equality we may then write

$$\begin{aligned} \frac{v(x) - (K - \mathrm{e}^{x^*})}{x - x^*} &= \frac{v(x) + K(\mathbb{E}(\mathrm{e}^{\underline{X}_{\mathbf{e}_q}}) - 1)}{x - x^*} \\ &= -K\frac{(\mathrm{e}^{x-x^*} - 1)}{x - x^*}\mathbb{E}\left(\mathrm{e}^{\underline{X}_{\mathbf{e}_q}}\mathbf{1}_{(-\underline{X}_{\mathbf{e}_q} > x-x^*)}\right) \\ &\quad +K\frac{\mathbb{E}\left((\mathrm{e}^{\underline{X}_{\mathbf{e}_q}} - 1)\mathbf{1}_{(-\underline{X}_{\mathbf{e}_q} \le x-x^*)}\right)}{x - x^*}. \end{aligned}$$

To simplify notations let us call A_x and B_x the last two terms, respectively. It is clear that

$$\lim_{x\downarrow x^*} A_x = -K\mathbb{E}\left(\mathrm{e}^{\underline{X}_{\mathbf{e}_q}}\mathbf{1}_{(-\underline{X}_{\mathbf{e}_q}>0)}\right). \tag{9.11}$$

On the other hand, we have that

$$\begin{aligned}
B_x &= K\frac{\mathbb{E}\left((\mathrm{e}^{\underline{X}_{\mathbf{e}_q}}-1)\mathbf{1}_{(0<-\underline{X}_{\mathbf{e}_q}\leq x-x^*)}\right)}{x-x^*} \\
&= K\int_{0+}^{x-x^*}\frac{\mathrm{e}^{-z}-1}{x-x^*}\mathbb{P}(-\underline{X}_{\mathbf{e}_q}\in \mathrm{d}z) \\
&= K\frac{\mathrm{e}^{x^*-x}-1}{x-x^*}\mathbb{P}(0<-\underline{X}_{\mathbf{e}_q}\leq x-x^*) \\
&\quad +\frac{K}{x-x^*}\int_0^{x-x^*}\mathrm{e}^{-z}\mathbb{P}(0<-\underline{X}_{\mathbf{e}_q}\leq z)\mathrm{d}z,
\end{aligned}$$

where in the first equality we have removed the possible atom at zero from the expectation by noting that $\exp\{\underline{X}_{\mathbf{e}_q}\}-1=0$ on $\{\underline{X}_{\mathbf{e}_q}=0\}$. This leads to $\lim_{x\downarrow x^*}B_x=0$. Using the expression for e^{x^*} we see that $v'(x^*+)=-\mathrm{e}^{x^*}+K\mathbb{P}(-\underline{X}_{\mathbf{e}_q}=0)$ which equals $-\mathrm{e}^{x^*}$ if and only if $\mathbb{P}(-\underline{X}_{\mathbf{e}_q}=0)=0$; in other words, if and only if 0 is regular for $(-\infty,0)$. □

Let us now discuss intuitively the dichotomy of continuous and smooth fit as a mathematical principle. In order to make the arguments more visible, we will specialise to the case that X is a spectrally negative Lévy process; in which case v and x^* are given in Corollary 9.3. We start by looking in closer detail at the analytic properties of the candidate solution v_y at its boundary point y.

Returning to the candidate solutions (v_y,τ_y^-) for $y\leq\log K$ we have again from Exercise 8.7 that

$$\begin{aligned}
v_y(x) &= K\left(Z^{(q)}(x-y)-W^{(q)}(x-y)\frac{q}{\Phi(q)}\right) \\
&\quad -\mathrm{e}^x\left(Z_1^{(p)}(x-y)-W_1^{(p)}(x-y)\frac{p}{\Phi_1(p)}\right) \\
&= KZ^{(q)}(x-y)-\mathrm{e}^xZ_1^{(p)}(x-y)+W^{(q)}(x-y)\frac{p}{\Phi_1(p)}\left(e^y-K\frac{q\Phi_1(p)}{\Phi(q)p}\right),
\end{aligned}$$

where $p=q-\psi(1)$ and the second equality follows from (8.25) and specifically the fact that $\mathrm{e}^xW_1^{(p)}(x)=W^{(q)}(x)$. Thanks to the analytical properties of the scale functions, we may observe that v_y is continuous everywhere except possibly at y. Indeed at the point y we find

$$v_y(y-)=(K-\mathrm{e}^y)$$

and

$$v_y(y+) = v_y(y-) + W^{(q)}(0)\frac{p}{\Phi_1(p)}\left(\mathrm{e}^y - K\frac{q\Phi_1(p)}{\Phi(q)p}\right). \tag{9.12}$$

Recall that $W^{(q)}(0) = 0$ if and only if X is of unbounded variation and otherwise $W^{(q)}(0) = 1/d$ where d is the drift in the usual decomposition of X; see (8.1) and Lemma 8.6. As X is spectrally negative, 0 is regular for $(-\infty, 0)$ if and only if X is of unbounded variation. We see then that v_y is continuous whenever 0 is regular for $(-\infty, 0)$ and otherwise there is, in general, a discontinuity at y. Specifically if $y < x^*$ then there is a negative jump at y; if $y > x^*$ then there is a positive jump at y and if $y = x^*$ then there is continuity at y.

Next we compute the derivative of v_y as follows. For $x < y$ we have $v_y'(x) = -\mathrm{e}^x$. For $x > y$, again using the fact that $\mathrm{e}^x W_1^{(p)}(x) = W^{(q)}(x)$, we have

$$\begin{aligned} v_y'(x) &= KqW^{(q)}(x-y) - \mathrm{e}^y pW^{(q)}(x-y) \\ &\quad -\mathrm{e}^x Z_1^{(p)}(x-y) + W^{(q)\prime}(x-y)\frac{p}{\Phi_1(p)}\left(\mathrm{e}^y - K\frac{q\Phi_1(p)}{\Phi(q)p}\right). \end{aligned}$$

We see then that

$$\begin{aligned} v_y'(y+) &= v_y'(y-) + W^{(q)}(0)(Kq - \mathrm{e}^y p) \\ &\quad +W^{(q)\prime}(0+)\frac{p}{\Phi_1(p)}\left(\mathrm{e}^y - K\frac{q\Phi_1(p)}{\Phi(q)p}\right). \end{aligned} \tag{9.13}$$

Recall from Exercise 8.5 (ii) that

$$W^{(q)\prime}(0+) = \begin{cases} 2/\sigma^2 & \text{if } \Pi(-\infty,0) = \infty \\ (\Pi(-\infty,0)+q)/d^2 & \text{if } \Pi(-\infty,0) < \infty \end{cases}$$

where σ is the Gaussian coefficient, Π is the Lévy measure of X and $d > 0$ is the drift in the case that X has bounded variation. We adopt the understanding that $1/0 = \infty$.

Figs. 9.2 and 9.1 sketch what one can expect to see in the shape of v_y by perturbing the value y about x^* for the cases of unbounded variation and bounded variation with infinite Lévy measure. With these diagrams in mind we may now intuitively understand the appearance of smooth or continuous fit as a principle via the following reasoning.

For the case 0 is irregular for $(-\infty, 0)$ for X. In general v_y has a discontinuity at y. When $y < x^*$, thanks to the analysis of (9.12) we know the function v_y does not upper bound the gain function $(K - \mathrm{e}^x)^+$ due to a negative jump at y and hence τ_y^- is not an admissible strategy in this regime of y. On the other hand, from (9.7) and (9.8) if $y \geq x^*$, v_y upper bounds the gain function. From (9.12) we see that there is a discontinuity in v_y at y when $y > x^*$ and continuity when $y = x^*$. By bringing y down

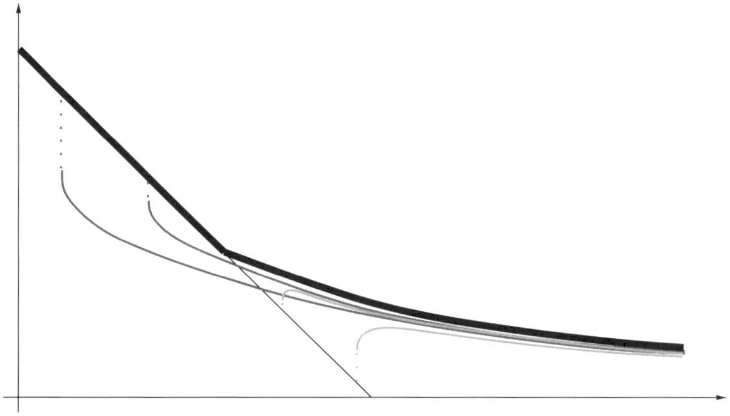

Fig. 9.1. A sketch of the functions $v_y(\log x)$ for different values of y when X is of bounded variation and $\Pi(-\infty, 0) = \infty$. Curves which do not upper bound the function $(K - x)^+$ correspond to examples of $v_y(\log x)$ with $y < x^*$. Curves which are lower bounded by $(K - x)^+$ correspond to examples of $v_y(\log x)$ with $y > x^*$. The unique curve which upper bounds the gain with continuous fit corresponds to $v_y(\log x)$ with $y = x^*$.

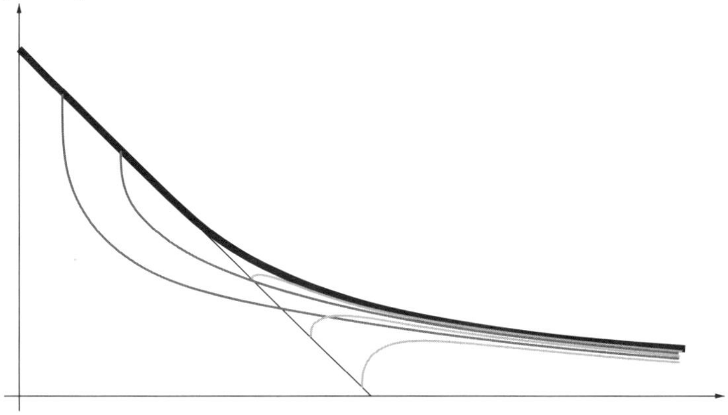

Fig. 9.2. A sketch of the functions $v_y(\log x)$ for different values of y when X is of unbounded variation and $\sigma = 0$. Curves which do not upper bound the function $(K - x)^+$ correspond to examples of $v_y(\log x)$ with $y < x^*$. Curves which are lower bounded by $(K - x)^+$ correspond to examples of $v_y(\log x)$ with $y > x^*$. The unique curve which upper bounds the gain with smooth fit corresponds to $v_y(\log x)$ with $y = x^*$.

to x^* it turns out that the function v_y is pointwise optimised. Here then we experience *a principle of continuous fit* and from (9.13) it transpires there is no smooth fit.

For the case 0 is regular for $(-\infty, 0)$ for X. All curves v_y are continuous. There is in general however a discontinuity in the first derivative of v_y at the point y. When $y < x^*$ the function v_y cannot upper bound the gain function $(K - e^y)^+$ as $v_y'(y+) < v_y'(y-)$ and hence τ_y^- is not an admissible strategy in this regime of y. As before, if $y \geq x^*$, v_y upper bounds the gain function. Again from (9.13) we see that there is a discontinuity in v_y' at y if $y > x^*$ and otherwise it is smooth when $y = x^*$. It turns out this time that by bringing y down to x^* the gradient $v_y'(y+)$ becomes equal to $v_y'(y-)$ and the function v_y is pointwise optimised. We experience then in this case *a principle of smooth fit* instead.

Whilst the understanding that smooth fit appears in the solutions of optimal stopping problems as a principle dates back to Mikhalevich (1958), the idea that continuous fit appears in certain classes of optimal stopping problems *as a principle* first appeared for the first time only recently in the work of Peskir and Shiryaev (2000, 2002).

9.4 The Novikov–Shiryaev Optimal Stopping Problem

The following family of optimal stopping problems was recently solved by Novikov and Shiryaev (2004) in an analogous random walk setting. Consider

$$v_n(x) = \sup_{\tau \in \mathcal{T}} \mathbb{E}_x(\mathrm{e}^{-q\tau}(X_\tau^+)^n), \tag{9.14}$$

where $\mathcal{T}$ is the set of $\mathbb{F}$-stopping times and it is assumed that X is any Lévy process, $q > 0$ and we may choose n to be any strictly positive integer. The solution we shall give is based on the arguments of Novikov and Shiryaev (2004)[4]. We first need to introduce a special class of polynomials based on cumulants of specified random variables.

Recall that if a non-negative random variable Y has characteristic function $\phi(\theta) = \mathbb{E}(\mathrm{e}^{i\theta Y})$ then its cumulant generating function is defined by $\log \phi(\theta)$. If Y has up to n moments then it is possible to make a Taylor expansion of the cumulant generating function up to order n plus an error term. In that case, the coefficients $\{\kappa_1, ..., \kappa_n\}$ are called the first n *cumulants*. If the first n cumulants are finite, then they may be written it terms of the first n moments. For example,

$$\begin{aligned}
\kappa_1 &= \mu_1, \\
\kappa_2 &= \mu_2 - \mu_1^2, \\
\kappa_3 &= 2\mu_1^3 - 3\mu_1\mu_2 + \mu_3, \\
&\dots
\end{aligned}$$

where $\mu_1, \mu_2, ...$ are the first, second, third, etc. moments of Y.

For a concise overview of cumulant generating functions, the reader is referred to Lukacs (1970).

[4]The continuous time arguments are also given in Kyprianou and Surya (2005)

Definition 9.5 (Appell Polynomials). *Suppose that Y is a non-negative random variable with nth cumulant given by κ_n for $n = 1, 2, ...$ Then define the Appell polynomials iteratively as follows. Take $Q_0(x) = 1$ and assuming that $|\kappa_n| < \infty$ (equivalently, Y has an nth moment) given $Q_{n-1}(x)$ we define $Q_n(x)$ via*

$$\frac{\mathrm{d}}{\mathrm{d}x}Q_n(x) = nQ_{n-1}(x). \tag{9.15}$$

This defines Q_n up to a constant. To pin this constant down we insist that $\mathbb{E}(Q_n(Y)) = 0$. The first three Appell polynomials are given for example by

$$Q_0(x) = 1,\ Q_1(x) = x - \kappa_1,\ Q_2(x) = (x - \kappa_1)^2 - \kappa_2,$$

$$Q_3(x) = (x - \kappa_1)^3 - 3\kappa_2(x - \kappa_1) - \kappa_3,$$

under the assumption that $\kappa_3 < \infty$. See also Schoutens (2000) for further details of Appell polynomials.

In the following theorem, we shall work with the Appell polynomials generated by the random variable $Y = \overline{X}_{\mathbf{e}_q}$ where as usual, for each $t \in [0, \infty)$, $\overline{X}_t = \sup_{s\in[0,t]} X_s$ and $\mathbf{e}_q$ is an exponentially distributed random variable which is independent of X.

Theorem 9.6. *Fix $n \in \{1, 2, ...\}$. Assume that*

$$\int_{(1,\infty)} x^n \Pi(dx) < \infty. \tag{9.16}$$

Then $Q_n(x)$ has finite coefficients and there exists $x_n^* \in [0, \infty)$ being the largest root of the equation $Q_n(x) = 0$. Let

$$\tau_n^* = \inf\{t \geq 0 : X_t \geq x_n^*\}.$$

Then τ_n^* is an optimal strategy to (9.14). Further,

$$v_n(x) = \mathbb{E}_x(Q_n(\overline{X}_{\mathbf{e}_q})\mathbf{1}_{(\overline{X}_{\mathbf{e}_q} \geq x_n^*)}).$$

Similarly to the McKean optimal stopping problem, we can establish a necessary and sufficient criterion for the occurrence of smooth fit. Once again, it boils down to the underlying path regularity.

Theorem 9.7. *For each $n = 1, 2, ...$ the solution to the optimal stopping problem in Theorem 9.6 is convex, in particular exhibiting continuous fit at x_n^*, and*

$$v_n'(x_n^*-) = v_n'(x_n^*+) - Q_n'(x_n^*)\mathbb{P}(\overline{X}_{\mathbf{e}_q} = 0).$$

Hence there is smooth fit at x_n^* if and only if 0 is regular for $(0, \infty)$ for X.

The proofs of the last two theorems require some preliminary results given in the following lemmas.

Lemma 9.8 (Mean value property). *Fix $n \in \{1, 2, ...\}$ Suppose that Y is a non-negative random variable satisfying $\mathbb{E}(Y^n) < \infty$. Then if Q_n is the nth Appell polynomial generated by Y, we have that*

$$\mathbb{E}(Q_n(x + Y)) = x^n$$

for all $x \in \mathbb{R}$.

Proof. Note the result is trivially true for $n = 1$. Next suppose the result is true for Q_{n-1}. Then using dominated convergence we have from (9.15) that

$$\frac{\mathrm{d}}{\mathrm{d}x}\mathbb{E}(Q_n(x + Y)) = \mathbb{E}\left(\frac{\mathrm{d}}{\mathrm{d}x}Q_n(x + Y)\right) = n\mathbb{E}(Q_{n-1}(x + Y)) = nx^{n-1}.$$

Solving together with the requirement that $\mathbb{E}(Q_n(Y)) = 0$ we have the required result. □

Lemma 9.9 (Fluctuation identity). *Fix $n \in \{1, 2, ...\}$ and suppose that*

$$\int_{(1,\infty)} x^n \Pi(\mathrm{d}x) < \infty.$$

Then for all $a > 0$ and $x \in \mathbb{R}$,

$$\mathbb{E}_x(\mathrm{e}^{-qT_a^+} X_{T_a^+}^n \mathbf{1}_{(T_a^+<\infty)}) = \mathbb{E}_x(Q_n(\overline{X}_{\mathbf{e}_q})\mathbf{1}_{(\overline{X}_{\mathbf{e}_q}\geq a)}),$$

where $T_a^+ = \inf\{t \geq 0 : X_t \geq a\}$.

Proof. On the event $\{T_a^+ < \mathbf{e}_q\}$ we have that $\overline{X}_{\mathbf{e}_q} = X_{T_a^+} + S$ where S is independent of $\mathcal{F}_{T_a^+}$ and has the same distribution as $\overline{X}_{\mathbf{e}_q}$. It follows that

$$\mathbb{E}_x(Q_n(\overline{X}_{\mathbf{e}_q})\mathbf{1}_{(\overline{X}_{\mathbf{e}_q}\geq a)}|\mathcal{F}_{T_a^+}) = \mathbf{1}_{(T_a^+<\mathbf{e}_q)}h(X_{T_a^+}),$$

where $h(x) = \mathbb{E}_x(Q_n(\overline{X}_{\mathbf{e}_q})) = x^n$ and the last equality follows from Lemma 9.8 with $Y = \overline{X}_{\mathbf{e}_q}$. Note also that by Exercise 7.1 the integral condition on Π implies that $\mathbb{E}(\overline{X}_{\mathbf{e}_q}^n) < \infty$ which has been used in order to apply Lemma 9.8. We see, by taking expectations again in the previous calculation, that

$$\mathbb{E}_x(Q_n(\overline{X}_{\mathbf{e}_q})\mathbf{1}_{(\overline{X}_{\mathbf{e}_q}\geq a)}) = \mathbb{E}_x(e^{-qT_a^+} X_{T_a^+}^n \mathbf{1}_{(T_a^+<\infty)})$$

as required. □

Lemma 9.10 (Largest positive root). *Fix $n \in \{1, 2, ...\}$ and suppose that*

$$\int_{(1,\infty)} x^n \Pi(\mathrm{d}x) < \infty.$$

Suppose that Q_n is generated by $\overline{X}_{\mathbf{e}_q}$. Then Q_n has a unique strictly positive root x_n^* such that $Q_n(x)$ is negative on $[0, x_n^*)$ and positive and increasing on $[x_n^*, \infty)$.

Proof. First note that the statement of the lemma is clearly true for $Q_1(x) = x - \kappa_1$. We proceed then by induction and assume that the result is true for Q_{n-1}.

The first step is to prove that $Q_n(0) \leq 0$. Let

$$\eta(a,n) = \mathbb{E}(e^{-qT_a^+} X^n_{T_a^+} \mathbf{1}_{(T_a^+ < \infty)}),$$

where $T_a^+ = \inf\{t \geq 0 : X_t \geq a\}$ and note that $\eta(a,n) \geq 0$ for all $a \geq 0$ and $n = 1, 2, ...$ On the other hand

$$\begin{aligned}\eta(a,n) &= \mathbb{E}(Q_n(\overline{X}_{\mathbf{e}_q})\mathbf{1}_{(\overline{X}_{\mathbf{e}_q} \geq a)}) \\ &= -\mathbb{E}(Q_n(\overline{X}_{\mathbf{e}_q})\mathbf{1}_{(\overline{X}_{\mathbf{e}_q} < a)}) \\ &= -\mathbb{P}(\overline{X}_{\mathbf{e}_q} < a)Q_n(0) \\ &\quad + \mathbb{E}((Q_n(0) - Q_n(\overline{X}_{\mathbf{e}_q}))\mathbf{1}_{(\overline{X}_{\mathbf{e}_q} < a)}),\end{aligned}$$

where the first equality follows from Lemma 9.9 and the second by Lemma 9.8. Since by definition

$$Q_n(x) = Q_n(0) + n\int_0^x Q_{n-1}(u)\mathrm{d}y \tag{9.17}$$

for all $x \geq 0$ we have the estimate

$$\left|\mathbb{E}_x((Q_n(0) - Q_n(\overline{X}_{\mathbf{e}_q}))\mathbf{1}_{(\overline{X}_{\mathbf{e}_q} < a)})\right| \leq na \sup_{y\in[0,a]} |Q_{n-1}(y)|\mathbb{P}(\overline{X}_{\mathbf{e}_q} < a)$$

which tends to zero as $a \downarrow 0$. We have in conclusion that

$$0 \leq \lim_{a\downarrow 0} \eta(a,n) \leq -\lim_{a\downarrow 0} \mathbb{P}(\overline{X}_{\mathbf{e}_q} < a)Q_n(0) + o(a)$$

and hence necessarily $Q_n(0) \leq 0$.

Under the induction hypothesis for Q_{n-1}, we see from (9.17), together with the fact that $Q_n(0) \leq 0$, that Q_n is negative and decreasing on $[0, x^*_{n-1})$. The point x^*_{n-1} corresponds to the minimum of Q_n thanks to the positivity and monotonicity of $Q_{n-1}(u)$ for $x > x^*_{n-1}$. In particular, $Q_n(x)$ tends to infinity from its minimum point and hence there must be a unique strictly positive root of the equation $Q_n(x) = 0$. □

We are now ready to move to the proofs of the main theorems of this section.

Proof (of Theorem 9.6). Fix $n \in \{1, 2, ...\}$. Thanks to (9.16), $\mathbb{E}(X_1) \in [-\infty, \infty)$ and hence the Strong Law of Large Numbers given in Exercise 7.2 implies that (9.2) is automatically satisfied. Indeed, if $q > 0$ then $(X_t^+)^n$ grows no faster than Ct^n for some constant $C > 0$.

Define

$$v_n^a(x) = \mathbb{E}_x(e^{-qT_a^+}(X_{T_a^+}^+)^n \mathbf{1}_{(T_a^+<\infty)}), \tag{9.18}$$

where as usual $T_a^+ = \inf\{t \geq 0 : X_t \geq a\}$. From Lemma 9.9 we know that

$$v_n^a(x) = \mathbb{E}_x(Q_n(\overline{X}_{\mathbf{e}_q})\mathbf{1}_{(\overline{X}_{\mathbf{e}_q} \geq a)}).$$

Again referring to the discussion following Lemma 9.1 we consider pairs (v_n^a, T_a^+) for $a > 0$ to be a class of candidate solutions to (9.14). Our goal then is to verify with the help of Lemma 9.1 that the candidate pair (v_n^a, T_a^+) solve (9.14) for some $a > 0$.

Lower bound (i). We need to prove that $v_n^a(x) \geq (x^+)^n$ for all $x \in \mathbb{R}$. Note that this statement is obvious for $x \in (-\infty, 0) \cup (a, \infty)$ just from the definition of v_n^a. Otherwise when $x \in (0, a)$ we have, using the mean value property in Lemma 9.8, that for all $x \in \mathbb{R}$,

$$\begin{aligned} v_n^a(x) &= \mathbb{E}_x(Q_n(\overline{X}_{\mathbf{e}_q})\mathbf{1}_{(\overline{X}_{\mathbf{e}_q} > a)}) \\ &= x^n - \mathbb{E}(Q_n(x + \overline{X}_{\mathbf{e}_q})\mathbf{1}_{(x+\overline{X}_{\mathbf{e}_q} < a)}). \end{aligned} \tag{9.19}$$

From Lemma 9.10 and specifically the fact that $Q_n(x) \leq 0$ on $(0, x_n^*]$ it follows that, provided

$$a \leq x_n^*,$$

we have in (9.19) that $v_n^a(x) \geq (x^+)^n$.

Supermartingale property (ii). Provided

$$a \geq x_n^*$$

we have almost surely that

$$Q_n(\overline{X}_{\mathbf{e}_q})\mathbf{1}_{(\overline{X}_{\mathbf{e}_q} \geq a)} \geq 0.$$

On the event that $\{\mathbf{e}_q > t\}$ we have $\overline{X}_{\mathbf{e}_q}$ is equal in distribution to $(X_t + S) \vee \overline{X}_t$ where S is independent of $\mathcal{F}_t$ and equal in distribution to $\overline{X}_{\mathbf{e}_q}$. In particular $\overline{X}_{\mathbf{e}_q} \geq X_t + S$. It now follows that

$$\begin{aligned} v_n^a(x) &\geq \mathbb{E}_x(\mathbf{1}_{(\mathbf{e}_q > t)} Q_n(\overline{X}_{\mathbf{e}_q})\mathbf{1}_{(\overline{X}_{\mathbf{e}_q} \geq a)}) \\ &\geq \mathbb{E}_x(\mathbf{1}_{(\mathbf{e}_q > t)} \mathbb{E}_x(Q_n(X_t + S)\mathbf{1}_{(X_t + S \geq a)} | \mathcal{F}_t)) \\ &= \mathbb{E}_x(\mathrm{e}^{-qt} v_n^a(X_t)). \end{aligned}$$

From this inequality together with the Markov property, it is easily shown as in the McKean optimal stopping problem that $\{\mathrm{e}^{-qt} v_n^a(X_t) : t \geq 0\}$ is a supermartingale. Right continuity follows again from the right continuity of the paths of X together with the right continuity of v_n^a which is evident from (9.19).

We now see that the unique choice $a = x_n^*$ allows all the conditions of Lemma 9.1 to be satisfied thus giving the solution to (9.14). □

Note that the case $q = 0$ can be dealt with in essentially the same manner. In that case it is necessary to assume that $\limsup_{t\uparrow\infty} X_t < \infty$ and if working with the gain function $(x^+)^n$ for $n = 1, 2, \ldots$, then one needs to assume that

$$\int_{(1,\infty)} x^{n+1}\Pi(dx) < \infty.$$

Note the power in the above integral is $n+1$ and not n as one must now deal with the nth moments of $\overline{X}_\infty$; see Exercise 7.1.

Proof (of Theorem 9.7). In a similar manner to the proof of Theorem 9.4 it is straighforward to prove that v is convex and hence continuous.

To establish when there is a smooth fit at x_n^* we compute as follows. For $x < x_n^*$,

$$\begin{aligned}\frac{v_n(x_n^*) - v_n(x^n)}{x_n^* - x} &= \frac{(x_n^*)^n - x^n}{x_n^* - x} + \frac{\mathbb{E}_x(Q_n(\overline{X}_{\mathbf{e}_q})\mathbf{1}_{(\overline{X}_{\mathbf{e}_q} < x_n^*)})}{x_n^* - x} \\ &= \frac{(x_n^*)^n - x^n}{x_n^* - x} + \frac{\mathbb{E}_x((Q_n(\overline{X}_{\mathbf{e}_q}) - Q_n(x_n^*))\mathbf{1}_{(\overline{X}_{\mathbf{e}_q} < x_n^*)})}{x_n^* - x},\end{aligned}$$

where the final equality follows because $Q_n(x_n^*) = 0$. Clearly

$$\lim_{x\uparrow x_n^*} \frac{(x_n^*)^n - x^n}{x_n^* - x} = v_n'(x_n^*+).$$

However,

$$\begin{aligned}&\frac{\mathbb{E}_x((Q_n(\overline{X}_{\mathbf{e}_q}) - Q_n(x_n^*))\mathbf{1}_{(\overline{X}_{\mathbf{e}_q} < x_n^*)})}{x_n^* - x} \\ &\quad= \frac{\mathbb{E}_x((Q_n(\overline{X}_{\mathbf{e}_q}) - Q_n(x))\mathbf{1}_{(x<\overline{X}_{\mathbf{e}_q} < x_n^*)})}{x_n^* - x} \\ &\qquad - \frac{\mathbb{E}_x((Q_n(x_n^*) - Q_n(x))\mathbf{1}_{(\overline{X}_{\mathbf{e}_q} < x_n^*)})}{x_n^* - x}\end{aligned} \tag{9.20}$$

where in the first term on the right-hand side we may restrict the expectation to $\{x < \overline{X}_{\mathbf{e}_q} < x_n^*\}$ as, under $\mathbb{P}_x$, the possible atom of $\overline{X}_{\mathbf{e}_q}$ at x gives zero mass to the expectation. Denote by A_x and B_x the two expressions on the right hand side of (9.20). We have that

$$\lim_{x\uparrow x_n^*} B_x = -Q_n'(x_n^*)\mathbb{P}(\overline{X}_{\mathbf{e}_q} = 0).$$

Integration by parts also gives

$$A_x = \int_{(0,x_n^*-x)} \frac{Q_n(x+y) - Q_n(x)}{x_n^* - x} \mathbb{P}(\overline{X}_{\mathbf{e}_q} \in \mathrm{d}y)$$
$$= \frac{Q_n(x_n^*) - Q_n(x)}{x_n^* - x} \mathbb{P}(\overline{X}_{\mathbf{e}_q} \in (0, x_n^* - x))$$
$$- \frac{1}{x_n^* - x} \int_0^{x_n^*-x} \mathbb{P}(\overline{X}_{\mathbf{e}_q} \in (0, y]) Q_n'(x+y)\mathrm{d}y.$$

Hence it follows that

$$\lim_{x \uparrow x_n^*} A_x = 0.$$

In conclusion we have that

$$\lim_{x \uparrow x_n^*} \frac{v_n(x_n^*) - v_n(x)}{x_n^* - x} = v_n'(x_n^*+) - Q_n'(x_n^*)\mathbb{P}(\overline{X}_{\mathbf{e}_q} = 0)$$

which concludes the proof. □

9.5 The Shepp–Shiryaev Optimal Stopping Problem

Suppose that X is a Lévy process. Then consider the optimal stopping problem

$$v(x) = \sup_{\tau \in \mathcal{T}} \mathbb{E}(\mathrm{e}^{-q\tau + (\overline{X}_\tau \vee x)}) \tag{9.21}$$

where $q > 0$, $\mathcal{T}$ is the set of $\mathbb{F}$-stopping times which are almost surely finite and $x \geq 0$. This optimal stopping problem was proposed and solved by Shepp and Shiryaev (1993) for the case that X is a linear Brownian motion and q is sufficiently large. Like the McKean optimal stopping problem, (9.21) appears in the context of an option pricing problem. Specifically, it addresses the problem of the optimal time to sell a risky asset for the minimum of either e^x or its running maximum when the risky asset follows the dynamics of an exponential linear Brownian motion and in the presence of discounting. In Avram et al. (2004) a solution was given to (9.21) in the case that X is a general spectrally negative Lévy process (and again q is sufficiently large). In order to keep to the mathematics that has been covered earlier on in this text, we give an account of a special case of that solution here. Specifically we deal with the case that X has bounded variation which we shall write in the usual form

$$X_t = dt - S_t, \tag{9.22}$$

where $d > 0$ and S is a driftless subordinator with Lévy measure Π (concentrated on $(0, \infty)$). We shall further assume that Π has no atoms. As we shall use scale functions in our solution, the latter condition will ensure that they are at least continuously differentiable; see Exercise 8.4. Our objective is the theorem below. Note that we use standard notation from Chap. 8.

Theorem 9.11. *Suppose that X is as stated in the the above paragraph, having Laplace exponent ψ. Suppose that $q > \psi(1)$. Define*

$$x^* = \inf\{x \geq 0 : Z^{(q)}(x) \leq qW^{(q)}(x)\}.$$

Then for each $x \geq 0$, the solution to (9.21) is given by the pair

$$v(x) = \mathrm{e}^x Z^{(q)}(x^* - x)$$

and

$$\tau^* = \inf\{t > 0 : Y_t^x > x^*\},$$

where $Y^x = \{Y_t^x : t \geq 0\}$ is the process X reflected in its supremum when initiated from $x \geq 0$ so that $Y_t^x = (x \vee \overline{X}_t) - X_t$.

Before moving to the proof of Theorem 9.11, let us consider the nature of (9.21) and its solution in a little more detail.

Firstly it needs to be pointed out that due to the involvement of the running supremum in the formulation of the problem, one may consider (9.21) as an optimal stopping problem which potentially concerns the three dimensional Markov process $\{(t, X_t, \overline{X}_t) : t \geq 0\}$. Nonetheless, as with the previous two examples of optimal stopping problems it is possible to reduce the dimension of the problem to just one. As indeed one notes from the statement of Theorem 9.11, the solution is formulated in terms of the process Y^x.

The way to do this was noted by Shepp and Shiryaev (1994) in a follow-up article to their original contribution. Recalling the method of change of measure described in Sects. 3.3 and 8.1 (see specifically Corollary 3.11), for each $\tau \in \mathcal{T}$ we may write

$$\mathbb{E}(\mathrm{e}^{-q\tau + (\overline{X}_\tau \vee x)}) = \mathbb{E}^1(\mathrm{e}^{-\alpha\tau + Y_\tau^x}),$$

where

$$\alpha = q - \psi(1)$$

and $Y^x = (x \vee \overline{X}) - X$ is the process reflected in its supremum and issued from x. Hence our objective is now to solve the optimal stopping problem

$$\sup_{\tau \in \mathcal{T}} \mathbb{E}^1(\mathrm{e}^{-\alpha\tau + Y_\tau^x}) \tag{9.23}$$

which is now based on the two dimensional Markov process $\{(t, Y_t^x) : t \geq 0\}$. Again, arguments along the lines of those in the paragraphs following Lemma 9.1 suggest that a suitable class of candidate solutions to (9.23) are of the form $(v_a(x), \overline{\sigma}_a^x)$ with $a \geq 0$ where for each $x \geq 0$

$$v_a(x) = \mathbb{E}^1(\mathrm{e}^{-\alpha\overline{\sigma}_a^x + Y_{\overline{\sigma}_a^x}^x}) \tag{9.24}$$

and

$$\overline{\sigma}_a^x = \inf\{t > 0 : Y_t^x \geq a\}.$$

(Note that the latter is $\mathbb{P}^1$-almost surely finite). In addition, one may also intuitively understand how a threshold strategy for Y^x is optimal for (9.21) as follows. The times at which Y^x is zero correspond to times at which $\overline{X}$, and hence the gain, is increasing. The times at which Y^x takes large values correspond to the times at which X is far from its running supremum. At such moments, one must wait for the process to return to its maximum before there is an increase in the gain. Exponential discounting puts a time penalty on waiting too long suggesting that one should look for a threshold strategy in which one should stop if X moves too far from its maximum.

Secondly let us consider the optimal threshold x^*. Define the function $f(x) = Z^{(q)}(x) - qW^{(q)}(x)$. Differentiating we have

$$\begin{aligned} f'(x) &= q(W^{(q)}(x) - W^{(q)\prime}(x)) \\ &= qe^{\Phi(q)x}((1-\Phi(q))W_{\Phi(q)}(x)) - W'_{\Phi(q)}(x) \end{aligned} \tag{9.25}$$

where, in the second equality we have used (8.25). Since

$$\psi'_{\Phi(q)}(0+) = \psi'(\Phi(q)) > 0,$$

we know that under $\mathbb{P}^{\Phi(q)}$ the process X drifts to infinity and hence recall from (8.7) that this implies that $W_{\Phi(q)}(x)$ is monotone increasing. In particular $W'_{\Phi(q)}(x) > 0$ for all $x > 0$. On the other hand, the assumption that $q > \psi(1)$ implies that $\Phi(q) > 1$. Hence the right-hand side of (9.25) is strictly negative for all $x > 0$. Further we may check with the help of Exercise 8.5 (i) that

$$\lim_{x\uparrow\infty} \frac{f(x)}{qW^{(q)}(x)} = \frac{1}{\Phi(q)} - 1 < 0.$$

From (8.25) again, it is clear that the denominator on the left-hand side above tends to infinity.

Note that $f(0+) = 1 - qW^{(q)}(0)$. Since X has bounded variation we know that $W^{(q)}(0) = d^{-1}$. Hence if $q \geq d$ then $f(0+) < 0$ so that necessarily $x^* = 0$. When $q < d$, we have that $f(0+) > 0$, $f'(x) < 0$ for all $x > 0$ and $f(\infty) = -\infty$. It follows then that there is a unique solution in $[0, \infty)$ to $f(x) = 0$, namely x^*, which is necessarily strictly positive.

Note also that as the solution can be expressed in terms of functions whose analytic properties are understood sufficiently well that, we may establish easily the situation with regard to smooth or continuous fit. For each $x > 0$, the process Y^x has the same small-time path behaviour as $-X$ and hence $\mathbb{P}(\overline{\sigma}^x_x = 0) = 0$ as $\mathbb{P}(\tau_0^- = 0) = 0$. This property is independent of $x > 0$ (the point $x = 0$ needs to be considered as a special case on account of reflection).

Corollary 9.12. *When $q < d$, the solution to (9.21) is convex and hence exhibits continuous fit at x^*. Further it satisfies*

$$v'(x^*-) = v'(x^*+) - \frac{q}{d}e^{x^*}$$

showing that there is no smooth fit at x^*.

Proof. The first part follows in a similar manner to the proof of convexity in the previous two optimal stopping problem. After a straightforward differentiation of the solution to (9.21), recalling that $W(0+) = 1/d$, the last part of the corollary follows. □

Proof (of Theorem 9.11). As indicated above, we consider candidate solutions of the form $(v_a, \overline{\sigma}_a^x)$. We can develop the right-hand side of v_a in terms of scale functions with the help of Theorem 8.10 (i). However, before doing so, again taking into account the conclusion of Lemma 8.4, let us note that the analogue of the scale function $W_{-1}^{(q)}$ when working under the measure $\mathbb{P}^1$ can be calculated as equal to

$$[W_1^{(q)}]_{-1}(x) = \mathrm{e}^x W_1^{(q+\psi_1(-1))}.$$

However, we also know that $\psi_1(-1) = \psi(1-1) - \psi(1) = -\psi(1)$ and hence applying Lemma 8.4 again we have further that

$$[W_1^{(q)}]_{-1}(x) = \mathrm{e}^x \mathrm{e}^{-x} W^{(q)}(x) = W^{(q)}(x).$$

This means that we may read out of Theorem 8.10 (i) the identity

$$v_a(x) = e^x \left(Z^{(q)}(a-x) - W^{(q)}(a-x) \frac{qW^{(q)}(a) - Z^{(q)}(a)}{W^{(q)\prime}(a) - W^{(q)}(a)} \right). \tag{9.26}$$

Revisiting the proof of Lemma 9.1 one sees that in fact that requirement that X is a Lévy process is not necessary and one may replace X by Y^x without affecting any of the arguments or conclusions. (Here we mean that $\mathcal{T}$ is still the family of stopping times with respect to the underlying Lévy process). Let us continue then to check the conditions of the aforementioned modified version of Lemma 9.1.

Lower bound (i). We need to show that $v_a(s) \geq \mathrm{e}^x$. The assumption $q > \psi(1)$ implies that $\Phi(q) > 1$ and hence

$$W^{(q)\prime}(a) - W^{(q)}(a) > W^{(q)\prime}(a) - \Phi(q)W^{(q)}(a).$$

On the other hand, from (8.25) we may compute

$$W^{(q)\prime}(a) - \Phi(q)W^{(q)}(a) = \mathrm{e}^{\Phi(q)a} W'_{\Phi(q)}(a) > 0,$$

where the inequality is, for example, due to (8.18). Together with the properties of $Z^{(q)}(a) - qW^{(q)}(a)$ as a function of a, we see that the coefficient

$$\frac{qW^{(q)}(a) - Z^{(q)}(a)}{W^{(q)\prime}(a) - W^{(q)}(a)}$$

is strictly positive when $a > x^*$ and non-positive when $a \in [0, x^*]$.

Recalling that $Z^{(q)}(x) \geq 1$, we conclude that $v_a(x) \geq \mathrm{e}^x$ when $a \in [0, x^*]$. On the other hand, suppose that $a > x^*$, then

$$v_a(a-) = \mathrm{e}^a - W^{(q)}(0)\frac{qW^{(q)}(a) - Z^{(q)}(a)}{W^{(q)\prime}(a) - W^{(q)}(a)} < \mathrm{e}^a, \tag{9.27}$$

showing that in order to respect the lower bound we necessarily must take $a \leq x^*$.

Supermartingale property (ii). We know that the function $v_a(x)$ is differentiable with continuous first derivative on $(0, a)$. Further, the right derivative at zero exists and is equal to zero. To see this, simply compute

$$\begin{aligned} v_a'(0+) &= \mathrm{e}^a \left(Z^{(q)}(a) - W^{(q)}(a)\frac{qW^{(q)}(a) - Z^{(q)}(a)}{W^{(q)\prime}(a) - W^{(q)}(a)} \right. \\ &\qquad \left. -qW^{(q)}(a) + W^{(q)\prime}(a)\frac{qW^{(q)}(a) - Z^{(q)}(a)}{W^{(q)\prime}(a) - W^{(q)}(a)} \right) \\ &= 0. \end{aligned}$$

Next recall from Sect. 8.1 that under $\mathbb{P}^1$, X remains a Lévy process of bounded variation with the same drift but now with exponentially tilted Lévy measure so that in the form (9.22), $\Pi(\mathrm{d}y)$ becomes $\mathrm{e}^{-y}\Pi(\mathrm{d}y)$. Applying the change of variable formula in the spirit of Exercise 4.2 and 4.1 we see that, despite the fact that the first derivative of v_a is not well defined at a,

$$\begin{aligned} \mathrm{e}^{-\alpha t} v_a(Y_t^x) &= v_a(x) - \alpha \int_0^t \mathrm{e}^{-\alpha s} v_a(Y_s^x)\mathrm{d}s \\ &\quad + \int_0^t \mathrm{e}^{-\alpha s}(v_a(a-) - v_a(a+))\mathrm{d}L_t^a \\ &\quad -\mathrm{d}\int_0^t \mathrm{e}^{-\alpha s} v_a'(Y_s^x)\mathrm{d}s + \int_0^t \mathrm{e}^{-\alpha s} v_a'(Y_s^x)\mathrm{d}(x \vee \overline{X}_s) \\ &\quad + \int_{[0,t]}\int_{(0,\infty)} \mathrm{e}^{-\alpha s}(v_a(Y_{s-}^x + y) - v_a(Y_{s-}^x))\mathrm{e}^{-y}\Pi(\mathrm{d}y)\mathrm{d}s \\ &\quad + M_t. \end{aligned} \tag{9.28}$$

$\mathbb{P}$-almost surely, where $\{L_t^a : t \geq 0\}$ counts the number of crossings of the process Y^x over the level a. Further, for $t \geq 0$,

$$\begin{aligned} M_t &= \int_{[0,t]}\int_{(0,\infty)} \mathrm{e}^{-\alpha s}(v_a(Y_{s-}^x + y) - v_a(Y_{s-}^x))N^1(\mathrm{d}s \times \mathrm{d}x) \\ &\quad - \int_{[0,t]}\int_{(0,\infty)} \mathrm{e}^{-\alpha s}(v_a(Y_{s-}^x + y) - v_a(Y_{s-}^x))\mathrm{e}^{-y}\Pi(\mathrm{d}y)\mathrm{d}s, \end{aligned}$$

where N^1 is the counting measure associated with the Poisson point process of jumps of the subordinator given in the representation (9.22) of X under $\mathbb{P}^1$.

In the third integral of (9.28), the process $x \vee \overline{X}_s$ increases only when $Y_s^x = 0$. Since $v_a'(0+) = 0$, then the third integral is equal to zero. Note also that it can be proved in a straightforward way with the help of the compensation formula in Theorem 4.4 that the process $M := \{M_t : t \geq 0\}$ is a martingale.

The Markov property together with (9.24) imply that

$$\mathbb{E}^1(\mathrm{e}^{-\alpha\overline{\sigma}_a^x + Y^x_{\overline{\sigma}_a^x}} | \mathcal{F}_t) = \mathrm{e}^{-\alpha(\overline{\sigma}_a^x \wedge t) + Y^x_{(\overline{\sigma}_a^x \wedge t)}}.$$

Hence considering the left-hand side of (9.28) on $\{t < \overline{\sigma}_a^x\}$, we deduce with the help of Exercise 4.3 that

$$\mathcal{L}^1 v_a(x) := \int_{(0,\infty)} (v_a(x+y) - v_a(x))\mathrm{e}^{-y}\Pi(\mathrm{d}y) - \mathrm{d}v_a'(x) - \alpha v_a(x) = 0 \quad (9.29)$$

for all $x \in (0, a)$. We also see by inspection $v_a(x) = e^x$ for all $x > a$. In that case, from the definition of the Laplace exponent ψ we know that the expression on the left-hand side of (9.29) satisfies

$$\begin{aligned}\mathcal{L}^1 v_a(x) &= \mathrm{e}^x \int_{(0,\infty)} (1 - \mathrm{e}^{-y})\Pi(\mathrm{d}y) - \mathrm{d}e^x - \alpha e^x \\ &= -\mathrm{e}^x(\psi(1) + \alpha) \\ &= -q\mathrm{e}^x < 0\end{aligned}$$

for $x > a$. In conclusion we have shown that $\mathcal{L}^1 v_a(x) \leq 0$ on $x \in (0,\infty)\backslash\{a\}$.

If $a \geq x^*$, then from (9.27) $v_a(a-) - v_a(a+) \leq 0$. Reconsidering (9.28) in this light, we see that when $a \geq x^*$, the process $\{\mathrm{e}^{-\alpha t} v_a(X_t) : t \geq 0\}$ is the difference of a martingale M and a non-decreasing adapted process thus making it a supermartingale. Right continuity can be seen from (9.28).

In conclusion, all properties of Lemma 9.1 are satisfied uniquely when $a = x^*$, thus concluding the proof. □

The semi-explicit nature of the functions $\{v_a : a \geq 0\}$ again gives the opportunity to show graphically how continuous fit occurs by perturbing the function v_a about the value $a = x^*$. See Fig. 9.3.

9.6 Stochastic Games

Suppose that H, G are two continuous functions mapping $\mathbb{R}$ to $[0,\infty)$ satisfying $H(x) > G(x)$ for all $x \in \mathbb{R}$ and let $\mathcal{T}$ be a family of $\mathbb{F}$-stopping times. For $q \geq 0$, $\sigma, \tau \in \mathcal{T}$ define

$$\Theta^q_{\tau,\sigma} = \mathrm{e}^{-q\tau}G(X_\tau)\mathbf{1}_{(\tau \leq \sigma)} + \mathrm{e}^{-q\sigma}H(X_\sigma)\mathbf{1}_{(\sigma < \tau)}.$$

Consider the existence of a function $v : \mathbb{R} \mapsto [0,\infty)$ and stopping times $\tau^*, \sigma^* \in \mathcal{T}$ which satisfy the relation

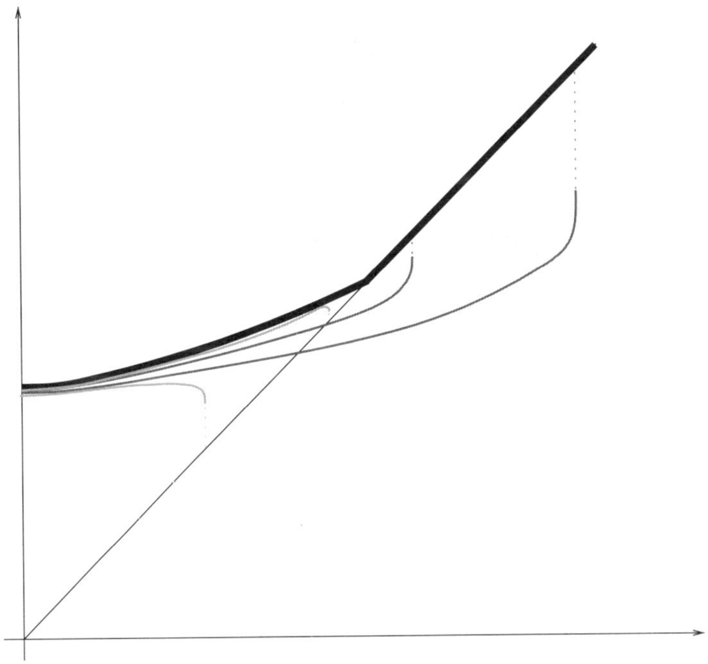

Fig. 9.3. A sketch of the functions $v_a(\log x)$ for different values of a when X is of bounded variation and $\Pi(-\infty,0)=\infty$. Curves which do not upper bound the diagonal correspond to $v_a(\log x)$ for $a > x^*$. Curves which are lower bounded by the diagonal correspond to $v_a(\log x)$ for $0 < a < x^*$. The unique curve which upper bounds the diagonal with continuous fit corresponds to $v_a(\log x)$ with $y = x^*$.

$$v(x) = \sup_{\tau\in\mathcal{T}} \inf_{\sigma\in\mathcal{T}} \mathbb{E}_x(\Theta^q_{\tau,\sigma}) = \inf_{\sigma\in\mathcal{T}} \sup_{\tau\in\mathcal{T}} \mathbb{E}_x(\Theta^q_{\tau,\sigma}) = \mathbb{E}_x(\Theta^q_{\tau^*,\sigma^*}). \tag{9.30}$$

The mathematical object (9.30) is called a stochastic game on account of its interpretation in the following context. In a two player game, player one agrees to pay player two an amount $G(X_\tau)$ when player two decides to *call payment* at any chosen time $\tau \in \mathcal{T}$. However, player one also reserves the right to *force payment* to the amount $H(X_\sigma)$ at any chosen time $\sigma \in \mathcal{T}$. If both players decide on payment at the same time, then the payment $G(X_\sigma)$ is made. The second player plays according to the policy of maximising their expected discounted gain, whilst the first player has the policy of minimising the expected discounted gain. (Note however it is not necessary that the optimal strategies of the two players are to stop the game at almost surely finite stopping times).

The problem of characterising the triple (v,τ^*,σ^*) has many similarities with establishing solutions to optimal stopping problems. Consider for example the following sufficient criteria for the triple (v,τ^*,σ^*) to characterise the saddle point of (9.30).

Lemma 9.13. *Consider the stochastic game (9.30) under the assumptions that for all $x \in \mathbb{R}$,*

$$\mathbb{E}_x(\sup_{t\geq 0} \mathrm{e}^{-qt} H(X_t)) < \infty \tag{9.31}$$

and

$$\mathbb{P}_x(\exists \lim_{t\uparrow\infty} \mathrm{e}^{-qt} G(X_t) < \infty) = 1. \tag{9.32}$$

Suppose that $\tau^* \in \mathcal{T}$ and $\sigma^* \in \mathcal{T}$ are candidate optimal strategies for the stochastic game (9.30) and let $v^*(x) = \mathbb{E}_x(\Theta^q_{\tau^*,\sigma^*})$. Then the triple (v^*, τ^*, σ^*) is a solution to (9.30) if

(i) $v^*(x) \geq G(x)$,
(ii) $v^*(x) \leq H(x)$
(iii) $v^*(X_{\tau^*}) = G(X_{\tau^*})$ almost surely on $\{\tau^* < \infty\}$,
(iv) $v^*(X_{\sigma^*}) = H(X_{\sigma^*})$ almost surely on $\{\sigma^* < \infty\}$,
(v) the process $\{\mathrm{e}^{-q(t\wedge\tau^*)} v^*(X_{t\wedge\tau^*}) : t \geq 0\}$ is a right continuous submartingale and
(vi) the process $\{\mathrm{e}^{-q(t\wedge\sigma^*)} v^*(X_{t\wedge\sigma^*}) : t \geq 0\}$ is a right continuous supermartingale.

Proof. From the supermartingale property (vi), Doob's Optional Stopping Theorem, (iv) and (i), we know that for any $\tau \in \mathcal{T}$,

$$\begin{aligned} v^*(x) &\geq \mathbb{E}_x(\mathrm{e}^{-q(t\wedge\tau\wedge\sigma^*)} v^*(X_{t\wedge\tau\wedge\sigma^*})) \\ &\geq \mathbb{E}_x(\mathrm{e}^{-q(t\wedge\tau)} G(X_{t\wedge\tau}) \mathbf{1}_{(\sigma^*\geq t\wedge\tau)} + \mathrm{e}^{-q\sigma^*} H(X_{\sigma^*}) \mathbf{1}_{(\sigma^* < t\wedge\tau)}). \end{aligned}$$

It follows that

$$v^*(x) \geq \mathbb{E}_x(\Theta^q_{\tau,\sigma^*}) \geq \inf_{\sigma\in\mathcal{T}} \mathbb{E}_x(\Theta^q_{\tau,\sigma}), \tag{9.33}$$

where the first inequality is a result of Fatou's Lemma and (9.32). Since $\tau \in \mathcal{T}$ is arbitrary we have on the one hand, from the first inequality in (9.33),

$$v^*(x) \geq \sup_{\tau\in\mathcal{T}} \mathbb{E}_x(\Theta^q_{\tau,\sigma^*}) \geq \inf_{\sigma\in\mathcal{T}} \sup_{\tau\in\mathcal{T}} \mathbb{E}_x(\Theta^q_{\tau,\sigma}) \tag{9.34}$$

and on the other hand, from the second inequality in (9.33),

$$v^*(x) \geq \sup_{\tau\in\mathcal{T}} \inf_{\sigma\in\mathcal{T}} \mathbb{E}_x(\Theta^q_{\tau,\sigma}). \tag{9.35}$$

Now using (v), Doob's Optimal Stopping Theorem, (iii), (ii) and (9.31) we have for any $\sigma \in \mathcal{T}$,

$$\begin{aligned} v^*(x) &\leq \mathbb{E}_x(\mathrm{e}^{-q(t\wedge\tau^*\wedge\sigma)} v^*(X_{t\wedge\tau^*\wedge\sigma})) \\ &= \mathbb{E}_x(\mathrm{e}^{-q\tau^*} v^*(X_{\tau^*}) \mathbf{1}_{(\tau^*\leq t\wedge\sigma)} + \mathrm{e}^{-q(t\wedge\sigma)} v^*(X_{t\wedge\sigma}) \mathbf{1}_{(\tau^* > t\wedge\sigma)}) \\ &\leq \mathbb{E}_x(\mathrm{e}^{-q\tau^*} G(X_{\tau^*}) \mathbf{1}_{(\tau^*\leq t\wedge\sigma)} + \mathrm{e}^{-q(t\wedge\sigma)} H(X_{t\wedge\sigma}) \mathbf{1}_{(\tau^* > t\wedge\sigma)}). \end{aligned}$$

Taking limits as $t \uparrow \infty$ and applying the Dominated Convergence Theorem with the help of (9.31), (9.32) and the non-negativity of G, we have

$$v^*(x) \leq \mathbb{E}_x(\Theta^q_{\tau^*,\sigma}) \leq \sup_{\tau\in\mathcal{T}} \mathbb{E}_x(\Theta^q_{\tau,\sigma}). \tag{9.36}$$

Since $\sigma \in \mathcal{T}$ is arbitrary, we have on the one hand, from the first inequality in (9.36),

$$v^*(x) \leq \inf_{\sigma\in\mathcal{T}} \mathbb{E}_x(\Theta^q_{\tau^*,\sigma}) \leq \sup_{\tau\in\mathcal{T}} \inf_{\sigma\in\mathcal{T}} \mathbb{E}_x(\Theta^q_{\tau,\sigma}) \tag{9.37}$$

and on the other hand, from the second inequality in (9.36),

$$v^*(x) \leq \inf_{\sigma\in\mathcal{T}} \sup_{\tau\in\mathcal{T}} \mathbb{E}_x(\Theta^q_{\tau,\sigma}). \tag{9.38}$$

From (9.34), (9.35), (9.37) and (9.38), we see that

$$v^*(x) = \inf_{\sigma\in\mathcal{T}} \sup_{\tau\in\mathcal{T}} \mathbb{E}_x(\Theta^q_{\tau,\sigma}) = \sup_{\tau\in\mathcal{T}} \inf_{\sigma\in\mathcal{T}} \mathbb{E}_x(\Theta^q_{\tau,\sigma}).$$

Finally from its definition

$$v^*(x) = \mathbb{E}_x(\Theta^q_{\tau^*,\sigma^*})$$

and hence the saddle point is achieved with the strategies (τ^*, σ^*). □

To the author's knowledge there is an extremely limited base of published literature giving concrete examples of stochastic games driven by Lévy processes.[5] We give an example here which is an excerpt from Baurdoux and Kyprianou (2005). Aspects of the general theory can be found in one of the original papers on the topic, Dynkin (1969).

Suppose that X is a spectrally negative Lévy process of bounded variation which drifts to infinity and denote as usual its Laplace exponent by ψ. That is to say, it has the same structure as in (9.22) with $d > \mathbb{E}(S_1) > 0$. Consider the stochastic game (9.30) when $G(x) = (K - \mathrm{e}^x)^+$, $H(x) = G(x) + \delta$ for fixed $\delta > 0$ and $\mathcal{T}$ consists of all $\mathbb{F}$-stopping times.

This stochastic game is very closely related to the McKean optimal stopping problem. We may think of it as modeling the optimal time to sell to an agent a risky asset for a fixed price K whose dynamics follow those of an exponential Lévy process. However, we have in addition now that the purchasing agent may also demand a forced purchase; in which case the agent must pay a supplementary penalty δ for forcing the purchase. Note, however, if the constant δ is too large (for example greater than K) then it will never be optimal for the agent to force a purchase in which case the stochastic game will have the same solution as the analogous optimal stopping problem. That is to say the optimal stopping problem (9.3) with $q = 0$.

[5]See for example Gapeev and Kühn (2005) in print

Suppose however that δ is a very small number. In this case, based on the experience of the above optimal stopping problems, it is reasonable to assume that one may optimise the expected gain by opting to sell the risky asset once the value of X drops below a critical threshold where the gain function G is large. The level of the threshold is determined by the fact that X drifts to infinity (and hence may never reach the threshold) as well as the behaviour of the purchasing agent. The latter individual can minimise the expected gain by stopping X in a region of $\mathbb{R}$ where the gain function H is small. This is clearly the half line $(\log K, \infty)$. As the gain G is identically zero on this interval, it turns out to be worthwhile to stop in the aforementioned interval, paying the penalty δ, rather than waiting for the other player to request a purchase requiring a potentially greater pay out by the agent.

These ideas are captured in the following theorem.

Theorem 9.14. *Under the assumptions above, the stochastic game (9.30) has the following two regimes of solutions.*

(a) Define x^* by

$$\mathrm{e}^{x^*} = K\frac{\psi'(0+)}{\psi(1)} \tag{9.39}$$

and note by strict convexity of ψ that $x^* < \log K$. Suppose that

$$\delta \geq K\psi(1)\int_0^{\log(\psi(1)/\psi'(0+))} \mathrm{e}^{-y}W(y)\mathrm{d}y,$$

then

$$v(x) = K - \mathrm{e}^x + \psi(1)\mathrm{e}^x \int_0^{x-x^*} \mathrm{e}^{-y}W(y)\mathrm{d}y, \tag{9.40}$$

where the saddle point strategies are given by

$$\tau^* = \inf\{t > 0 : X_t < x^*\} \text{ and } \sigma^* = \infty.$$

That is to say, (v, τ^*) is also the solution to the optimal stopping problem

$$v(x) = \sup_{\tau\in\mathcal{T}} \mathbb{E}_x((K - \mathrm{e}^{X_\tau})^+).$$

(b) Suppose that

$$\delta < K\psi(1)\int_0^{\log(\psi(1))/\psi'(0+)} \mathrm{e}^{-y}W(y)\mathrm{d}y$$

then

$$v(x) = \begin{cases} K - \mathrm{e}^x + \psi(1)\mathrm{e}^x \int_0^{x-z^*} \mathrm{e}^{-y}W(y)\mathrm{d}y & \text{for } x < \log K \\ \delta & \text{for } x \geq \log K, \end{cases} \tag{9.41}$$

where $z^* \in (0, K)$ is the unique solution of the equation

$$K\psi(1)\int_0^{\log K - z} \mathrm{e}^{-y} W(y)\mathrm{d}y = \delta \tag{9.42}$$

and the saddle point strategies are given by

$$\tau^* = \inf\{t > 0 : X_t < z^*\} \text{ and } \sigma^* = \inf\{t > 0 : X_t > \log K\}.$$

Proof. For both parts (a) and (b) we shall resort to checking conditions given in Lemma 9.13. Note that for the stochastic game at hand, conditions (9.31) and (9.32) are both satisfied. The first condition is trivially satisfied and the second follows from the fact that $\lim_{t\uparrow\infty} X_t = \infty$ and $H(x) = \delta$ for all $x > \log K$.

(a) As indicated in the statement of the Theorem, the proposed solution necessarily solves the McKean optimal stopping problem. Under the present circumstances of spectral negativity and bounded variation, the solution to this optimal stopping problem is given in Corollary 9.3. It is left to the reader to check that this solution corresponds to the function given on the right-hand side of (9.40).

Let v^* be the expression given on the right-hand side of (9.40). We now proceed to show that v^*, τ^* and σ^* fulfill the conditions (i) – (vi) of Lemma 9.13.

Bounds (i) and (ii). Since by definition of W, $\int_0^\infty e^{-y} W(y)\mathrm{d}y = 1/\psi(1)$ we may apply l'Hôpital's rule and compute

$$\begin{aligned}
\lim_{x\uparrow\infty} v^*(x) &= K - \lim_{x\uparrow\infty} \frac{1 - \psi(1)\int_0^{x-x^*} \mathrm{e}^{-y} W(y)\mathrm{d}y}{\mathrm{e}^{-x}} \\
&= K - \lim_{x\uparrow\infty} \psi(1)\mathrm{e}^{x^*} W(x - x^*) \\
&= K - \frac{\psi(1)}{\psi'(0+)}\mathrm{e}^{x^*} \\
&= 0,
\end{aligned}$$

where the final equality follows from (8.15) and (9.39). Hence, with the assumption on δ, which reworded says $\delta \geq v^*(\log K)$, and the convexity of v^* (cf. Theorem 9.4) we have

$$(K - \mathrm{e}^x)^+ \leq v^*(x) \leq (K - \mathrm{e}^x)^+ + \delta.$$

Hence conditions (i) and (ii) of Lemma 9.13 are fulfilled.

Equality with the gain properties (iii) and (iv). Property (iii) is trivially satisfied by virtue of the fact that v^* solves (9.3) with $q = 0$. As $\sigma^* = \infty$ the statement (iv) is empty.

Supermartingale and subartingale properties (v) and (vi). From the proof of Theorem 9.2 we have seen that $\{v^*(X_t) : t \geq 0\}$ is a supermartingale with right continuous paths. For the submartingale property, we actually need to show that $\{v^*(X_{t\wedge\tau^*}) : t \geq 0\}$ is a martingale. However this follows by the Strong Markov Property since for all $t \geq 0$,

$$\begin{aligned}\mathbb{E}_x((K - \mathrm{e}^{X_{\tau^*}})|\mathcal{F}_{t\wedge\tau^*}) &= \mathbb{E}_{x'}(K - \mathrm{e}^{X_{\tau^*}}) \text{ where } x' = X_{t\wedge\tau^*} \\ &= v^*(X_{t\wedge\tau^*}).\end{aligned}$$

(b) Technically speaking, one reason why we may not appeal to the solution to (9.3) with $q = 0$ as the solution to the problem at hand is because the condition on δ now implies that the solution to part (a) is no longer bounded above by $(K - e^x)^+ + \delta$, at least at the point $x = \log K$.

The right-hand side of (9.41) can easily be checked using the fluctuation identities in Chap. 8 to be equal to

$$v^*(x) = \mathbb{E}_x((K - \exp\{X_{\tau^-_{z^*}}\})\mathbf{1}_{(\tau^-_{z^*} < \tau^+_{\log K})}) + \delta\mathbb{E}_x(\mathbf{1}_{(\tau^-_{z^*} > \tau^+_{\log K})}) \qquad (9.43)$$

and we check now whether the triple (v^*, τ^*, σ^*) fulfill the conditions (i) – (vi) of Lemma 9.13.

Bounds (i) and (ii). The lower bound is easily established since by (9.43) we have that $v^*(x) \geq 0$ and from (9.41) $v^*(x) \geq K - \mathrm{e}^x$. Hence $v^*(x) \geq (K - \mathrm{e}^x)^+$. For the upper bound, note that $v^*(x) = H(x)$ for $x \geq \log K$. On $x < \log K$ we also have from (9.41) that

$$\begin{aligned}v^*(x) - (K - \mathrm{e}^x) &= \frac{\mathrm{e}^x}{K} K\psi(1) \int_0^{x - z^*} \mathrm{e}^{-y} W(y) \mathrm{d}y \\ &\leq K\psi(1) \int_0^{\log K - z^*} \mathrm{e}^{-y} W(y)\mathrm{d}y \\ &= \delta\end{aligned}$$

thus confirming that $v^*(x) \leq H(x)$.

Equality with the gain properties (iii) and (iv). This is clear by inspection.

Supermartingale properties (v). By considering the representation (9.43) the martingale property is confirmed by a similar technique to the one used in part (a) and applying the Strong Markov Property to deduce that

$$v^*(X_{t\wedge\tau^*\wedge\sigma^*}) = \mathbb{E}_x((K - \mathrm{e}^{X_{\tau^-_{z^*}}})\mathbf{1}_{(\tau^-_{z^*} < \tau^+_{\log K})} + \delta\mathbf{1}_{(\tau^-_{z^*} > \tau^+_{\log K})}|\mathcal{F}_t).$$

Note that in establishing the above equality, it has been used that $v^*(X_{\tau^-_{z^*}}) = (K - \mathrm{e}^{X_{\tau^-_{z^*}}})$ and $v^*(X_{\tau^+_{\log K}}) = \delta$. Noting that v^* is a continuous function which is also C^1 function on $\mathbb{R}\backslash\{\log K\}$ we may apply the change of variable formula in the spirit of Exercise 4.1 and obtain

$$v^*(X_t) = v^*(x) + \int_0^t \mathcal{H}v^*(X_{s-})\mathrm{d}s + M_t,\ t \geq 0 \qquad (9.44)$$

$\mathbb{P}_x$ almost surely where $\{M_t : t \geq 0\}$ is a right continuous martingale,

$$\mathcal{H}v^*(x) = d\frac{\mathrm{d}v^*}{\mathrm{d}x}(x) + \int_{(-\infty,0)} (v^*(x+y) - v^*(x))\Pi(\mathrm{d}y)$$

and Π is the Lévy measure of X and d is its drift. The details of this calculation are very similar in nature to those of a related calculation appearing in the proof of Theorem 9.11 and hence are left to the reader.

Since $\{v^*(X_t) : t < \tau^* \wedge \sigma^*\}$ is a martingale we deduce from Exercise 4.3 that $\mathcal{H}v^*(x) = 0$ for all $x \in (z^*, \log K)$. A straightforward calculation also shows that for $x < z^*$, where $v^*(x) = K - e^x$,

$$\mathcal{H}v^*(x) = -\mathrm{e}^x \left(d + \int_{(-\infty,0)} (\mathrm{e}^y - 1)\Pi(\mathrm{d}y) \right) = -\mathrm{e}^x \psi(1) < 0.$$

Reconsidering (9.44) we see then that on $\{t < \sigma^*\}$ the Lebesgue integral is a non-increasing adapted process and hence $\{v^*(X_t) : t < \sigma^*\}$ is a right continuous supermaringale.

Submartingale property (vi). For $x > \log K$, where $v^*(x) = \delta$, our aim is to show that

$$\mathcal{H}v^*(x) = \int_{(-\infty,0)} (v^*(x+y) - \delta)\Pi(\mathrm{d}y) \geq 0$$

as a consequence of the fact that

$$v^*(x) \geq \delta \text{ on } (0, \log K). \qquad (9.45)$$

If this is the case then in (9.44) we have on $\{t < \tau^*\}$ that the Lebesgue integral is a non-decreasing adapted process and hence $\{v^*(X_t) : t < \tau^*\}$ is a right continuous submartingale. Showing that (9.45) holds turns out to be quite difficult and we prove this in the lemma immediately below.

In conclusion, the conditions of Lemma 9.13 are satisfied by the triple (v^*, τ^*, σ^*) thus establishing part (b) of the theorem. □

Lemma 9.15. *The function v^* in the proof of part (b) of the above theorem is strictly greater than δ for all $x < \log K$.*

Proof. We assume the notation from the proof of part (b) of the above theorem. As a first step we show that v^* together with the stopping time τ^* solves the optimal stopping problem

$$f(x) = \sup_{\tau \in \mathcal{T}} \mathbb{E}_x(G(X^*_\tau)), \tag{9.46}$$

where $\{X^*_t : t \geq 0\}$ is the process X stopped on first entry to $(\log K, \infty)$, so that $X^*_t = X_{t\wedge\sigma^*}$, and

$$G(x) = (K - \mathrm{e}^x)^+ \mathbf{1}_{(x \leq \log K)} + \delta \mathbf{1}_{(x > \log K)}.$$

Recalling the remark at the end of Sect. 9.1, we note from the lower bound (i), equality with the gain property (iii), martingale property and supermartingale property (vi) established in the proof of part (b) of the above theorem that (v, τ^*) solves the optimal stopping problem (9.46).

Next define for each $q \geq 0$ the solutions to the McKean optimal stopping problem

$$g^{(q)}(x) = \sup_{\tau \in \mathcal{T}} \mathbb{E}_x(\mathrm{e}^{-q\tau}(K - \mathrm{e}^{X_\tau})^+)$$

for $x \in \mathbb{R}$. The associated optimal stopping time for each $q \geq 0$ is given by $\tau^*_q = \inf\{t > 0 : X_t < x^*_p\}$. From Theorem 9.2 and Corollary 9.3 we know that

$$\mathrm{e}^{x^*_q} = K\mathbb{E}(\mathrm{e}^{\underline{X}_{\mathbf{e}_q}}) = K\frac{q}{\Phi(q)}\frac{\Phi(q) - 1}{q - \psi(1)}.$$

Hence the constant x^*_q is continuous and strictly increasing to $\log K$ as q tends to infinity. It may also be verified from Corollary 9.3 that

$$\begin{aligned} g^{(q)}(\log K) = Kq &\int_0^{\log K - x^*_q} (1 - \mathrm{e}^{-y}) W^{(q)}(y) \mathrm{d}y \\ &+ K\psi(1) \int_0^{\log K - x^*_q} \mathrm{e}^{-y} W^{(q)}(y) \mathrm{d}y. \end{aligned}$$

We see then that $g^{(q)}(\log K)$ is continuous in q. From Theorem 9.2 we also know that for each fixed $x \in \mathbb{R}$, $g^{(q)}(x) \downarrow 0$ as $q \uparrow \infty$. Thanks to these facts we may deduce that there exists a solution to the equation $g^{(q)}(\log K) = \delta$. Henceforth we shall assume that q solves the latter equation.

The Strong Markov Property implies that

$$\begin{aligned} g^{(q)}(x) &= \mathbb{E}_x(\mathrm{e}^{-q\tau^*_q}(K - \mathrm{e}^{X_{\tau^*_q}})) \\ &= \mathbb{E}_x(\mathrm{e}^{-q\tau^*_q}(K - \mathrm{e}^{X_{\tau^*_q}})\mathbf{1}_{(\tau^*_q < \tau^+_{\log K})} + \mathrm{e}^{-q\tau^+_{\log K}} g^{(q)}(\log K)\mathbf{1}_{(\tau^*_q > \tau^+_{\log K})}) \\ &= \mathbb{E}_x(\mathrm{e}^{-q\tau^*_q}(K - \mathrm{e}^{X_{\tau^*_q}})\mathbf{1}_{(\tau^*_q < \tau^+_{\log K})} + \mathrm{e}^{-q\tau^+_{\log K}} \delta\mathbf{1}_{(\tau^*_q > \tau^+_{\log K})}). \end{aligned}$$

Hence as v^* solves (9.46) we have

$$\begin{aligned} v^*(x) &= \sup_{\tau \in \mathcal{T}} \mathbb{E}_x((K - \mathrm{e}^{X_\tau})^+\mathbf{1}_{(\tau < \sigma^*)} + \delta\mathbf{1}_{(\tau \geq \sigma^*)}) \\ &\geq \mathbb{E}_x((K - \mathrm{e}^{X_{\tau^*_q}})\mathbf{1}_{(\tau^*_q < \sigma^*)} + \delta\mathbf{1}_{(\tau^*_q \geq \sigma^*)}) \\ &\geq g^{(q)}(x). \end{aligned}$$

We know from Theorem 9.4 that $g^{(q)}$ is convex and hence the last inequality ensures that $v^*(x) \geq \delta$ for all $x < \log K$ as required. □

Exercises

9.1. The following exercise is based on Novikov and Shiryaev (2004). Suppose that X is any Lévy process and either

$$q > 0 \text{ or } q = 0 \text{ and } \lim_{t\uparrow\infty} X_t = -\infty.$$

Consider the optimal stopping problem

$$v(x) = \sup_{\tau\in\mathcal{T}} \mathbb{E}_x(\mathrm{e}^{-q\tau}(1 - \mathrm{e}^{-(X_\tau)^+})) \tag{9.47}$$

where $\mathcal{T}$ is the set of $\mathbb{F}$-stopping times.

(i) For $a > 0$, prove the identity

$$\mathbb{E}_x\left(\mathrm{e}^{-qT_a^+}\left(1 - \mathrm{e}^{-X_{T_a^+}}\right)\mathbf{1}_{(T_a^+<\infty)}\right) = \mathbb{E}_x\left(\left(1 - \frac{\mathrm{e}^{-\overline{X}_{\mathbf{e}_q}}}{\mathbb{E}(\mathrm{e}^{-\overline{X}_{\mathbf{e}_q}})}\right)\mathbf{1}_{(\overline{X}_{\mathbf{e}_q}\geq a)}\right),$$

where $T_a^+ = \inf\{t \geq 0 : X_t \geq a\}$.

(ii) Show that the solution to (9.47) is given by the pair $(v_{x^*}, T_{x^*}^+)$ where $v_{x^*}(x)$ is equal to the left-hand side of the identity in part (i) with $a = x^*$ where

$$\mathrm{e}^{-x^*} = \mathbb{E}(\mathrm{e}^{-\overline{X}_{\mathbf{e}_q}}).$$

(iii) Show that there is smooth fit at x^* if and only if 0 is regular for $(0, \infty)$ for X and otherwise continuous fit.

9.2. This exercise is based on Baurdoux and Kyprianou (2005). Suppose that X is a bounded variation spectrally negative Lévy process with characteristic exponent $\psi(\theta) = d\theta - \int_{(0,\infty)}(1-\mathrm{e}^{-\theta x})\Pi(\mathrm{d}x)$ for $\theta \geq 0$ where d is the drift and Π is the Lévy measure of the pure jump subordinator in the decomposition of X. Consider the stochastic game which requires one to find $v(x)$ satisfying

$$v(x) = \sup_{\tau\in\mathcal{T}} \inf_{\sigma\in\mathcal{T}} \mathbb{E}(\mathrm{e}^{-q\tau+(x\vee\overline{X}_\tau)}\mathbf{1}_{(\tau\leq\sigma)} + \mathrm{e}^{-q\sigma}(\mathrm{e}^{(x\vee\overline{X}_\sigma)} + \delta\mathrm{e}^{X_\sigma})\mathbf{1}_{(\tau>\sigma)}), \tag{9.48}$$

where $\delta > 0$, $x \geq 0$, $d > q > \psi(1) > 0$ and the supremum and infimum are interchangeable.

(i) Let x^* be the constant defined in Theorem 9.11. Show that if $1 + \delta \geq Z^{(q)}(x^*)$ the solution to (9.48) is the same as the solution given in Theorem 9.11.

(ii) Henceforth assume that $1 + \delta < Z^{(q)}(x^*)$. Show that there exists a unique solution to equation $Z^{(q)}(z^*) = 1 + \delta$ and further that $z^* < x^*$. Consider the function $f(x) = Z^{(q)}(x) - qW^{(q)}(x)$ defined in the discussion preceding Theorem 9.11. Using the analytic properties of f discussed there show that function $\mathrm{e}^x Z^{(q)}(z^* - x)$ is convex satisfying $v'(0+) > 0$ and $v'(z^*) = \mathrm{e}^{z^*}(1 - q/d)$.

(iii) Show that the solution to (9.48) is given by

$$v(x) = \mathrm{e}^x Z^{(q)}(z^* - x),$$

where z^* is given in the previous part. Further the saddle point is achieved with the stopping times

$$\sigma^* = \inf\{t > 0 : Y_t^x = 0\} \text{ and } \tau^* = \inf\{t > 0 : Y_t^x > z^*\},$$

where $Y^x = (x \vee \overline{X}) - X$ is the process reflected in its supremum issued from x. Note that there is continuous fit at z^* but not smooth fit.

10

Continuous-State Branching Processes

Our interest in continuous-state branching processes will be in exposing their intimate relationship with spectrally positive Lévy processes. A flavour for this has already been given in Sect. 1.3.4 where it was shown that compound Poisson process killed on exiting $(0, \infty)$ can be time changed to obtain a continuous-time Bienaymé-Galton-Watson process. The analogue of this path transformation in fuller generality consists of time changing the path of a spectrally positive Lévy process killed on exiting $(0, \infty)$ to obtain a process equal in law to the path of a Markov process which observes the so-called *branching property* (defined in more detail later) and vice versa. The latter process is what we refer to as the continuous-state branching process. The time change binding the two processes together is called the Lamperti transform, named after the foundational work of Lamperti (1967a,b).[1]

Having looked closely at the Lamperti transform we shall give an account of a number of observations concerning the long term behaviour as well as conditioning on survival of continuous-state branching processes. Thanks to some of the results in Chap. 8 semi-explicit results can be obtained.

10.1 The Lamperti Transform

A $[0, \infty)$-valued strong Markov process $Y = \{Y_t : t \geq 0\}$ with probabilities $\{P_x : x \geq 0\}$ is called a continuous-state branching process if it has paths that are right continuous with left limits and its law observes the branching process given in Definition 1.14. Another way of phrasing the branching property is that for all $\theta \geq 0$ and $x, y \geq 0$,

$$E_{x+y}(\mathrm{e}^{-\theta Y_t}) = E_x(\mathrm{e}^{-\theta Y_t})E_y(\mathrm{e}^{-\theta Y_t}). \tag{10.1}$$

Note from the above equality that after an iteration we may always write for each $x > 0$,

[1] See also Silverstein (1968).

$$E_x(\mathrm{e}^{-\theta Y_t}) = E_{x/n}(\mathrm{e}^{-\theta Y_t})^n \tag{10.2}$$

showing that Y_t is infinitely divisible for each $t > 0$. If we define for $\theta, t \geq 0$,

$$g(t, \theta, x) = -\log E_x(\mathrm{e}^{-\theta Y_t}),$$

then (10.2) implies that for any positive integer m,

$$g(t, \theta, m) = ng(t, \theta, m/n) \text{ and } g(t, \theta, m) = mg(t, \theta, 1)$$

showing that for $x \in \mathbb{Q} \cap [0, \infty)$,

$$g(t, \theta, x) = xu_t(\theta), \tag{10.3}$$

where $u_t(\theta) = g(t, \theta, 1) \geq 0$. From (10.1) we also see that for $0 \leq z < y$, $g(t, \theta, z) \leq g(t, \theta, y)$ which implies that $g(t, \theta, x-)$ exists as a left limit and is less than or equal to $g(t, \theta, x+)$ which exists as a right limit. Thanks to (10.3), both left and right limits are the same so that for all $x > 0$

$$E_x(\mathrm{e}^{-\theta Y_t}) = \mathrm{e}^{-xu_t(\theta)}. \tag{10.4}$$

The Markov property in conjunction with (10.4) implies that for all $x > 0$ and $t, s, \theta \geq 0$,

$$\mathrm{e}^{-xu_{t+s}(\theta)} = E_x\left(E(\mathrm{e}^{-\theta Y_{t+s}}|Y_t)\right) = E_x\left(\mathrm{e}^{-Y_t u_s(\theta)}\right) = \mathrm{e}^{-xu_t(u_s(\theta))}.$$

In other words the Laplace exponent of Y obeys the semi-group property

$$u_{t+s}(\theta) = u_t(u_s(\theta)).$$

The first significant glimpse one gets of Lévy processes in relation to the above definition of a continuous-state branching process comes with the following result for which we offer no proof on account of technicalities (see however Exercise 1.11 for intuitive motivation and Chap. II of Le Gall (1999) or Silverstein (1968) for a proof).

Theorem 10.1. *For $t, \theta \geq 0$, suppose that $u_t(\theta)$ is the Laplace functional given by (10.4) of some continuous-state branching process. Then it is differentiable in t and satisfies*

$$\frac{\partial u_t}{\partial t}(\theta) + \psi(u_t(\theta)) = 0 \tag{10.5}$$

with initial condition $u_0(\theta) = \theta$ where for $\lambda \geq 0$,

$$\psi(\lambda) = -q - a\lambda + \frac{1}{2}\sigma^2\lambda^2 + \int_{(0,\infty)} (\mathrm{e}^{-\lambda x} - 1 + \lambda x \mathbf{1}_{(x<1)})\Pi(\mathrm{d}x) \tag{10.6}$$

and in the above expression, $q \geq 0$, $a \in \mathbb{R}$, $\sigma \geq 0$ and Π is a measure supported in $(0, \infty)$ satisfying $\int_{(0,\infty)}(1 \wedge x^2)\Pi(\mathrm{d}x) < \infty$.

Note that for $\lambda \geq 0$, $\psi(\lambda) = \log \mathbb{E}(\mathrm{e}^{-\lambda X_1})$ where X is either a spectrally positive Lévy process[2] or a subordinator, killed independently at rate $q \geq 0$.[3] Otherwise said, ψ is the Laplace exponent of a killed spectrally negative Lévy process or the negative of the Laplace exponent of a killed subordinator. From Sects. 8.1 and 5.5, respectively, we know for example that ψ is convex, infinitely differentiable on $(0,\infty)$, $\psi(0) = q$ and $\psi'(0+) \in [-\infty,\infty)$. Further, if X is a (killed) subordinator, then $\psi(\infty) < 0$ and otherwise we have that $\psi(\infty) = \infty$.

For each $\theta > 0$ the solution to (10.5) can be uniquely identified by the relation

$$-\int_{\theta}^{u_t(\theta)} \frac{1}{\psi(\xi)} \mathrm{d}\xi = t. \tag{10.7}$$

(This is easily confirmed by elementary differentiation, note also that the lower delimiter implies that $u_0(\theta) = \theta$ by letting $t \downarrow 0$.)

From the discussion earlier we may deduce that *if a continuous-state branching process exists*, then it is associated with a particular function $\psi : [0,\infty) \mapsto \mathbb{R}$ given by (10.6). Formally speaking, we shall refer to all such ψ as *branching mechanisms*. We will now state without proof the Lamperti transform which, amongst other things, shows that for every branching mechanism ψ there exists an associated continuous-state branching process.

Theorem 10.2. *Let ψ be any given branching mechanism.*

(i) Suppose that $X = \{X_t : t \geq 0\}$ is a Lévy process with no negative jumps, initial position $X_0 = x$, killed at an independent exponentially distributed time with parameter $q \geq 0$. Further, $\psi(\lambda) = \log \mathbb{E}_x(\mathrm{e}^{-\lambda(X_1 - x)})$. Define for $t \geq 0$,

$$Y_t = X_{\theta_t \wedge \tau_0^-},$$

where $\tau_0^- = \inf\{t > 0 : X_t < 0\}$ and

$$\theta_t = \inf\{s > 0 : \int_0^s \frac{\mathrm{d}u}{X_u} > t\}$$

then $Y = \{Y_t : t \geq 0\}$ is a continuous-state branching process with branching mechanism ψ and initial value $Y_0 = x$.

(ii) Conversely suppose that $Y = \{Y_t : t \geq 0\}$ is a continuous-state branching process with branching mechanism ψ, such that $Y_0 = x > 0$. Define for $t \geq 0$,

$$X_t = Y_{\varphi_t},$$

[2] Recall that our definition of spectrally positive processes excludes subordinators. See the discussion following Lemma 2.14.

[3] As usual, we understand the process X killed at rate q to mean that it is killed after an independent and exponentially distributed time with parameter q. Further $q = 0$ means there is no killing.

where

$$\varphi_t = \inf\{s > 0 : \int_0^s Y_u \mathrm{d}u > t\}.$$

Then $X = \{X_t : t \geq 0\}$ is a Lévy process with no negative jumps, killed at the minimum of the time of the first entry into $(-\infty, 0)$ and an independent and exponentially distributed time with parameter $q \geq 0$, with initial position $X_0 = x$ and satisfying $\psi(\lambda) = \log \mathbb{E}(\mathrm{e}^{-\lambda X_1})$.

It can be shown that a general continuous-state branching process appears as the result of an asymptotic re-scaling (in time and space) of the continuous time Bienaymé–Galton–Watson process discussed in Sect. 1.3.4; see Jirina (1958). Roughly speaking the Lamperti transform for continuous-state branching processes then follows as a consequence of the analogous construction being valid for the continuous time Bienaymé–Galton–Watson process; recall the discussion in Sect. 1.3.4.

10.2 Long-term Behaviour

For the forthcoming discussion it will be useful to recall the definition of a Bienaymé–Galton–Watson process; the process for which continuous-state branching processes are a natural analogue in continuous time and space. The latter is a discrete time Markov chain $Z = \{Z_n : n = 0, 1, 2, ...\}$ with state space $\{0, 1, 2, ...\}$. The quantity Z_n is to be thought of as the size of the n-th generation of some asexually reproducing population. The process Z has probabilities $\{P_x : x = 0, 1, 2, ...\}$ such that, under P_x, $Z_0 = x$ and

$$Z_n = \sum_{i=1}^{Z_{n-1}} \xi_i^{(n)} \tag{10.8}$$

for $n = 1, 2, ...$ where for each $n \geq 1$, $\{\xi_i^{(n)} : i = 1, 2, ...\}$ are independent and identically distributed on $\{0, 1, 2, ...\}$.

Without specifying anything further about the common distribution of the offspring there are two events which are of immediate concern for the Markov chain Z; explosion and absorption. In the first case it is not clear whether or not the event $\{Z_n = \infty\}$ has positive probability for some $n \geq 1$ (the latter could happen if, for example, the offspring distribution has no moments). When $P_x(Z_n < \infty) = 1$ for all $n \geq 1$ we say the process is conservative (in other words there is no explosion). In the second case, we note from the definition of Z that if $Z_n = 0$ for some $n \geq 1$ then $Z_{n+m} = 0$ for all $m \geq 0$ which makes 0 an absorbing state. As Z_n is to be thought of as the size of the nth generation of some asexually reproducing population, the event $\{Z_n = 0 \text{ for some } n > 0\}$ is referred to as extinction.

In this section we consider the analogues of conservative behaviour and extinction within the setting of continuous-state branching processes. In addition we shall examine the laws of the supremum and total progeny process of continuous-state branching processes. These are the analogues of

$$\sup_{n\geq 0} Z_n \text{ and } \{\sum_{0\leq k\leq n} Z_k : n \geq 0\}$$

for the Bienaymé–Galton–Watson process. Note in the latter case, total progeny is interpreted as the total number of offspring to date.

10.2.1 Conservative Processes

A continuous-state branching process $Y = \{Y_t : t \geq 0\}$ is said to be *conservative* if for all $t > 0$, $P(Y_t < \infty) = 1$. The following result is taken from Grey (1974).

Theorem 10.3. *A continuous-state branching process with branching mechanism ψ is conservative if and only if*

$$\int_{0+} \frac{1}{|\psi(\xi)|} \mathrm{d}\xi = \infty.$$

A necessary condition is, therefore, $\psi(0) = 0$ and a sufficient condition is $\psi(0) = 0$ and $|\psi'(0+)| < \infty$ (equivalently $q = 0$ and $\int_{[1,\infty)} x\Pi(\mathrm{d}x) < \infty$).

Proof. From the definition of $u_t(\theta)$, a continuous-state branching process is conservative if and only if $\lim_{\theta\downarrow 0} u_t(\theta) = 0$ since, for each $x > 0$,

$$P_x(Y_t < \infty) = \lim_{\theta\downarrow 0} E_x(\mathrm{e}^{-\theta Y_t}) = \exp\{-x \lim_{\theta\downarrow 0} u_t(\theta)\},$$

where the limits are justified by monotonicity. However, note from (10.7) that as $\theta \downarrow 0$,

$$t = -\int_\theta^\delta \frac{1}{\psi(\xi)} \mathrm{d}\xi + \int_{u_t(\theta)}^\delta \frac{1}{\psi(\xi)} \mathrm{d}\xi,$$

where $\delta > 0$ is sufficiently small. However, as the left-hand side is independent of θ we are forced to conclude that $\lim_{\theta\downarrow 0} u_t(\theta) = 0$ if and only if

$$\int_{0+} \frac{1}{|\psi(\xi)|} \mathrm{d}\xi = \infty.$$

Note that $\psi(\theta)$ may be negative in the neighbourhood of the origin and hence the absolute value is taken in the integral.

From this condition and the fact that ψ is a smooth function, one sees immediately that a *necessary* condition for a continuous-state branching process to be conservative is that $\psi(0) = 0$; in other words the "killing rate" $q = 0$. It is also apparent that a *sufficient* condition is that $q = 0$ and that $|\psi'(0+)| < \infty$ (so that ψ is locally linear passing through the origin). Due to the fact that $\psi(\lambda) = \log \mathbb{E}(\mathrm{e}^{-\lambda X_1})$ where X is a Lévy process with no negative jumps, it follows that the latter condition is equivalent to $\mathbb{E}|X_1| < \infty$ where X is the Lévy processes with no negative jumps associated with ψ. Hence by Theorem 3.8 it is also equivalent to $\int_{[1,\infty)} x\Pi(\mathrm{d}x) < \infty$. □

Henceforth we shall assume for all branching mechanisms that $q = 0$.

10.2.2 Extinction Probabilities

Thanks to the representation of continuous-state branching processes given in Theorem 10.2 (i), it is clear that the latter processes observe the fundamental property that if $Y_t = 0$ for some $t > 0$, then $Y_{t+s} = 0$ for $s \geq 0$. Let $\zeta = \inf\{t > 0 : Y_t = 0\}$. The event $\{\zeta < \infty\} = \{Y_t = 0 \text{ for some } t > 0\}$ is thus referred to as *extinction* in line with terminology used for the Bienaymé–Galton–Watson process.

This can also be seen from the branching property (10.1). By taking $y = 0$ there we see that P_0 must be the measure that assigns probability one to the processes which is identically zero. Hence by the Markov property, once in state zero, the process remains in state zero.

Note from (10.4) that $u_t(\theta)$ is continuously differentiable in $\theta > 0$ (since by dominated convergence, the same is true of the left-hand side of the aforementioned equality). Differentiating (10.4) in $\theta > 0$ we find that for each $x, t > 0$,

$$E_x(Y_t \mathrm{e}^{-\theta Y_t}) = x\frac{\partial u_t}{\partial \theta}(\theta)\mathrm{e}^{-xu_t(\theta)} \tag{10.9}$$

and hence taking limits as $\theta \downarrow 0$ we obtain

$$E_x(Y_t) = x\frac{\partial u_t}{\partial \theta}(0+) \tag{10.10}$$

so that both sides of the equality are infinite at the same time. Differentiating (10.5) in $\theta > 0$ we also find that

$$\frac{\partial}{\partial t}\frac{\partial u_t}{\partial \theta}(\theta) + \psi'(u_t(\theta))\frac{\partial u_t}{\partial \theta}(\theta) = 0.$$

Standard techniques for first-order differential equations then imply that

$$\frac{\partial u_t}{\partial \theta}(\theta) = c\mathrm{e}^{-\int_0^t \psi'(u_s(\theta))\mathrm{d}s} \tag{10.11}$$

where $c > 0$ is a constant. Inspecting (10.9) as $t \downarrow 0$ we see that $c = 1$. Now taking limits as $\theta \downarrow 0$ and recalling that for each fixed $s > 0$, $u_s(\theta) \downarrow 0$ it is straightforward to deduce from (10.10) and (10.11) that

$$E_x(Y_t) = x\mathrm{e}^{-\psi'(0+)t}, \tag{10.12}$$

where we understand the left-hand side to be infinite whenever $\psi'(0+) = -\infty$. Note that from the definition $\psi(\theta) = \log \mathbb{E}(\mathrm{e}^{-\theta X_1})$ where X is a Lévy process with no negative jumps, we know that ψ is convex and $\psi(0+) \in [-\infty, \infty)$ (cf. Sects. 5.5 and 8.1). Hence in particular to obtain (10.12), we have used dominated convergence in the integral in (10.11) when $|\psi'(0+)| < \infty$ and monotone convergence when $\psi'(0+) = -\infty$.

This leads to the following classification of continuous-state branching processes.

Definition 10.4. *A continuous-state branching process with branching mechanism ψ is called*

(i) subcritical, if $\psi'(0+) > 0$,
(ii) critical, if $\psi'(0+) = 0$ and
(iii) supercritical, if $\psi'(0+) < 0$.

The use of the terminology "criticality" refers then to whether the process will, on average, decrease, remain constant or increase. The same terminology is employed for Bienaymé–Galton–Watson processes where now the three cases in Definition 10.4 correspond to the mean of the offspring distribution being strictly less than, equal to and strictly greater than unity, respectively. The classic result due to the scientists after which the latter process is named states that there is extinction with probability 1 if and only if the mean offspring size is less than or equal to unity (see Chap. I of Athreya and Ney (1972) for example). The analogous result for continuous-state branching processes might therefore read that there is extinction with probability one if and only if $\psi'(0+) \geq 0$. However, here we encounter a subtle difference for continuous-state branching processes as the following simple example shows: In the representation given by Theorem 10.2, take $X_t = 1 - t$ corresponding to $Y_t = \mathrm{e}^{-t}$. Clearly $\psi(\lambda) = \lambda$ so that $\psi'(0+) = 1 > 0$ and yet $Y_t > 0$ for all $t > 0$.

Extinction is characterised by the following result due to Grey (1974); see also Bingham (1976).

Theorem 10.5. *Suppose that Y is a continuous-state branching process with branching mechanism ψ. Let $p(x) = P_x(\zeta < \infty)$.*

(i) If $\psi(\infty) < 0$, then for all $x > 0$, $p(x) = 0$
(ii) Otherwise, when $\psi(\infty) = \infty$, $p(x) > 0$ for some (and then for all) $x > 0$ if and only if

$$\int^{\infty} \frac{1}{\psi(\xi)} \mathrm{d}\xi < \infty$$

in which case $p(x) = \mathrm{e}^{-\Phi(0)x}$ where $\Phi(0) = \sup\{\lambda \geq 0 : \psi(\lambda) = 0\}$.

Proof. (i) If $\psi(\lambda) = \log \mathbb{E}(\mathrm{e}^{-\lambda X_1})$ where X is a subordinator, then clearly from the path representation given in Theorem 10.2 (i), extinction occurs with probability zero. From the discussion following Theorem 10.1, the case that X is a subordinator is equivalent to $\psi(\lambda) < 0$ for all $\lambda > 0$.

(ii) Since for $s, t > 0$, $\{Y_t = 0\} \subseteq \{Y_{t+s} = 0\}$ we have by monotonicity that for each $x > 0$,

$$P_x(Y_t = 0) \uparrow p(x) \tag{10.13}$$

as $t \uparrow \infty$. Hence $p(x) > 0$ if and only if $P_x(Y_t = 0) > 0$ for some $t > 0$. Since $P_x(Y_t = 0) = \mathrm{e}^{-xu_t(\infty)}$, we see that $p(x) > 0$ for some (and then all) $x > 0$ if and only if $u_t(\infty) < \infty$ for some $t > 0$.

Fix $t > 0$. Taking limits in (10.7) as $\theta \uparrow \infty$ we see that if $u_t(\infty) < \infty$, then it follows that

$$\int^{\infty} \frac{1}{\psi(\xi)} \mathrm{d}\xi < \infty. \tag{10.14}$$

Conversely, if the above integral holds, then again taking limits in (10.7) as $\theta \uparrow \infty$ it must necessarily hold that $u_t(\infty) < \infty$.

Finally, assuming (10.14), we have learnt that

$$\int_{u_t(\infty)}^{\infty} \frac{1}{\psi(\xi)} \mathrm{d}\xi = t. \tag{10.15}$$

From (10.13) and the fact that $u_t(\infty) = -x^{-1} \log P_x(Y_t = 0)$ we see that $u_t(\infty)$ decreases as $t \uparrow \infty$ to the largest constant $c \geq 0$ such that $\int_c^\infty 1/\psi(\xi)\mathrm{d}\xi$ becomes infinite. Appealing to the convexity and smoothness of ψ, the constant c must necessarily correspond to a root of ψ in $[0, \infty)$, at which point it will behave linearly and thus cause $\int_c^\infty 1/\psi(\xi)\mathrm{d}\xi$ to blow up. There are at most two such points, and the largest of these is described precisely by $c = \Phi(0) \in [0, \infty)$ (see Sect. 8.1). In conclusion,

$$p(x) = \lim_{t \uparrow \infty} \mathrm{e}^{-xu_t(\infty)} = \mathrm{e}^{-\Phi(0)x}$$

as required. □

On account of the convexity of ψ we also recover the following corollary to part (ii) of the above theorem.

Corollary 10.6. *For a continuous-state branching process with branching mechanism ψ satisfying $\psi(\infty) = \infty$ and*

$$\int^{\infty} \frac{1}{\psi(\xi)} \mathrm{d}\xi < \infty,$$

we have $p(x) < 1$ for some (and then for all) $x > 0$ if and only if $\psi'(0+) < 0$.

To summarise the conclusions of Theorem 10.5 and Corollary 10.6, we have the following cases for the extinction probability $p(x)$:

Condition	$p(x)$
$\psi(\infty) < 0$	0
$\psi(\infty) = \infty,\ \int^{\infty} \psi(\xi)^{-1}\mathrm{d}\xi = \infty$	0
$\psi(\infty) = \infty,\ \psi'(0+) < 0,\ \int^{\infty} \psi(\xi)^{-1}\mathrm{d}\xi < \infty$	$\mathrm{e}^{-\Phi(0)x} \in (0,1)$
$\psi(\infty) = \infty,\ \psi'(0+) \geq 0,\ \int^{\infty} \psi(\xi)^{-1}\mathrm{d}\xi < \infty$	1

10.2.3 Total Progeny and the Supremum

Thinking of a continuous-state branching process, $\{Y_t : t \geq 0\}$ as the continuous time, continuous-state analogue of the Bienaymé–Galton–Watson process, it is reasonable to refer to

$$J_t := \int_0^t Y_u \mathrm{d}u$$

as the total progeny until time $t \geq 0$.

In this section our main goal, given in the theorem below, is to provide distributional identities for $J_{T_a^+}$ where

$$T_a^+ = \inf\{t > 0 : Y_t > a\}$$

and $\sup_{s \leq \zeta} Y_s$. To ease the statement of the main result, let us first recall the following notation. As noted above, for any branching mechanism ψ, when $\psi(\infty) = \infty$ (in other words when the Lévy process associated with ψ is not a subordinator) we have that ψ is the Laplace exponent of a spectrally negative Lévy process. Let $\Phi(q) = \sup\{\theta \geq 0 : \psi(\theta) = q\}$ (cf. Sect. 8.1). Associated with the latter are the scale functions $W^{(q)}$ and $Z^{(q)}$; see Sect. 8.2. In particular,

$$\int_0^{\infty} \mathrm{e}^{-\beta x} W^{(q)}(x)\mathrm{d}x = \frac{1}{\psi(\beta) - q} \text{ for } \beta > \Phi(q)$$

and $Z^{(q)}(x) = 1 + q\int_0^x W^{(q)}(y)\mathrm{d}y$. Following the notational protocol of Chap. 8 we denote $W^{(0)}$ by W.

Theorem 10.7. *Let $Y = \{Y_t : t \geq 0\}$ be a continuous-state branching process with branching mechanism ψ satisfying $\psi(\infty) = \infty$.*

(i) For each $a \geq x > 0$ and $q \geq 0$,

$$E_x(\mathrm{e}^{-q\int_0^{T_a^+} Y_s \mathrm{d}s} \mathbf{1}_{(T_a^+ < \zeta)}) = Z^{(q)}(a - x) - W^{(q)}(a - x)\frac{Z^{(q)}(a)}{W^{(q)}(a)}.$$

(ii) For each $a \geq x > 0$ and $q \geq 0$,

$$E_x(\mathrm{e}^{-q\int_0^{\zeta} Y_s \mathrm{d}s} \mathbf{1}_{(\zeta < T_a^+)}) = \frac{W^{(q)}(a - x)}{W^{(q)}(a)}.$$

Proof. Suppose now that X is the Lévy process mentioned in Theorem 10.2 (ii). Write in the usual way $\tau_a^+ = \inf\{t > 0 : X_t > a\}$ and $\tau_0^- = \inf\{t > 0 : X_t < 0\}$. Then a little thought shows that

$$\tau_a^+ = \int_0^{T_a^+} Y_s \mathrm{d}s \text{ and } \tau_0^- = \int_0^{\zeta} Y_s \mathrm{d}s.$$

The proof is now completed by invoking Theorem 8.1 (iii) for the process X. Note that X is a spectrally positive Lévy process and hence to implement the aforementioned result, which applies to spectrally negative processes, one must consider the problem of two-sided exit from $[0, a]$ of $-X$ when $X_0 = a-x$. □

We conclude this section by noting the following two corollaries of Theorem 10.7 which, in their original form, are due to Bingham (1976).

Corollary 10.8. *Under the assumptions of Theorem 10.7 we have for each* $a \geq x > 0$,

$$P_x(\sup_{s\leq\infty} Y_s \leq a) = \frac{W(a-x)}{W(a)}.$$

In particular

$$P_x(\sup_{s\leq\infty} Y_s < \infty) = \mathrm{e}^{-\varPhi(0)x}$$

and the right-hand side is equal to unity if and only if Y *is not supercritical.*

Proof. The first part is obvious by taking limits as $q \downarrow 0$ in Theorem 10.7 (i). The second part follows by taking limits as $a \uparrow \infty$ and making use of Exercise 8.5 (i). Recall that $\varPhi(0) > 0$ if and only if $\psi'(0+) < 0$. □

Corollary 10.9. *Under the assumptions of Theorem 10.7 we have for each* $x > 0$ *and* $q \geq 0$,

$$E_x(\mathrm{e}^{-q\int_0^\zeta Y_s \mathrm{d}s}) = \mathrm{e}^{-\varPhi(q)x}.$$

Proof. The proof is again a simple consequence of Theorem 10.7 (i) by taking limits as $a \uparrow \infty$ and then applying the conclusion of Exercise 8.5 (i). □

10.3 Conditioned Processes and Immigration

In the classical theory of Bienaymé–Galton–Watson processes where the offspring distribution is assumed to have finite mean, it is well understood that by taking a critical or subcritical process (for which extinction occurs with probability one) and conditioning it in the long term to remain positive uncovers a beautiful relationship between a martingale change of measure and processes with immigration; cf. Athreya and Ney (1972) and Lyons et al. (1995). Let us be a little more specific.

A Bienaymé–Galton–Watson process with immigration is defined as the Markov chain $Z^* = \{Z_n^* : n = 0, 1, ...\}$ where $Z_0^* = z \in \{0, 1, 2, ...\}$ and for $n = 1, 2, ...,$

$$Z_n^* = Z_n + \sum_{k=1}^{n} Z_{n-k}^{(k)}, \tag{10.16}$$

where now $Z = \{Z_n : n \geq 0\}$ has law P_z and for each $k = 1, 2, ..., n$, $Z_{n-k}^{(k)}$ is independent and equal in distribution to numbers in the $(n-k)$th generation of (Z, P_{η_k}) where it is assumed that the initial numbers, η_k, are, independently of everything else, randomly distributed according to the probabilities $\{p_k^* : k = 0, 1, 2, ...\}$. Intuitively speaking one may see the process Z^* as a variant of the Bienaymé–Galton–Watson process, Z, in which, from the first and subsequent generations, there is a generational stream of immigrants $\{\eta_1, \eta_2, ...\}$ each of whom initiates an independent copy of (Z, P_1).

Suppose now that Z is a Bienaymé–Galton–Watson process with probabilities $\{P_x : x = 1, 2, ...\}$ as described above. For any event A which belongs to the sigma algebra generated by the first n generations, it turns out that for each $x = 0, 1, 2, ...$

$$P_x^*(A) := \lim_{m\uparrow\infty} P_x(A|Z_k > 0 \text{ for } k = 0, 1, ..., n + m)$$

is well defined and further,

$$P_x^*(A) = E_x(\mathbf{1}_A M_n),$$

where $M_n = m^{-n} Z_n/Z_0$ and $m = E_1(Z_1)$ which is assumed to be finite. It is not difficult to show that $E_x(Z_n) = xm^n$ and that $\{M_n : n \geq 0\}$ is a martingale using the iteration (10.8). What is perhaps more intriguing is that the new process (Z, P_x^*) can be identified in two different ways:

1. The process $\{Z_n - 1 : n \geq 0\}$ under P_x^* can be shown to have the same law as a Bienaymé–Galton–Watson process with immigration having $x-1$ initial ancestors. The immigration probabilities satisfy $p_k^* = (k+1)p_{k+1}/m$ for $k = 0, 1, 2, ...$ where $\{p_k : k = 0, 1, 2, ...\}$ is the offspring distribution of the original Bienaymé–Galton–Watson process and immigrants initiate independent copies of Z.
2. The process Z under P_x^* has the same law as $x - 1$ initial individuals each initiating independently a Bienaymé–Galton–Watson process under P_1 together with one individual initiating an independent immortal genealogical line of descent, *the spine*, along which individuals reproduce with the tilted distribution $\{kp_k/m : k = 1, 2, ...\}$. The offspring of individuals on the spine who are not themselves part of the spine initiate copies of a Bienaymé–Galton–Watson process under P_1. By subtracting off individuals on the spine from the aggregate population, one observes a Bienaymé–Galton–Watson process with immigration described in (1).

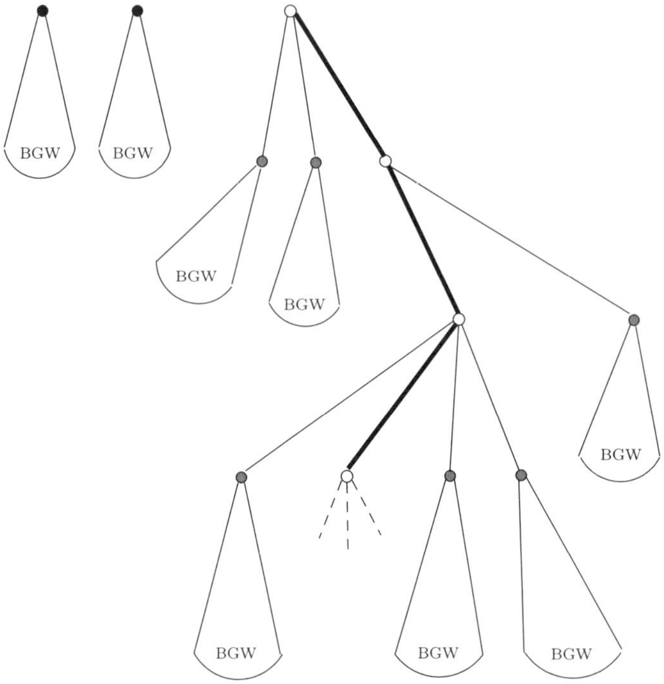

Fig. 10.1. Nodes shaded in *black* initiate Bienaymé–Galton–Watson processes under P_1. Nodes in *white* are individuals belonging to the immortal genealogical line of descent known as the spine. Nodes shaded in *grey* represent the offspring of individuals on the spine who are not themselves members of the spine. These individuals may also be considered as "immigrants".

Effectively, taking the second interpretation above to hand, the change of measure has adjusted the statistics on just one genealogical line of descent to ensure that it, and hence the whole process itself, is immortal. See Fig. 10.1.

Our aim in this section is to establish the analogue of these ideas for critical or subcritical continuous-state branching processes. This is done in Sect. 10.3.2. However, we first address the issue of how to condition a spectrally positive Lévy process to stay positive. Apart from as being a useful comparison for the case of conditioning a continuous-state branching process, there are reasons to believe that the two classes of conditioned processes might be connected through a Lamperti-type transform on account of the relationship given in Theorem 10.2. This is the very last point we address in Sect 10.3.2.

10.3.1 Conditioning a Spectrally Positive Lévy Process to Stay Positive

It is possible to talk of conditioning any Lévy process to stay positive and this is now a well understood and well documented phenomenon; also for the case of random walks. See Bertoin (1993), Bertoin and Doney (1994b), Chaumont (1994, 1996), Konstantopoulos and Richardson (2002), Duquesne (2003) and Chaumont and Doney (2005) to name but some of the most recent additions to the literature; see also Lambert (2000) who considers conditioning a spectrally negative Lévy process to stay in a strip. We restrict our attention to the case of spectrally positive Lévy processes; in part because this is what is required for the forthcoming discussion and in part because this facilitates the mathematics.

Suppose that $X = \{X_t : t \geq 0\}$ is a spectrally positive Lévy process with $\psi(\lambda) = \log \mathbb{E}(\mathrm{e}^{-\lambda X_1})$ for all $\lambda \geq 0$. (So as before, ψ is the Laplace exponent of the spectrally negative process $-X$). First recall from Theorem 3.12 that for all $x > 0$,

$$\mathbb{E}_x(\mathrm{e}^{-q\tau_0^-}\mathbf{1}_{(\tau_0^-<\infty)}) = \mathrm{e}^{-\Phi(q)x}, \tag{10.17}$$

where, as usual, $\tau_0^- = \inf\{t > 0 : X_t < 0\}$ and Φ is the right inverse of ψ. In particular when $\psi'(0+) < 0$, so that $\lim_{t\uparrow\infty} X_t = \infty$, we have that $\Phi(0) > 0$ and $\mathbb{P}(\tau_0^- = \infty) = 1 - \mathrm{e}^{-\Phi(0)x}$. In that case, for any $A \in \mathcal{F}_t$, we may simply apply Bayes' formula and the Markov property, respectively, to deduce that for all $x > 0$,

$$\begin{aligned}
\mathbb{P}_x^{\uparrow}(A) &:= \mathbb{P}_x(A|\tau_0^- = \infty) \\
&= \frac{\mathbb{E}_x\left(\mathbf{1}_{(A,t<\tau_0^-)}\mathbb{P}(\tau_0^- = \infty|\mathcal{F}_t)\right)}{\mathbb{P}_x(\tau_0^- = \infty)} \\
&= \mathbb{E}_x\left(\mathbf{1}_{(A,t<\tau_0^-)}\frac{1-\mathrm{e}^{-\Phi(0)X_t}}{1-\mathrm{e}^{-\Phi(0)x}}\right)
\end{aligned}$$

thus giving sense to "conditioning X to stay positive". If however $\psi'(0+) \geq 0$, in other words, $\liminf_{t\uparrow\infty} X_t = -\infty$, then the above calculation is not possible as $\Phi(0) = 0$ and it is less clear what it means to condition the process to stay positive. The sense in which this may be understood is given in Chaumont (1994).

Theorem 10.10. *Suppose that* $\mathbf{e}_q$ *is an exponentially distributed random variable with parameter* q *independent of* X. *Suppose that* $\psi'(0+) \geq 0$. *For all* $x, t > 0$ *and* $A \in \mathcal{F}_t$,

$$\mathbb{P}_x^{\uparrow}(A) := \lim_{q\downarrow 0} \mathbb{P}_x(A, t < \mathbf{e}_q|\tau_0^- > \mathbf{e}_q)$$

exists and satisfies

$$\mathbb{P}_x^{\uparrow}(A) = \mathbb{E}_x(\mathbf{1}_{(A,t<\tau_0^-)}\frac{X_t}{x}).$$

Proof. Again appealing to Bayes' formula followed by the Markov property in conjunction with the lack of memory property and (10.17), we have

$$\begin{aligned}\mathbb{E}_x(\Lambda, t < \mathbf{e}_q | \tau_0^- > \mathbf{e}_q) &= \frac{\mathbb{P}_x(\Lambda,\ t < \mathbf{e}_q,\ \tau_0^- > \mathbf{e}_q)}{\mathbb{P}_x(\tau_0^- > \mathbf{e}_q)} \\ &= \frac{\mathbb{E}_x(\mathbf{1}_{(\Lambda, t < \mathbf{e}_q \wedge \tau_0^-)}\mathbb{E}(\tau_0^- > \mathbf{e}_q | \mathcal{F}_t))}{\mathbb{E}_x(1 - \mathrm{e}^{-q\tau_0^-})} \\ &= \mathbb{E}_x\left(\mathbf{1}_{(\Lambda, t<\tau_0^-)}\mathrm{e}^{-qt}\frac{1 - \mathrm{e}^{-\Phi(q)X_t}}{1 - \mathrm{e}^{-\Phi(q)x}}\right). \qquad (10.18)\end{aligned}$$

Under the assumption $\psi'(0+) \geq 0$, we know that $\Phi(0) = 0$ and hence by l'Hôpital's rule

$$\lim_{q \downarrow 0} \frac{1 - \mathrm{e}^{-\Phi(q)X_t}}{1 - \mathrm{e}^{-\Phi(q)x}} = \frac{X_t}{x}. \qquad (10.19)$$

Noting also that for all q sufficiently small,

$$\frac{1 - \mathrm{e}^{-\Phi(q)X_t}}{1 - \mathrm{e}^{-\Phi(q)x}} \leq \frac{\Phi(q)X_t}{1 - \mathrm{e}^{-\Phi(q)x}} \leq C\frac{X_t}{x},$$

where $C > 1$ is a constant. The condition $\psi'(0+) \geq 0$ also implies that for all $t > 0$, $\mathbb{E}(|X_t|) < \infty$ (see Sect. 8.1) and hence by dominated convergence we may take limits in (10.18) as $q \downarrow 0$ and apply (10.19) to deduce the result. □

It is interesting to note that, whilst $\mathbb{P}_x^{\uparrow}$ is a probability measure for each $x > 0$, when $\psi'(0+) < 0$, this is not necessarily the case when $\psi'(0+) \geq 0$. The following lemma gives a precise account.

Lemma 10.11. *Fix $x > 0$. When $\psi'(0+) = 0$, $\mathbb{P}_x^{\uparrow}$ is a probability measure and when $\psi'(0+) > 0$, $\mathbb{P}_x^{\uparrow}$ is a sub-probability measure.*

Proof. All that is required to be shown is that for each $t > 0$, $\mathbb{E}_x(\mathbf{1}_{(t<\tau_0^-)}X_t) = x$ for $\mathbb{P}_x^{\uparrow}$ to be a probability measure and $\mathbb{E}_x(\mathbf{1}_{(t<\tau_0^-)}X_t) < x$ for a sub-probability measure. To this end, recall from the proof of Theorem 10.10 that

$$\begin{aligned}\mathbb{E}_x(\mathbf{1}_{(t<\tau_0^-)}\frac{X_t}{x}) &= \lim_{q\downarrow 0}\mathbb{P}_x(\ t < \mathbf{e}_q | \tau_0^- > \mathbf{e}_q) \\ &= 1 - \lim_{q\downarrow 0}\mathbb{P}_x(\mathbf{e}_q \leq t | \tau_0^- > \mathbf{e}_q) \\ &= 1 - \lim_{q\downarrow 0}\int_0^t \frac{q\mathrm{e}^{-qu}}{1 - \mathrm{e}^{-\Phi(q)x}}\mathbb{P}_x(\tau_0^- > u)\mathrm{d}u \\ &= 1 - \lim_{q\downarrow 0}\frac{q}{\Phi(q)x}\int_0^t \mathrm{e}^{-qu}\mathbb{P}_x(\tau_0^- > u)\mathrm{d}u \\ &= 1 - \lim_{q\downarrow 0}\frac{\psi'(0+)}{x}\int_0^t \mathrm{e}^{-qu}\mathbb{P}_x(\tau_0^- > u)\mathrm{d}u.\end{aligned}$$

It is now clear that when $\psi'(0+) = 0$ the right-hand side above is equal to unity and otherwise is strictly less than unity thus distinguishing the case of a probability measure from a sub-probability measure. □

Note that when $\mathbb{E}_x(\mathbf{1}_{(t<\tau_0^-)}X_t) = x$, an easy application of the Markov property implies that $\{\mathbf{1}_{(t<\tau_0^-)}X_t/x : t \geq 0\}$ is a unit mean $\mathbb{P}_x$-martingale so that $\mathbb{P}_x^\uparrow$ is obtained by a martingale change of measure. Similarly when $\mathbb{E}_x(\mathbf{1}_{(t<\tau_0^-)}X_t) \leq x$, the latter process is a supermartingale.

On a final note, the reader may be curious as to how one characterises spectrally positive Lévy processes, and indeed a general Lévy process, to stay positive when the initial value $x = 0$. In general, this is a non-trivial issue, but possible by considering the weak limit of the measure $\mathbb{P}_x^\uparrow$ as measure on the space of paths that are right continuous with left limits. The interested reader should consult Chaumont and Doney (2005) for the most recent and up to date account.

10.3.2 Conditioning a (sub)Critical Continuous-State Branching Process to Stay Positive

Let us now progress to conditioning of continuous-state branching processes to stay positive, following closely Chap. 3 of Lambert (2001). We continue to adopt the notation of Sect. 10.1. Our interest is restricted to the case that there is extinction with probability one for all initial values $x > 0$. According to Corollary 10.6 this corresponds to $\psi(\infty) = \infty$, $\psi'(0+) \geq 0$ and

$$\int^\infty \frac{1}{\psi(\xi)} \mathrm{d}\xi < \infty$$

and henceforth we assume that these conditions are in force. For notational convenience we also set

$$\rho = \psi'(0+).$$

Theorem 10.12. *Suppose that $Y = \{Y_t : t \geq 0\}$ is a continuous-state branching process with branching mechanism ψ satisfying the above conditions. For each event $A \in \sigma(Y_s : s \leq t)$, and $x > 0$,*

$$P_x^\uparrow(A) := \lim_{s\uparrow\infty} P_x(A|\zeta > t+s)$$

is well defined as a probability measure and satisfies

$$P_x^\uparrow(A) = E_x(\mathbf{1}_A \mathrm{e}^{\rho t}\frac{Y_t}{x}).$$

In particular, $P_x^\uparrow(\zeta < \infty) = 0$ and $\{\mathrm{e}^{\rho t}Y_t : t \geq 0\}$ is a martingale.

Proof. From the proof of Theorem 10.5 we have seen that for $x > 0$,

$$P_x(\zeta \le t) = P_x(Y_t = 0) = \mathrm{e}^{-xu_t(\infty)},$$

where $u_t(\theta)$ satisfies (10.15). Crucial to the proof will be the convergence

$$\lim_{s\uparrow\infty} \frac{u_s(\infty)}{u_{t+s}(\infty)} = \mathrm{e}^{\rho t} \tag{10.20}$$

for each $t > 0$ and hence we first show that this result holds.

To this end note from (10.15) that

$$\int_{u_{t+s}(\infty)}^{u_s(\infty)} \frac{\rho}{\psi(\xi)} \mathrm{d}\xi = \rho t.$$

On the other hand, recall from the proof of Theorem 10.5 that $u_t(\theta)$ is decreasing to $\Phi(0) = 0$ as $t \downarrow 0$. Hence, since $\lim_{\xi\downarrow 0} \psi(\xi)/\xi = \psi'(0+) = \rho$, it follows that

$$\log \frac{u_s(\infty)}{u_{t+s}(\infty)} = \int_{u_{t+s}(\infty)}^{u_s(\infty)} \frac{1}{\xi} \mathrm{d}\xi = \int_{u_{t+s}(\infty)}^{u_s(\infty)} \frac{\psi(\xi)}{\xi\rho} \frac{\rho}{\psi(\xi)} \mathrm{d}\xi \to \rho t,$$

as $s \uparrow \infty$ thus proving the claim.

With (10.20) in hand we may now proceed to note that

$$\lim_{s\uparrow\infty} \frac{1 - \mathrm{e}^{-Y_t u_s(\infty)}}{1 - \mathrm{e}^{-xu_{t+s}(\infty)}} = \frac{Y_t}{x} \mathrm{e}^{\rho t}.$$

In addition,

$$\frac{1 - \mathrm{e}^{-Y_t u_s(\infty)}}{1 - \mathrm{e}^{-xu_{t+s}(\infty)}} \le \frac{Y_t u_s(\infty)}{1 - \mathrm{e}^{-xu_{t+s}(\infty)}} \le C \frac{Y_t}{x} \mathrm{e}^{\rho t}$$

for some $C > 1$. Hence we may now apply the Markov property and then the Dominated Convergence Theorem to deduce that

$$\begin{aligned}
\lim_{s\uparrow\infty} P_x(A|\zeta > t+s) &= \lim_{s\uparrow\infty} E_x\left(\mathbf{1}_{(A,\zeta>t)} \frac{P_{Y_t}(\zeta > s)}{P_x(\zeta > t+s)}\right) \\
&= \lim_{s\uparrow\infty} E_x\left(\mathbf{1}_{(A,\zeta>t)} \frac{1 - \mathrm{e}^{-Y_t u_s(\infty)}}{1 - \mathrm{e}^{-xu_{t+s}(\infty)}}\right) \\
&= \lim_{s\uparrow\infty} E_x(\mathbf{1}_{(A,\zeta>t)} \frac{Y_t}{x} \mathrm{e}^{\rho t}).
\end{aligned}$$

Note that we may remove the qualification $\{t < \zeta\}$ from the indicator on the right-hand side above as $Y_t = 0$ on $\{t \ge \zeta\}$. To show that $P_x^{\uparrow}$ is a probability measure it suffices to show that for each $x, t > 0$, $E_x(Y_t) = \mathrm{e}^{-\rho t} x$. However, the latter was already proved in (10.12). A direct consequence of this is that $P_x^{\uparrow}(\zeta > t) = 1$ for all $t \ge 0$ which implies that $P_x^{\uparrow}(\zeta < \infty) = 0$.

The fact that $\{e^{\rho t}Y_t : t \geq 0\}$ is a martingale follows in the usual way from consistency of Radon–Nikodym densities. Alternatively, it follows directly from (10.12) by applying the Markov property as follows. For $0 \leq s \leq t$,

$$E_x(e^{\rho t}Y_t|\sigma(Y_u : u \leq s)) = e^{\rho s}E_{Y_s}(e^{\rho(t-s)}Y_{t-s}) = e^{\rho s}Y_s,$$

which establishes the martingale property. □

Note that in older literature, the process $(Y, P_x^{\uparrow})$ is called the Q-process. See for example Athreya and Ney (1972).

We have thus far seen that conditioning a (sub)critical continuous-state branching process to stay positive can be performed mathematically in a similar way to conditioning a spectrally positive Lévy processes to stay positive. Our next objective is to show that, in an analogous sense to what has been discussed for Bienaymé–Galton–Watson processes, the conditioned process has the same law as a continuous-state branching process with immigration. Let us spend a little time to give a mathematical description of the latter.

In general we define a Markov process $Y^* = \{Y_t^* : t \geq 0\}$ with probabilities $\{\mathbf{P}_x : x \geq 0\}$ to be a continuous-state branching process with branching mechanism ψ and immigration mechanism ϕ if it is $[0, \infty)$-valued and has paths that are right continuous with left limits and for all $x, t > 0$ and $\theta \geq 0$

$$\mathbf{E}_x(e^{-\theta Y_t^*}) = \exp\{-xu_t(\theta) - \int_0^t \phi(u_{t-s}(\theta))ds\}, \tag{10.21}$$

where $u_t(\theta)$ is the unique solution to (10.5) and ϕ is the Laplace exponent of any subordinator. Specifically, for $\theta \geq 0$,

$$\phi(\theta) = d\theta + \int_{(0,\infty)} (1 - e^{-\theta x})\Lambda(dx),$$

where Λ is a measure concentrated on $(0, \infty)$ satisfying $\int_{(0,\infty)}(1 \wedge x)\Lambda(dx) < \infty$.

It is possible to see how the above definition plays an analogous role to (10.16) by considering the following sample calculations (which also show the existence of continuous-state branching processes with immigration). Suppose that $S = \{S_t : t \geq 0\}$ under $\mathbb{P}$ is a pure jump subordinator[4] with Laplace exponent $\phi(\theta)$ (hence $d = 0$). Now define a process

$$Y_t^* = Y_t + \int_{[0,t]} \int_{(0,\infty)} Y_{t-s}^{(x)} N(ds \times dx),\ t \geq 0,$$

where N is the Poisson random measure associated with the jumps of S, Y_t is a continuous-state branching process and for each (s, x) in the support of

[4]Examples of such processes when $\phi(\lambda) = c\lambda^\alpha$ for $\alpha \in (0, 1)$ and $c > 0$ are considered by Etheridge and Williams (2003).

N, $Y_{t-s}^{(x)}$ is an independent copy of the process (Y, P_x) at time $t-s$. Note that since S has a countable number of jumps, the integral above is well defined and for the forthcoming calculations it will be more convenient to write the expression for Y_t^* in the form

$$Y_t^* = Y_t + \sum_{u \le t} Y_{t-u}^{(\Delta S_u)},$$

where $\Delta S_u = S_u - S_{u-}$ so that $\Delta S_u = 0$ at all but a countable number of $u \in [0,t]$. We immediately see that $Y^* = \{Y_t^* : t \ge 0\}$ is a natural analogue of (10.16) where now the subordinator S_t plays the role of $\sum_{i=1}^n \eta_i$, the total number of immigrants in Z^* up to and including generation n. Let us proceed further to compute its Laplace exponent. If $\mathbf{P}_x$ is the law of Y^* when $Y_0^* = Y_0 = x$ then, with $\mathbf{E}_x$ as the associated expectation operator, for all $\theta \ge 0$,

$$\mathbf{E}_x(\mathrm{e}^{-\theta Y_t^*}) = \mathbf{E}_x\left(\mathrm{e}^{-\theta Y_t} \prod_{v \le t} E(\mathrm{e}^{-\theta Y_{t-v}^{\Delta S_v}} | S)\right),$$

where the interchange of the product and the conditional expectation is a consequence of monotone convergence.[5] Continuing this calculation we have

$$\begin{aligned}
\mathbf{E}_x(\mathrm{e}^{-\theta Y_t^*}) &= E_x(\mathrm{e}^{-\theta Y_t})\mathbb{E}\left(\prod_{v \le t} E_{\Delta S_v}(\mathrm{e}^{-\theta Y_{t-v}})\right)\\
&= \mathrm{e}^{-x u_t(\theta)}\mathbb{E}\left(\prod_{v \le t} \mathrm{e}^{-\Delta S_v u_{t-v}(\theta)}\right)\\
&= \mathrm{e}^{-x u_t(\theta)}\mathbb{E}\left(\mathrm{e}^{-\sum_{v \le t} \Delta S_v u_{t-v}(\theta)}\right)\\
&= \mathrm{e}^{-x u_t(\theta)}\mathbb{E}\left(\mathrm{e}^{-\int_{[0,t]}\int_{(0,\infty)} x u_{t-s}(\theta) N(\mathrm{d}s \times \mathrm{d}x)}\right)\\
&= \exp\{-x u_t(\theta) - \int_{[0,t]}\int_{(0,\infty)} (1 - \mathrm{e}^{-x u_{t-s}(\theta)})\mathrm{d}s\Lambda(\mathrm{d}x)\}\\
&= \exp\{-x u_t(\theta) - \int_0^t \phi(u_{t-s}(\theta))\mathrm{d}s\},
\end{aligned}$$

where the penultimate equality follows from Theorem 2.7 (ii).

---

[5]Note that for each $\varepsilon > 0$, the Lévy–Itô decomposition tells us that

$$\mathbf{E}(1 - \prod_{u \le t} \mathbf{1}_{(\Delta S_u > \varepsilon)}\mathrm{e}^{-\theta Y_{t-u}^{\Delta S_u}} | S) = 1 - \prod_{u \le t} \mathbf{1}_{(\Delta S_u > \varepsilon)}\mathbf{E}(\mathrm{e}^{-\theta Y_{t-u}^{\Delta S_u}} | S)$$

due to there being a finite number of independent jumps greater than ε. Now take limits as $\varepsilon \downarrow 0$ and apply monotone convergence.

Allowing a drift component in ϕ introduces some lack of clarity with regard to a path-wise construction of Y^* in the manner shown above (and hence its existence). Intuitively speaking, if d is the drift of the underlying subordinator, then the term $d\int_0^t u_{t-s}(\theta)\mathrm{d}s$ which appears in the Laplace exponent of (10.21) may be thought of as due to a "continuum immigration" where, with rate d, in each $\mathrm{d}t$ an independent copy of $(Y, P_\cdot)$ immigrates with infinitesimally small initial value. The problem with this intuitive picture is that there are an uncountable number of immigrating processes which creates measurability problems when trying to construct the "aggregate integrated mass" that has immigrated up to time t. Nonetheless, Lambert (2002) gives a path-wise construction with the help of excursion theory and Itô synthesis; a technique which goes beyond the scope of this text. Returning to the relationship between processes with immigration and conditioned processes, we see that the existence of a process Y^* with an immigration mechanism containing drift can otherwise be seen from the following lemma.

Lemma 10.13. *Fix $x > 0$. Suppose that (Y, P_x) is a continuous-state branching process with branching mechanism ψ. Then $(Y, P_x^\uparrow)$ has the same law as a continuous-state branching process with branching mechanism ψ and immigration mechanism ϕ where for $\theta \geq 0$,*

$$\phi(\theta) = \psi'(\theta) - \rho.$$

Proof. Fix $x > 0$. Clearly $(Y, \mathbb{P}_x^\uparrow)$ has paths that are right continuous with left limits as for each $t > 0$, when restricted to $\sigma(Y_s : s \leq t)$ we have $\mathbb{P}_x^\uparrow << \mathbb{P}_x$. Next we compute the Laplace exponent of Y_t under $\mathbb{P}^\uparrow$ making use of (10.4),

$$\begin{aligned}
\mathbb{E}_x^\uparrow(\mathrm{e}^{-\theta Y_t}) &= \mathbb{E}_x(\mathrm{e}^{\rho t}\frac{Y_t}{x}\mathrm{e}^{-\theta Y_t}) \\
&= -\frac{\mathrm{e}^{\rho t}}{x}\frac{\partial}{\partial\theta}\mathbb{E}_x(\mathrm{e}^{-\theta Y_t}) \\
&= -\frac{\mathrm{e}^{\rho t}}{x}\frac{\partial}{\partial\theta}\mathrm{e}^{-x u_t(\theta)} \\
&= \mathrm{e}^{\rho t}\mathrm{e}^{-x u_t(\theta)}\frac{\partial u_t}{\partial\theta}(\theta).
\end{aligned} \tag{10.22}$$

Recall from (10.11) that

$$\frac{\partial u_t}{\partial\theta}(\theta) = \mathrm{e}^{-\int_0^t \psi'(u_s(\theta))\mathrm{d}s} = \mathrm{e}^{-\int_0^t \psi'(u_{t-s}(\theta))\mathrm{d}s},$$

in which case we may identify with the help of (10.6),

$$\begin{aligned}
\phi(\theta) &= \psi'(\theta) - \rho \\
&= \sigma^2\theta + \int_{(0,\infty)} (1 - \mathrm{e}^{-\theta x})x\Pi(\mathrm{d}x).
\end{aligned}$$

The latter is the Laplace exponent of a subordinator with drift σ^2 and Lévy measure $x\Pi(\mathrm{d}x)$. □

Looking again to the analogy with conditioned Bienaymé–Galton–Watson processes, it is natural to ask if there is any way to decompose the conditioned process in some way as to identify the analogue of the genealogical line of descent, earlier referred to as the spine, along which copies of the original process immigrate. This is possible, but again somewhat beyond the scope of this text. We refer the reader instead to Duquesne (2006) and Lambert (2002).

Finally, as promised earlier, we show the connection between $(X, \mathbb{P}_x^{\uparrow})$ and $(Z, P_x^{\uparrow})$ for each $x > 0$. We are only able to make a statement for the case that $\psi'(0+) = 0$.

Lemma 10.14. *Suppose that* $Y = \{Y_t : t \geq 0\}$ *is a continuous-state branching process with branching mechanism* ψ. *Suppose further that* $X = \{X_t : t \geq 0\}$ *is a spectrally positive Lévy process with no positive jumps with Laplace exponent* $\psi(\theta) = \log \mathbb{E}(\mathrm{e}^{-\theta X_1})$ *for* $\theta \geq 0$. *Fix* $x > 0$. *If* $\psi'(0+) = 0$ *and*

$$\int^{\infty} \frac{1}{\psi(\xi)} \mathrm{d}\xi < \infty,$$

then

(i) the process $\{X_{\theta_t} : t \geq 0\}$ *under* $\mathbb{P}_x^{\uparrow}$ *has the same law as* $(Y, P_x^{\uparrow})$ *where*

$$\theta_t = \inf\{s > 0 : \int_0^s \frac{1}{X_u} \mathrm{d}u > t\},$$

(ii) the process $\{Y_{\varphi_t} : t \geq 0\}$ *under* $P_x^{\uparrow}$ *has the same law as* $(X, \mathbb{P}_x^{\uparrow})$ *where*

$$\varphi_t = \inf\{s > 0 : \int_0^s Y_u \mathrm{d}u > t\}.$$

Proof. Note that the condition $\psi'(0+) = 0$ necessarily excludes the case that X is a subordinator.

(i) It is easy to show that θ_t is a stopping time with respect to $\{\mathcal{F}_t : t \geq 0\}$ of X. Using Theorem 10.10 and the Lamperti transform we have that if $F(X_{\theta_s} : s \leq t)$ is a non-negative measurable functional of X, then for each $x > 0$,

$$\begin{aligned}
\mathbb{E}_x^{\uparrow}(F(X_{\theta_s} : s \leq t)\mathbf{1}_{(\theta_t < \infty)}) &= \mathbb{E}_x(\frac{X_{\theta_t}}{x} F(X_{\theta_s} : s \leq t)\mathbf{1}_{(\theta_t < \tau_0^-)}) \\
&= E_x(\frac{Y_t}{x} F(Y_s : s \leq t)\mathbf{1}_{(t < \zeta)}) \\
&= E_x^{\uparrow}(F(Y_s : s \leq t)).
\end{aligned}$$

(ii) The proof of the second part is a similar argument and left to the reader. □

10.4 Concluding Remarks

It would be impossible to complete this chapter without mentioning that the material presented above is but the tip of the iceberg of a much grander theory of continuous time branching processes. Suppose in the continuous time Bienaymé–Galton–Watson process we allowed individuals to independently move around according to some Markov process then we would have an example of a *spatial Markov branching particle process*. If continuous-state branching processes are the continuous-state analogue of continuous time Bienaymé–Galton–Watson process then what is the analogue of a spatial Markov branching particle process?

The answer to this question opens the door to the world of measure valued diffusions (or superprocesses) which, apart from its implicit probabilistic and mathematical interest, has many consequences from the point of view of mathematical biology, genetics and statistical physics. The interested reader is referred to the excellent monographs of Etheridge (2000), Le Gall (1999) and Duquesne and Le Gall (2002) for an introduction.

Exercises

10.1. This exercise is due to Prof. A.G. Pakes. Suppose that $Y = \{Y_t : t \geq 0\}$ is a continuous-state branching process with branching mechanism

$$\psi(\theta) = c\theta - \int_{(0,\infty)} (1 - \mathrm{e}^{\theta x})\lambda F(\mathrm{d}x),$$

where $c, \lambda > 0$ and F is a probability distribution concentrated on $(0, \infty)$. Assume further that $\psi'(0+) > 0$ (hence Y is subcritical).

(i) Show that Y survives with probability one.
(ii) Show that for all t sufficiently large, $Y_t = \mathrm{e}^{-ct}\Delta$ where Δ is a positive random variable.

10.2. This exercise is based in part on Chaumont (1994). Suppose that X is a spectrally positive Lévy process with Laplace exponent $\psi(\theta) = \log \mathbb{E}(\mathrm{e}^{-\theta X_1})$ for $\theta \geq 0$. Assume that $\psi'(0+) \geq 0$.

(i) Show using the Wiener–Hopf factorisation that for each $x, q > 0$ and continuous, compactly supported $f : [0, \infty) \to [0, \infty)$,

$$\mathbb{E}^{\uparrow}_x \left(\int_0^\infty \mathrm{e}^{-qt} f(X_t)\mathrm{d}t \right)$$
$$= \frac{\Phi(q)}{qx} \int_0^\infty \mathrm{d}y \; \mathrm{e}^{-\Phi(q)y} \mathbf{1}_{(y<x)} \cdot \int_{[0,\infty)} \mathbb{P}(\overline{X}_{\mathbf{e}_q} \in \mathrm{d}z) \cdot f(x+z-y)(x+z-y).$$

(ii) Hence show the following identity holds for the potential density of the process conditioned to stay positive

$$\int_0^\infty \mathrm{d}t \cdot \mathbb{P}_x^\uparrow(X_t \in \mathrm{d}y) = \frac{y}{x}\{W(y) - W(y-x)\}\mathrm{d}y,$$

where W is the scale function defined in Theorem 8.1.

(iii) Show that when $\psi'(0+) = 0$ (in which case it follows from Lemma 10.11 that $\mathbb{P}_x^\uparrow$ is a probability measure for each $x > 0$) we have

$$\mathbb{P}_x^\uparrow(\tau_z^+ < \tau_y^-) = 1 - \frac{y}{x}\frac{W(x-y)}{W(z-y)},$$

where $0 \leq y < x < z < \infty$ and

$$\tau_z^+ = \inf\{t > 0 : X_t > z\} \text{ and } \tau_y^- = \inf\{t > 0 : X_t < y\}.$$

Hence deduce that for all $x > 0$,

$$\mathbb{P}_x^\uparrow(\liminf_{t\uparrow\infty} X_t = \infty) = 1.$$

10.3. This exercise is taken from Lambert (2001). Suppose that Y is a conservative continuous-state branching process with branching mechanism ψ (we shall adopt the same notation as the main text in this chapter). Suppose that $\psi'(\infty) = \infty$ (so that the underlying Lévy process is not a subordinator), $\int^\infty \psi(\xi)^{-1}\mathrm{d}\xi < \infty$ and $\rho := \psi'(0+) \geq 0$.

(i) Using (10.7) show that one may write for each $t, x > 0$ and $\theta \geq 0$,

$$E_x^\uparrow(\mathrm{e}^{-\theta Y_t}) = \mathrm{e}^{-xu_t(\theta)+\rho t}\frac{\psi(u_t(\theta))}{\psi(\theta)}$$

which is a slightly different representation to (10.21) used in the text.

(ii) Assume that $\rho = 0$. Show that for each $x > 0$

$$P_x^\uparrow(\lim_{t\uparrow\infty} Y_t = \infty) = 1.$$

(Hint: you may use the conclusion of Exercise 10.2 (iii)).

(iii) Now assume that $\rho > 0$. Recalling that the convexity of ψ implies that $\int_{(1,\infty)} x\Pi(\mathrm{d}x) < \infty$ (cf. Sect. 8.1), show that

$$0 \leq \int_0^\theta \frac{\psi(\xi) - \rho\xi}{\xi^2}\mathrm{d}\xi = \frac{1}{2}\sigma^2\theta + \int_{(0,\infty)} x\Pi(\mathrm{d}x) \cdot \int_0^{\theta x}\left(\frac{\mathrm{e}^{-\lambda} - 1 + \lambda}{\lambda^2}\right)\mathrm{d}\lambda.$$

Hence using the fact that $\psi(\xi) \sim \rho\xi$ as $\xi \downarrow 0$ show that

$$\int^\infty x\log x\Pi(\mathrm{d}x) < \infty$$

if and only if

$$0 \leq \int_{0+}\left(\frac{1}{\rho\xi} - \frac{1}{\psi(\xi)}\right)\mathrm{d}\xi < \infty.$$

(iv) Keeping with the assumption that $\rho > 0$ and $x > 0$, show that

$$Y_t \stackrel{P_x^\uparrow}{\to} \infty$$

if $\int^\infty x \log x \Pi(\mathrm{d}x) = \infty$ and otherwise Y_t converges in distribution under $P_x^\uparrow$ as $t \uparrow \infty$ to a non-negative random variable Y_∞ with Laplace transform

$$E_x^\uparrow(\mathrm{e}^{-\theta Y_\infty}) = \frac{\rho\theta}{\psi(\theta)} \exp\left\{-\rho \int_0^\theta \left(\frac{1}{\rho\xi} - \frac{1}{\psi(\xi)}\right) \mathrm{d}\xi\right\}.$$

Epilogue

The applications featured in this book have been chosen specifically because they exemplify, utilise and have stimulated many different aspects of the mathematical subtleties which together are commonly referred to as the fluctuation theory of Lévy processes. There are of course many other applications of Lévy processes which we have not touched upon. The literature in this respect is vast.

None the less, let us mention a few topics with a few key references for the interested reader. The list is by no means exhaustive but merely a selection of current research activities at the time of writing.

Stable and stable-like processes. Stable processes and variants thereof are a core class of Lévy processes which offer the luxury of a higher degree of mathematical tractability in a wide variety of problems. This is in part due to their inherent scaling properties. Samorodnitsky and Taqqu (1994) provides an excellent starting point for further reading.

Stochastic control. The step from optimal stopping problems driven by diffusions to optimal stopping problems driven by processes with jumps comes hand in hand with the movement to stochastic control problems driven by jump processes. Recent progress is summarised in Øksendal and Sulem (2004).

Financial mathematics. In Sect. 2.7.3 we made some brief remarks concerning how properties of Lévy processes may be used to one's advantage while modelling risky assets. This picture is far from complete as, at the very least, we have made no reference to the more substantial and effective stochastic volatility models. The use of such models increases the mathematical demands on subtle financial issues such as hedging, completeness, exact analytic pricing, measures of risk and so on. Whilst solving some problems in mathematical finance, the use of Lévy processes also creates many problems. The degree of complexity of the latter now supports a large and vibrant community of researchers engaged in many new and interesting forms of mathematical theories. The reader is again referred to Boyarchenko and Levendorskii (2002a), Schoutens (2003), Cont and Tankov (2004) and Barndorff–Nielsen and Shephard (2005). See also Kyprianou et al. (2005).

Regenerative sets and combinatorics. By sampling points independently from an exponential distribution and grouping them in a way that is determined by a pre-specified regenerative set on $[0, \infty)$, such as is given by the range of a subordinator, one may describe certain combinatorial sampling formulae. See Chap. 9 of Kingman (1993) or Gnedin and Pitman (2005).

Stochastic differential equations driven by Lévy processes. There is a well established theory for existence, uniqueness and characterisation of the solution to stochastic differential equations driven by Brownian motion which crop up in countless scenarios within the physical and engineering sciences (cf. Øksendal (2003)). It is natural to consider analogues of these equations where now the driving source of randomness is a Lévy process. Applebaum (2004) offers a recent treatment.

Continuous-time time series models. Lévy processes are the continuous-time analogue of random walks. What then are the continuous time analogues of time series models, particularly those that are popular in mathematical finance such as GARCH processes? The answer to this question has been addressed in very recent literature such as Klüppelberg et al. (2004, 2006) and Brockwell et al. (2005). Lévy processes play an important role here.

Lévy Copulas. The method of using copulas to build in certain parametric dependencies in multivariate distributions from their marginals is a well established theory. See for example the up-to-date account in Nelson (2006). Inspired by this methodology, a limited volume of recent literature has proposed to address the modelling of multi-dimensional Lévy processes by working with copulas on the Lévy measure. The foundational ideas are to be found in Tankov (2003) and Kallsen and Tankov (2004).

Lévy-type processes and pseudodifferential operators. Jacob (2001, 2002, 2005) summarises the analysis of Markov processes through certain pseudodifferential operators. The latter are intimately related to the infinitesimal generator of the underlying process via complex analysis.

Fractional Lévy processes. The concept of fractional Brownian motion also has its counterpart for Lévy processes; see Samorodnitsky and Taqqu (1994). Interestingly, whilst fractional Brownian motion has at least two consistent definitions in the form of stochastic integrals with respect to Brownian motion (the harmonisable representation and the moving-average representation), the analogues of these two definitions for fractional Lévy processes throw out subtle differences. See for example Benassi, Cohen and Istas (2002, 2004).

Quantum Independent Increment Processes. Lévy processes have also been introduced in quantum probability, where they can be thought of as an abstraction of a "noise" perturbing a quantum system. The first examples arose in models of quantum systems coupled to a heat bath and in von Waldenfels' investigations of light emission and absorption. The algebraic structure underlying the notions of increment and independence in this setting was developed by Accardi, Schürmann and von Waldenfels. For an introduction to the subject and a description of the latest research in this area, see Applebaum et al. (2005) and Barndorff-Nielsen et al. (2006).

Lévy networks. These systems can be thought of multi-dimensional Lévy processes reflected on the boundary of the positive orthant of $\mathbb{R}^d$ which appear as limiting models of communication networks with traffic process of unconventional (i.e., long-range dependent) type. See, for example, Harrison and Williams (1987), Kella (1993) and Konstantopoulos et al. (2004). The justification of these as limits can be found, for example, in Konstantopoulos and Lin (1998) and Mikosch et al. (2002). Although Brownian stochastic networks have, in some cases, stationary distributions which can be simply described, this is not the case with more general Lévy networks. The area of multidimensional Lévy processes is a challenging field of research.

Fragmentation theory. The theory of fragmentation has recently undergone a surge of interest thanks to its close relation to other probability models which have also enjoyed an equal amount of attention. Closely related to classical spatial branching processes, a core class of fragmentation processes model the way in which an object of unit total mass *fragments* in continuous time so that at any moment there are always a countably infinite number of pieces and further, up to an infinite number of fragmentations may occur over any finite time horizon. In a way that is familiar to Lévy processes, the construction of fragmentation processes is done with the help of Poisson point processes. In addition there are many embedded probabilistic structures which are closely related to special classes of Lévy processes. We refer to the forthcoming monograph of Bertoin (2006) for an indication of the state of the art.

Solutions

Chapter 1

1.1 For each $i = 1, \ldots, n$ let $X^{(i)} = \{X_t^{(i)} : t \geq 0\}$ be independent Lévy processes and define the process $X = \{X_t : t \geq 0\}$ by

$$X_t = \sum_{i-1}^{n} X_t^{(i)}.$$

Note that the first two conditions in Definition 1.1 are automatically satisfied. For $0 \leq s \leq t < \infty$ it is clear that $X_t - X_s$ is independent of $\{X_u^{(i)} : u \leq s\}$ for each $i = 1, \ldots, n$ and hence is independent of $\{X_u : u \leq s\}$. Finally $X_t - X_s = \sum_{i=1}^{n} X_t^{(i)} - X_s^{(i)} \overset{d}{=} \sum_{i=1}^{n} X_{t-s}^{(i)} \overset{d}{=} X_{t-s}$.

1.2 (i) Recall the negative Binomial distribution with parameter $c \in \{1, 2, ...\}$ and $p \in (0,1)$ is considered to be the distribution one obtains by summing c independent geometric distributions. Let $q = 1 - p$. An easy calculation shows that $\mathbb{E}(\mathrm{e}^{\mathrm{i}\theta \mathbf{\Gamma}_p}) = p/(1 - q\mathrm{e}^{\mathrm{i}\theta})$ for $\theta \in \mathbb{R}$ and hence if $\mathbf{\Lambda}_{c,p}$ is a negative Binomial with parameters c, p as above, then $\mathbb{E}(\mathrm{e}^{\mathrm{i}\theta \mathbf{\Lambda}_{c,p}}) = p^c/(1 - q\mathrm{e}^{\mathrm{i}\theta})^c$ and the probabilities of $\mathbf{\Lambda}_{x,p}$ are given by

$$\mathbb{P}(\mathbf{\Lambda}_{c,p} = k) = \binom{-c}{k} p^c (-q)^k = (k!)^{-1}(-c)(-c-1)...(-c-k+1)p^c(-q)^k,$$

where k runs through the non-negative integers. One may easily confirm that the restriction on c can be relaxed to $c > 0$ in the given analytical description of the negative Binomial distribution. It is now clear that $\mathbf{\Gamma}_p$ is infinitely divisible since

$$\mathbb{E}(\mathrm{e}^{\mathrm{i}\theta \mathbf{\Gamma}_p}) = \mathbb{E}(\mathrm{e}^{\mathrm{i}\theta \mathbf{\Lambda}_{1/n,p}})^n = \left(\frac{p}{1 - q\mathrm{e}^{\mathrm{i}\theta}}\right).$$

(ii) This follows by showing that $\mathbb{E}(\exp\{\mathrm{i}\theta S_{\mathbf{\Gamma}_p}\}) = \mathbb{E}(\exp\{\mathrm{i}\theta S_{\mathbf{\Lambda}_{1/n,p}}\})^n$ which is a straightforward exercise.

1.3 (i) Using Fubini's theorem we have for $\infty > b > a > 0$,

$$\int_0^\infty \frac{f(ax) - f(bx)}{x} \mathrm{d}x = -\int_0^\infty \int_a^b f'(yx)\mathrm{d}y\mathrm{d}x$$
$$= -\int_a^b \frac{1}{y}(f(\infty) - f(0))\mathrm{d}y$$
$$= (f(0) - f(\infty))\log(b/a).$$

(ii) Choosing $f(x) = \mathrm{e}^{-x}$, $a = \alpha > 0$, and $b = \alpha - z$, where $z < 0$,

$$\int_0^\infty (1 - \mathrm{e}^{zx})\frac{\beta}{x}\mathrm{e}^{-\alpha x}\mathrm{d}x = \log\left((1 - z/\alpha)^\beta\right) \tag{S.1}$$

from which the first claim follows.

One should use the convention that $1/(1-z/\alpha)^\beta = \exp\{-\beta\log(1-z/\alpha)\}$ where an appropriate branch of the logarithm function is taken thus showing that the right-hand side of (S.1) is analytic. Further, to show that $\int_0^\infty (1 - \mathrm{e}^{zx})\frac{\beta}{x}\mathrm{e}^{-\alpha x}\mathrm{d}x$ is analytic on $\Re z < 0$, one may estimate

$$\left|\int_0^\infty (1 - \mathrm{e}^{zx})\frac{\beta}{x}\mathrm{e}^{-\alpha x}\mathrm{d}x\right| \leq 2\beta\int_1^\infty \mathrm{e}^{-\alpha x}\mathrm{d}x + \beta\int_0^1 \sum_{k\geq 1}\frac{|z|^k}{k!}x^{k-1}\mathrm{e}^{-\alpha x}\mathrm{d}x$$

from which one may easily show with the help of Fubini's Theorem that the left-hand side is finite. The fact that $\int_0^\infty (1 - \mathrm{e}^{zx})\frac{\beta}{x}\mathrm{e}^{-\alpha x}\mathrm{d}x$ is analytic now follows again from an expansion of e^{zx} together with Fubini's Theorem; specifically

$$\int_0^\infty (1 - \mathrm{e}^{zx})\frac{\beta}{x}\mathrm{e}^{-\alpha x}\mathrm{d}x = \beta\sum_{k\geq 1}\frac{(-z)^k}{k!}\int_0^\infty x^{k-1}\mathrm{e}^{-\alpha x}\mathrm{d}x.$$

The Identity Theorem tells us that if two functions are analytic on the same domain and they agree on a set which has a limit point in that domain then the two functions are identical. Since both sides of (S.1) are analytic on $\{w \in \mathbb{C} : \Re w < 0\}$ and agree on real $z < 0$, there is equality on $\{w \in \mathbb{C} : \Re w < 0\}$. Equality on $\Re w = 0$ follows since the limit as $\Re w \uparrow 0$ exists on the left-hand side of (S.1) and hence the limit on the right-hand side exists and both limits are equal.

1.4 (i) Integration by parts shows that

$$\int_0^\infty (\mathrm{e}^{-ur} - 1)r^{-\alpha-1}\mathrm{d}r = -\frac{u}{\alpha}\int_0^\infty \mathrm{e}^{-ur}r^{-\alpha}\mathrm{d}r = -\frac{1}{\alpha}u^\alpha\Gamma(1-\alpha), \tag{S.2}$$

where the second equality follows from substitution $t = ur$ in the integral appearing in the first equality. Now using the fact that $\Gamma(1-\alpha) = -\alpha\Gamma(-\alpha)$ the claim follows. Analytic extension may be performed in a similar manner to the calculations in the solution to Exercise 1.3.

To establish (1.7) from (1.9) with $\alpha \in (0,1)$ first choose $\sigma = 0$ and $a = \eta - \int_{\mathbb{R}} x\mathbf{1}_{(|x|<1)}\Pi(\mathrm{d}x)$, where Π is given by (1.9). Our task is to prove that

$$\int_{\mathbb{R}} (1 - \mathrm{e}^{\mathrm{i}\theta x})\Pi(\mathrm{d}x) = c|\theta|^\alpha(1 - \mathrm{i}\beta\tan(\pi\alpha/2)\mathrm{sgn}\,\theta).$$

We have

$$
\begin{aligned}
&\int_{\mathbb{R}} (1 - \mathrm{e}^{\mathrm{i}\theta x}) \Pi(\mathrm{d}x) \\
&\quad = -c_1 \Gamma(-\alpha)|\theta|^\alpha \mathrm{e}^{-\mathrm{i}\pi\alpha \mathrm{sgn}\theta/2} - c_2 \Gamma(-\alpha)|\theta|^\alpha \mathrm{e}^{\mathrm{i}\pi\alpha\mathrm{sgn}\theta/2} \\
&\quad = -\Gamma(-\alpha)\cos(\pi\alpha/2)|\theta|^\alpha \\
&\qquad \times (c_1 + c_2 - \mathrm{i}c_1 \tan(\pi\alpha/2)\mathrm{sgn}\theta + \mathrm{i}c_2 \tan(\pi\alpha/2)\mathrm{sgn}\theta) \\
&\quad = -(c_1 + c_2)\Gamma(-\alpha)\cos(\pi\alpha/2)(1 - \mathrm{i}\beta\tan(\pi\alpha/2)\mathrm{sgn}\theta).
\end{aligned}
\tag{S.3}
$$

The required representation follows by replacing $-\Gamma(-\alpha)\cos(\pi\alpha/2)c_i$ by another constant (also called c_i) for $i = 1, 2$ and then setting $\beta = (c_1 - c_2)/(c_1 + c_2)$ and $c = (c_1 + c_2)$.

(ii) The first part is a straightforward computation. Fourier inversion allows one to write

$$
\frac{1}{2\pi} \int_{\mathbb{R}} 2 \left(\frac{1 - \cos\theta}{\theta^2} \right) \mathrm{e}^{\mathrm{i}\theta x} \mathrm{d}\theta = 1 - |x|.
$$

Choosing $x = 0$ and using symmetry to note that $\int_{\mathbb{R}} (1 - \cos\theta)/\theta^2 \mathrm{d}\theta = 2\int_0^\infty (1 - \cos\theta)/\theta^2 \mathrm{d}\theta$ the second claim follows. Now note that

$$
\begin{aligned}
&\int_0^\infty (1 - \mathrm{e}^{\mathrm{i}rz} + \mathrm{i}rz\mathbf{1}_{(r<1)}) \frac{1}{r^2} \mathrm{d}r \\
&\quad = \int_0^\infty (1 - \cos zr) \frac{1}{r^2} - \mathrm{i} \int_0^{1/z} \frac{1}{r^2} (\sin zr - zr) \mathrm{d}r \\
&\qquad - \mathrm{i} \int_{1/z}^\infty \frac{1}{r^2} \sin rz \mathrm{d}r + \mathrm{i} \int_{1/z}^1 \frac{1}{r^2} zr \mathrm{d}r \\
&\quad = \frac{\pi}{2} z - \mathrm{i}z \left(\int_0^1 \frac{1}{r^2} (\sin r - r) \mathrm{d}r + \int_1^\infty r^{-2} \sin r \mathrm{d}r - \log z \right) \\
&\quad = \frac{\pi}{2} z + \mathrm{i}z \log z - ikz
\end{aligned}
$$

for an obvious choice of the constant k. The complex conjugate of this equality is the identity

$$
\int_0^\infty (1 - \mathrm{e}^{-\mathrm{i}rz} - \mathrm{i}rz\mathbf{1}_{(r<1)}) \frac{1}{r^2} \mathrm{d}r = \frac{\pi}{2} z - \mathrm{i}z \log z + \mathrm{i}kz.
$$

To obtain (1.8), let Π be given as in (1.9) and set $\alpha = 1$, $\sigma = 0$, and $a = \eta + (c_1 - c_2)\mathrm{i}k$ and note that for that

$$
\begin{aligned}
&\int_{\mathbb{R}} (1 - \mathrm{e}^{\mathrm{i}x\theta} + \mathrm{i}x\theta\mathbf{1}_{(|x|<1)}) \Pi(\mathrm{d}x) \\
&\quad = c_1 \int_0^\infty (1 - \mathrm{e}^{\mathrm{i}r\theta} + \mathrm{i}r\theta\mathbf{1}_{(r<1)}) \frac{1}{r^2} \mathrm{d}r \\
&\qquad + c_2 \int_{-\infty}^0 (1 - \mathrm{e}^{\mathrm{i}r\theta} + \mathrm{i}r\theta\mathbf{1}_{(r<1)}) \frac{1}{r^2} \mathrm{d}r \\
&\quad = (c_1 + c_2) \frac{\pi}{2} |\theta| + \mathrm{sgn}\theta (c_1 - c_2) \mathrm{i}|\theta| \log|\theta|
\end{aligned}
$$

$$
\begin{aligned}
&-\mathrm{sgn}\theta(c_1-c_2)\mathrm{i}k|\theta| \\
&= c|\theta|\left(1+\frac{2}{\pi}\beta\mathrm{sgn}\theta\log|\theta|\right)-(c_1-c_2)\mathrm{i}k\theta,
\end{aligned}
$$

where $c=(c_1+c_2)\pi/2$ and $\beta=(c_1-c_2)/(c_1+c_2)$.

(iii) The suggested integration by parts is straightforward. Following similar reasoning to the previous two parts of the question we can establish (1.7) for $\alpha\in(1,2)$ by taking $\sigma=0$, $a=\eta+\int_{\mathbb{R}}x\mathbf{1}_{|x|>1}\Pi(\mathrm{d}x)$, where Π is given by (1.9). Note that one easily confirms that the last integral converges as $\alpha>1$. Further, note that

$$
\begin{aligned}
&\int_{\mathbb{R}}(1-\mathrm{e}^{\mathrm{i}\theta x}+\mathrm{i}\theta x)\Pi(\mathrm{d}x) \\
&= c_1\int_0^\infty(1-\mathrm{e}^{\mathrm{i}\theta x}+i\theta x)\frac{1}{x^{1+\alpha}}\mathrm{d}x+c_2\int_0^\infty(1-\mathrm{e}^{-\mathrm{i}\theta x}-\mathrm{i}\theta x)\frac{1}{x^{1+\alpha}}\mathrm{d}x \\
&= c_1\int_0^\infty(1-\mathrm{e}^{\mathrm{i}\mathrm{sgn}\theta z}+\mathrm{i}\mathrm{sgn}\theta z)|\theta|^\alpha\frac{1}{z^{1+\alpha}}\mathrm{d}z \\
&\quad +c_2\int_0^\infty(1-\mathrm{e}^{-\mathrm{i}\mathrm{sgn}\theta z}-\mathrm{i}\mathrm{sgn}\theta z)|\theta|^\alpha\frac{1}{z^{1+\alpha}}\mathrm{d}z \\
&\quad -c_1\Gamma(-\alpha)|\theta|^\alpha\mathrm{e}^{-\mathrm{i}\pi\alpha\mathrm{sgn}\theta/2}-c_2\Gamma(-\alpha)|\theta|^\alpha\mathrm{e}^{\mathrm{i}\pi\alpha\mathrm{sgn}\theta/2}.
\end{aligned}
$$

The right-hand side above is the same as (S.3) and the calculation thus proceeds in the same way as it does there.

1.5 Let $M_t=\exp\{\mathrm{i}\theta X_t+\Psi(\theta)t\}$. Clearly $\{M_t: t\geq 0\}$ is adapted to the filtration $\mathcal{F}_t=\sigma(X_s:s\leq t)$ and

$$
\mathbb{E}(|M_t|)\leq\mathrm{e}^{\Re\Psi(\theta)}=\exp\left\{\int_{\mathbb{R}}(1-\cos\theta x)\Pi(\mathrm{d}x)\right\}\leq\exp\left\{c\int_{\mathbb{R}}(1\wedge x^2)\Pi(\mathrm{d}x)\right\}
$$

for some sufficiently large $c>0$. Stationary independent increments also implies that for $0\leq s\leq t<\infty$,

$$
\mathbb{E}(M_t|\mathcal{F}_s)=M_s\mathbb{E}(\mathrm{e}^{\mathrm{i}\theta(X_t-X_s)}|\mathcal{F}_s)\mathrm{e}^{-\Psi(\theta)(t-s)}=M_s\mathbb{E}(\mathrm{e}^{\mathrm{i}\theta X_{t-s}})\mathrm{e}^{\Psi(\theta)(t-s)}=M_s.
$$

1.6 (i) Similar arguments to those given in the solution to Exercise 1.5 show that $\{\exp\{\lambda B_t-\lambda^2t/2\}: t\geq 0\}$ is a martingale. We have from Doob's Optimal Stopping Theorem that

$$
1=\mathbb{E}(\mathrm{e}^{\lambda B_{t\wedge\tau_s}-\frac{1}{2}\lambda^2(t\wedge\tau_s)})=\mathbb{E}(\mathrm{e}^{\lambda(B_{t\wedge\tau_s}+b(t\wedge\tau_s))-(\frac{1}{2}\lambda^2+b\lambda)(t\wedge\tau_s)}).
$$

Since $B_{t\wedge\tau_s}+b(t\wedge\tau_s)\leq s$ for all $t\geq 0$ and $b>0$, which implies that $\lim_{t\uparrow\infty}B_t=\infty$ and hence that $\tau_s<\infty$, it follows with the help of the Dominated Convergence Theorem and the continuity of the paths of Brownian motion with drift that

$$
1=\lim_{t\uparrow\infty}\mathbb{E}(\mathrm{e}^{\lambda B_{t\wedge\tau_s}-\frac{1}{2}\lambda^2(t\wedge\tau_s)})=\mathbb{E}(\mathrm{e}^{\lambda s-(\frac{1}{2}\lambda^2+b\lambda)\tau_s})
$$

as required. By setting $q=\lambda^2/2+b\lambda$ we deduce that

$$
\mathbb{E}(\mathrm{e}^{-q\tau_s})=\mathrm{e}^{-s\left(\sqrt{b^2+2q}-b\right)}.
$$

Both right-hand side and left-hand side can be shown to be analytical functions when we replace q by $a - \mathrm{i}\theta$ for $a > 0$ and $\theta \in \mathbb{R}$ and hence they agree on this parameter range. Taking limits as a tends to zero confirms that both functions agree when we replace q by $\mathrm{i}\theta$ with θ as above.

(ii) When $\Pi(\mathrm{d}x) = (2\pi x^3)^{-1/2}\mathrm{e}^{-xb^2/2}$ on $x > 0$, using Exercise 1.4 (i)

$$
\begin{aligned}
&\int_0^\infty (1 - \mathrm{e}^{\mathrm{i}\theta x})\Pi(\mathrm{d}x)\\
&= \int_0^\infty \frac{1}{\sqrt{2\pi x^3}}(1 - \mathrm{e}^{\mathrm{i}\theta - b^2 x/2})\mathrm{d}x - \int_0^\infty \frac{1}{\sqrt{2\pi x^3}}(1 - \mathrm{e}^{-b^2 x/2})\mathrm{d}x\\
&= -\frac{\Gamma\left(-\frac{1}{2}\right)}{2\pi}\left(\frac{b^2}{2} - \mathrm{i}\theta\right)^{1/2} + \frac{\Gamma\left(-\frac{1}{2}\right)}{2\pi}\left(\frac{b^2}{2}\right)^{1/2}\\
&= (b^2 - 2\mathrm{i}\theta)^{1/2} - b.
\end{aligned}
$$

From the Lévy–Khintchine formula we clearly require $\sigma = 0$ and the above calculation indicates that $a = \int_{(0,1)} x\Pi(\mathrm{d}x)$.

(iii) With $\mu_s(\mathrm{d}x) = s(2\pi x^3)^{-1/2}\exp\{sb - (s^2x^{-1} + b^2x)/2\}\mathrm{d}x$,

$$
\begin{aligned}
\int_0^\infty \mathrm{e}^{-\lambda x}\mu_s(\mathrm{d}x) &= \mathrm{e}^{bs - s\sqrt{b^2+2\lambda}} \int_0^\infty \frac{s}{\sqrt{2\pi x^3}}\mathrm{e}^{-\frac{1}{2}\left(\frac{s}{\sqrt{x}} - \sqrt{(b^2+2\lambda)x}\right)^2}\mathrm{d}x\\
&= \mathrm{e}^{bs - s\sqrt{b^2+2\lambda}} \int_0^\infty \sqrt{\frac{2\lambda + b^2}{2\pi u}}\mathrm{e}^{-\frac{1}{2}\left(\frac{s}{\sqrt{u}} - \sqrt{(b^2+2\lambda)u}\right)^2}\mathrm{d}u,
\end{aligned}
$$

where the second equality follows from the substitution $sx^{-1/2} = ((2\lambda + b^2)u)^{1/2}$. Adding the last two integrals together and dividing by two gives

$$
\int_0^\infty \mathrm{e}^{-\lambda x}\mu_s(\mathrm{d}x) = \frac{1}{2}\int_0^\infty \left(\frac{s}{\sqrt{2\pi x^3}} + \sqrt{\frac{2\lambda + b^2}{2\pi x}}\right)\mathrm{e}^{-\frac{1}{2}\left(\frac{s}{\sqrt{x}} - \sqrt{(b^2+2\lambda)x}\right)^2}\mathrm{d}x.
$$

making the substitution $\eta = sx^{-1/2} - \sqrt{(b^2 + 2\lambda)x}$ yields

$$
\int_0^\infty \mathrm{e}^{-\lambda x}\mu_s(\mathrm{d}x) = \mathrm{e}^{bs - s\sqrt{b^2+2\lambda}} \int_{\mathbb{R}} \frac{1}{\sqrt{2\pi}}\mathrm{e}^{-\frac{1}{2}\eta^2}\mathrm{d}\eta = \mathrm{e}^{bs - s\sqrt{b^2+2\lambda}}.
$$

1.7 Note by definition $\tau = \{\tau_s : s \geq 0\}$ is also the inverse of the process $\{\overline{B}_t : t \geq 0\}$, where $\overline{B}_t = \sup_{s\leq t} B_s$. The latter is continuous and $\overline{B}_t > 0$ for all $t > 0$ hence τ satisfies the first two conditions of Definition 1.1. The Strong Markov Property, the fact that $B_{\tau_s} = s$ and spatial homogeneity of Brownian motion implies that $\{B_{\tau_s+t} - s : t \geq 0\}$ is independent of $\{B_u : u \leq \tau_s\}$. Further, this implies that for each $q \geq 0$, $\tau_{s+q} - \tau_s$ is equal in distribution to τ_q and independent of $\{\tau_u : u \leq s\}$. Similar analysis to the solution of Exercise 1.6 centred around an application of Doob's Optimal Stopping Theorem with the stopping time τ_s to the exponential martingale shows that

$$
\mathbb{E}(\mathrm{e}^{-q\tau_s}) = \mathrm{e}^{-\sqrt{2q}s}.
$$

Note however from the first part of the solution to part (i) of Exercise 1.4 we see that the above equality implies that τ is a stable process with index $\alpha = 1/2$ whose Lévy measure is supported on $(0, \infty)$, in other words, $\beta = 1$.

1.8 (i) Is self-explanatory.

(ii) Let $T_0 = 0$ and define recursively for $n = 1, 2 \ldots,$ $T_n = \inf\{k > T_{n-1} : S_k > S_{T_{n-1}}\}$ and let $H_n = S_{T_n}$ if $T_n < \infty$. The indexes T_n are called the strong ascending ladder times and H_n are the ladder heights. It is straightforward to prove that T_n are stopping times. Note that for each $n \geq 1$, from the Strong Markov Property, $H_n - H_{n-1}$ has the same distribution as $S_{T_0^+}$.

$$
\begin{aligned}
&P(S_{T_0^-} \in dx) \\
&\quad = P(S_1 \in \mathrm{d}x) + \sum_{n\geq 1} P(S_1 > 0, ..., S_n > 0, S_{n+1} \in \mathrm{d}x) \\
&\quad = P(S_1 \in \mathrm{d}x) + \sum_{n\geq 1} \int_{(0,\infty)} P(S_1 > 0, ..., S_n > 0, S_n \in \mathrm{d}y)\mu(\mathrm{d}x - y) \\
&\quad = P(S_1 \in \mathrm{d}x) \\
&\qquad + \int_{(0,\infty)} \sum_{n\geq 1} P(S_n \in \mathrm{d}y, S_n > S_j \text{ for all } j = 0, ..., n-1)\mu(\mathrm{d}x - y).
\end{aligned}
$$

The sum in the final equality sums over all indexes of the random walk which correspond to a ladder time. Hence

$$
\begin{aligned}
P(S_{T_0^-} \in dx) &= P(S_1 \in \mathrm{d}x) + \int_{(0,\infty)} \sum_{n\geq 1} P(H_n \in \mathrm{d}y)\mu(\mathrm{d}x - y) \\
&= \int_{(0,\infty)} V(\mathrm{d}y)\mu(\mathrm{d}x - y),
\end{aligned}
$$

where V is given in the question.

(iii) Note that

$$
\mu(z, \infty) = P(\mathbf{e}_\beta > z + \xi_1) = E\left(\mathrm{e}^{-\beta(z+\xi)}\right) = \mathrm{e}^{-\beta z} \int_0^\infty \mathrm{e}^{-\beta u} F(\mathrm{d}u)
$$

and

$$
\mu(-\infty, z) = P(\xi \geq \mathbf{e}_\beta + z) = E(\overline{F}(z + \mathbf{e}_\beta)).
$$

(iv) Recall that the net profit condition implies that $\lim_{n\uparrow\infty} S_n = \infty$ which in turn implies that $P(T_0^+ < \infty) = 1$. The lack of memory condition together with the fact that upward jumps in S are exponentially distributed implies that $S_{T_0^+}$ is exponentially distributed with parameter β (in particular the latter is a proper distribution because first passage above the origin is guaranteed with probability one). From this it follows that for each $n = 1, 2, ...,$

$$
P(H_n \in \mathrm{d}x) = \frac{1}{(n-1)!}\beta^n x^{n-1}\mathrm{e}^{-\beta x}\mathrm{d}x
$$

(in other words a gamma distribution with parameters n and β). Returning to the definition of V, we see that $V(\mathrm{d}y) = \delta_0(dy) + \beta\mathrm{d}y$ and hence

$$
\begin{aligned}
&P(-S_{T_0^-} > x) \\
&= E\left(\overline{F}(\mathbf{e}_\beta + x) + \int_0^\infty \beta \overline{F}(\mathbf{e}_\beta + x + y)\mathrm{d}y\right) \\
&= E\left(\overline{F}(\mathbf{e}_\beta + x) + \int_x^\infty \beta \overline{F}(\mathbf{e}_\beta + z)\mathrm{d}z\right).
\end{aligned}
$$

Continuing we have

$$
\begin{aligned}
&P(-S_{T_0^-} > x) \\
&= E\left(\overline{F}(\mathbf{e}_\beta + x) + \int_{x+\mathbf{e}_\beta}^\infty \beta \overline{F}(z)\mathrm{d}y\right) \\
&= \int_0^\infty \mathrm{d}u \cdot \beta \mathrm{e}^{-\beta u} E\left(\overline{F}(u + x) + \int_{x+u}^\infty \beta \overline{F}(z)\mathrm{d}y\right) \\
&= \int_x^\infty \beta \mathrm{e}^{-\beta(u-x)} \overline{F}(u)\mathrm{d}u + \int_x^\infty \beta \mathrm{d}z \cdot \overline{F}(z) \int_0^{z-x} \beta \mathrm{e}^{-\beta u}\mathrm{d}u \\
&= \beta \int_x^\infty \overline{F}(z)\mathrm{d}z.
\end{aligned}
$$

1.9 (i) The Laplace exponent may be computed directly using the Poisson distribution of N_1 in a similar manner to the calculations in Sect. 1.2.2. One may compute directly the second derivative $\psi''(\theta) = \lambda \int_{(0,\infty)} x^2 \mathrm{e}^{-\theta x} F(\mathrm{d}x) > 0$ showing that ψ is strictly convex and in particular continuous on $(0, \infty)$. It is clear that $\psi(0) = 0$ by taking limits as $\theta \downarrow 0$ on both sides of the equality $\mathbb{E}(\mathrm{e}^{\theta X_1}) = \mathrm{e}^{\psi(\theta)}$ (note $X_1 \leq 1$). Further it is also clear that $\psi(\infty) = \infty$ since $\mathrm{e}^{\psi(\theta)} \geq \mathbb{E}(\mathrm{e}^{\theta X_1}; X_1 > 0)$. On account of convexity, there is one root to the equation $\psi(\theta) = 0$ (that is $\theta = 0$) in $[0, \infty)$, when $\psi'(0+) \geq 0$ and otherwise there are two distinct roots in $[0, \infty)$ (the smallest of these two being $\theta = 0$).

(ii) The martingle properties follow from similar arguments to those given in the solution to Exercise 1.5 (with the help of Doob's Optimal Stopping Theorem). In particular, when sampling at the stopping time $t \wedge \tau_x^+$ and taking limits as t tends to infinity,

$$
1 = \mathbb{E}(\mathrm{e}^{\theta^* X_{\tau_x^+} - \psi(\theta^*)\tau_x^+} \mathbf{1}_{(\tau_x^+ < \infty)}) = \mathrm{e}^{\theta^* x} \mathbb{P}(\tau_x^+ < \infty) = \mathrm{e}^{\theta^* x} \mathbb{P}(\overline{X}_\infty > x).
$$

Note this shows that $\overline{X}_\infty$ is exponentially distributed with parameter θ^*. In the case that $\theta^* = 0$ we interpret this to mean that $\mathbb{P}(\overline{X}_\infty = \infty) = 1$.

(iii) Note that

$$
\{W_s = 0\} = \{\overline{X}_s = X_s \text{ and } \overline{X}_s > w\}.
$$

Note also that on the latter event $\mathrm{d}s = \mathrm{d}X_s = \mathrm{d}\overline{X}_s$. Hence

$$
\int_0^t \mathbf{1}_{(W_s = 0)}\mathrm{d}s = \int_0^t \mathbf{1}_{(\overline{X}_s = X_s, \overline{X}_s > w)}\mathrm{d}\overline{X}_s = \int_0^t \mathbf{1}_{(\overline{X}_s > w)}\mathrm{d}\overline{X}_s - (\overline{X}_t \quad w) \vee 0.
$$

(iv) A direct computation shows that $\psi'(0+) = 1 - \lambda\mu$. Hence if $\lambda\mu \leq 1$ then $\theta^* = 0$ and $\overline{X}_\infty = \infty$ almost surely and hence from the previous part, $I = \infty$ almost surely.

(v) When $\lambda\mu > 1$ we have that $\mathbb{P}(\tau_w^+ = \infty) = \mathbb{P}(\overline{X}_\infty \leq w) = 1 - \mathrm{e}^{-\theta^* w}$. Since $I = 0$ on $\{\tau_w^+ = \infty\}$, it follows that

$$\mathbb{P}(I \in \mathrm{d}x, \tau_w^+ = \infty | W_0 = w) = (1 - \mathrm{e}^{-\theta^* w})\delta_0(\mathrm{d}x).$$

On the other hand, on the event $\{\tau_w^+ < \infty\}$, since τ_w^+ is a stopping time, and $\overline{X}_\infty$ is exponentially distributed, the Strong Markov Property and the Lack of Memory Property imply that

$$\begin{aligned}\mathbb{P}(I \in \mathrm{d}x, \tau_w^+ < \infty | W_0 = w) &= \mathbb{P}\left(\overline{X}_\infty - w \in \mathrm{d}x, \overline{X}_\infty > w | W_0 = w\right) \\ &= \theta^* \mathrm{e}^{-\theta^*(x+w)}\mathrm{d}x.\end{aligned}$$

1.10 (i) This is a straightforward exercise in view of Exercise 1.5.

(ii) Using Doob's Optimal Stopping Theorem we have for the bounded stopping time $t \wedge \tau_x^+$ that

$$\mathbb{E}(\mathrm{e}^{\theta X_{t\wedge\tau_x^+} - (\sigma^2\theta^2/2 + \gamma\theta)(t\wedge\tau_x^+)}) = 1.$$

Setting $\theta = (\sqrt{\gamma^2 + 2\sigma^2 q} - \gamma)/\sigma^2$ and letting $t \uparrow \infty$, it follows from the fact that $X_{t\wedge\tau_x^+} \leq x$ for all $t \geq 0$ and Dominated Convergence that,

$$\mathbb{E}(\mathrm{e}^{-\tau_x^+}) = \mathrm{e}^{-x(\sqrt{\gamma^2+2\sigma^2 q}-\gamma)/\sigma^2}.$$

Now note that by considering the process $-X$, we may simply write down an expression for $\mathbb{E}(\mathrm{e}^{-q\tau_{-a}^-})$ by considering the expression above for $\mathbb{E}(\mathrm{e}^{-q\tau_a^+})$ but with γ replaced by $-\gamma$.

(iii) This is a straightforward exercise.

1.11 (i) From the branching property one easily deduces that, Y_t under P_y is equal in law to the independent sum $\sum_{i=1}^y Y_t^{(i)}$, where $Y_t^{(i)}$ has the same distribution as Y_t under P_1. From this it follows immediately that $E_y(\mathrm{e}^{-\phi Y_t}) = \mathrm{e}^{-y u_t(\phi)}$, where $u_t(\phi) = -\log E_1(\mathrm{e}^{-\phi Y_t})$. Note in particular that since at least one of the y individuals initiating the branching process may survive until time t with probability $\mathrm{e}^{-\lambda t}$, it follows that $u_t(\phi) \geq \mathrm{e}^{-\phi y + \lambda t} > 0$.

(ii) Applying the Markov property we have for all $y = 1, 2, ..., \phi > 0$ and $t, s \geq 0$,

$$E_y(\mathrm{e}^{-\phi Y_{t+s}}) = E_y(E_{Y_s}(\mathrm{e}^{-\phi Y_t})) = E_y(\mathrm{e}^{-u_t(\phi)Y_s})$$

and hence $u_{t+s}(\phi) = u_s(u_t(\phi))$.

(iii) Under the assumption $\pi_0 = 0$, we have as $h \downarrow 0$,

$$P_1(Y_h = 1 + i) = \lambda\pi_i h + o(h) \quad \text{for} \quad i = -1, 1, 2, ...$$

From this it follows that

$$\lim_{h\downarrow 0} \frac{\mathbb{E}_1(\mathrm{e}^{-\phi Y_h}) - \mathrm{e}^{-\phi}}{h} = \sum_{i=-1,1,2,...} \mathrm{e}^{-\phi(i+1)}\lambda\pi_i - \mathrm{e}^{-\phi} = -\mathrm{e}^{-\phi}\psi(\phi).$$

Hence reconsidering the left-hand side above, it follows that

$$\left.\frac{\partial u_t(\phi)}{\partial t}\right|_{t=0} = \psi(\phi).$$

Now use part (ii) and write

$$\frac{\partial u_t(\phi)}{\partial t} = \lim_{h\downarrow 0} \frac{u_h(u_t(\phi)) - u_t(\phi)}{h} = \psi(u_t(\phi)).$$

Chapter 2

2.1 We prove (i) and (ii) together. Similar calculations to those in 1.2.1 give

$$E(s^{\theta N_i}) = \mathrm{e}^{-\lambda_i(1-s)}$$

for $0 < s < 1$. Independence and Dominated Convergence thus implies that

$$E(s^{\sum_{i\geq 1} N_i}) = \lim_{n\uparrow\infty} E(s^{\sum_{i=1}^n N_i}) = \mathrm{e}^{-\lim_{n\uparrow\infty}\sum_{i=1}^n \lambda_i(1-s)}.$$

It now follows that $E(s^{\sum_{i\geq 1} N_i}) = 0$ if and only if $P(\sum_{i\geq 1} N_i = \infty) = 1$ if and only if $\sum_{i=1}^n \lambda_i = \infty$. Further, when $\sum_{i=1}^n \lambda_i < \infty$ $\mathbb{P}(\sum_{i\geq 1} N_i < \infty) = \lim_{s\uparrow 1}\exp\{-\sum_{i=1}^\infty \lambda_i(1-s)\} = 1$.

2.2 (i) Suppose that $\{S_i : i = 1, 2, ..., n\}$ are independent and exponentially distributed. The joint density of $(S_1, ..., S_n)$ is given by $\lambda^n \mathrm{e}^{-\lambda(s_1+\cdots+s_n)}$ for $s_1 > 0, ..., s_n > 0$. Using the classical density transform, the density of the partial sums $(T_1 = S_1, T_2 = S_1 + S_2, ..., T_n = S_1 + \cdots + S_n)$ is given by $\lambda^n \mathrm{e}^{-\lambda t_n}$ for $t_1 \leq t_2 \leq \cdots \leq t_n$. If these partial sums are the arrival times of a Poisson process, then for $A \in \mathcal{B}([0,\infty)^n)$,

$$P((T_1, ..., T_n) \in A \text{ and } N_t = n)$$
$$= \int_{(t_1,\ldots,t_n)\in A} \mathbf{1}_{(t_1\leq t_2\leq\cdots\leq t_n\leq t)} \lambda^n \mathrm{e}^{-\lambda t_n} \mathrm{d}t_1\cdots\mathrm{d}t_n. \tag{S.4}$$

Dividing by $P(N_t = n) = \mathrm{e}^{-\lambda t}(\lambda t)^n/n!$ gives the conclusion of part (i)

(ii) The right-hand side of (S.4) is the density of the distribution of an ordered independent sample from a uniform distribution on $[0, t]$ Specifically, n independent samples from the latter distribution has density $1/t^n$ and there are $n!$ ways of ordering them.

2.3 Note that on $[0,\infty)$, $1 - \mathrm{e}^{-\phi y} \leq 1 \wedge \phi y \leq (1 \vee \phi)(1 \wedge y)$. Hence for all $\phi > 0$,

$$\int_S (1 - \mathrm{e}^{-\phi f(x)})\eta(\mathrm{d}x) \leq (1 \vee \phi)\int_S (1 \wedge f(x))\eta(\mathrm{d}x)$$

and the claim is now obvious.

2.4 (i) Let us suppose that there exists a function $f : [0,1] \to \mathbb{R}$ such that $\lim_{n\uparrow\infty}\sup_{x\in[0,1]}|f_n(x) - f(x)|$, where $\{f_n : n = 1, 2, ...\}$ is a sequence in $D[0,1]$. To show right continuity note that for all $x \in [0,1)$ and $\varepsilon \in (0, 1-x)$,

$$\begin{aligned}|f(x+\varepsilon) - f(x)| \leq\ & |f(x+\varepsilon) - f_n(x+\varepsilon)| \\ & +|f_n(x+\varepsilon) - f_n(x)| \\ & +|f_n(x) - f(x)|.\end{aligned}$$

Each of the three terms on the right-hand side can be made arbitrarily small on account of the fact that $f_n \in D[0,1]$ or by the convergence of f_n to f by choosing ε sufficiently small or n sufficiently large, respectively.

For the existence of a left limit, note that it suffices to prove that for each $x \in (0,1]$, $f(x-\varepsilon)$ is a Cauchy sequence with respect to the distance metric $|\cdot|$ as $\varepsilon \downarrow 0$. To this end note that for $x \in (0,1]$ and $\varepsilon, \eta \in (0,x)$,

$$\begin{aligned}|f(x-\varepsilon) - f(x-\eta)| \leq & |f(x-\varepsilon) - f_n(x-\varepsilon)| \\ & + |f_n(x-\varepsilon) - f_n(x-\eta)| \\ & + |f_n(x-\eta) - f(x-\eta)|.\end{aligned}$$

Once again, each of the three terms on the right-hand side can be made arbitrarily small by choosing either n sufficiently large or η, ε sufficiently small.

(ii) Suppose that Δ_c is a countable set for each $c > 0$, then since

$$\Delta = \bigcup_{n\geq 1} \Delta_{1/n}$$

it follows that Δ is also a countable set.

Suppose then for contradiction that, for a given $c > 0$, the set Δ_c has an accumulation point, say x. This means there exists a sequence $y_n \to x$ such that for each $n \geq 1$, $y_n \in \Delta_c$. From this sequence we may assume without loss of generality that there exists as subsequence $x_n \uparrow x$ (otherwise if this fails the forthcoming argument may be performed for a subsequence $x_n \downarrow x$ to the function $g(x) = f(x-)$ which is left continuous with right limits but has the same set of discontinuities). Now suppose that N is sufficiently large so that for a given $\delta > 0$, $|x_m - x_n| < \delta$ for all $m, n > N$. We have

$$f(x_n) - f(x_m) = [f(x_n) - f(x_n-)] + [f(x_n-) - f(x_m)]$$

and so it cannot happen that $|f(x_n) - f(x_m)| \to 0$ as $n, m \uparrow \infty$ as $|f(x_n) - f(x_n-)| > c$ and yet by left continuity at each x_n we may make $|f(x_n-) - f(x_m)|$ arbitrarily small. This means that $\{f(x_n) : n \geq 1\}$ is not a Cauchy sequence which is a contradiction as the limit $f(x-)$ exists.

In conclusion, for each $c > 0$ there can be no accumulation points in Δ_c and thus the latter is at most countable with the implication that Δ is countable.

2.5 (i) The function f^{-1} jumps precisely when f is piece-wise constant and this clearly happens on the dyadic rationals in $[0,1]$ (that is to say $\mathbb{Q}_2 \cap [0,1]$). This is easily seen by simply reflecting the graph of f about the diagonal. This also shows that jumps of f^{-1} and g are necessarily positive. Since the set $\mathbb{Q} \cap [0,1]$ is dense in $[0,1]$ then so are the jumps of g. The function g has bounded variation since it is the difference of two monotone functions. Finally g is right continuous with left limits because the same is true of f^{-1} as an inverse function.

(ii) Note that for any $y > 0$, for each $k = 1, 2, 3, ...$, there are either an even or odd number of jumps of magnitude 2^{-k} in $[0, y]$. In the case that there are an even number of jumps, they make net contribution of zero to the value $f(y) = \sum_{s\in[0,y]\cap\mathbb{Q}_2} J(x)$. When there are an odd number of jumps they make a net contribution of magnitude 2^{-k} to $f(y)$. Hence we can upper estimate $|f(y)|$ by $\sum_{k\geq 1} 2^{-k} < \infty$. Since f only changes value at dyadic rationals, it follows that f is bounded $[0,1]$.

Using similar reasoning to the above, by fixing $x \in [0,1]$ which is not a dyadic rational, one may show that for all integers $n > 0$, there exists a sufficiently large integer $N > 0$ such that $|f(x) - f(y)| \leq \sum_{k \geq N} 2^{-k}$ whenever $|x - y| < 2^{-n}$. Further, $N \uparrow \infty$ as $n \uparrow \infty$. In other words, when there is no jump, there is a continuity point in f.

If on the other hand, there is a jump at $x \in [0,1]$ (in other words, x is a dyadic rational), then a similar argument to the above shows that there is right continuity (consider in that case $x < y < x + 2^{-n}$). Further, the existence of a left limit can be established by showing in a similar way that $|f(x) - J(x) - f(y)| \leq \sum_{k \geq N} 2^{-k}$ for $x - 2^{-n} < y < x$, where $N \uparrow \infty$ as $n \uparrow \infty$.

To show that f has unbounded variation, note that the total variation over $[0,1]$ is given by

$$\sum_{s \in [0,1] \cap \mathbb{Q}_2} |J(s)| = \frac{1}{2} \sum_{n=1}^{\infty} 2^n \cdot 2^{-n} = \infty.$$

Right continuity and left limits, follows from the definition of f and the fact that there are a countable number of jumps.

2.6 (i) Apply Theorem 2.7 to the case that $S = [0, \infty) \times \mathbb{R}$ and the intensity measure is $\mathrm{d}t \times \Pi(\mathrm{d}x)$. According to the latter, by taking $f(s, x) = x^n$ we have that for each $t > 0$, $\int_{[0,t]} \int_{\mathbb{R}} f(s, x) N(\mathrm{d}s \times \mathrm{d}x) < \infty$ if and only if $\int_{[0,t]} \int_{\mathbb{R}} (1 \wedge |x|^n) \mathrm{d}s \times \Pi(\mathrm{d}x) < \infty$; in other words, if and only if $\int_{\mathbb{R}} (1 \wedge |x|^n) \Pi(\mathrm{d}x) < \infty$. Note however that necessarily $\int_{(-1,1)} x^2 \Pi(\mathrm{d}x) < \infty$ and this implies that $\int_{(-1,1)} |x|^n \Pi(\mathrm{d}x) < \infty$. In conclusion, a necessary and sufficient condition is $\int_{|x| \geq 1} |x|^n \Pi(\mathrm{d}x) < \infty$.

(ii) The proof follows very closely the proof of Lemma 2.9.

2.7 (i) The probability that $\sup_{0 \leq s \leq t} |X_s - X_{s-}|$ is greater than a is one minus the probability that there is no jump of magnitude a or greater. In other words $1 - \mathbb{P}(N([0,t] \times \{\mathbb{R} \backslash (-a, a)\}) = 0)$. The latter probability is equal to $\exp\{-\int_{[0,t]} \int_{\mathbb{R} \backslash (-a,a)} \mathrm{d}s \times \Pi(\mathrm{d}x)\} = \exp\{-t\Pi(\mathbb{R} \backslash (-a, a))\}$. (Note that this also shows that $T_a := \inf\{t > 0 : |X_t - X_{t-}| > a\}$ is exponentially distributed).

(ii) The paths of X are continuous if and only if for all $a > 0$, $\mathbb{P}(\sup_{0 \leq s \leq t} |X_s - X_{s-}| > a) = 0$ if and only if $\Pi(\mathbb{R} \backslash (-a, a)) = 0$ (from the previous part) if and only if $\Pi = 0$.

(iii) and (iv) Obviously a compound Poisson process with drift is piece-wise linear. Clearly, if $\sigma > 0$ then paths cannot be piece-wise linear. Suppose now that $\Pi(\mathbb{R}) = \infty$. In that case, $\lim_{a \downarrow 0} \mathbb{P}(T_a > t) = 0$ showing that $\lim_{a \downarrow 0} T_a = 0$ almost surely. Hence 0 is a limiting point of jump times. Stationary and independent increments and the latter imply that any $s \in \mathbb{Q} \cap (0, \infty)$ (or indeed any $s > 0$) is a limit point of jump times from the right. In conclusion, paths are piece-wise linear if and only if $\Pi(\mathbb{R}) < \infty$ and $\sigma = 0$ and when $\Pi(\mathbb{R}) = \infty$ then the jumps are dense in $[0, \infty)$.

2.8 We have seen that any Lévy process of bounded variation may be written in the form

$$X_t = dt + \int_{[0,t]} \int_{\mathbb{R}} x N(\mathrm{d}s \times \mathrm{d}x), \ t \geq 0,$$

where N is the Poisson random measure associated with the jumps. However, we may also decompose the right-hand side so that

$$X_t = \left\{(d \vee 0) + \int_{[0,t]} \int_{(0,\infty)} xN(\mathrm{d}s \times \mathrm{d}x)\right\} - \left\{|d \wedge 0| + \int_{[0,t]} \int_{(-\infty,0)} |x|N(\mathrm{d}s \times \mathrm{d}x)\right\}.$$

Note that both integrals converge on account of Theorem 2.7 (i) as the assumption that X is of bounded variation implies that both $\int_{(0,\infty)}(1 \wedge x)\Pi(\mathrm{d}x)$ and $\int_{(-\infty,0)}(1 \wedge |x|)\Pi(\mathrm{d}x)$ are finite. As N has independent counts on disjoint domains, it follows that the two integrals on the right-hand side above are independent. The processes in curly brackets above clearly have monotone paths and hence the claim follows.

2.9 (i) The given process is a Lévy process by Lemma 2.15. Also by the same Lemma its characteristic exponent is computable as $\Xi \circ \Lambda$, where $\Lambda(\theta) = \sigma^2\theta^2/2 + c\theta$ and $\Xi(\theta) = \beta \log(1 - i\theta/\alpha)$. Hence the given expression for Ψ follows with a minimal amount of algebra.

(ii) It is not clear apriori that the Variance Gamma process as defined in part (i) of the question is a process of bounded variation. However, one notes the factorisation

$$\left(1 - \mathrm{i}\frac{\theta c}{\alpha} + \frac{\sigma^2\theta^2}{2\alpha}\right) = \left(1 - \frac{\mathrm{i}\theta}{\alpha^{(1)}}\right) \times \left(1 - \frac{\mathrm{i}\theta}{\alpha^{(2)}}\right)$$

and hence Ψ decomposes into the sum of terms of the form $\beta \log(1 - i\theta/\alpha^{(i)})$ for $i = 1, 2$. This shows that X has the same law as the difference of two independent gamma subordinators with the given parameters.

2.10 Increasing the dimension of the space on which the Poisson random measure of jumps is defined has little effect to the given calculations for the one dimensional case presented. Using almost identical proofs to the one dimensional case one may show that $\int_{\mathbb{R}}(1 - \mathrm{e}^{\mathrm{i}\theta \cdot \mathbf{x}} + \mathrm{i}\theta \cdot \mathbf{x}\mathbf{1}_{(|\mathbf{x}|<1)})\Pi(\mathrm{d}\mathbf{x})$ is the characteristic exponent of a pure jump Lévy process consisting of a compound Poisson process whose jumps are greater than or equal to unity in magnitude plus an independent process of compensated jumps of magnitude less than unity. Add to this a linear drift in the direction $\mathbf{a}$ and an independent d-dimensional Brownian motion and one recovers the characteristic exponent given in (2.28).

2.11 As a subordinator is non-negative valued, the Lapalce exponent $\Phi(q)$ exists for all $q \geq 0$. This means that the characteristic exponent $\Psi(\theta) = -\log \mathbb{E}(\mathrm{e}^{\mathrm{i}\theta X_1})$, initially defined for $\theta \in \mathbb{R}$, can now be analytically extended to allow for the case that $\Im\theta > 0$. In that case, plugging $\theta = \mathrm{i}q$ with $q \geq 0$ into the Lévy–Khinchine formula for subordinators (2.23), one obtains

$$\Phi(q) = dq + \int_{(0,\infty)} (1 - \mathrm{e}^{-qx})\Pi(\mathrm{d}x).$$

Suppose temporarily that we truncate the mass of Π within an ϵ-neighbourhood of the origin and replace $\Pi(\mathrm{d}x)$ by $\Pi_\epsilon(\mathrm{d}x) = \mathbf{1}_{(x>\epsilon)}\Pi(\mathrm{d}x)$. In particular, by doing this, note that $\Pi_\epsilon(0,\infty) < \infty$. Integration by parts yields

$$\int_{(a,A)} (1 - e^{-qx}) \Pi_\epsilon(dx) = -(1 - e^{-qA}) \Pi_\epsilon(A, \infty) + (1 - e^{-qa}) \Pi_\epsilon(a, \infty)$$

$$+q \int_a^A e^{-qx} \Pi_\epsilon(x, \infty) dx.$$

Note however the fact that $\Pi_\epsilon(0, \infty) < \infty$ implies both that $(1 - e^{-qA}) \Pi_\epsilon(A, \infty) \to 0$ as $A \uparrow \infty$ and that $(1 - e^{-qa}) \Pi_\epsilon(a, \infty) \to 0$ as $a \downarrow 0$. We have then,

$$\int_{(\epsilon,\infty)} (1 - e^{-qx}) \Pi_\epsilon(dx) = \int_{(0,\infty)} (1 - e^{-qx}) \Pi_\epsilon(dx) = q \int_0^\infty e^{-qx} \Pi_\epsilon(x, \infty) dx.$$

Now talking limits as $\epsilon \downarrow 0$ and appealing to the Monotone Convergence Theorem on the right-hand side above we finally conclude that

$$\Phi(q) = dq + q \int_{(0,\infty)} e^{-qx} \Pi(x, \infty) dx.$$

The final part of the question is now a simple application of the Dominated Convergence Theorem.

2.12 In all parts of this exercise, one ultimately appeals to Lemma 2.15.

(i) Let $B = \{B_t : t \geq 0\}$ be a standard Brownian motion and $\tau = \{\tau_t : t \geq 0\}$ be an independent stable subordinator with index $\alpha/2$, where $\alpha \in (0, 2)$. Then for $\theta \in \mathbb{R}$,

$$\mathbb{E}(e^{i\theta B_{\tau_t}}) = \mathbb{E}(e^{-\frac{1}{2}\theta^2 \tau_t}) = e^{-C|\theta|^\alpha}$$

showing that $\{B_{\tau_t} : t \geq 0\}$ has the law of a symmetric stable process of index α.

(ii) The subordinator τ has Laplace exponent

$$-\log \mathbb{E}(e^{-q\tau_1}) = \frac{2}{a} \left(\frac{q}{q + a^2} \right)$$

for $q \geq 0$; this computation follows with the help of (1.3). To see this, note that X_1 has the same distribution as an independent Poisson sum of independent exponentially distributed random variables where the Poisson distribution has parameter $2/a^2$ and the exponential distributions have parameter a^2. Similarly to the first part of the question we have

$$-\log \mathbb{E}(e^{i\theta\sqrt{2}B_{\tau_1}}) = -\log \mathbb{E}(e^{-\theta^2 \tau_1}) = \frac{2}{a} \left(\frac{\theta^2}{\theta^2 + a^2} \right).$$

To show that the right-hand side is the characteristic exponent of the given compound Poisson process note that

$$\int_{(0,\infty)} (1 - e^{i\theta x}) e^{-ax} dx = \frac{1}{a} - \frac{1}{a - i\theta} = -\frac{1}{a} \left(\frac{i\theta}{a - i\theta} \right)$$

and hence

$$-\log \mathbb{E}(e^{i\theta X_1}) = -\frac{1}{a} \left(\frac{i\theta}{a - i\theta} \right) + \frac{1}{a} \left(\frac{i\theta}{a + i\theta} \right) = \frac{2}{a} \left(\frac{\theta^2}{\theta^2 + a^2} \right).$$

(iii) We note that

$$\mathbb{E}(e^{i\theta\sigma B_{N_1}}) = \mathbb{E}(e^{-\frac{1}{2}\sigma^2\theta^2 N_1}) = \exp\{2\lambda(e^{-\sigma^2\theta^2/2} - 1)\}.$$

It is straightforward to show that the latter is equal to $\mathbb{E}(e^{i\theta X_1})$ using (1.3).

Chapter 3

3.1 (i) We handle the case that $t_2 > t > t_1 \geq 0$, the remaining cases $t \geq t_2 > t_1 \geq 0$ and $t_2 > t_1 \geq t \geq 0$ are handled similarly. Note that when $t_2 > u \geq t > t_1$, using Exercise 1.5 we have

$$\begin{aligned}\mathbb{E}(\mathrm{e}^{\mathrm{i}\theta_1 X_{t_1} + \mathrm{i}\theta_2 X_{t_2}}|\mathcal{F}_u) &= \mathrm{e}^{\mathrm{i}\theta_1 X_{t_1}}\mathbb{E}(\mathrm{e}^{\mathrm{i}\theta_2 X_{t_2} + \Psi(\theta_2)t_2}|\mathcal{F}_u)\mathrm{e}^{-\Psi(\theta_2)t_2} \\ &= \mathrm{e}^{\mathrm{i}\theta_1 X_{t_1}}\mathrm{e}^{\mathrm{i}\theta_2 X_u + \Psi(\theta_2)u}\mathrm{e}^{-\Psi(\theta_2)t_2}.\end{aligned}$$

In particular, considering the right-hand side above for $u = t$ and using right continuity, the limit as $u \downarrow t$ satisfies the right expression. Note that this method can be used to prove the same statement when $\theta_1 X_{t_1} + \theta_2 X_{t_2}$ is replaced by $\sum_{i=1}^n \theta_i X_{t_i}$ for $\theta_i \in \mathbb{R}$ and $t_i \geq 0$ for $i = 1, ..., n$.

(ii) This is a consequence of standard measure theory (see for example Theorem 4.6.3 of Durrett (2004)) and part (i).

(iii) The conclusion of (ii) implies in particular that for all cylinder sets $A = \{X_{t_1} \in A_1, ..., X_{t_n} \in A_n\}$ with $t_i \geq 0$ and $A_i \in \mathcal{B}(\mathbb{R})$ for $i = 1, ..., n$,

$$\mathbb{P}(A|\mathcal{F}_{t+}) = \mathbb{P}(A|\mathcal{F}_t) \tag{S.5}$$

almost surely where $\mathcal{F}_{t+} = \bigcap_{s>t}\mathcal{F}_s$. Using the fact that the above cylinder set together with null sets of $\mathbb{P}$ generate members of the filtration $\mathbb{F}$ we deduce that (S.5) holds for any $A \in \mathcal{F}_{t+}$. As the left-hand side of (S.5) in that case is also equal to $\mathbf{1}_A$ almost surely and $\mathbb{F}$ is completed by the null sets of $\mathbb{P}$ it follows that $\mathcal{F}_{t+} \subseteq \mathcal{F}_t$. The inclusion $\mathcal{F}_t \subseteq \mathcal{F}_{t+}$ is obvious and hence the filtration $\mathbb{F}$ is right continuous.

3.2 It suffices to prove the result for the first process as the second process is the same as the first process when one replaces X by $-X$. To this end, define for each $y \geq 0$, $Y_t^y = (y \vee \overline{X}_t) - X_t$ and let $\widetilde{X}_u = X_{t+u} - X_t$ for any $u \geq 0$. Note that for $t, s \geq 0$,

$$\begin{aligned}(y \vee \overline{X}_{t+s}) - X_{t+s} &= \left(y \vee \overline{X}_t \vee \sup_{u\in[t,t+s]} X_u\right) - X_t - \widetilde{X}_s \\ &= \left[(y \vee \overline{X}_t - X_t) \vee \left(\sup_{u\in[t,t+s]} X_u - X_t\right)\right] - \widetilde{X}_s \\ &= \left[Y_t^y \vee \sup_{u\in[0,s]} \widetilde{X}_u\right] - \widetilde{X}_s.\end{aligned}$$

From the right-hand side above, it is clear that the law of Y_{t+s}^y depends only on Y_t^y and $\{\widetilde{X}_u : u \in [0,s]\}$, the latter being independent of $\mathcal{F}_t$. Hence $\{Y_t^y : t \geq 0\}$ is a Markov process.

To upgrade to a strong Markov process, one uses a standard technique that takes account of the right continuity of the paths of $\{Y_t^y : t \geq 0\}$. This involves approximating a given stopping time τ by a discretised version $\tau^{(n)} \geq \tau$ defined in (3.2) and then proving with the help of the Markov property that for each $H \in \mathcal{F}_\tau$ and $0 \leq t_1 \leq t_2 \leq \cdots \leq t_n < \infty$,

$$\mathbb{P}\left(H \cap \bigcap_{i=1}^n \{Y_{\tau^{(n)}+t_i}^y \in A_i\}\right) = \int_{[0,\infty)} \mathbb{P}\left(\bigcap_{i=1}^n \{Y_{t_i}^z \in A_i\}\right) p(y, H, \mathrm{d}z),$$

where $p(y, H, \mathrm{d}z) = \mathbb{P}(H \cap \{Y^y_{\tau^{(n)}} \in \mathrm{d}z\})$. Then since $\tau^{(n)} \downarrow \tau$ as $n \uparrow \infty$ and Y^y is right continuous, by taking limits as $n \uparrow \infty$ the same equality above is valid when $\tau^{(n)}$ is replaced by τ thus exhibiting the Strong Markov Property.

3.3 The solution to this exercise is taken from Sect. 25 of Sato (1999).

(i) To prove that $g(x) \le a_g \mathrm{e}^{b_g|x|}$ for $a_g, b_g > 0$ use submultiplicativity to show that for integer n chosen so that $|x| \in (n-1, n]$,

$$g(x) = g\left(\sum_{i=1}^{n} x/n\right) \le c^{n-1} g(x/n)^n.$$

Since g is bounded on compacts, we may define $a_g = \max\{c^{-1}, \sup_{|x|\in[0,1]} g(x)\} \in (0,\infty)$ and further set $b_g = a_g c$. We now have from the previous inequality that

$$g(x) \le a_g (c a_g)^{n-1} \le a_g (c a_g)^{|x|} = a_g \mathrm{e}^{b_g |x|}.$$

Now suppose that $\mathbb{E}(g(X_t)) < \infty$ for all $t > 0$. Decomposing as in the proof of Theorem 3.6, submultiplicativity and independence in the Lévy–Itô decomposition imply that

$$\mathbb{E}(g(X_t)) = \int_{\mathbb{R}} \int_{\mathbb{R}} g(x+y) \mathrm{d}F_2(y) \mathrm{d}F_{1,3}(x),$$

where F_2 is the distribution of $X^{(2)}_t$ and $F_{1,3}$ is the distribution of $X^{(1)}_t + X^{(3)}_t$. The last equality implies that for at least one $x \in \mathbb{R}$, $\int_{\mathbb{R}} g(x+y)\mathrm{d}F_2(y) < \infty$. Next note from the submulitplicative property that $g(y) \le c g(-x) g(x+y) \le c a_g \mathrm{e}^{b_g|x|} g(x+y)$ and hence $\mathbb{E}(g(X^{(2)}_t)) \le c a_g \mathrm{e}^{b_g|x|} \int_{\mathbb{R}} g(x+y)\mathrm{d}F_2(y) < \infty$. Recalling that $X^{(2)}$ is a compound Poisson process, a similar calculation to the one in (3.7) shows that $\int_{|y|\ge 1} g(y)\Pi(\mathrm{d}y) < \infty$.

Now suppose that $\int_{|y|\ge 1} g(y)\Pi(\mathrm{d}y) < \infty$. We can write

$$\mathbb{E}(g(X_t)) \le c\mathbb{E}(g(X^{(2)}_t))\mathbb{E}(g(X^{(1)}_t + X^{(3)}_t)).$$

Note however that since $X^{(2)}$ is a compound Poisson process then one may compute using submultiplicativity

$$\begin{aligned}\mathbb{E}(g(X^{(2)}_t)) &= \sum_{k\ge 0} \mathrm{e}^{-\lambda t} \frac{(\lambda t)^k}{k!} \mathbb{E}\left(g\left(\sum_{i=1}^{k} \xi_i\right)\right) \\ &\le \sum_{k\ge 0} \mathrm{e}^{-\lambda t} \frac{c^{k-1} t^k}{k!} \left(\int_{|x|>1} g(x)\Pi(\mathrm{d}x)\right)^k < \infty,\end{aligned}$$

where $\{\xi_i : i = 1, 2, ...\}$ are independent with common distribution given by $\Pi(\mathbb{R}\backslash(-1,1))^{-1}\Pi|_{\mathbb{R}\backslash(-1,1)}$. Note also that

$$\mathbb{E}(g(X^{(1)}_t + X^{(3)}_t)) \le a_g \mathbb{E}\left(\mathrm{e}^{b_g|X^{(1)}_t + X^{(3)}_t|}\right)$$

and the right-hand side can be shown to be finite in a similar way to (3.8) in the proof of Theorem 3.6. In conclusion it follows that $\mathbb{E}(g(X_t)) < \infty$.

(ii) Suppose that $h(x)$ is a positive increasing function on $\mathbb{R}$ such that for $x \le b$ it is constant and for $x > b$, $\log h(x)$ is concave. Then writing $f(x) = \log h(x)$ it follows for $u, v \ge b$ that

$$f(u+b) - f(u) \le f(2b) - f(b)$$
$$f(u+v) - f(v) \le f(u+b) - f(b)$$

and hence

$$f(u+v) \le f(2b) - 2f(b) + f(u) + f(v).$$

This implies that h is submultiplicative. As h is an increasing function, it also follows that

$$h(|x+y|) \le h(|x| + |y|) \le ch(|x|)h(|y|),$$

where $c > 0$ is a constant.

Note that the function $x \vee 1$ fits the description of h and hence, together with $|x| \vee 1$ is submultiplicative. The discussion preceding Theorem 3.8 shows then that $(x^\alpha \vee 1)$ and $|x|^\alpha \vee 1$ are also submultiplicative. One may handle the remaining functions similarly.

(iii) Apply the conclusion of Theorem 3.8 to the Lévy measure of a stable process.

3.4 First note that any Lévy process, X, with no Gaussian component may always be written up to a linear drift as the difference to two spectrally positive processes. Indeed if X has characteristic exponent given in Theorem 1.6 then one may always write $X = X^{\mathrm{up}} - X^{\mathrm{down}}$, where X^{up} and X^{down} have characteristic exponents

$$\Psi^{\mathrm{up}}(\theta) = \mathrm{i}a\theta + \int_{(0,\infty)} (1 - \mathrm{e}^{\mathrm{i}\theta x} + \mathrm{i}\theta x \mathbf{1}_{(x<1)})\Pi(\mathrm{d}x)$$

and

$$\Psi^{\mathrm{down}}(\theta) = \int_{(0,\infty)} (1 - \mathrm{e}^{\mathrm{i}\theta x} + \mathrm{i}\theta x \mathbf{1}_{(x<1)})\nu(\mathrm{d}x),$$

respectively, where Π is the Lévy measure of X and $\nu(x,\infty) = \Pi(-\infty, x)$. Then (i) is easily verified by showing that the total mass of the Lévy measure is finite, (ii) and (iii) are a matter of definition and (iv) follows from Theorem 3.9.

3.5 Note that for $\beta > 0$,

$$\psi(\beta) = -a\beta + \frac{1}{2}\sigma^2\beta^2 + \int_{x<-1} (\mathrm{e}^{\beta x} - 1)\Pi(\mathrm{d}x) + \int_{0>x>-1} (\mathrm{e}^{\beta x} - 1 - \beta x)\Pi(\mathrm{d}x)$$

and since for all $x < -1$, $|\mathrm{e}^{\beta x} - 1| < 1$ and for $0 > x > -1$ we have that

$$|\mathrm{e}^{\beta x} - 1 - \beta x| \le \sum_{k\ge 2} \frac{\beta^k}{k!}|x|^k \le \mathrm{e}^{\beta} x^2$$

then recalling the restriction $\int_{(-\infty,0)} (1 \wedge x^2)\Pi(\mathrm{d}x) < \infty$ one may appeal to a standard argument of dominated convergence to differentiate under the integral and obtain

$$\psi'(\beta) = -a + \sigma^2\beta + \int_{(-\infty,0)} x(\mathrm{e}^{\beta x} - \mathbf{1}_{(x>-1)})\Pi(\mathrm{d}x).$$

Using similar arguments one may use dominated convergence and the integrability condition on Π to deduce that for $n \geq 2$,

$$\frac{\mathrm{d}^n\psi}{\mathrm{d}\beta^n}(\beta) = \sigma^2\mathbf{1}_{(n=2)} + \int_{(-\infty,0)} x^n \mathrm{e}^{\beta x}\Pi(\mathrm{d}x).$$

Note in particular that $\psi''(\beta) > 0$ showing strict convexity. The fact that $\psi(0) = 0$ is trivial from the definition of ψ and since $\mathrm{e}^{\psi(\beta)} \geq \mathbb{E}(\mathrm{e}^{\beta X_1}\mathbf{1}_{(X_1>0)})$ it follows that $\psi(\infty) = \infty$.

3.6 (i) Recall the notation $\mathbb{P}_x(\cdot) = \mathbb{P}(\cdot|X_0 = x)$ for any $x \in \mathbb{R}$. Now using spectral negativity, stationary independent increments of the subordinator $\{\tau_x^+ : x \geq 0\}$ and the lack of memory property we have

$$\begin{aligned}
\mathbb{P}(\overline{X}_{\mathbf{e}_q} > x + y) &= \mathbb{P}(\tau_{x+y}^+ < \mathbf{e}_q) \\
&= \mathbb{E}(\mathbf{1}_{(\tau_x^+ < \mathbf{e}_q)}\mathbb{P}_{X_{\tau_x^+}}(\tau_{x+y}^+ < \mathbf{e}_q)) \\
&= \mathbb{E}(\mathbf{1}_{(\tau_x^+ < \mathbf{e}_q)}\mathbb{P}_x(\tau_{x+y}^+ < \mathbf{e}_q)) \\
&= \mathbb{P}(\tau_x^+ < \mathbf{e}_q)\mathbb{P}(\tau_y^+ < \mathbf{e}_q) \\
&= \mathbb{P}(\overline{X}_{\mathbf{e}_q} > x)\mathbb{P}(\overline{X}_{\mathbf{e}_q} > y).
\end{aligned}$$

(ii) Write $f(x) = \mathbb{P}(\overline{X}_{\mathbf{e}_q} > x)$ for all $x \geq 0$. Note that as the tail of a distribution function, f is right continuous. Further, it fulfils the relation $f(x + y) = f(x)f(y)$ for $x, y \geq 0$. The last relation iterates so that for integer $n > 0$, $f(nx) = f(x)^n$. In the usual way one shows that this implies that for positive integers p, q we have that $f(p/q) = f(1/q)^p$ and $f(1) = f(1/q)^q$ and hence for rational $x > 0$, $f(x) = f(1)^x = \exp\{-\theta x\}$, where $\theta = -\log f(1) \geq 0$. Right continuity of f implies that f is identically equal to the aforementioned exponential function on $[0, \infty)$. In conclusion, $\overline{X}_{\mathbf{e}_q}$ is exponentially distributed and in particular, by choosing q sufficiently large, for any $\beta \geq 0$,

$$\mathbb{E}(\mathrm{e}^{\beta \overline{X}_{\mathbf{e}_q}}) = \int_0^\infty q\mathrm{e}^{-qt}\mathbb{E}(\mathrm{e}^{\beta \overline{X}_t})\mathrm{d}t < \infty \tag{S.6}$$

showing in particular that $\mathbb{E}(\mathrm{e}^{\beta X_t}) \leq \mathbb{E}(\mathrm{e}^{\beta \overline{X}_t}) < \infty$. Note the last inequality follows on account of (S.6) and the continuity of $\{\beta \overline{X}_t : t \geq 0\}$.

(iii) From Theorem 3.12 we also have that

$$\begin{aligned}
\mathbb{P}(\overline{X}_{\mathbf{e}_q} > x) &= \mathbb{P}(\tau_x^+ < \mathbf{e}_q) \\
&= \mathbb{E}\left(\int_0^\infty \mathbf{1}_{(t>\tau_x^+)}q\mathrm{e}^{-qt}\mathrm{d}t\right) \\
&= \mathbb{E}(\mathrm{e}^{-q\tau_x^+}) = \mathrm{e}^{-\Phi(q)x}.
\end{aligned}$$

Hence we deduce that $\overline{X}_{\mathbf{e}_q}$ is exponentially distributed with parameter $\Phi(q)$. Letting $q \downarrow 0$ we know that $\overline{X}_{\mathbf{e}_q}$ converges in distribution to $\overline{X}_\infty$ and hence the latter is infinite if and only if $\Phi(0) = 0$. If $\Phi(0) > 0$ then it is clear that $\overline{X}_\infty$ is exponentially distributed with parameter $\Phi(0)$. From Exercise 3.5 we know that ψ is convex and hence there are at most two non-negative roots of the equation $\psi(\beta) = 0$. There is a unique root at zero if and only if $\psi'(0) = \mathbb{E}(X_1) \geq 0$ and otherwise the largest root is strictly positive.

3.7 We prove (i) and (ii) together. Consider the characteristic exponent developed for $\theta \in \mathbb{R}$ as follows,

$$\begin{aligned}\Psi(\theta) &= c\theta^{\alpha}\left(1 - \mathrm{i}\tan\frac{\pi\alpha}{2}\right) \\ &= c(\cos(\pi\alpha/2))^{-1}\theta^{\alpha}(\cos(\pi\alpha/2) - \mathrm{i}\sin(\pi\alpha/2)) \\ &= c(\cos(\pi\alpha/2))^{-1}(\theta/\mathrm{i})^{\alpha} \\ &= c(\cos(\pi\alpha/2))^{-1}(-\theta\mathrm{i})^{\alpha}.\end{aligned}$$

According to Theorem 3.6 $\mathbb{E}(\mathrm{e}^{-\gamma X_1}) < \infty$ (this is obvious in the case that $\alpha \in (0,1)$ as then $\mathbb{P}(X_1 \geq 0) = 1$ since X is a subordinator) and hence one may analytically extend $\Psi(\theta)$ to the complex plane where $\Im\theta \geq 0$. By taking $\theta = \mathrm{i}\gamma$ we have from above the required result.

Chapter 4

4.1 (i) Recall that necessarily X must be the difference of a linear drift with rate, say, $d > 0$ and a pure jump subordinator. It follows from Lemma 4.11 that $\mathbb{P}(X_t > 0$ for all sufficiently small $t > 0) = 1$ and hence $\mathbb{P}(\tau^{\{0\}} > 0) = 1$ where $\tau^{\{0\}} = \inf\{t > 0 : X_t = 0\}$. The latter is a stopping time and hence applying the Strong Markov Property at this time it follows from the strict positivity of $\tau^{\{0\}}$ that there can be at most a finite number of visits to the origin in any finite time horizon with probability one. The claim concerning $\{L_t^0 : t \geq 0\}$ is now obvious.

(ii) It requires a straightforward review of the proof of change of variable formula to see that the time horizon $[0,t]$ may be replaced by any finite time horizon of random length. Hence if T_n is the time of the n-th visit of X to 0 for $n = 0, 1, 2, ...$ with $T_0 = 0$, we easily deduce that on $\{t \geq T_{n-1}\}$, taking account of the possible discontinuity in f at 0,

$$\begin{aligned}f(X_{t\wedge T_n}) &= f(X_{T_{n-1}}) + d\int_{T_{n-1}}^{t\wedge T_n} f'(X_s)\mathrm{d}s \\ &\quad + \int_{(T_{n-1},t\wedge T_n)}\int_{(-\infty,0)} (f(X_{s-}+x) - f(X_{s-}))N(\mathrm{d}s\times\mathrm{d}x) \\ &\quad + (f(X_{T_n}) - f(X_{T_n-}))\mathbf{1}_{(t>T_n)}.\end{aligned}$$

The claim follows by supposing that $T_{n-1} \leq t < T_n$ and then using the above identity and the simple telescopic sum

$$f(X_t) = f(X_0) + \sum_{i=1}^{n-1}\{f(X_{T_i}) - f(X_{T_{i-1}})\} + \{f(X_t) - f(X_{T_{n-1}})\}$$

and noting that

$$\int_0^t (f(X_s) - f(X_{s-}))\mathrm{d}L_t^0 = \sum_{i=1}^{n-1}(f(X_{T_i}) - f(X_{T_i-})).$$

4.2 Construct the process $X^{(\varepsilon)}$ and define N as in the proof of Theorem 4.2 and consider the telescopic sum

$$f(t, X_t^\varepsilon) = f(0, X_0^\varepsilon) + \sum_{i=1}^{N}(f(\overline{X}_{T_i}, X_{T_i}^\varepsilon) - f(\overline{X}_{T_{i-1}}, X_{T_{i-1}}^\varepsilon)) \\ +(f(\overline{X}_t, X_t^\varepsilon) - f(\overline{X}_{T_N}, X_{T_N}^\varepsilon)).$$

One may now proceed as in the proof of Theorem 4.2 to apply regular Lebesgue–Steiltjes calculus between the jumps of the process $X^{(\varepsilon)}$ (note in particular that $\{\overline{X}_t : t \geq 0\}$ has non-decreasing paths and is continuous) taking care to add in the increments $f(\overline{X}_{T_i}, X_{T_i-}^\varepsilon + \xi_i) - f(\overline{X}_{T_i}, X_{T_i-}^\varepsilon)$, where $\{\xi_i : i = 1, 2, ..\}$ are the successive jumps of $X^{(\varepsilon)}$ before taking the limit as $\varepsilon \downarrow 0$.

4.3 (i) It is a straightforward exercise to prove that $y \mapsto \int_{\mathbb{R}}(f(x+y) - f(y))\Pi(dx)$ is a continuous function. It follows that $T_n = \inf\{t \geq 0 : X_t \in B_n\}$, where B_n is an open set and hence by Theorem 3.3 it is also true that T_n is a stopping time.

(ii) This is an application of Corollary 4.6 when one takes $\phi(s, x) = \mathrm{e}^{-\lambda s}(f(X_{s-} + x) - f(X_{s-}))\mathbf{1}_{(t \leq T_n)}$ (note in particular one uses the fact that $\int_{\mathbb{R}} |\phi(s, x)|\Pi(\mathrm{d}x) \leq n$).

(iii) The Change of Variable Formula allows us to write on $\{t < T_n\}$,

$$\mathrm{e}^{-\lambda t} f(X_t) = \int_0^t \mathrm{e}^{-\lambda s}\left\{d\frac{\partial f}{\partial y}(X_s) + \int_{\mathbb{R}}(f(X_s + x) - f(X_s))\Pi(\mathrm{d}x) - \lambda f(X_s)\right\}\mathrm{d}s + M_t.$$

It is known that a non-zero Lebesgue integral cannot be a martingale (cf. Exercise 3.16 of Lamberton and Lapeyre (1996)) and hence, since $T_n \uparrow \infty$ almost surely, we are forced to deduce that

$$d\frac{\partial f}{\partial y}(X_s) + \int_{\mathbb{R}}(f(X_s + x) - f(X_s))\Pi(\mathrm{d}x) = \lambda f(X_s)$$

for Lebesgue almost every $0 \leq s \leq t$ with probability one. As the support of the distribution of X_s is $\mathbb{R}$ for all $s > 0$, we may conclude with the help of continuity of

$$y \mapsto d\frac{\partial f}{\partial y}(y) + \int_{\mathbb{R}}(f(y + x) - f(y))\Pi(\mathrm{d}x) - \lambda f(y)$$

that the latter is identically equal to zero on $\mathbb{R}$.

(iv) This is clear from the conclusion of the previous part.

4.4 Starting with an approximation to ϕ via (4.8) one may establish this identity in a similar way to the proof of Theorem 4.4, although the calculations are somewhat more technically voluminous.

4.5 (i) The Lévy–Itô decomposition tells us that we may write for each $t \geq 0$,

$$X_t^{(\varepsilon)} = -at + \int_{[0,t]}\int_{|x|\geq 1} xN(\mathrm{d}s \times \mathrm{d}x) + \int_{[0,t]}\int_{\varepsilon<|x|<1} xN(\mathrm{d}s \times \mathrm{d}x) - t\int_{\varepsilon<|x|<1} x\Pi(\mathrm{d}x),$$

where N is the Poisson random measure associated with the jumps of X. The given expression follows by the change of variable formula which now says

$$
\begin{aligned}
f(t, X_t^{(\varepsilon)}) &= f(0, X_0^{(\varepsilon)}) + \int_0^t \frac{\partial f}{\partial s}(s, X_s^{(\varepsilon)})\mathrm{d}s \\
&\quad + \left(-a - \int_{\varepsilon<|x|<1} x\Pi(\mathrm{d}x)\right)\int_0^t \frac{\partial f}{\partial x}(s, X_s^{(\varepsilon)})\mathrm{d}s \\
&\quad + \int_{[0,t]}\int_{|x|>\varepsilon} (f(s, X_{s-}^{(\varepsilon)} + x) - f(s, X_{s-}^{(\varepsilon)}) - x\frac{\partial f}{\partial x}(s, X_{s-}^{(\varepsilon)}))N(\mathrm{d}s \times \mathrm{d}x) \\
&\quad + \int_{[0,t]}\int_{|x|>\varepsilon} x\frac{\partial f}{\partial x}(s, X_{s-}^{(\varepsilon)})N(\mathrm{d}s \times \mathrm{d}x).
\end{aligned}
$$

Hence we may reasonably interpret

$$
\int_{[0,t]}\int_{|x|\geq 1} x\frac{\partial f}{\partial x}(s, X_{s-}^{(\varepsilon)})N(\mathrm{d}s \times \mathrm{d}x) - a\int_0^t \frac{\partial f}{\partial x}(s, X_s^{(\varepsilon)})\mathrm{d}s = \int_0^t \frac{\partial f}{\partial x}(s, X_{s-}^{(\varepsilon)})\mathrm{d}X_s^{(2)}
$$

as the process $X^{(2)}$ has a discrete set of jump times. Further, we identify

$$
M_t^{(\varepsilon)} = \int_{[0,t]}\int_{1>|x|>\varepsilon} x\frac{\partial f}{\partial x}(s, X_{s-}^{(\varepsilon)})N(\mathrm{d}s\times\mathrm{d}x) - \left(\int_{\varepsilon<|x|<1} x\Pi(\mathrm{d}x)\right)\int_0^t \frac{\partial f}{\partial x}(s, X_s^{(\varepsilon)})\mathrm{d}s
$$

which is a martingale with respect to $\{\mathcal{F}_t : t \geq 0\}$ by Corollary 4.6. Its paths are clearly right continuous by definition (note that there are a discrete set of jump times for $M^{(\varepsilon)}$) and further, by Exercise 4.4 it is also a square integrable martingale. (Note, it has implicitly been used so far that the first derivative of f in x is uniformly bounded). Note that on account of there being a discrete set of jumps in $M^{(\varepsilon)}$ we may also write it

$$
M_t^{(\varepsilon)} = \int_0^t \frac{\partial f}{\partial x}(s, X_{s-}^{(\varepsilon)})\mathrm{d}X_s^{(3,\varepsilon)},
$$

where $X^{(3,\varepsilon)}$ is a Lévy process with characteristic exponent

$$
\Psi^{(3,\varepsilon)}(\theta) = \int_{\varepsilon<|x|<1} (1 - \mathrm{e}^{\mathrm{i}\theta x} + \mathrm{i}\theta x)\Pi(\mathrm{d}x).
$$

(ii) Using the notation of Chap. 2, for fixed $T > 0$, considering $\{M_t^{(\varepsilon)} : t \in [0, T]\}$ as an element of $\mathcal{M}_T$, we may appeal again to Exercise 4.4 to deduce that for any $0 < \varepsilon < \eta < 1$,

$$
\begin{aligned}
&||M^{(\varepsilon)} - M^{(\eta)}|| = \\
&\quad \mathbb{E}\left(\int_{[0,T]}\int_{|x|<1} x^2\left\{\frac{\partial f}{\partial x}(s, X_s^{(\varepsilon)})\mathbf{1}_{(|x|>\varepsilon)} - \frac{\partial f}{\partial x}(s, X_s^{(\eta)})\mathbf{1}_{(|x|>\eta)}\right\}^2 \mathrm{d}s\Pi(\mathrm{d}x)\right)
\end{aligned}
$$

which tends to zero as $\varepsilon, \eta \downarrow 0$ on account of the boundedness of the first derivative of f in x and the necessary condition for X to be a Lévy process, $\int_{(-1,1)} x^2\Pi(\mathrm{d}x) < \infty$.

(iii) We know from the Lévy–Itô decomposition that $X^{(\varepsilon)}$ converges uniformly on $[0, T]$ with probability one along some deterministic subsequence, say $\{\varepsilon_n : n = 1, 2, ...\}$, to X. Similar reasoning also shows that thanks to the result in part (ii) there exists a subsubsequence, say $\epsilon = \{\epsilon_n : n = 1, 2, ...\}$, of the latter subsequence along which $M^{(\varepsilon)}$ converges uniformly along $[0, T]$ with probability one to its limit, say M. Hence from part (i) we may say that uniformly on $[0, T]$,

$$
\int_0^{\cdot} \frac{\partial f}{\partial x}(s, X_{s-}^{(\varepsilon)})\mathrm{d}X_s^{(2)} + \int_0^{\cdot} \frac{\partial f}{\partial x}(s, X_{s-}^{(\varepsilon)})\mathrm{d}X_s^{(3,\varepsilon)}
$$

has an almost sure limit as $\varepsilon \downarrow 0$ which is equal to

$$\int_0^{\cdot} \frac{\partial f}{\partial x}(s, X_{s-})\mathrm{d}X_s^{(2)} + M$$

and since X is the sum of $X^{(2)}$ and the limit of $X^{(3,\varepsilon)}$ we may reasonably denote the above limit by

$$\int_0^{\cdot} \frac{\partial f}{\partial x}(s, X_{s-})\mathrm{d}X_s.$$

For the other terms in (4.20), note that continuity of f ensures that $f(t, X_t^{(\varepsilon)}) \to f(t, X_t)$ as $\varepsilon \downarrow 0$ (along ϵ). Further, it can be shown using the assumptions of boundedness that for all $0 \leq s \leq t$ and $y \in \mathbb{R}$,

$$\left| f(s, y + x) - f(s, y) - x\frac{\partial f}{\partial x}(s, y) \right| \leq Cx^2$$

for some constant $C > 0$. Hence by almost sure dominated convergence it follows that the limit of

$$\int_{[0,t]} \int_{|x|>\varepsilon} (f(s, X_{s-}^{(\varepsilon)} + x) - f(s, X_{s-}^{(\varepsilon)}) - x\frac{\partial f}{\partial x}(s, X_{s-}^{(\varepsilon)}))N(\mathrm{d}s \times \mathrm{d}x)$$

is equal to

$$\int_{[0,t]} \int_{\mathbb{R}} (f(s, X_{s-} + x) - f(s, X_{s-}) - x\frac{\partial f}{\partial x}(s, X_{s-}))N(\mathrm{d}s \times \mathrm{d}x)$$

as $\varepsilon \downarrow 0$. Note in particular that we use Theorem 2.7 and the integrability condition $\int_{\mathbb{R}}(1 \wedge x^2)\Pi(\mathrm{d}x) < \infty$.

(iv) See the technique used in Exercise 4.3.

4.6 (i) The queue remains empty for a period of time that is exponentially distributed. Starting with the initial workload from the first arriving customer, which is independent of the subsequent evolution of the workload and has distribution F, the period B is equal to the time it takes for the workload to become zero again. The latter has the same distribution as $\tau_x^+ = \inf\{t > 0 : X_t > x\}$ when x is independently randomised using the distribution F. Note that X denotes the underlying spectrally positive Lévy process which drives W and whose Laplace exponent is given by

$$\log \mathbb{E}(\mathrm{e}^{\theta X_1}) = \psi(\theta) = \theta - \int_{(0,\infty)} (1 - \mathrm{e}^{-\theta x})\lambda F(\mathrm{d}x)$$

for $\theta \geq 0$. With the help of Theorem 3.12 we can thus deduce that for $\beta \geq 0$,

$$\mathbb{E}(\mathrm{e}^{-\beta B}) = \int_{(0,\infty)} \mathbb{E}(\mathrm{e}^{-\beta\tau_x^+})F(\mathrm{d}x) = \int_{(0,\infty)} \mathrm{e}^{-\Phi(\beta)x}F(\mathrm{d}x) = \hat{F}(\Phi(\beta)),$$

where $\Phi(\beta)$ is the largest root of the equation $\psi(\theta) = \beta$.

(ii) The case that $\rho = \lambda\int_{(0,\infty)} xF(\mathrm{d}x) > 1$ corresponds to the case that $\mathbb{E}(X_1) < 0$ which implies also that $\Phi(0) > 0$, in other words, following similar arguments to those given in Sect. 1.3.1 (involving the Strong Law of Large Numbers), $\lim_{t\uparrow\infty} X_t = -\infty$. In turn this implies that the last moment at which X is at its supremum is finite almost surely. The process X undergoes a finite number of excursions from the maximum before entering an excursion which never returns to the previous maximum. Note that by the Strong Markov Property, excursions from the maximum are independent and between them the process X spends independent and exponentially distributed lengths of time drifting upwards during which time, $X = \overline{X}$. If we set $p = \mathbb{P}(B = \infty)$ then the above description of excursions implies that the number of finite excursions is distributed according to a geometric random variable with parameter p.

Note now that $p = 1 - \lim_{\beta \downarrow 0} \mathbb{E}(\mathrm{e}^{-\beta B}) = 1 - \hat{F}(\Phi(0))$. According to the above discussion

$$\int_0^\infty \mathbf{1}_{(W_t=0)}\mathrm{d}t = \sum_{k=1}^{\Gamma_p+1} \mathbf{e}_\lambda^{(k)},$$

where Γ_p is a geometric random variable with parameter p and $\{\mathbf{e}_\lambda^{(k)} : k = 1, 2, ...\}$ is a sequence of independent random variables (also independent of Γ_p) each of which is exponentially distributed with parameter λp. One may compute the moment generating function of the random variable on the right-hand side above to find that it has the same law as an exponentially distributed random variable with parameter λp. Note however that by definition we have that $0 = \Phi(0) - \int_{(0,\infty)}(1-\mathrm{e}^{-\Phi(0)x})\lambda F(\mathrm{d}x) = \Phi(0) - \lambda p$. Hence as required $\int_0^\infty \mathbf{1}_{(W_t=0)}\mathrm{d}t$ is exponentially distributed with parameter $\Phi(0)$.

(iii) From its definition we have $\rho = 1 - \psi'(0) = \lambda/\mu$. Recalling that the convolution of an exponential distribution is a gamma distribution, we have from Theorem 4.10, the stationary distribution is given by

$$\begin{aligned}
&\left(1 - \frac{\lambda}{\mu}\right)\left(\delta_0(\mathrm{d}x) + \mathbf{1}_{(x>0)}\sum_{k=1}^\infty \lambda^k x^{k-1}\frac{\mathrm{e}^{-\mu x}}{(k-1)!}\mathrm{d}x\right) \\
&= \left(1 - \frac{\lambda}{\mu}\right)\left(\delta_0(\mathrm{d}x) + \mathbf{1}_{(x>0)}\lambda \mathrm{e}^{-\mu x + \lambda x}\mathrm{d}x\right).
\end{aligned}$$

4.7 (i) Recall that $\mathcal{E}_t(\alpha) = \exp\{\alpha X_t - \psi(\alpha)t\}$ and note that

$$\begin{aligned}
\mathrm{d}\mathcal{E}_t(\alpha) &= \mathcal{E}_{t-}(\alpha)\big(\alpha\,\mathrm{d}X_t - \psi(\alpha)\,\mathrm{d}t\big) + \frac{1}{2}\alpha^2\mathcal{E}_{t-}(\alpha)\,\mathrm{d}[X,X]_t^c \\
&\quad + \{\triangle\mathcal{E}_t(\alpha) - \alpha\mathcal{E}_{t-}(\alpha)\triangle X_t\}.
\end{aligned}$$

Note also that

$$\begin{aligned}
\mathrm{d}M_t &= \psi(\alpha)\mathrm{e}^{-\alpha Z_{t-}}\,\mathrm{d}t + \alpha\mathrm{e}^{-\alpha Z_{t-}}\,\mathrm{d}Z_t - \frac{1}{2}\alpha^2\mathrm{e}^{-\alpha Z_{t-}}\,\mathrm{d}[X,X]_t^c \\
&\quad - \{\triangle\mathrm{e}^{-\alpha Z_t} + \alpha\triangle Z_t\} - \alpha\,\mathrm{d}\overline{X}_t \\
&= \mathrm{e}^{-\alpha\overline{X}_t + \psi(\alpha)t}\Big[\psi(\alpha)\mathcal{E}_{t-}(\alpha)\,\mathrm{d}t + \alpha\mathcal{E}_{t-}(\alpha)\big(\mathrm{d}\overline{X}_t - \mathrm{d}X_t\big) \\
&\qquad - \frac{1}{2}\alpha^2\mathcal{E}_{t-}(\alpha)\,\mathrm{d}[X,X]_t^c \\
&\qquad - \mathcal{E}_{t-}(\alpha)\{\mathrm{e}^{\alpha\triangle X_t} - 1 - \alpha\triangle X_t\} \\
&\qquad - \alpha\mathrm{e}^{\alpha\overline{X}_t - \psi(\alpha)t}\,\mathrm{d}\overline{X}_t\Big] \\
&= \mathrm{e}^{-\alpha\overline{X}_t + \psi(\alpha)t}\Big\{-\mathrm{d}\mathcal{E}_t(\alpha) + \alpha\big(\mathcal{E}_t(\alpha) - \mathrm{e}^{\alpha\overline{X}_t - \psi(\alpha)t}\big)\,\mathrm{d}\overline{X}_t\Big\},
\end{aligned}$$

where $Z_t = \overline{X}_t = X_t$ and we have used that $\overline{X}_{t-} = \overline{X}_t$. Since $\overline{X}_t = X_t$ if and only if $\overline{X}_t$ increases, we may write

$$\begin{aligned} \mathrm{d}M_t &= \mathrm{e}^{-\alpha \overline{X}_t + \psi(\alpha)t}\Big\{-\mathrm{d}\mathcal{E}_t(\alpha) + \alpha\big(\mathcal{E}_t(\alpha) - \mathrm{e}^{\alpha \overline{X}_t - \psi(\alpha)t}\big)\mathbf{1}_{(\overline{X}_t = X_t)}\,\mathrm{d}\overline{X}_t\Big\} \\ &= -\mathrm{e}^{-\alpha \overline{X}_t + \psi(\alpha)t}\,\mathrm{d}\mathcal{E}_t(\alpha) \end{aligned}$$

showing that M_t is a local martingale since $\mathcal{E}_t(\alpha)$ is a martingale.

(ii) For $q > 0$, we know from Exercise 3.6 (iii) that $\overline{X}_{\mathbf{e}_q}$ is exponentially distributed with parameter $\Phi(q)$. Hence

$$E\big(\overline{X}_{\mathbf{e}_q}\big) = \int_0^\infty q\mathrm{e}^{-qt}\, E\big(\overline{X}_t\big)\,\mathrm{d}t = \frac{1}{\Phi(q)} < \infty$$

and hence, since $\overline{X}_t$ is a continuous increasing process, we have $E\big(\overline{X}_t\big) < \infty$ for all t.

(iii) Now note by the positivity of the process Z and again since $\overline{X}$ increases,

$$E\Big(\sup_{s\le t}|M_s|\Big) \le \psi(\alpha)t + 2 + \alpha\, E\big(\overline{X}_t\big) < \infty$$

for each finite $t > 0$.

4.8 (i) The Change of Variable Formula has been given for real valued functions. Nonetheless, one may apply the change of variable formula to both real and imaginary parts. Similarly one may derive a version of the compensation formula (4.8) for complex valued functions.

The proof that the given process M is a martingale follows a similar line to the proof of Theorem 4.7 by applying the appropriate version of the change of variable formula to $\exp\{\mathrm{i}\alpha(X_t - \overline{X}_t) + \mathrm{i}\beta\overline{X}_t\}$.

(ii) Similar techniques to those used in the proof of Theorem 4.8 show that

$$\mathbb{E}\left(\int_0^{\mathbf{e}_q} \mathrm{e}^{\mathrm{i}\alpha(X_s - \overline{X}_s) + \mathrm{i}\beta\overline{X}_s}\mathrm{d}s\right) = \frac{1}{q}\mathbb{E}\left(\mathrm{e}^{\mathrm{i}\alpha(X_{\mathbf{e}_q} - \overline{X}_{\mathbf{e}_q}) + \mathrm{i}\beta\overline{X}_s}\right).$$

Further,

$$\begin{aligned} \mathbb{E}\left(\int_0^{\mathbf{e}_q} \mathrm{e}^{\mathrm{i}\alpha(X_s - \overline{X}_s) + \mathrm{i}\beta\overline{X}_s}\mathrm{d}\overline{X}_s\right) &= \mathbb{E}\left(\int_0^{\mathbf{e}_q} \mathrm{e}^{\mathrm{i}\beta\overline{X}_s}\mathrm{d}\overline{X}_s\right) \\ &= \mathbb{E}\left(\int_0^\infty q\mathrm{e}^{-uq}\mathrm{d}u \cdot \int_0^\infty \mathbf{1}_{(s<u)}\mathrm{e}^{\mathrm{i}\beta\overline{X}_s}\mathrm{d}\overline{X}_s\right) \\ &= \mathbb{E}\left(\int_0^\infty \mathrm{e}^{-qs}\mathrm{e}^{\mathrm{i}\beta\overline{X}_s}\mathrm{d}\overline{X}_s\right) \\ &= \mathbb{E}\left(\int_0^\infty \mathrm{e}^{-q\tau_x^+ + \mathrm{i}\beta x}\mathbf{1}_{(\tau_x^+<\infty)}\mathrm{d}x\right) \\ &= \int_0^\infty \mathrm{e}^{\mathrm{i}\beta x}\mathrm{e}^{-\Phi(q)x}\mathrm{d}x \\ &= \frac{1}{\Phi(q) - \mathrm{i}\beta}, \end{aligned}$$

where the first equality uses the fact that $X_s - \overline{X}_s = 0$ on the times that $\overline{X}$ increases, the third equality is a result of an application of Fubini's Theorem, the fourth equality is the result of a change of variable $s \mapsto \tau_x^+$, (noting in particular that due to spectral negativity, $\overline{X}_{\tau_x^+} = x$) and finally the fifth equality uses Fubini's Theorem and Theorem 3.12.

Using the equality $\mathbb{E}(M_{\mathbf{e}_q}) = 0$ which follows from the fact that M is a martingale and rearranging terms on the left-hand side of the latter equality with the help of the above two observations gives the required result.

(iii) Note that (4.21) factors as follows

$$\frac{\Phi(q)}{\Phi(q) - \mathrm{i}\beta} \times \frac{q}{\Phi(q)} \frac{\Phi(q) - \mathrm{i}\alpha}{q + \Psi(\alpha)}.$$

From Exercise 3.6 (iii) we know that $\overline{X}_{\mathbf{e}_q}$ is exponentially distributed with parameter $\Phi(q)$ and so $\mathbb{E}(\mathrm{e}^{\mathrm{i}\beta \overline{X}_{\mathbf{e}_q}})$ is equal to the first factor above. Standard theory of Fourier transforms lead us to the conclusion that $\overline{X}_{\mathbf{e}_q}$ and $(\overline{X}_{\mathbf{e}_q} - X_{\mathbf{e}_q})$ are independent.

4.9 (i) By writing X as the difference of two subordinators (which is possible as it is a process of bounded variation, see Exercise 2.8) the limit follows.

(ii) As with the case of a spectrally negative Lévy process of bounded variation, as $d > 0$, part (i) implies that, almost surely, for all sufficiently small $t > 0$ it is the case that $X_t > 0$; in which case $\mathbb{P}(\tau_0^+ > 0) = 1$.

(iii) By inspection one sees that the proof of the Pollaczek–Khintchine formula in Sect. 4.6 is still valid under the current circumstances in light of the conclusion of part (ii).

Chatper 5

5.1 The proof follows verbatim the proof of Lemma 5.5 (ii) by substituting U for V and Π for F and further, using Corollary 5.3 in place of Theorem 5.1. It should also be noted that $\mathbb{E}(\tau_x^+) = U(x)$ and hence as before $U(x+y) \leq U(x) + U(y)$. Finally note, that from the first part of the theorem we have that

$$\begin{aligned} 1 - \lim_{x \uparrow \infty} \mathbb{P}(X_{\tau_x^+} = x) &= \lim_{x \uparrow \infty} \mathbb{P}(X_{\tau_x^+} - x > 0, x - X_{\tau_x^+ -} \geq 0) \\ &= \frac{1}{\mu} \int_0^\infty \Pi(y, \infty) \mathrm{d}y. \end{aligned} \tag{S.7}$$

If $\Phi(\theta)$ is the Laplace exponent of X then from (5.1) it is straightforward to deduce that $\Phi'(0) = \mu = d + \int_0^\infty \Pi(y,\infty)\mathrm{d}y$. Hence from (S.7) we conclude that $\lim_{x \uparrow \infty} \mathbb{P}(X_{\tau_x^+} = x) = d/\mu$.

5.2 (i) From Sect. 4.6 we know that for the process Y, $\mathbb{P}(\sigma_0^+ > 0) = 1$ and hence the times of new maxima form a discrete set. Note that $\overline{Y}_{\sigma_x^+ -}$ corresponds to the undershoot of the last maximum of Y prior to first passage over x and $Y_{\sigma_x^+}$ corresponds to the next maximum thereafter. If we define an auxilliary process, say $X = \{X_t : t \geq 0\}$, which has positive and independent jumps which are equal in distribution to $Y_{\sigma_0^+}$, then according to the decomposition of the path of Y into its excursions from the maximum, the pair $(Y_{\sigma_x^+} - x, x - \overline{Y}_{\sigma_x^+ -})$ conditional on the event $\sigma_0^+ < \infty$ is equal in law to the overshoot and undershoot of X when first passing x. It suffices then to take X as a compound Poisson subordinator with the aforementioned jump distribution and thanks to Corollary 4.12 the latter is given by $(d - \psi'(0))^{-1}\nu(x,\infty)\mathrm{d}x$.

(ii) An application of Theorem 5.5 (ii) shows that provided $\mathbb{E}(Y_{\sigma_0^+}) < \infty$,

$$\begin{aligned}&\lim_{x\uparrow\infty} \mathbb{P}(Y_{\sigma_x^+} - x \in \mathrm{d}u, x - \overline{Y}_{\sigma_x^+-} \in \mathrm{d}y | \sigma_x^+ < \infty)\\ &= \frac{1}{\mathbb{E}(Y_{\sigma_0^+})}\mathbb{P}(Y_{\sigma_0^+} \in \mathrm{d}u + y)\mathrm{d}y\end{aligned}$$

which gives the required result once one notes that $\mathbb{E}(Y_{\sigma_0^+})$ is finite if and only if $\int_0^\infty x\nu(x,\infty)\mathrm{d}x < \infty$.

(iii) When modelling a risk process by $-Y$, the previous limit is the asymptotic joint distribution of the deficit at ruin and the wealth prior to ruin conditional on ruin occurring when starting from an arbitrary large capital.

5.3 Note from Theorem 5.6 that

$$\begin{aligned}&\lim_{x\uparrow\infty} \mathbb{P}(X_{\tau_x^+} - X_{\tau_x^+-} \in \mathrm{d}z)\\ &= \lim_{x\uparrow\infty} \int_{[0,z]} \mathbb{P}(X_{\tau_x^+} - x \in \mathrm{d}z - y, x - X_{\tau_x^+-}) \in \mathrm{d}y\\ &= \lim_{x\uparrow\infty} \int_{[0,z]} U(x - \mathrm{d}y)\Pi(\mathrm{d}z)\\ &= \lim_{x\uparrow\infty} (U(x) - U(x - z))\Pi(\mathrm{d}z)\\ &= \lim_{x\uparrow\infty} \frac{1}{\mu} z\Pi(\mathrm{d}z),\end{aligned}$$

where in the final equality we have appealed to Corollary 5.3 (i) (and hence used the assumption on U). The last part is an observation due to Winter (1989). We have with the help of Fubini's Theorem,

$$\begin{aligned}\mathbb{P}((1-U)Z > u, UZ > y) &= \mathbb{P}\left(Z > y + u, \frac{y}{Z} < U < 1 - \frac{z}{Z}\right)\\ &= \iint \mathbf{1}_{(z>y+u)}\mathbf{1}_{(\theta\in(y/z,1-u/z))}\mathbf{1}_{(\theta\in(0,1))}\mathrm{d}\theta \cdot \frac{1}{\mu}z\Pi(\mathrm{d}z)\\ &= \int \mathbf{1}_{(z>y+u)}\frac{z-u-y}{z}\frac{z}{\mu}\Pi(\mathrm{d}z)\\ &= \frac{1}{\mu}\int \mathbf{1}_{(z>y+u)}\left\{\int \mathbf{1}_{(y+u<s<z)}\mathrm{d}s\right\}\Pi(\mathrm{d}z)\\ &= \frac{1}{\mu}\iint \mathbf{1}_{(z>s)}\mathbf{1}_{(s>y+u)}\Pi(\mathrm{d}z)\mathrm{d}s\\ &= \frac{1}{\mu}\int_{y+u}^\infty \Pi(s,\infty)\mathrm{d}s\\ &= \mathbb{P}(V > u, W > y).\end{aligned}$$

5.4 An analogue of Theorem 4.4 is required for the case of a Poisson random measure N which is defined on $([0,\infty)\times\mathbb{R}^d, \mathcal{B}[0,\infty)\times\mathcal{B}(\mathbb{R}), dt\times\mathbf{\Pi})$, where $d\in\{1,2,...\}$ (although our attention will be restricted to the case that $d=2$). Following the proof of Theorem 4.4 one deduces that if $\phi:[0,\infty)\times\mathbb{R}^d\times\Omega\to[0,\infty)$ is a random time-space function such that:

(i) As a trivariate function $\phi=\phi(t,x)[\omega]$ is measurable.
(ii) For each $t\geq 0$ $\phi(t,x)[\omega]$ is $\mathcal{F}_t\times\mathcal{B}(\mathbb{R}^d)$-measurable.
(iii) For each $x\in\mathbb{R}^d$, with probability one, $\{\phi(t,x)[\omega]: t\geq 0\}$ is a left continuous process.

Then for all $t\geq 0$,

$$\mathbb{E}\left(\int_{[0,t]}\int_{\mathbb{R}^d}\phi(s,x)N(\mathrm{d}s\times\mathrm{d}x)\right)=\mathbb{E}\left(\int_0^t\int_{\mathbb{R}^d}\phi(s,x)\mathrm{d}s\mathbf{\Pi}(\mathrm{d}x)\right) \tag{S.8}$$

with the understanding that the right-hand side is infinite if and only if the left-hand side is.

From this the required result follows from calculations similar in nature to those found in the proof of Theorem 5.6.

5.5 (i) First suppose that $\alpha,\beta,x>0$. Use the Strong Markov property and note that

$$\begin{aligned}&\mathbb{E}\left(\mathrm{e}^{-\beta X_{\mathbf{e}_\alpha}}\mathbf{1}_{(X_{\mathbf{e}_\alpha}>x)}\right)\\&=\mathbb{E}\left(\mathrm{e}^{-\beta X_{\mathbf{e}_\alpha}}\mathbf{1}_{(\tau_x^+<\mathbf{e}_\alpha)}\right)\\&=\mathbb{E}\left(\mathbf{1}_{(\tau_x^+<\mathbf{e}_\alpha)}\mathrm{e}^{-\beta X_{\tau_x^+}}\mathbb{E}\left(\mathrm{e}^{-\beta(X_{\mathbf{e}_\alpha}-X_{\tau_x^+})}\Big|\mathcal{F}_{\tau_x^+}\right)\right).\end{aligned}$$

Now, conditionally on $\mathcal{F}_{\tau_x^+}$ and on the event $\{\tau_x^+<\mathbf{e}_\alpha\}$ the random variables $X_{\mathbf{e}_\alpha}-X_{\tau_x^+}$ and $X_{\mathbf{e}_\alpha}$ have the same distribution thanks to the lack of memory property of $\mathbf{e}_\alpha$ and the stationary and independent increments of X. Hence

$$\mathbb{E}\left(\mathrm{e}^{-\beta X_{\mathbf{e}_\alpha}}\mathbf{1}_{(X_{\mathbf{e}_\alpha}>x)}\right)=\mathbb{E}\left(\mathrm{e}^{-\alpha\tau_x^+-\beta X_{\tau_x^+}}\right)\mathbb{E}\left(\mathrm{e}^{-\beta X_{\mathbf{e}_\alpha}}\right).$$

Noting that $\mathbb{P}(X_{\mathbf{e}_\alpha}\in\mathrm{d}x)=\alpha U^{(\alpha)}(\mathrm{d}x)$ we have

$$\mathbb{E}\left(\mathrm{e}^{-\beta X_{\mathbf{e}_\alpha}}\mathbf{1}_{(X_{\mathbf{e}_\alpha}>x)}\right)=\int_{(x,\infty)}\alpha\mathrm{e}^{-\beta z}U^{(\alpha)}(\mathrm{d}z)\text{ and }\mathbb{E}\left(\mathrm{e}^{-\beta X_{\mathbf{e}_\alpha}}\right)=\frac{\alpha}{\alpha+\Phi(\beta)}$$

and hence (5.20) follows. For the case that at least one of α,β or x is zero, simply take limits on both sides of (5.20).

(ii) Since for $q>0$,

$$\int_{[0,\infty)}\mathrm{e}^{-qy}U^{(\alpha)}(\mathrm{d}y)=\frac{1}{\alpha+\Phi(q)}, \tag{S.9}$$

we have by taking Laplace transforms in (5.20) and applying Fubini's Theorem that when $q>\beta$,

$$\int_0^\infty \mathrm{e}^{-qx}\mathbb{E}\left(\mathrm{e}^{-\alpha\tau_x^+ - \beta(X_{\tau_x^+}-x)}\right)\mathrm{d}x$$
$$= (\alpha+\Phi(\beta))\int_{[0,\infty)}\int_0^\infty \mathbf{1}_{(z>x)}\mathrm{e}^{-\beta z}\mathrm{e}^{-(q-\beta)x}\mathrm{d}xU^{(\alpha)}(\mathrm{d}z)$$
$$= \frac{(\alpha+\Phi(\beta))}{q-\beta}\int_{[0,\infty)}(\mathrm{e}^{-\beta z}-\mathrm{e}^{-qz})U^{(\alpha)}(\mathrm{d}z)$$
$$= \frac{1}{q-\beta}\left(1-\frac{\alpha+\Phi(\beta)}{\alpha+\Phi(q)}\right). \tag{S.10}$$

(iii) Recall from Exercise 2.11 that

$$\lim_{\beta\uparrow\infty}\frac{\Phi(\beta)}{\beta} = d.$$

By taking $\beta\uparrow\infty$ and then $\alpha\downarrow 0$ in (S.10) we see with the help of the above limit and dominated convergence that

$$\int_0^\infty \mathrm{e}^{-qx}\mathbb{E}(\mathrm{e}^{-\alpha\tau_x^+}\mathbf{1}_{(X_{\tau_x^+}=x)})\mathrm{d}x = \frac{d}{\alpha+\Phi(q)}.$$

It is now immediate that if $d=0$ then $\mathbb{E}(\mathrm{e}^{-\alpha\tau_x^+}\mathbf{1}_{(X_{\tau_x^+}=x)}) = 0$ Lebesgue almost everywhere and hence by Theorem 5.9, this can be upgraded to everywhere. Also, from the above Laplace transform and (S.9) it follows that if $d>0$ then $U^{(\alpha)}(\mathrm{d}x) = u^{(\alpha)}(x)\mathrm{d}x$, where Lebesgue almost everywhere $u^{(\alpha)}(x) := d^{-1}\mathbb{E}(\mathrm{e}^{-\alpha\tau_x^+}\mathbf{1}_{(X_{\tau_x^+}=x)})$. Since we may think of the latter expectation as the probability that X visits the point x when killed independently at rate α, Theorem 5.9 tells us that it is continuous strictly positive on $(0,\infty)$ and hence we take $u^{(\alpha)}$ to be this version.

(iv) From Lemma 5.11 we know that $\mathbb{E}(\mathrm{e}^{-\alpha\tau_x^+}\mathbf{1}_{(X_{\tau_x^+}=x)}) \to 1$ as $x\downarrow 0$. This implies that $u^{(\alpha)}(0+)$ exists and is equal to $1/d$.

5.6 (i) From Exercise 1.4 (i) we see that the Lévy measure necessarily takes the form

$$\Pi(\mathrm{d}x) = \frac{x^{-(1+\alpha)}}{-\Gamma(-\alpha)}\mathrm{d}x.$$

(ii) From (S.9) we know that for $q>0$,

$$\int_{[0,\infty)}\mathrm{e}^{-qx}U(\mathrm{d}x) = \frac{1}{q^\alpha}.$$

On the other hand, by definition of the Gamma function

$$\int_0^\infty \mathrm{e}^{-t}\frac{t^{\alpha-1}}{\Gamma(\alpha)}\mathrm{d}t = 1$$

and hence with the change of variable $t=qx$ one sees the agreement of the measures $U(\mathrm{d}x)$ and $\Gamma(\alpha)^{-1}x^{\alpha-1}\mathrm{d}x$ on $(0,\infty)$ via their Laplace transforms.

(iii) The identity is obtained by filling in (5.7) with the explicit expressions for U and Π.

(iv) Recall from Exercise 2.11 that the drift component is obtained by taking the limit $\lim_{\theta\uparrow\infty}\theta^{\alpha}/\theta$ and hence is equal to zero. Creeping occurs if and only if a drift is present and hence stable subordinators do not creep.

5.7 (i) Using the conclusion of Theorem 5.6 we compute for $\beta,\gamma\geq 0$,

$$
\begin{aligned}
&\mathbb{E}(\mathrm{e}^{-\beta X_{\tau_x^+-}-\gamma X_{\tau_x^+}}\mathbf{1}_{(X_{\tau_x^+}>x)})\\
&=\iint \mathbf{1}_{(0\leq y<x)}\mathbf{1}_{(v>x-y)}\mathrm{e}^{-\beta y-\gamma(y+v)}U(\mathrm{d}y)\Pi(\mathrm{d}v)\\
&=\int_{[0,x)}\mathrm{e}^{-(\beta+\gamma)y}U(\mathrm{d}y)\left\{\int_{(x-y,\infty)}\mathrm{e}^{-\gamma v}\Pi(\mathrm{d}v)\right\}.
\end{aligned}
$$

Noting that the latter is a convolution, it follows for $q>\gamma$,

$$
\begin{aligned}
&\int_0^\infty \mathrm{d}x\cdot\mathrm{e}^{-qx}\mathbb{E}(\mathrm{e}^{-\beta X_{\tau_x^+-}-\gamma(X_{\tau_x^+}-x)}\mathbf{1}_{(X_{\tau_x^+}>x)})\\
&=\int_0^\infty \mathrm{e}^{-(q-\gamma)x}\mathrm{e}^{-(\beta+\gamma)x}U(\mathrm{d}x)\\
&\quad\times\int_0^\infty \mathrm{d}x\cdot\mathrm{e}^{-(q-\gamma)x}\int_{(x,\infty]}\mathrm{e}^{-\gamma v}\Pi(\mathrm{d}v)\\
&=\frac{1}{\Phi(q+\beta)}\int_0^\infty \mathrm{d}x\cdot\mathrm{e}^{-(q-\gamma)x}\int_{(x,\infty]}\mathrm{e}^{-\gamma v}\Pi(\mathrm{d}v),
\end{aligned}
$$

where the final equality uses (S.9). Fubini's theorem shows that the integral on the right-hand side is equal to

$$
\begin{aligned}
&\int_{(0,\infty)}\int_0^\infty \mathbf{1}_{(v>x)}\mathrm{e}^{-(q-\gamma)x}\mathrm{e}^{-\gamma v}\mathrm{d}x\Pi(\mathrm{d}v)\\
&=\frac{1}{q-\gamma}\int_{(0,\infty)}\Pi(\mathrm{d}v)\{1-\mathrm{e}^{-qv}-(1-\mathrm{e}^{-\gamma v})\}\\
&=\frac{1}{q-\gamma}\{\Phi(q)-\Phi(\gamma)-d(q-\gamma)\}.
\end{aligned}
$$

In conclusion we have that

$$
\begin{aligned}
&\int_0^\infty \mathrm{d}x\cdot\mathrm{e}^{-qx}\mathbb{E}(\mathrm{e}^{-\beta X_{\tau_x^+-}-\gamma(X_{\tau_x^+}-x)}\mathbf{1}_{(X_{\tau_x^+}>x)})\\
&=\frac{1}{q-\gamma}\frac{\Phi(q)-\Phi(\gamma)}{\Phi(q+\beta)}-\frac{d}{\Phi(q+\beta)}.
\end{aligned}
$$

(ii) Theorem 5.9 (ii) and Lemma 5.8 also tell us that

$$
\int_0^\infty \mathrm{d}x\cdot\mathrm{e}^{-qx}\mathbb{E}(\mathrm{e}^{-\beta X_{\tau_x^+-}-\gamma(X_{\tau_x^+}-x)}\mathbf{1}_{(X_{\tau_x^+}=x)})=\frac{d}{\Phi(q+\beta)}
$$

and hence

$$\int_0^\infty \mathrm{d}x \cdot \mathrm{e}^{-qx}\mathbb{E}(\mathrm{e}^{-\beta X_{\tau_x^+-}-\gamma(X_{\tau_x^+}-x)}) = \frac{1}{q-\gamma}\frac{\Phi(q)-\Phi(\gamma)}{\Phi(q+\beta)}.$$

The required expression follows by making the change of variables $x \mapsto tx$, $q \mapsto q/t$, $\beta \mapsto \beta/t$, $\gamma \mapsto \gamma/t$.

(iii) Suppose that Φ is slowly varying at zero with index 0. Note with the help of the Dominated Convergence Theorem,

$$\int_0^\infty \mathrm{d}x \cdot \mathrm{e}^{-qx} \lim_{t\uparrow\infty} \mathbb{E}(\mathrm{e}^{-\beta(X_{\tau_{tx}^+-}/t)-\gamma(X_{\tau_{tx}^+}-tx)/t}) = \frac{1}{q}\mathbf{1}_{(\gamma=0)}.$$

By considering the cases that $q = 0$ and $q > 0$ separately we see then that for some $x > 0$ (in fact almost every $x > 0$), $X_{\tau_{tx}^+-}/t$ tends in distribution to zero as $t \uparrow \infty$ and $X_{\tau_{tx}^+}/t$ tends in distribution to infinity as $t \uparrow \infty$. In particular this implies the same limiting apply when $x = 1$ (by dividing both expressions by x and then making the change of variable $tx \mapsto x$). The other cases are handled similarly.

(iv) Note that $(X_{\tau_t^+-}/t, (X_{\tau_t^+} - t)/t)$ has a limiting distribution as t tends to infinity (resp. zero) if and only if $(X_{\tau_{tx}^+-}/t, (X_{\tau_{tx}^+} - tx)/t)$ has a limiting distribution as t tends to infinity (resp. zero) for all $x > 0$. Note however from the identity in part (ii) and the information provided on regularly varying functions in the question, one sees by first considering the case that $\gamma = 0$ and then the case that $\beta = 0$ as t tends to infinity (resp. zero) that a limiting distribution exists if and only if Φ is regularly varying at zero (infinity resp.). As discussed earlier in Sect. 5.5, if Φ is regularly varying with index α then necessarily $\alpha \in [0, 1]$. Appealing to the conclusion of part (iii) one thus concludes that there is a non-trivial limiting distribution of $(X_{\tau_t^+-}/t, (X_{\tau_t^+} - t)/t)$ as t tends to infinity (resp. zero) if and only if $\alpha \in (0, 1)$.

5.8 (i) Introduce the auxilliary process $\widetilde{X} = \{\widetilde{X}_t : t \geq 0\}$ which is the subordinator whose Laplace exponent is given by $\Phi(q) - \eta$. Define also $\widetilde{\tau}_x^+ = \inf\{t > 0 : \widetilde{X}_t > x\}$. We have

$$\begin{aligned}
U(x,\infty) &= \mathbb{E}\left(\int_0^\infty \mathbf{1}_{(X_t > x)}\mathrm{d}t\right) \\
&= \mathbb{E}\left(\int_0^\infty \mathrm{e}^{-\eta t}\mathbf{1}_{(\widetilde{X}_t > x)}\mathrm{d}t\right) \\
&= \mathbb{E}\left(\int_{\widetilde{\tau}_x^+}^\infty \mathrm{e}^{-\eta t}\mathbf{1}_{(\widetilde{X}_t > x)}\mathrm{d}t\right) \\
&= \mathbb{E}\left(\mathrm{e}^{-\eta\widetilde{\tau}_x^+}\int_0^\infty \mathrm{e}^{-\eta t}\mathrm{d}t\right) \\
&= \frac{1}{\eta}\mathbb{P}(\tau_x^+ < \infty).
\end{aligned}$$

(ii) Now note with the help of (5.20) that

$$\begin{aligned}\mathbb{E}(\mathrm{e}^{-\beta(X_{\tau_x^+}-x)}|\tau_x^+<\infty) &= \frac{\mathbb{E}(\mathrm{e}^{-\beta(X_{\tau_x^+}-x)}\mathbf{1}_{(\tau_x^+<\infty)})}{\mathbb{P}(\tau_x^+<\infty)}\\ &= \frac{\mathbb{E}(\mathrm{e}^{-\beta(\widetilde{X}_{\tilde{\tau}_x^+}-x)}\mathrm{e}^{-\eta\tilde{\tau}_x^+})}{\mathbb{E}(\mathrm{e}^{-\eta\tilde{\tau}_x^+})}\\ &= \frac{[\eta+(\Phi(\beta)-\eta)]\int_{(x,\infty)}\mathrm{e}^{-\beta(y-x)}\widetilde{U}^{(\eta)}(\mathrm{d}y)}{\eta\widetilde{U}^{(\eta)}(x,\infty)},\end{aligned}$$

where $\widetilde{U}^{(\eta)}$ is the η-potential of $\widetilde{X}$. However, it is straightforward to prove from its definition that $\widetilde{U}^{(\eta)}=U$, the potential measure of X, from which the required expression follows.

(iii) Integration by parts yields

$$\mathbb{E}(\mathrm{e}^{-\beta(X_{\tau_x^+}-x)}|\tau_x^+<\infty) = \frac{\Phi(\beta)}{\eta}\left(1-\int_0^\infty \beta\mathrm{e}^{-\beta y}\frac{U(y+x,\infty)}{U(x,\infty)}\mathrm{d}y\right).$$

As it has been assumed that U belongs to class $\mathcal{L}^{(\alpha)}$ it follows by dominated convergence that

$$\lim_{x\uparrow\infty}\mathbb{E}(\mathrm{e}^{-\beta(X_{\tau_x^+}-x)}|\tau_x^+<\infty) = \frac{\Phi(\beta)}{\eta}\left(\frac{\alpha}{\alpha+\beta}\right).$$

(iv) Note that

$$\begin{aligned}G(0,\infty) &= \frac{1}{\eta}\left(\Phi(-\alpha)+\int_{(0,\infty)}(\mathrm{e}^{\alpha y}-1)\Pi(\mathrm{d}y)\right)\\ &= \frac{1}{\eta}(\Phi(-\alpha)+\eta-\alpha\mathrm{d}-\Phi(-\alpha))\\ &= 1-\frac{d\alpha}{\eta}.\end{aligned}$$

We also have using integration by parts

$$\begin{aligned}\int_{(0,\infty)}\mathrm{e}^{-\beta x}G(\mathrm{d}x) &= 1-\frac{\alpha d}{\eta}-\beta\int_0^\infty \mathrm{e}^{-\beta x}G(x,\infty)\mathrm{d}x\\ &= 1-\frac{\alpha d}{\eta}-\frac{\beta}{\eta}\int_0^\infty \mathrm{d}x\cdot\mathrm{e}^{-(\beta+\alpha)x}\int_{(0,\infty)}\mathbf{1}_{(y>x)}(\mathrm{e}^{\alpha y}-\mathrm{e}^{\alpha x})\Pi(\mathrm{d}y).\end{aligned}$$

Further, starting with an application of Fubini's Theorem, the last integral can be developed as follows

$$\begin{aligned}&\int_0^\infty \mathrm{d}x\cdot\mathrm{e}^{-(\beta+\alpha)x}\int_{(0,\infty)}\mathbf{1}_{(y>x)}(\mathrm{e}^{\alpha y}-\mathrm{e}^{\alpha x})\Pi(\mathrm{d}y)\\ &= \int_{(0,\infty)}\Pi(\mathrm{d}y)\cdot\left\{\mathrm{e}^{\alpha y}\frac{(1-\mathrm{e}^{-(\alpha+\beta)y})}{(\alpha+\beta)}-\frac{(1-\mathrm{e}^{-\beta y})}{\beta}\right\}\\ &= \frac{1}{(\alpha+\beta)}\int_{(0,\infty)}(\mathrm{e}^{\alpha y}-1)\Pi(\mathrm{d}y)-\frac{\alpha}{\beta(\alpha+\beta)}\int_{(0,\infty)}(1-\mathrm{e}^{-\beta y})\Pi(\mathrm{d}y)\\ &= \frac{1}{(\alpha+\beta)}[-\Phi(-\alpha)+\eta-\alpha d]-\frac{\alpha}{\beta(\alpha+\beta)}[\Phi(\beta)-\eta-d\beta].\end{aligned}$$

Putting the pieces together we find

$$\int_{(0,\infty)} \mathrm{e}^{-\beta x} G(\mathrm{d}x) = \frac{\alpha \Phi(\beta)}{\eta(\beta+\alpha)} - \frac{\alpha d}{\eta}$$

as required.

(v) Taking account of the atom at the origin, we now have that

$$\int_{[0,\infty)} \mathrm{e}^{-\beta x} G(\mathrm{d}x) = \frac{\alpha \Phi(\beta)}{\eta(\beta+\alpha)}$$

which agrees with the Laplace transform of the limiting conditional distribution of the overshoot from part (iii) and hence G represents the overshoot distribution. In particular, $\mathrm{d}\alpha/\eta$ is the asymptotic conditional probability of creeping across the barrier.

5.9 Recall from the Lévy–Itô decomposition and specifically Exercise 2.8, the process X can be written as the difference of two pure jump subordinators, say $X = X^u - X^d$. As they are independent we may condition on the path of one of the subordinators and note that for $x \geq 0$,

$$\mathbb{P}(\inf\{t > 0 : X_t = x\} < \infty) = \mathbb{E}[\mathbb{P}(\inf\{t > 0 : X_t^u = x + X_t^d\} < \infty | X^d)].$$

Now consider any non-decreasing right continuous path $g : [0,\infty) \to [0,\infty)$ whose range is a countable set. From Theorem 5.9 we know that X^u hits a given non-negative level at any time in $[0,\infty)$ with probability zero. Therefore, it hits a given non-negative level at any given countable subset of time in $[0,\infty)$ also with probability zero. That is to say

$$\mathbb{P}(\inf\{t > 0 : X_t^u = g(t)\} < \infty) = 0.$$

Referring back to (S.11) we now see that $\mathbb{P}(\inf\{t > 0 : X_t^u = x + X_t^d\} < \infty | X^d)$ and hence $\mathbb{P}(\inf\{t > 0 : X_t = x\} < \infty)$ is equal to zero.

Chapter 6

6.1 Irrespective of path variation, any symmetric Lévy process which is not a compound Poisson process has the property that 0 is regular for both $(0,\infty)$ and $(-\infty, 0)$. To see why, it suffices to note that if this were not true then by symmetry it would be the case that 0 is irregular for both $(0,\infty)$ and $(-\infty,0)$.

Here are two examples of a Lévy process of bounded variation for which 0 is irregular for $(0,\infty)$. Firstly consider a spectrally positive Lévy process of bounded variation. The claim follows from Lemma 4.11. For the second example consider the difference of two stable subordinators with indices $0 < \alpha < \alpha' < 1$, where the subtracted subordinator is of index α'. Now check using the second integral test in Theorem 6.5 that

$$\int_{(0,1)} \frac{x \Pi(\mathrm{d}x)}{\int_0^x \Pi(-\infty,-y)\mathrm{d}y} = O\left(\int_0^1 \frac{x \cdot x^{-(1+\alpha)}\mathrm{d}x}{\int_0^x y^{-\alpha'}\mathrm{d}y}\right) = O\left(\int_0^1 \frac{1}{x^{1+\alpha'-\alpha}}\mathrm{d}x\right)$$

and apply the conclusion of the same theorem to deduce there is irregularity of $(0,\infty)$. Clearly there must be regularity of 0 for $(-\infty,0)$ as otherwise we would have a compound Poisson process which is not the case.

6.2 (i) By assumption, $\lim_{q\uparrow\infty}\Phi(q)=\infty$, and we have with the help of Exercise 2.11,

$$d=\lim_{q\uparrow\infty}\frac{\Phi(q)}{q}=\lim_{q\uparrow\infty}\frac{\Phi(q)}{\psi(\Phi(q))}=\lim_{\theta\uparrow\infty}\frac{\theta}{\psi(\theta)}=\lim_{\theta\uparrow\infty}\frac{1}{\psi'(\theta)},$$

where the last equality follows by L'Hôpital's rule. However, we know from Exercise 3.5 that ψ, which necessarily takes the form

$$\psi(\theta)=-a\theta+\frac{1}{2}\sigma^2\theta^2+\int_{(-\infty,0)}(\mathrm{e}^{\theta x}-1-\theta x\mathbf{1}_{(x>-1)})\Pi(\mathrm{d}x),$$

is infinitely differentiable. Hence

$$d=\lim_{\theta\uparrow\infty}[-a+\sigma^2\theta+\int_{(-\infty,0)}x(\mathrm{e}^{\theta x}-\mathbf{1}_{(x>-1)})\Pi(\mathrm{d}x)]^{-1}=0,$$

where in the last equality we have used the fact that either $\sigma>0$ or $\int_{(-1,0)}|x|\Pi(\mathrm{d}x)=\infty$.

(ii) As τ^+ is a pure jump subordinator, we have from the proof of Theorem 4.11 that

$$\lim_{x\downarrow 0}\frac{\tau^+_x}{x}=0$$

and hence as $X_{\tau^+_x}=x$ it follows that

$$\limsup_{t\downarrow 0}\frac{X_t}{t}\geq\limsup_{x\downarrow 0}\frac{X_{\tau^+_x}}{\tau^+_x}=\infty.$$

The latter implies that there exists a sequence of random times $\{t_n:n\geq 0\}$ with $t_n\downarrow 0$ as $n\uparrow\infty$ such that almost surely $X_{t_n}>0$. This in turn implies regularity of 0 for $(0,\infty)$. If τ^+ has finite jump measure then $\mathbb{P}(\tau^+_0>0)=1$ which implies that X is irregular for $(0,\infty)$. Hence necessarily τ^+ has a jump measure with infinite mass.

(iii) From the Wiener–Hopf factorisation given in Sect. 6.5.2 we have that

$$\mathbb{E}\left(\mathrm{e}^{\theta\underline{X}_{\mathbf{e}_q}}\right)=\frac{q}{\Phi(q)}\frac{\Phi(q)-\theta}{q-\psi(\theta)}.$$

Taking limits as $\theta\uparrow\infty$ and recalling from part (i) that $\lim_{\theta\uparrow\infty}\theta\psi(\theta)^{-1}=0$ gives us

$$\mathbb{P}(\underline{X}_{\mathbf{e}_q}=0)=0.$$

Finally note that $\mathbb{P}(\underline{X}_{\mathbf{e}_q}=0)>0$ if and only if $\mathbb{P}(\tau^+_0>0)>0$ and hence it must be the case that $\mathbb{P}(\tau^+_0=0)=1$, in other words, 0 is regular for $(0,\infty)$.

6.3 Suppose that $N=\{N_t:t\geq 0\}$ is a Poisson process with rate $\lambda\rho$ and $\{\xi_n:n=1,2,...\}$ are independent (also of N) and identically distributed and further, $\mathbf{e}_{\lambda(1-\rho)}$

is an independent exponentially distributed random variable with parameter $\lambda(1-\rho)$, then

$$-\log \mathbb{E}\left(\mathrm{e}^{-\theta \sum_{i=1}^{N_1} \xi_i} \mathbf{1}_{(1<\mathbf{e}_{\lambda(1-\rho)})}\right) = -\log \mathbb{E}\left(\mathrm{e}^{-\theta \sum_{i=1}^{N_1} \xi_i} \mathrm{e}^{-\lambda(1-\rho)}\right)$$
$$= \lambda(1-\rho) + \lambda\rho(1 - \mathbb{E}(\mathrm{e}^{-\theta\xi_1})).$$

On the other hand, suppose that $\widetilde{N} = \{\widetilde{N}_t : t \geq 0\}$ is an independent Poisson process with rate λ and $\mathbf{\Gamma}_{1-\rho}$ is a geometric distribution with parameter $1-\rho$, then

$$-\log \mathbb{E}\left(\mathrm{e}^{-\theta \sum_{i=1}^{\widetilde{N}_t} \xi_i} \mathbf{1}_{(\widetilde{N}_t \leq \mathbf{\Gamma}_{1-\rho})}\right) = -\log \mathbb{E}\left(\mathrm{e}^{-\theta \sum_{i=1}^{\widetilde{N}_t} \xi_i} \rho^{\widetilde{N}_t}\right)$$
$$= \lambda - \lambda\rho\mathbb{E}(\mathrm{e}^{-\theta\xi_1})$$

and hence the required result follows.

6.4 Note that $\int_0^\infty \mathbf{1}_{(\overline{X}_t - X_t = 0)}\mathrm{d}t > 0$ if and only if $\int_0^\infty \mathbb{P}(\overline{X}_t - X_t = 0)\mathrm{d}t > 0$ if and only if $\int_0^\infty \mathbb{P}(\underline{X}_t = 0)\mathrm{d}t > 0$ (by duality) if and only if 0 is irregular for $(-\infty, 0)$. Similarly $\int_0^\infty \mathbf{1}_{(X_t - \underline{X}_t = 0)}\mathrm{d}t > 0$ if and only if 0 is irregular for $(0,\infty)$. This excludes all Lévy processes except for compound Poisson processes.

6.5 (i) We know that the Laplace exponent of a spectrally negative Lévy process satisfies

$$\psi(\theta) = -\log \mathbb{E}(\mathrm{e}^{\theta X_1}) = -a\theta + \frac{1}{2}\sigma^2\theta^2 + \int_{(-\infty,0)} (\mathrm{e}^{\theta x} - 1 - \theta x \mathbf{1}_{(x>-1)})\Pi(\mathrm{d}x).$$

Taking account of Exercise 3.5 we may compute $\mathbb{E}(X_1) = \psi'(0+)$ and hence

$$\mathbb{E}(X_1) = -a + \lim_{\theta\downarrow 0} \int_{(-\infty,0)} x(\mathrm{e}^{\theta x} - \mathbf{1}_{(x>-1)})\Pi(\mathrm{d}x) = -a + \int_{(-\infty,-1)} x\Pi(\mathrm{d}x).$$

Integration by parts now implies that $\int_{-\infty}^{-1} \Pi(-\infty, x)\mathrm{d}x < \infty$.

(ii) As it is assumed that $\mathbb{E}(X_1) > 0$, we know from Corollary 3.14 that the inverse local time is a subordinator and hence the ascending ladder height process is a subordinator. In fact, taking $L = \overline{X}$, we have already seen that $L_t^{-1} = \tau_t^+ = \inf\{s > 0 : X_s > t\}$ and hence with this choice of local time, $H_t = t$. It follows that $\kappa(0, -\mathrm{i}\theta) = \mathrm{i}\theta$ and so from Theorem 6.16 (iv) we have up to a multiplicative constant that

$$\begin{aligned}
\widehat{\kappa}(0, \mathrm{i}\theta) &= \frac{-\psi(\mathrm{i}\theta)}{\mathrm{i}\theta} \\
&= a - \frac{1}{2}\mathrm{i}\sigma^2\theta - \frac{1}{\mathrm{i}\theta}\int_{(-\infty,0)} (\mathrm{e}^{\mathrm{i}\theta x} - 1 - \mathrm{i}\theta x \mathbf{1}_{(x>-1)})\Pi(\mathrm{d}x) \\
&= \left(-a + \int_{(-\infty,-1)} x\Pi(\mathrm{d}x)\right) - \frac{1}{2}\mathrm{i}\sigma^2\theta + \int_{(-\infty,0)} (1 - \mathrm{e}^{\mathrm{i}\theta x})\Pi(-\infty, x)\mathrm{d}x,
\end{aligned}$$

where the final equality follows by integration by parts. Note that for $\beta \in \mathbb{R}$, $\kappa(0, -\mathrm{i}\beta)$ is the characteristic exponent of the ascending ladder height process. Hence from the right-hand side above we see that $\kappa(0,-\mathrm{i}\beta)$ is the characteristic exponent of a killed subordinator where the killing rate is $\mathbb{E}(X_1)$, the drift is $\sigma^2/2$ and the Lévy measure is given by $\Pi(-\infty,-x)\mathrm{d}x$ on $(0,\infty)$.

6.6 (i) A spectrally negative stable process of index $\alpha\in(1,2)$ has Laplace exponent defined, up to a multiplicative constant, by $\psi(\theta)=\theta^\alpha$ (see Exercise 3.7). From Theorem 3.12 and Corollary 3.14 we have that the first passage time process $\{\tau_x^+ : x\geq 0\}$ is a subordinator with Laplace exponent $\theta^{1/\alpha}$. However the latter process is also the inverse local time at the maximum and hence up to a multiplicative constant $\kappa(\theta,0)=\theta^{1/\alpha}$. According to the calculations in Sect. 6.5.3 the inverse local time at the maximum is a stable subordinator with index ρ. It follows then that $\rho=1/\alpha$.

(ii) With the help of Fubini's theorem we have for p sufficiently large that

$$\begin{aligned}\int_0^\infty p\mathrm{e}^{-pt}\sum_{n=0}^\infty\frac{(-\theta t^{1/\alpha})^n}{\Gamma(1+n/\alpha)}\mathrm{d}t &= \sum_{n=0}^\infty(-\theta)^n\int_0^\infty\frac{\mathrm{e}^{-pt}t^{n/\alpha}}{\Gamma(1+n/\alpha)}\frac{p^{1+n/\alpha}}{p^{n/\alpha}}\mathrm{d}t\\ &=\sum_{n=0}^\infty\frac{(-\theta)^n}{p^{n/\alpha}}\\ &=\frac{p^{1/\alpha}}{p^{1/\alpha}+\theta}.\end{aligned}$$

We know from Sect. 6.5.2 that $\kappa(p,\theta)=p^{1/\alpha}+\theta$ and from the Wiener–Hopf factorisation $\mathbb{E}(\mathrm{e}^{-\theta\overline{X}_{\mathbf{e}_p}})=\kappa(p,0)/\kappa(p,\theta)$. It follows that

$$\mathbb{E}(\mathrm{e}^{-\theta\overline{X}_t})=\sum_{n=0}^\infty\frac{(-\theta t^{1/\alpha})^n}{\Gamma(1+n/\alpha)}$$

for Lebesgue almost every $t\geq 0$. As $t\mapsto\overline{X}_t$ and hence, by dominated convergence $t\mapsto\mathbb{E}(\mathrm{e}^{-\theta\overline{X}_t})$ are continuous, the last equality is in fact valid for all $t\geq 0$.

6.7 (i) For $\alpha,x>0$, $\beta\geq 0$,

$$\begin{aligned}&\mathbb{E}\left(\mathrm{e}^{-\beta\overline{X}_{\mathbf{e}_\alpha}}\mathbf{1}_{(\overline{X}_{\mathbf{e}_\alpha}>x)}\right)\\ &=\mathbb{E}\left(\mathrm{e}^{-\beta\overline{X}_{\mathbf{e}_\alpha}}\mathbf{1}_{(\tau_x^+<\mathbf{e}_\alpha)}\right)\\ &=\mathbb{E}\left(\mathbf{1}_{(\tau_x^+<\mathbf{e}_\alpha)}\mathrm{e}^{-\beta X_{\tau_x^+}}\mathbb{E}\left(\mathrm{e}^{-\beta(\overline{X}_{\mathbf{e}_\alpha}-X_{\tau_x^+})}\Big|\mathcal{F}_{\tau_x^+}\right)\right).\end{aligned}$$

Now, conditionally on $\mathcal{F}_{\tau_x^+}$ and on the event $\tau_x^+<\mathbf{e}_\alpha$ the random variables $\overline{X}_{\mathbf{e}_\alpha}-X_{\tau_x^+}$ and $\overline{X}_{\mathbf{e}_\alpha}$ have the same distribution thanks to the lack of memory property of $\mathbf{e}_\alpha$ and the strong Markov property. Hence, we have the factorisation

$$\mathbb{E}\left(\mathrm{e}^{-\beta\overline{X}_{\mathbf{e}_\alpha}}\mathbf{1}_{(\overline{X}_{\mathbf{e}_\alpha}>x)}\right)=\mathbb{E}\left(\mathrm{e}^{-\alpha\tau_x^+-\beta X_{\tau_x^+}}\right)\mathbb{E}\left(\mathrm{e}^{-\beta\overline{X}_{\mathbf{e}_\alpha}}\right).$$

(ii) By taking Laplace transforms of both sides of (6.33) and using Fubini's Theorem, we can write for $q>0$,

$$
\begin{aligned}
&\int_0^\infty \mathrm{e}^{-qx}\mathbb{E}\left(\mathrm{e}^{-\alpha\tau_x^+ - \beta(X_{\tau_x^+}-x)}\mathbf{1}_{(\tau_x^+<\infty)}\right)\mathrm{d}x\\
&= \frac{1}{\mathbb{E}\left(\mathrm{e}^{-\beta\overline{X}_{\mathbf{e}_\alpha}}\right)}\int_0^\infty \mathrm{e}^{-qx}\mathbb{E}\left(\mathrm{e}^{-\beta(\overline{X}_{\mathbf{e}_\alpha}-x)}\mathbf{1}_{(\overline{X}_{\mathbf{e}_\alpha}>x)}\right)\mathrm{d}x\\
&= \frac{1}{\mathbb{E}\left(\mathrm{e}^{-\beta\overline{X}_{\mathbf{e}_\alpha}}\right)}\int_0^\infty \mathrm{e}^{-qx}\int_0^\infty \mathbf{1}_{(y>x)}\mathrm{e}^{-\beta(y-x)}\mathbb{P}\left(\overline{X}_{\mathbf{e}_\alpha}\in\mathrm{d}y\right)\mathrm{d}x\\
&= \frac{1}{\mathbb{E}\left(\mathrm{e}^{-\beta\overline{X}_{\mathbf{e}_\alpha}}\right)}\int_0^\infty \mathrm{e}^{-\beta y}\int_0^\infty \mathbf{1}_{(y>x)}\mathrm{e}^{-qx+\beta x}\mathrm{d}x\mathbb{P}\left(\overline{X}_{\mathbf{e}_\alpha}\in\mathrm{d}y\right)\\
&= \frac{1}{(q-\beta)\mathbb{E}\left(\mathrm{e}^{-\beta\overline{X}_{\mathbf{e}_\alpha}}\right)}\int_0^\infty \left(\mathrm{e}^{-\beta y}-\mathrm{e}^{-qy}\right)\mathbb{P}\left(\overline{X}_{\mathbf{e}_\alpha}\in\mathrm{d}y\right)\\
&= \frac{\mathbb{E}\left(\mathrm{e}^{-\beta\overline{X}_{\mathbf{e}_\alpha}}\right)-\mathbb{E}\left(\mathrm{e}^{-q\overline{X}_{\mathbf{e}_\alpha}}\right)}{(q-\beta)\mathbb{E}\left(\mathrm{e}^{-\beta\overline{X}_{\mathbf{e}_\alpha}}\right)}.
\end{aligned}
$$

The required statement follows by recalling from the Wiener–Hopf factorisation that $\mathbb{E}(\mathrm{e}^{-\theta\overline{X}_{\mathbf{e}_\alpha}}) = \kappa(\alpha,0)/\kappa(\alpha,\theta)$.

6.8 (i) Suppose first that 0 is regular for $(0,\infty)$. This implies that for $p>0$, $\mathbb{P}(\overline{X}_{\mathbf{e}_p}=0)=\lim_{\beta\uparrow\infty}\mathbb{E}(\mathrm{e}^{-\beta\overline{X}_{\mathbf{e}_p}})=0$. However, from the Wiener–Hopf factorisation we know that

$$
\mathbb{E}(\mathrm{e}^{-\beta\overline{X}_{\mathbf{e}_p}}) = \exp\left\{-\int_0^\infty\int_{[0,\infty)}(1-\mathrm{e}^{-\beta x})\mathrm{e}^{-pt}\frac{1}{t}\mathbb{P}(X_t\in\mathrm{d}x)\mathrm{d}t\right\}.
$$

Note that the second integral can be taken over $(0,\infty)$ as the integrand is zero when $x=0$. Hence regularity implies that $\lim_{\beta\uparrow\infty}\int_0^\infty\int_{(0,\infty)}(1-\mathrm{e}^{-\beta x})\mathrm{e}^{-pt}t^{-1}\mathbb{P}(X_t\in\mathrm{d}x)\mathrm{d}t=\infty$. Monotone convergence implies that the latter limit is equal to

$$
\int_0^\infty \mathrm{e}^{-pt}t^{-1}\mathbb{P}(X_t>0)\mathrm{d}t.
$$

As $\int_1^\infty \mathrm{e}^{-pt}t^{-1}\mathrm{d}t<\infty$ we have that $\int_0^1\mathrm{e}^{-pt}t^{-1}\mathbb{P}(X_t>0)\mathrm{d}t=\infty$. As $\mathrm{e}^{-pt}\leq 1$ it follows that $\int_0^1 t^{-1}\mathbb{P}(X_t>0)\mathrm{d}t=\infty$.

Now suppose that $\int_0^1 t^{-1}\mathbb{P}(X_t>0)\mathrm{d}t=\infty$. Using the estimate $\mathrm{e}^{-pt}\geq\mathrm{e}^{-p}$ for $t\in(0,1)$ we have that $\int_0^\infty\mathrm{e}^{-pt}t^{-1}\mathbb{P}(X_t>0)\mathrm{d}t=\infty$. Reversing the arguments above, we see that $\mathbb{P}(\overline{X}_{\mathbf{e}_p}=0)=\lim_{\beta\uparrow\infty}\mathbb{E}(\mathrm{e}^{-\beta\overline{X}_{\mathbf{e}_p}})=0$ and hence 0 is regular for $(0,\infty)$.

(ii) We show equivalently that 0 is irregular for $[0,\infty)$ if and only if $\int_0^1 t^{-1}\mathbb{P}(X_t\geq 0)\mathrm{d}t<\infty$. To this end, suppose that 0 is irregular for $[0,\infty)$ so that $\lim_{\lambda\uparrow\infty}\mathbb{E}(\mathrm{e}^{-\lambda\overline{G}_{\mathbf{e}_p}}) = \mathbb{P}(\overline{G}_{\mathbf{e}_p}=0)>0$. From the Wiener–Hopf factorisation we have that

$$
\mathbb{E}(\mathrm{e}^{-\lambda\overline{G}_{\mathbf{e}_p}}) = \exp\left\{-\int_0^\infty(1-\mathrm{e}^{-\lambda t})\mathrm{e}^{-pt}\frac{1}{t}\mathbb{P}(X_t\geq 0)\mathrm{d}t\right\}
$$

and hence again appealing to monotone convergence $\int_0^\infty t^{-1}e^{-pt}\mathbb{P}(X_t \geq 0)dt < \infty$. Using the bound $e^{-pt} \geq e^{-p}$ on $(0,1)$ thus produces the conclusion that $\int_0^1 t^{-1}\mathbb{P}(X_t \geq 0)dt < \infty$.

Conversely suppose that $\int_0^1 t^{-1}\mathbb{P}(X_t \geq 0)dt < \infty$. Using the estimate $e^{-pt} \leq 1$ one may deduce that $\int_0^\infty t^{-1}e^{-pt}\mathbb{P}(X_t \geq 0)dt < \infty$. Reversing the arguments above, we have that $\lim_{\lambda\uparrow\infty}\mathbb{E}(e^{-\lambda \overline{G}_{\mathbf{e}_p}}) = \mathbb{P}(\overline{G}_{\mathbf{e}_p} = 0) > 0$ and thus irregularity of 0 for $[0,\infty)$.

6.9 (i) For $s \in (0,1)$ and $\theta \in \mathbb{R}$, let $q = 1-p$ and note that on the one hand,

$$
\begin{aligned}
E(s^{\mathbf{\Gamma}_p}e^{i\theta S_{\mathbf{\Gamma}_p}}) &= E(E(se^{i\theta S_1})^{\mathbf{\Gamma}_p}) \\
&= \sum_{k\geq 0} p(qsE(e^{i\theta S_1}))^k \\
&= \frac{p}{1-qsE(e^{i\theta S_1})}.
\end{aligned}
$$

On the other hand, with the help of Fubini's Theorem,

$$
\begin{aligned}
&\exp\left\{-\int_{\mathbb{R}}\sum_{n=1}^{\infty}(1-s^n e^{i\theta x})q^n\frac{1}{n}F^{*n}(dx)\right\} \\
&= \exp\left\{-\sum_{n=1}^{\infty}(1-s^n E(e^{i\theta S_n}))q^n\frac{1}{n}\right\} \\
&= \exp\left\{-\sum_{n=1}^{\infty}(1-s^n E(e^{i\theta S_1})^n)q^n\frac{1}{n}\right\} \\
&= \exp\left\{\log(1-q) - \log(1-sqE(e^{i\theta S_1}))\right\} \\
&= \frac{p}{1-qsE(e^{i\theta S_1})},
\end{aligned}
$$

where in the last equality we have appealed to the Mercator–Newton series expansion of the logarithm. Comparing the conclusions of the last two series of equalities, the required expression for $E(s^{\mathbf{\Gamma}_p}e^{i\theta S_{\mathbf{\Gamma}_p}})$ follows. Infinite divisibility follows from Exercise 2.10 (which implies the existence of infinitely divisible distributions in higher dimensions whose characteristic exponent is the natural analogue of the Lévy–Khintchine formula in one dimension) noting that the Lévy measure is given by

$$
\Pi(dy, dx) = \sum_{n=1}^{\infty}\delta_{\{n\}}(dy)F^{*n}(dx)\frac{1}{n}q^n.
$$

(ii) The path of the random walk may be broken into $\nu \in \{0,1,2,....\}$ finite excursions from the maximum followed by an additional excursion which straddles the random time $\mathbf{\Gamma}_p$. By the Strong Markov Property for random walks[6] and the lack

[6]The Strong Markov Property for random walks is the direct analogue of the same property for Lévy processes. In other words, for any $\{0,1,2,...\}$-valued random time τ satisfying $\{\tau < n\} \in \sigma(S_0, S_1, ..., S_n)$, the process $\{S_{\tau+k} - S_\tau : k = 1,2,...\}$ conditional on $\tau < \infty$ is independent of $\sigma(S_0, S_1, ..., S_n)$ and has the same law as S.

of memory property for the geometric distribution the finite excursions must have the same law, namely that of a random walk sampled on the time points $\{1, 2, ..., N\}$ conditioned on the event that $\{N \leq \mathbf{\Gamma}_p\}$ and ν is geometrically distributed with parameter $1 - P(N \leq \mathbf{\Gamma}_p)$. Hence we may write

$$(G, S_G) = \sum_{i=1}^{\nu} (N^{(i)}, H^{(i)}),$$

where the pairs $\{(N^{(i)}, H^{(i)}) : i = 1, 2, ...\}$ are independent having the same distribution as (N, S_N) conditioned on $\{N \leq \mathbf{\Gamma}_p\}$. Infinite divisibility follows as a consequence of part (i).

(iii) The independence of (G, S_G) and $(\mathbf{\Gamma}_p - G, S_{\mathbf{\Gamma}_p} - S_G)$ is immediate from the decomposition described in part (ii). Duality[7] for random walks implies that the latter pair is equal in distribution to (D, S_D).

(iv) We know that $(\mathbf{\Gamma}_p, S_{\mathbf{\Gamma}_p})$ may be written as the independent sum of (G, S_G) and $(\mathbf{\Gamma}_p - G, S_G - S_{\mathbf{\Gamma}_p})$ where the latter is equal in distribution to (D, S_D). Reviewing the proof of part (ii) when the strong ladder height is replaced by a weak ladder height, we see that $(\mathbf{\Gamma}_p - G, S_{\mathbf{\Gamma}_p} - S_G)$, like (G, S_G) is infinitely divisible (in that case one works with the stopping time $N' = \inf\{n > 0 : S_n \leq 0\}$; note the relationship between the inequality in the definition of N' and the max in the definition of D). Further, (G, S_G) is supported on $\{1, 2, ...\} \times [0, \infty)$ and $(\mathbf{\Gamma}_p - G, S_{\mathbf{\Gamma}_p} - S_G)$ is supported on $\{1, 2, ...\} \times (-\infty, 0]$. This means that $E(s^G \mathrm{e}^{\mathrm{i}\theta S_G})$ can be analytically extended to the upper half of the complex plane and $E(s_p^{\mathbf{\Gamma}} - G\mathrm{e}^{\mathrm{i}\theta S_{\mathbf{\Gamma}_p} - S_G})$ to the lower half of the complex plane. Taking account of the Lévy-Khinchine formula in higher dimensions (see the remarks in part (iii) in the light of Exercise 2.10), this forces the factorisation[8] of the expression for $E(s^{\mathbf{\Gamma}_p} \mathrm{e}^{\mathrm{i}\theta S_{\mathbf{\Gamma}_p}})$ in such a way that

$$E(s^G \mathrm{e}^{\mathrm{i}\theta S_G}) = \exp\left\{ -\int_{(0,\infty)} \sum_{n=1}^{\infty} (1 - s^n \mathrm{e}^{\mathrm{i}\theta x}) q^n \frac{1}{n} F^{*n}(\mathrm{d}x) \right\} \tag{S.11}$$

and

$$E(s^{(\mathbf{\Gamma}_p - G)} \mathrm{e}^{\mathrm{i}\theta (S_{\mathbf{\Gamma}_p} - S_G)}) = \frac{E(s^{\mathbf{\Gamma}_p} \mathrm{e}^{\mathrm{i}\theta S_{\mathbf{\Gamma}_p}})}{E(s^G \mathrm{e}^{\mathrm{i}\theta S_G})}.$$

(v) Note that the path decomposition given in part (i) shows that

$$E(s^G \mathrm{e}^{\mathrm{i}\theta S_G}) = E(s^{\Sigma_{i=1}^{\nu} N^{(i)}} \mathrm{e}^{\mathrm{i}\theta \Sigma_{i=1}^{\nu} H^{(i)}}),$$

where the pairs $\{(N^{(i)}, H^{(i)}) : i = 1, 2, ...\}$ are independent having the same distribution as (N, S_N) conditioned on $\{N \leq \mathbf{\Gamma}_p\}$. Hence we have

---

[7] Duality for random walks is the same concept as for Lévy processes. In other words, for any $n = 0, 1, 2...$ (which may later be randomised with any independent distribution) note that the independence and common distribution of increments implies that $\{S_{n-k} - S_n : k = 0, 1, ..., n\}$ has the same law as $\{-S_k : k = 0, 1, ..., n\}$.

[8] Here we catch a glimpse again of the technique which has the taste of the analytical factorisation methods of Wiener and Hopf. It is from this point in the reasoning that the name "Wiener–Hopf factorisation for random walks" has emerged.

$$\begin{aligned}
E(s^G e^{i\theta S_G}) &= \sum_{k\geq 0} P(\mathbf{\Gamma}_p \leq N) P(\mathbf{\Gamma}_p > N)^k E(s^{\Sigma_{i=1}^k N^{(i)}} e^{i\theta \Sigma_{i=1}^k H^{(i)}}) \\
&= \sum_{k\geq 0} P(N > \mathbf{\Gamma}_p) P(N \leq \mathbf{\Gamma}_p)^k E(s^N e^{i\theta S_N} | N \leq \mathbf{\Gamma}_p)^k \\
&= \sum_{k\geq 0} P(N > \mathbf{\Gamma}_p) E(s^N e^{i\theta S_N} \mathbf{1}_{(N\leq \mathbf{\Gamma}_p)})^k \\
&= \sum_{k\geq 0} P(N > \mathbf{\Gamma}_p) E((qs)^N e^{i\theta S_N})^k \\
&= \frac{P(N > \mathbf{\Gamma}_p)}{1 - E((qs)^N e^{i\theta S_N})}.
\end{aligned}$$

Note in the fourth equality we use the fact that $P(\mathbf{\Gamma}_p \geq n) = q^n$.

The second equality to be proved follows by setting $s = 0$ in (S.11) to recover

$$P(N > \mathbf{\Gamma}_p) = \exp\left\{ -\int_{(0,\infty)} \sum_{n=1}^{\infty} \frac{q^n}{n} F^{*n}(\mathrm{d}x) \right\}.$$

Chapter 7

7.1 (i) As it is assumed that $\mathbb{E}((\max\{X_1, 0\})^n) < \infty$ we have that $\mathbb{E}(|X_t^K|^n) < \infty$ if and only if $\mathbb{E}((\max\{-X_1^K, 0\})^n) < \infty$ and by Exercise 3.3 the latter is automatic since $\int_{(-\infty,-1)} |x|^n \Pi(\mathrm{d}x) < \infty$.

(ii) First suppose that $q > 0$. The Wiener–Hopf factorisation gives us

$$\mathbb{E}(e^{-i\theta \overline{X}^K_{\mathbf{e}_q}}) = \mathbb{E}(e^{i\theta X^K_{\mathbf{e}_q}}) \frac{\widehat{\kappa}^K(q, i\theta)}{\widehat{\kappa}^K(q, 0)}, \tag{S.12}$$

where $\widehat{\kappa}^K$ is the Laplace–Fourier exponent of the bivariate descending ladder process and $\mathbf{e}_q$ is an independent and exponentially distributed random variable with mean $1/q$. Note that the descending ladder height process of $\widehat{H}$ cannot have jumps of size greater than K as X^K cannot jump downwards by more than K. Hence the Lévy measure of the descending ladder height process of X^K has bounded support which with the help of Exercise 3.3 implies that all moments of the aforementioned process exist. Together with the fact that $\mathbb{E}(|X_t^K|^n) < \infty$ for all $t > 0$ this implies that the right-hand side of (S.12) has a Maclaurin expansion up to order n. Specifically this means that $\mathbb{E}((\overline{X}^K_{\mathbf{e}_q})^n) < \infty$. Finally note that due to the truncation, $\overline{X}_{\mathbf{e}_q} \leq \overline{X}^K_{\mathbf{e}_q}$ and hence $\mathbb{E}(\overline{X}^n_{\mathbf{e}_q}) < \infty$.

(iii) Now suppose that $\limsup_{t\uparrow\infty} X_t < \infty$. In this case we may use the Wiener–Hopf factorisation for X^K in the form (up to a multiplicative constant)

$$\kappa^K(0, -i\theta) = \frac{\Psi^K(\theta)}{\widehat{\kappa}^K(0, i\theta)},$$

where κ^K and Ψ^K are obviously defined. The same reasoning in the previous paragraph shows that the Maclaurin expansion on the right-hand side above exists up to order n and hence the same is true of the left-hand side. We make the truncation level K large enough so that it is still the case that $\lim_{t\uparrow\infty} X^K_\infty = -\infty$. This is possible by choosing K sufficiently large so that $\mathbb{E}(X^K_1) < 0$.

We now have that $\widehat{\kappa}^K(0,0) = 0$ and that $\widehat{\kappa}(0,\mathrm{i}\theta)$ has an infinite Maclaurin expansion. The integral assumption implies that $\Psi^K(\theta)$ has Maclaurin expansion up to order n and as a matter of fact $\Psi^K(0) = 0$. It now follows that the ratio $\Psi^K(\theta)/\widehat{\kappa}^K(0,\mathrm{i}\theta)$ has a Maclaurin expansion up to order $n-1$. Since $\kappa^K(0,-\mathrm{i}\theta)$ is the cumulative generating function of the ascending ladder height process of X^K it follows that the aforementioned process has finite $(n-1)$th moments. Since $\overline{X}^K_\infty$ is equal in law to the ascending ladder height process of X^K stopped at an independent and exponentially distributed time, we have that $\mathbb{E}((\overline{X}^K_\infty)^n) < \infty$. Finally we have $\mathbb{E}(\overline{X}^n_\infty) < \infty$ similar to above $\overline{X}_\infty \leq \overline{X}^K_\infty$.

7.2 (i) From the stationary and independent increments of X we have that

$$\mathbb{E}(Y_n) = \mathbb{E}(Y_1) \leq \mathbb{E}(\max\{\overline{X}_1, -\underline{X}_1\}) \leq \mathbb{E}(\overline{X}_1) - \mathbb{E}(\underline{X}_1).$$

According to Exercise 7.1 the right-hand side above is finite when we assume that $\mathbb{E}(\max\{X_1,0\}) < \infty$ and $\mathbb{E}(\max\{-X_1,0\}) < \infty$, in other words $\mathbb{E}(|X_1|) < \infty$.

(ii) The Strong Law of Large Numbers now applies to the sequence of partial sums of $\{Y_1, Y_2, ...\}$ so that $\lim_{n\uparrow\infty} \sum_{i=1}^n Y_i/n = \mathbb{E}(Y_1)$. This shows in particular that $\lim_{n\uparrow\infty} Y_n/n = 0$ almost surely.

(iii) Let $[t]$ be the integer part of t. Write

$$\frac{X_t}{t} = \frac{\sum_{i=1}^{[t]}(X_i - X_{i-1})}{[t]}\frac{[t]}{t} + \frac{X_t - X_{[t]}}{[t]}\frac{[t]}{t}.$$

Note that the first term on the right-hand side converges almost surely to $\mathbb{E}(X_1)$ by stationary independent increments and the classical Strong Law of Large Numbers. The second term can be dominated in absolute value by $Y_{[t]}/[t]$ which from part (ii) tends almost surely to zero.

(iv) Now suppose that $\mathbb{E}(X_1) = \infty$. This implies that $\mathbb{E}(\max\{-X_1,0\}) < \infty$ and $\mathbb{E}(\max\{X_1,0\}) = \infty$. From Exercise 3.3 we know that this is equivalent to $\int_{(-\infty,-1)} |x|\Pi(\mathrm{d}x) < \infty$ and $\int_{(1,\infty)} x\Pi(\mathrm{d}x) = \infty$ where Π is the Lévy measure of X. For $K > 1$ define the adjusted Lévy process X^K from X so that all jumps which are greater than or equal to K are replaced by a jump of size precisely K. Note the latter process has Lévy measure given by

$$\Pi^K(\mathrm{d}x) = \Pi(\mathrm{d}x)\mathbf{1}_{(x<K)} + \Pi(K,\infty)\delta_K(\mathrm{d}x).$$

Since $\Pi(1,\infty) < \infty$ we have that $\int_{(1,\infty)} x\Pi^K(\mathrm{d}x) < \infty$ for all $K > 1$. Hence as we also have that $\int_{(-\infty,-1)} |x|\Pi^K(\mathrm{d}x) = \int_{(-\infty,-1)} |x|\Pi(\mathrm{d}x) < \infty$, we have that $\mathbb{E}(|X^K_1|)$ exists and is finite. Clearly $X^K_t \leq X_t$ for all $t \geq 0$ and $\mathbb{E}(X^K_1) \uparrow \mathbb{E}(X_1)$ as $K \uparrow \infty$. Hence by choosing K sufficiently large, it can be arranged that $\mathbb{E}(X^K_1) > 0$. Now applying the result from part (iii) we have that $\liminf_{t\uparrow\infty} X_t/t \geq \liminf_{t\uparrow\infty} X^K_t/t = \mathbb{E}(X^K_1)$. Since K may be taken arbitrarily large the result follows.

7.3 (i) Recall from the strict convexity ψ it follows that $\Phi(0) > 0$ if and only if $\psi'(0+) < 0$ and hence

$$\lim_{q\downarrow 0}\frac{q}{\Phi(q)}=\begin{cases}0 & \text{if } \psi'(0^+)<0\\ \psi'(0^+) & \text{if } \psi'(0^+)\geq 0.\end{cases}$$

The Wiener–Hopf factorisation for spectrally negative Lévy processes (see Sect. 6.5.2) gives us for $\beta\geq 0$,

$$\mathbb{E}(\mathrm{e}^{\beta \underline{X}_{\mathbf{e}_q}})=\frac{q}{\Phi(q)}\frac{\Phi(q)-\beta}{q-\psi(\beta)}.$$

By taking q to zero in the identity above we now have that

$$E\left(\mathrm{e}^{\alpha \underline{X}_\infty}\right)=\begin{cases}0 & \text{if } \psi'(0^+)<0\\ \psi'(0^+)\alpha/\psi(\alpha) & \text{if } \psi'(0^+)\geq 0.\end{cases}$$

(ii) From the other Wiener–Hopf factor we have for $\beta>0$

$$E\left(\mathrm{e}^{-\beta \overline{X}_{\mathbf{e}_q}}\right)=\frac{\Phi(q)}{\Phi(q)+\beta}$$

and hence by taking the limit of both sides as q tends to zero,

$$E\left(\mathrm{e}^{-\alpha \overline{X}_\infty}\right)=\begin{cases}\Phi(0)/(\beta+\Phi(0)) & \text{if } \psi'(0^+)<0\\ 0 & \text{if } \psi'(0^+)\geq 0.\end{cases}$$

(iii) The trichotomy in Theorem 7.1 together with the conclusions of parts (i) and (ii) give the required asymptotic behaviour. For example, when $\psi'(0+)<0$ we have $\overline{X}_\infty<\infty$ and $\underline{X}_\infty=-\infty$ and when compared against the trichotomy of asymptotic behaviour, this can only happen in the case of drifting to $-\infty$.

(iv) The given process has Laplace exponent given by $\psi(\theta)=c\theta^\alpha$ for some $c>0$ (see Exercise 3.7). Clearly $\psi'(0+)=0$ and hence spectrally negative stable processes of index $\alpha\in(1,2)$ necessarily oscillate according to the conclusion of part (iii).

7.4 (i) According to the Wiener–Hopf factorisation the ascending ladder height process is a stable subordinator with parameter $\alpha\rho$ (see Sect. 6.5.3). The latter process has no drift and hence X cannot creep upwards (see Exercise 3.7 and Lemma 7.10).

(ii) The measure $U(\mathrm{d}x)$ is the potential measure of the ascending ladder height process. As mentioned above this process is a stable subordinator with index $\alpha\rho$ and hence for $\theta>0$ (and up to a constant),

$$\int_{[0,\infty)}\mathrm{e}^{-\theta x}U(\mathrm{d}x)=\frac{1}{\theta^{\alpha\rho}}.$$

It is straightforward to check that the right-hand side agrees with the Laplace transform of the measure $\Gamma(\alpha\rho)^{-1}x^{\alpha\rho-1}\mathbf{1}_{(x>0)}\mathrm{d}x$. (Note one will need to make use of the definition $\Gamma(z)=\int_0^\infty t^{z-1}\mathrm{e}^{-t}\mathrm{d}t$ for $z>0$, or just revisit Exercise 5.6).

(iii) By simple considerations of symmetry, we deduce easily that $\widehat{U}(\mathrm{d}x):=\int_{[0,\infty)}\widehat{\mathcal{U}}(\mathrm{d}x,\mathrm{d}s)$ is identifiable up to a constant as $\Gamma(\alpha(1-\rho))^{-1}x^{\alpha(1-\rho)-1}\mathbf{1}_{(x>0)}\mathrm{d}x$. The required expression now follows from the quintuple law when marginalised to a triple law with the help of the expressions for $U(\mathrm{d}x)$, $\widehat{U}(\mathrm{d}x)$ and $\Pi(\mathrm{d}x)=x^{-(\alpha+1)}\mathrm{d}x$ (all up to a constant).

(iv) Since there is no creeping, there is no atom at zero in the overshoot distribution. The constant c must therefore be such that the given triple law is a probability distribution. We show that

$$\int_0^x \int_y^\infty \int_0^\infty \frac{(x-y)^{\alpha\rho-1}(v-y)^{\alpha(1-\rho)-1}}{(v+u)^{1+\alpha}} du\ dv\ dy = \frac{\pi}{\sin\alpha\rho\pi} \frac{\Gamma(\alpha\rho)\Gamma(\alpha(1-\rho))}{\Gamma(1+\alpha)} \tag{S.13}$$

to complete the exercise. We do this with the help of the Beta function,

$$\int_0^1 u^{p-1}(1-u)^{q-1}\mathrm{d}u = \frac{\Gamma(p)\Gamma(q)}{\Gamma(p+q)}$$

for $p, q > 0$.

First note that

$$\int_0^\infty \frac{1}{(v+u)^{1+\alpha}} du = \frac{1}{\alpha} v^{-\alpha}.$$

Next, changing variables with $s = y/v$ we have

$$\begin{aligned}\int_y^\infty (v-y)^{\alpha(1-\rho)-1} v^{-\alpha}\mathrm{d}v &= y^{-\alpha\rho}\int_0^1 (1-s)^{\alpha(1-\rho)-1}s^{\alpha\rho-1}\mathrm{d}s \\ &= \frac{\Gamma(\alpha(1-\rho))\Gamma(\alpha\rho)}{\Gamma(\alpha)}.\end{aligned}$$

Finally using the change of variables $t = y/x$ we have

$$\begin{aligned}\int_0^x (x-y)^{\alpha\rho-1}y^{-\alpha\rho}\mathrm{d}y &= \int_0^1 (1-t)^{\alpha\rho-1}t^{1-\alpha\rho-1}\mathrm{d}t \\ &= \frac{\Gamma(\alpha\rho)\Gamma(1-\alpha\rho)}{\Gamma(1)} \\ &= \frac{\pi}{\sin\alpha\rho\pi}.\end{aligned}$$

Gathering the constants from the above three integrals and recalling that $\Gamma(1+\alpha) = \alpha\Gamma(\alpha)$ completes the proof of (S.13)

7.5 For a given Lévy process, X, with the usual notation, by marginalising the quintuple law in Theorem 7.7 we have

$$\begin{aligned}&\mathbb{P}(\tau_x^+ - \overline{G}_{\tau_x^+-} \in \mathrm{d}t, \overline{G}_{\tau_x^+-} \in \mathrm{d}s, X_{\tau_x^+} - x \in \mathrm{d}u, x - \overline{X}_{\tau_x^+-} \in \mathrm{d}y) \\ &= \mathcal{U}(\mathrm{d}s, x-\mathrm{d}y)\int_{[y,\infty)} \widehat{\mathcal{U}}(\mathrm{d}t, \mathrm{d}v - y)\Pi(\mathrm{d}u + v) \\ &= \mathcal{U}(\mathrm{d}s, x-\mathrm{d}y)\int_{[0,\infty)} \widehat{\mathcal{U}}(\mathrm{d}t, \mathrm{d}\theta)\Pi(\mathrm{d}u + \theta + y)\end{aligned}$$

for $u > 0$, $y \in [0, x]$ and $s, t \geq 0$. In terms of the bivariate ascending ladder height process (L^{-1}, H), this quadruple is also equal in distribution to the quadruple $(\Delta L^{-1}_{T_x}, L^{-1}_{T_x-}, x - H_{T_x-}, H_{T_x} - x)$ where

$$T_x^+ = \inf\{t > 0 : H_t > x\}.$$

According to the conclusion of Exercise 5.4 we also have

$$\begin{aligned}&\mathbb{P}(\Delta L^{-1}_{T_x} \in \mathrm{d}t, L^{-1}_{T_x-} \in \mathrm{d}s, x - H_{T_x-} \in \mathrm{d}y, H_{T_x} - x \in \mathrm{d}u)\\ &= \mathcal{U}(\mathrm{d}s, x - \mathrm{d}y)\mathbf{\Pi}(\mathrm{d}t, \mathrm{d}u + y)\end{aligned}$$

for $u > 0$, $y \in [0, x]$ and $s, t \geq 0$. Comparing these two quadruple laws it follows that

$$\mathbf{\Pi}(\mathrm{d}t, \mathrm{d}u) = \int_{[0,\infty)} \widehat{\mathcal{U}}(\mathrm{d}t, \mathrm{d}\theta)\Pi(\mathrm{d}u + \theta).$$

When X is spectrally positive we have $\widehat{H}_t = t$ on $t < \widehat{L}_\infty$ and hence

$$\widehat{\mathcal{U}}(\mathrm{d}t, \mathrm{d}\theta) = \int_0^\infty \mathbb{P}(L^{-1}_s \in \mathrm{d}t, H_s \in \mathrm{d}\theta)\mathrm{d}s = \mathbb{P}(L^{-1}_\theta \in \mathrm{d}t)\mathrm{d}\theta$$

and the second claim follows.

7.6 Recall the Lévy–Khintchine formula for Ψ, the characteristic exponent of a general Lévy process X

$$\Psi(\theta) = \mathrm{i}a\theta + \frac{1}{2}\sigma^2\theta^2 + \int_{\mathbb{R}} (1 - \mathrm{e}^{\mathrm{i}\theta x} + \mathrm{i}\theta x \mathbf{1}_{(|x|<1)})\Pi(\mathrm{d}x)$$

for $\theta \in \mathbb{R}$ where $a \in \mathbb{R}$, $\sigma \geq 0$ and Π is a measure supported in $\mathbb{R}\backslash\{0\}$ such that $\int_{\mathbb{R}}(1 \wedge x^2)\Pi(\mathrm{d}x) < \infty$. It is clear that

$$\lim_{|\theta|\uparrow\infty} \frac{\mathrm{i}a\theta + \sigma^2\theta^2/2}{\theta^2} = \frac{\sigma^2}{2}$$

and we are therefore required to prove that

$$\lim_{|\theta|\uparrow\infty} \frac{1}{\theta^2}\int_{\mathbb{R}} (1 - \mathrm{e}^{\mathrm{i}\theta x} + \mathrm{i}\theta x \mathbf{1}_{(|x|<1)})\Pi(\mathrm{d}x) = 0. \tag{S.14}$$

Making use of the inequalities $|1 - \cos a| \leq 2(1 \wedge a^2)$ and $|a - \sin a| \leq 2(|a| \wedge |a|^3)$ one deduces that for all $|\theta|$ sufficiently large,

$$\left|\frac{(1 - \mathrm{e}^{\mathrm{i}\theta x} + \mathrm{i}\theta x \mathbf{1}_{(|x|<1)})}{\theta^2}\right| \leq 4(1 \wedge x^2).$$

Hence dominated convergence implies that the limit in (S.14) passes through the integral thus justifying the right-hand side.

According to Lemma 7.10, there is creeping upwards if and only if $\lim_{\beta\uparrow\infty}\kappa(0,\beta)/\beta > 0$. An analogous statement also holds for downward creeping. However, from the above and the Wiener–Hopf factorisation we have (up to a multiplicative constant)

$$\frac{\sigma^2}{2} = \lim_{|\theta|\uparrow\infty} \frac{\kappa(0, -\mathrm{i}\theta)}{\theta}\frac{\widehat{\kappa}(0, \mathrm{i}\theta)}{\theta}.$$

Creeping both upwards and downwards happens if and only if the right-hand side is non-zero and hence if and only if a Gaussian component is present.

(ii) Suppose that $d < 0$. By Theorem 6.5 we know that 0 is irregular for $[0,\infty)$ and hence the ascending ladder height process must be driftless with finite jump measure in which case creeping upwards is excluded. If $d = 0$ then by Exercise 5.9 X cannot hit points and therefore cannot creep. If $d > 0$ then from Sect. 6.1 we know that the local time at the maximum may be constructed in the form $L_t = \int_0^t \mathbf{1}_{(\overline{X}_s = X_s)} ds$, $t \geq 0$. In that case, Corollary 6.11 implies that the inverse local time process, L^{-1}, has a drift. Recalling that the ladder height process $H = X_{L^{-1}}$, since X has positive drift d it follows that one may identify a non-singular contribution (with respect to Lebesgue measure) to the expression for H equal to $d \times ct$ where c is the drift of L^{-1}. In other words, H has a drift and X creeps.

(iii) The proof of this part is contained in the proof of part (ii).

(iv) A spectrally negative Lévy process always creeps upwards by definition. If it has no Gaussian component then by part (i), since it is guaranteed to creep upwards, it cannot creep downwards.

(v) Any symmetric process must either creep in both directions or not at all (by symmetry). Since stable processes have no Gaussian component then any symmetric stable process cannot creep in either direction. A symmetric α-stable process has characteristic exponent $\Psi(\theta) = c|\theta|^\alpha$. Hence when $\alpha \in (1, 2)$ it is clear that

$$\int_{\mathbb{R}} \left(\frac{1}{1 + c|\theta|^\alpha} \right) d\theta < \infty.$$

As the latter process has unbounded variation it follows from Theorem 7.12 that symmetric α-stable processes with index $\alpha \in (1, 2)$ can hit all points but cannot creep.

7.7 (i) From Exercise 5.6 we know that when X is a stable subordinator with index $\alpha \in (0, 1)$ and $\tau_x^+ = \inf\{t > 0 : Y_t > x\}$ we have for $u \geq 0$

$$\begin{aligned} \frac{d}{du}\mathbb{P}(X_{\tau_x^+} - x \leq u) &= \frac{\alpha \sin \alpha\pi}{\pi} \int_0^x (x-y)^{\alpha-1}(y+u)^{-(\alpha+1)} dy \\ &= \frac{\alpha \sin \alpha\pi}{x\pi} \int_0^1 (1-\phi)^{\alpha-1} \left(\phi + \frac{u}{x}\right)^{-(\alpha+1)} d\phi, \end{aligned}$$

where in the second equality we have changed variables by $y = x\phi$. If we want to establish the required identity then, remembering that the ascending ladder height process is a stable subordinator with index $\alpha\rho$, we need to prove that the right-hand side above is equal to

$$\frac{d}{du}\Phi_\alpha\left(\frac{u}{x}\right) = \frac{\sin \alpha\pi}{\pi} \left(\frac{u}{x}\right)^{-\alpha} \left(1 + \frac{u}{x}\right)^{-1} \frac{1}{x}.$$

Equivalently we need to prove that for all $\theta > 0$

$$\theta^{-\alpha}(1+\theta)^{-1} = \alpha \int_0^1 (1-\phi)^{\alpha-1}(\phi+\theta)^{-(\alpha+1)} d\phi.$$

On the right-hand side above we can change variables via $(1+\theta)(1-u) = \phi + \theta$ to show that in fact our goal is to show that for all $\theta > 0$

$$\frac{\theta^{-\alpha}}{\alpha} = \int_0^{1/(1+\theta)} u^{\alpha-1}(1-u)^{-(\alpha+1)} du.$$

The quickest way to establish the above identity is to note that both sides are smooth and have the same derivatives as well as tending to 0 as $\theta \uparrow \infty$.

(ii) Note that

$$\begin{aligned} r(x,y) &= \mathbb{P}_x(X_{\tau_1^+} \leq 1+y) - \mathbb{P}_x(X_{\tau_1^+} \leq 1+y, \tau_0^- < \tau_1^+) \\ &= \Phi_{\alpha\rho}\left(\frac{y}{1-x}\right) - \int_{(0,\infty)} \mathbb{P}_x(-X_{\tau_0^-} \in \mathrm{d}z, \tau_0^- < \tau_1^+)\mathbb{P}_{-z}(X_{\tau_1^+} \leq 1+y) \\ &= \Phi_{\alpha\rho}\left(\frac{y}{1-x}\right) - \int_{(0,\infty)} l(x,\mathrm{d}z)\Phi_{\alpha\rho}\left(\frac{y}{1+z}\right). \end{aligned}$$

To obtain the second identity, one notes by considering $-X$ that one may replace r by l, ρ by $1-\rho$ and x by $1-x$.

(iii) It turns out to be easier to differentiate the equations in (ii) in the variable y and to check that the density of the given expression fits this equation. One transforms the solution for r into the solution for l by applying the same method at the end of part (ii) above.

Chapter 8

8.1 We know that for $x, q > 0$, $\mathbb{E}(\mathrm{e}^{-q\tau_x^+}) = \mathrm{e}^{-\Phi(q)x}$. As it has been assumed that $\psi'(0+) < 0$, it follows that $\Phi(0) > 0$ and hence taking limits as $q \uparrow \infty$ we have $\mathbb{P}(\tau_x^+ < \infty) = \mathrm{e}^{-\Phi(0)x}$. Now appealing to Bayes formula and then the Markov Property we have

$$\begin{aligned} \mathbb{P}(A|\tau_x^+ < \infty) &= \mathrm{e}^{\Phi(0)x}[\mathbb{P}(A, \infty > \tau_x^+ < t) + \mathbb{P}(A, \tau_x^+ \geq t)] \\ &= \mathrm{e}^{\Phi(0)x}[\mathbb{P}(A, \tau_x^+ < t) + \mathbb{E}(\mathbf{1}_{(A,\tau_x^+ \geq t)}\mathbb{P}_{X_t}(\tau_x^+ < \infty))] \\ &= \mathrm{e}^{\Phi(0)x}[\mathbb{P}(A, \tau_x^+ < t) + \mathbb{E}(\mathbf{1}_{(A,\tau_x^+ \geq t)}\mathrm{e}^{\Phi(0)(X_t - x)})] \\ &= \mathrm{e}^{\Phi(0)x}\mathbb{P}(A, \tau_x^+ < t) + \mathbb{P}^{\Phi(0)}(A, \tau_x^+ \geq t). \end{aligned}$$

The second term on the right-hand side tends to $\mathbb{P}^{\Phi(0)}(A)$ as $x \uparrow \infty$. We thus need to show that

$$\lim_{x\uparrow\infty} \mathrm{e}^{\Phi(0)x}\mathbb{P}(A, \tau_x^+ < t) = 0 \tag{S.15}$$

for all $t > 0$.

To this end, note that for $q > 0$

$$\mathrm{e}^{\Phi(0)x}\mathbb{P}(\tau_x^+ < \mathbf{e}_q) = \mathrm{e}^{\Phi(0)x}\mathbb{E}(\mathrm{e}^{-q\tau_x^+}) = \mathrm{e}^{-(\Phi(q)-\Phi(0))x},$$

where, as usual, $\mathbf{e}_q$ is a random variable independent of X which is exponentially distributed with parameter q. We know that Φ is strictly increasing (as ψ is strictly increasing) and hence $\lim_{x\uparrow\infty} \mathrm{e}^{\Phi(0)x}\mathbb{P}(A, \tau_x^+ < \mathbf{e}_q) = 0$. Since we can choose q arbitrarily small making $\mathbb{P}(\mathbf{e}_q > t) = \mathrm{e}^{-qt}$ arbitrarily close to 1 and since

$$\mathbb{P}(A, \tau_x^+ < \mathbf{e}_q) \geq \mathbb{P}(A, \tau_x^+ < t, t < \mathbf{e}_q)$$

the limit (S.15) follows.

8.2 (i) Note that for any spectrally negative process X, with the usual notation, integration by parts gives for $\beta > \Phi(q)$

$$\int_0^\infty \mathrm{e}^{-\beta x}\overline{W}^{(q)}(x)\mathrm{d}x = -\frac{1}{\beta}\lim_{x\uparrow\infty} \mathrm{e}^{-\beta x}\int_0^x W^{(q)}(y)\mathrm{d}y + \frac{1}{\beta}\int_0^\infty \mathrm{e}^{-\beta x}W^{(q)}(x)\mathrm{d}x.$$

The second term on the right-hand side is equal to $1/\beta(\psi(\beta)-q)$ and the first term is equal to zero since $W^{(q)}(x) = \mathrm{e}^{\Phi(q)x}W_{\Phi(q)}(x)$ and hence, writing $\beta = \Phi(q) + 2\varepsilon$ for some $\varepsilon > 0$,

$$\lim_{x\uparrow\infty} \mathrm{e}^{-\beta x}\int_0^x W^{(q)}(y)\mathrm{d}y \le \lim_{x\uparrow\infty} \mathrm{e}^{-\varepsilon x}\int_0^x \mathrm{e}^{-\varepsilon y}W_{\Phi(q)}(y)\mathrm{d}y.$$

We have used here that X under $\mathbb{P}^{\Phi(q)}$ drifts to infinity and hence $\int_0^\infty \mathrm{e}^{-\varepsilon y}W_{\Phi(q)}(y)\mathrm{d}y < \infty$. In particular using a standard geometric series expansion we have

$$\int_0^\infty \mathrm{e}^{-\beta x}\overline{W}^{(q)}(x)\mathrm{d}x = \frac{1}{\beta}\frac{1}{\beta^\alpha - q} = \frac{1}{\beta^{1+\alpha}}\sum_{k\ge 0}\left(\frac{q}{\beta^\alpha}\right)^k = \sum_{n\ge 1} q^{n-1}\beta^{-n\alpha-1}.$$

(ii) For $q > 0$ and $\beta > \Phi(q)$,

$$\int_0^\infty \mathrm{e}^{-\beta x}Z^{(q)}(x)\mathrm{d}x = \frac{1}{\beta} + \int_0^\infty \mathrm{e}^{-\beta x}\overline{W}^{(q)}(x)\mathrm{d}x = \sum_{n\ge 0} q^n\beta^{-\alpha n-1}.$$

On the other hand, with the help of Fubini's Theorem,

$$\begin{aligned}\int_0^\infty \mathrm{e}^{-\beta x}\sum_{n\ge 0} q^n\frac{x^{\alpha n}}{\Gamma(1+n\alpha)}\mathrm{d}x &= \sum_{n\ge 0}\frac{q^n}{\Gamma(1+n\alpha)}\frac{1}{\beta^{\alpha n+1}}\int_0^\infty \mathrm{e}^{-z}z^{\alpha n}\mathrm{d}z \\ &= \sum_{n\ge 0} q^n\beta^{-\alpha n-1} \\ &= E_\alpha(qx^\alpha).\end{aligned}$$

Continuity of $Z^{(q)}$ means that we can identify it as the given series. The case $q=0$ has $Z^{(q)} = 1$ by definition.

(iii) We have for $q > 0$,

$$W^{(q)}(x) = \frac{1}{q}\frac{\mathrm{d}}{\mathrm{d}x}Z^{(q)}(x) = \alpha x^{\alpha-1}E'_\alpha(qx^\alpha).$$

Since we know from Lemma 8.3 that $W^{(q)}(x)$ is continuous in q for each fixed x, it follows that $W(x) = \alpha x^{\alpha-1}$.

(iv) The function $W^{(q)}$ is obtained by a standard exercise in Laplace inversion of $(\beta^2/2 - q)^{-1}$ and $Z^{(q)}$ follows easily thereafter by its definition in terms of $W^{(q)}$.

8.3 (i) The assumption $\lim_{t\uparrow\infty} X_t$ implies that $\psi'(0+) > 0$ where ψ is the Laplace exponent of X. The latter can be written in the form

$$\psi(\theta) = d\theta - \theta\int_{(0,\infty)} \mathrm{e}^{-\theta x}\Pi(x,\infty)\mathrm{d}x$$

for $\theta \ge 0$ (see (8.1) and Exercise 2.11). Taking derivatives we have

$$\psi'(0+) = d - \int_0^\infty \Pi(x,\infty)\mathrm{d}x$$

thus showing that $d^{-1}\int_0^\infty \Pi(x,\infty)\mathrm{d}x < \infty$.

(ii) From (8.17) and the representation of ψ given above we now have for $\beta > 0$,

$$\begin{aligned}\int_{[0,\infty)} \mathrm{e}^{-\beta x} W(\mathrm{d}x) &= \frac{1}{d - \int_0^\infty \mathrm{e}^{-\beta x}\Pi(x,\infty)\mathrm{d}x} \\ &= \frac{1}{d}\sum_{k\geq 0}\left(\frac{1}{\mathrm{d}}\int_0^\infty \mathrm{e}^{-\beta x}\Pi(x,\infty)\mathrm{d}x\right)^k \\ &= \frac{1}{d}\sum_{k\geq 0}\int_0^\infty \mathrm{e}^{-\beta x}\nu^{*k}(\mathrm{d}x) \\ &= \frac{1}{d}\int_0^\infty \mathrm{e}^{-\beta x}\sum_{k\geq 0}\nu^{*k}(\mathrm{d}x),\end{aligned}$$

where the final equality follows by Fubini's Theorem and we understand $\nu^{*0}(\mathrm{d}x) = \delta_0(\mathrm{d}x)$. It follows that

$$W(\mathrm{d}x) = d^{-1}\sum_{n\geq 0}\nu^{*n}(\mathrm{d}x) \tag{S.16}$$

on $[0,\infty)$.

(iii) In the case that S is a compound Poisson subordinator with jump rate $\lambda > 0$ and jumps which are exponentially distributed with parameter μ we have that $\Pi(x,\infty) = \lambda\mathrm{e}^{-\mu x}$ and the condition $d^{-1}\int_0^\infty \Pi(x,\infty)\mathrm{d}x < 1$ implies that $\lambda < d\mu$. In addition for $n \geq 1$, $\nu^{*n}(\mathrm{d}x) = ((n-1)!)^{-1}(\lambda/d)^n x^{n-1}\mathrm{e}^{-\mu x}\mathrm{d}x$. Now note that

$$\begin{aligned}W(x) &= \frac{1}{d}\left(1 + \sum_{n\geq 1}\nu^{*n}[0,x]\right) \\ &= \frac{1}{d}\left(1 + \sum_{n\geq 1}\left(\frac{\lambda}{d}\right)^n \frac{1}{(n-1)!}\int_0^x y^{n-1}\mathrm{e}^{-\mu y}\mathrm{d}y\right) \\ &= \frac{1}{d}\left(1 + \int_0^x \mathrm{e}^{-\mu y}\frac{\lambda}{d}\sum_{k\geq 0}\frac{1}{k!}\left(\frac{\lambda y}{d}\right)^k \mathrm{d}y\right) \\ &= \frac{1}{d}\left(1 + \frac{\lambda}{d}\int_0^x \mathrm{e}^{-(\mu - d^{-1}\lambda)y}\mathrm{d}y\right) \\ &= \frac{1}{d}\left(1 + \frac{\lambda}{d\mu - \lambda}(1 - \mathrm{e}^{-(\mu - d^{-1}\lambda)x})\right).\end{aligned}$$

8.4 (i) From (S.16) Exercise 8.3 (ii) it is clear that for $x > 0$ (so that the term $\delta_0(\mathrm{d}x) = 0$) the measure $W(\mathrm{d}x)$ is absolutely continuous with respect to Lebesgue measure as the same is true of ν. The smoothness of the density is dictated by the term indexed $n = 1$ in (S.16) as all higher convolutions of ν have a density which can be expressed in terms of Lebesgue integrals of the form $\int_0^x \cdots \mathrm{d}y$. Indeed we see that W has a continuous density if and only if $\Pi(x,\infty)$ has no jumps, in other words, Π has no atoms.

(ii) To obtain a continuous kth derivative of W we need to impose conditions on the convolutions indexed $n = 1, ..., k-1$ in (S.16). Specifically the minimum required (which pertains to the term indexed $n = 1$) is that $\Pi(x, \infty)$ has a continuous $(n-1)$th derivative.

(iii) We may write $W^{(q)}(x) = \mathrm{e}^{\Phi(q)x} W_{\Phi(q)}(x)$ where $W_{\Phi(q)}$ plays the role of W for $(X, \mathbb{P}^{\Phi(q)})$ (note in the case that $q = 0$ and $\lim_{t\uparrow\infty} X_t = -\infty$ we have that $\Phi(0) > 0$). Recall in particular that under $\mathbb{P}^{\Phi(q)}$ the process X remains within the class of spectrally negative Lévy processes but now it drifts to ∞. Also, when decomposed into the difference of a positive drift and a pure jump subordinator, the associated Lévy measure of the subordinator is $\Pi_{\Phi(q)}(\mathrm{d}x) = \mathrm{e}^{\Phi(q)x}\Pi(\mathrm{d}x)$. Hence the result of part (ii) applies to $\Pi_{\Phi(q)}$. In turn this implies $W^{(q)}$ is n times differentiable with continuous derivatives if and only if $\Pi(x, \infty)$ is $n - 1$ times differentiable with continuous derivatives.

(iv) Finally, when $q = 0$ and X oscillates, from (8.24) we have that $W^{(q)}(x) = \sum_{k\geq 0} q^k W^{*(k+1)}(x)$ and hence considering the term indexed $k = 0$ we see that W is continuously differentiable if and only if $W^{(q)}$ is, which establishes the claim on account of the conclusion of part (ii).

8.5 (i) First fix $q > 0$. Taking limits in (8.9) as $a \uparrow \infty$ we must obtain agreement with (8.6) by dominated convergence. The first result follows immediately. The second result follows by taking limits as $a \uparrow \infty$ in (8.8) when x is replaced by $a - x$, applying dominated convergence again and comparing with the conclusion of Theorem 3.12. For both cases when $q = 0$, one may take limits as $q \downarrow$ in the preceding conclusions.

(ii) Integrating by parts wc havc for $q > 0$ and $\beta > \Phi(q)$,

$$
\begin{aligned}
\int_{[0,\infty)} \mathrm{e}^{-\beta x} W^{(q)}(\mathrm{d}x) &= W^{(q)}(\{0\}) + \int_{(0,\infty)} \mathrm{e}^{-\beta x} W^{(q)}(\mathrm{d}x) \\
&= \int_0^\infty \beta \mathrm{e}^{-\beta x} W^{(q)}(\{0\})\mathrm{d}x + \int_0^\infty \beta \mathrm{e}^{-\beta x} W^{(q)}(0, x]\mathrm{d}x \\
&= \beta \int_0^\infty \mathrm{e}^{-\beta x} W^{(q)}(x)\mathrm{d}x \\
&= \frac{\beta}{\psi(\beta) - q}.
\end{aligned}
$$

Noting further that $W^{(q)}$ is always differentiable we have

$$
\int_0^\infty \mathrm{e}^{-\beta x} W^{(q)\prime}(x)\mathrm{d}x + W^{(q)}(0) = \frac{\beta}{\psi(\beta) - q}.
$$

Next suppose that X has unbounded variation. In that case we know that $W^{(q)}(0) = 0$ and further that $W^{(q)}$ is differentiable. Hence

$$
W^{(q)\prime}(0) = \lim_{\beta\uparrow\infty} \int_0^\infty \beta \mathrm{e}^{-\beta x} W^{(q)\prime}(x)\mathrm{d}x = \lim_{\beta\uparrow\infty} \frac{\beta^2}{\psi(\beta) - q} = \frac{2}{\sigma^2},
$$

where the last equality follows from Exercise 7.6 (i).

Now suppose that X has bounded variation. In this case we have from Lemma 8.6 that $W^{(q)}(0) = d^{-1}$ where d is the drift and hence

$$
\begin{aligned}
W^{(q)\prime}(0) &= \lim_{\beta\uparrow\infty}\int_0^\infty \beta \mathrm{e}^{-\beta x} W^{(q)\prime}(x)\mathrm{d}x \\
&= \lim_{\beta\uparrow\infty} \frac{\beta^2}{\mathrm{d}\beta - \beta\int_0^\infty \mathrm{e}^{-\beta x}\Pi(-\infty,-x)\mathrm{d}x - q} - \beta W^{(q)}(0) \\
&= \lim_{\beta\uparrow\infty} \frac{\beta^2(1 - W^{(q)}(0)d + W^{(q)}(0)\int_0^\infty \mathrm{e}^{-\beta x}\Pi(-\infty,-x)\mathrm{d}x) + q\beta W^{(q)}(0)}{\mathrm{d}\beta - \int_0^\infty \beta \mathrm{e}^{-\beta x}\Pi(-\infty,-x)\mathrm{d}x - q} \\
&= \lim_{\beta\uparrow\infty} \frac{1}{d}\frac{\int_0^\infty \beta \mathrm{e}^{-\beta x}\Pi(-\infty,-x)\mathrm{d}x + q}{d - \int_0^\infty \mathrm{e}^{-\beta x}\Pi(-\infty,-x)\mathrm{d}x} \\
&= \frac{\Pi(-\infty,0)+q}{d^2}.
\end{aligned}
$$

In particular, if $\Pi(-\infty,0)=\infty$ then the right-hand side above is equal to ∞ and otherwise if $\Pi(-\infty,0)<\infty$ then $W^{(q)\prime}(0)$ is finite.

In conclusion $W^{(q)\prime}(0)$ is finite if and only if X has a Gaussian component or $\Pi(-\infty,0)<\infty$.

8.6 It has been established that a spectrally negative Lévy process creeps downwards if and only if it has a Gaussian component ($\sigma>0$). Hence if $\sigma=0$ then $\mathbb{P}(X_{\tau_x^+}=x)=0$. Without loss of generality we are thus left to deal with the case that $\sigma>0$.

We know that for all $x\leq 0$, $\mathbb{P}(X_{\tau_x^+}=x)=\mathbb{P}(\widehat{H}_{T^+_{-x}}=-x)$ where $\widehat{H}$ is the descending ladder height process and $T^+_{-x}=\inf\{t>0:\widehat{H}_t>-x\}$. According to Theorem 5.9 $\mathbb{P}(\widehat{H}_{T^+_{-x}}=-x)=\widehat{c}\widehat{u}(-x)$ where $\widehat{c}$ is the drift of $\widehat{H}$ and $\widehat{u}$ is the continuous version of the density of $\widehat{U}(y)=\mathbb{E}(\int_0^\infty \mathbf{1}_{(\widehat{H}_t\leq y)}\mathrm{d}t)$ when $\widehat{c}>0$ and otherwise equal to zero. Recalling from the Wiener–Hopf factorisation for spectrally negative processes that

$$
\begin{aligned}
\int_0^\infty \mathrm{e}^{-\beta y}\widehat{u}(y)\mathrm{d}y &= \int_{[0,\infty)} \mathrm{e}^{-\beta y}\widehat{U}(\mathrm{d}y) \\
&= \frac{1}{\widehat{\kappa}(0,\beta)} \\
&= \frac{\beta-\Phi(0)}{\psi(\beta)}
\end{aligned}
$$

for some appropriate normalisation of local time. (Recall that $\widehat{\kappa}(0,\infty)$ is the Laplace exponent of $\widehat{H}$). Since $\int_0^\infty \mathrm{e}^{-\beta x}W(x)\mathrm{d}x=\psi(\beta)^{-1}$ and $\int_0^\infty \mathrm{e}^{-\beta x}W'(x)\mathrm{d}x=\beta\psi(\beta)^{-1}$ for $\beta>\Phi(0)$ (see Exercise 8.5) it follows by continuity and Laplace inversion that

$$
\widehat{u}(y)=W'(y)-\Phi(0)W(y).
$$

With the particular normalisation of local time we have chosen, we can identify the drift $\widehat{c}$ by

$$
\widehat{c}=\lim_{\beta\uparrow\infty}\frac{\widehat{\kappa}(0,\beta)}{\beta}=\lim_{\beta\uparrow\infty}\frac{\psi(\beta)}{\beta^2}=\frac{\sigma^2}{2}.
$$

(Here we have used Exercise 7.6 (i)).

8.7 (i) By definition

$$\begin{aligned}\Phi_c(q) &= \sup\{\lambda \in \mathbb{R} : \psi_c(\lambda) = q\}\\ &= \sup\{\lambda \in \mathbb{R} : \psi(\lambda + c) = \psi(c) + q\}\\ &= \Phi(\psi(c) + q) - c.\end{aligned}$$

(ii) Appealing to an exponential change of measure we have for $x > 0$, $c \geq 0$ and $p > \psi(c) \vee 0$,

$$\begin{aligned}\mathbb{E}\left(\mathrm{e}^{-p\tau^-_{-x} + cX_{\tau^-_{-x}}}\mathbf{1}_{(\tau^-_{-x}<\infty)}\right) &= \mathbb{E}\left(\mathrm{e}^{-q\tau^-_{-x} + cX_{\tau^-_{-x}} - \psi(c)\tau^-_{-x}}\mathbf{1}_{(\tau^-_{-x}<\infty)}\right)\\ &= \mathbb{E}^c\left(\mathrm{e}^{-q\tau^-_{-x}}\mathbf{1}_{(\tau^-_{-x}<\infty)}\right)\\ &= \mathbb{E}^c_x\left(\mathrm{e}^{-q\tau^-_0}\mathbf{1}_{(\tau^-_0<\infty)}\right)\\ &= Z^{(q)}_c(x) - \frac{q}{\Phi_c(q)}W^{(q)}_c(x), \end{aligned}\tag{S.17}$$

where $q = p - \psi(c)$. The case that $p = \psi(c)$ is dealt with by taking limits as $p \downarrow \psi(c)$. For each fixed $c \geq 0$ the left-hand side is finite for all $p \geq 0$ and hence can be extended analytically to $\mathbb{C}^+ = \{z \in \mathbb{C} : \Re z \geq 0\}$. We also know that the functions $Z^{(q)}_c(x)$ and $W^{(q)}_c(x)$ can be extended analytically to $\mathbb{C}$ for fixed x. With regard to the factor $q/\Phi_c(q)$, note that it is a constant if $q = 0$ and otherwise

$$\frac{q}{\Phi_c(q)} = \frac{p - \psi(c)}{\Phi(p) - c}.$$

The term $\Phi(p)$ is the Laplace exponent of a (possibly killed) subordinator and up to the addition of a constant takes the form

$$\Phi(p) = dp + \int_{(0,\infty)} (1 - \mathrm{e}^{-px})\nu(\mathrm{d}x) \geq 0$$

for some measure ν. The term $q/\Phi_c(q)$ is thus also analytically extendable to $\mathbb{C}^+$. Since both left- and right-hand side of (S.17) are equal on a set of accumulation points on the interior of $\mathbb{C}^+$, the Identity Theorem of complex analysis allows us to conclude that they agree for $p \in \{z \in \mathbb{C} : \Re z > 0\}$. By taking limits as $p \downarrow 0$ in (S.17) we see that they also agree on $\mathbb{C}^+$.

(iii) Conditioning on $\mathcal{F}_{\tau^-_{-x}}$ we have

$$\begin{aligned}&\mathbb{E}\left(\mathbb{E}\left(\mathrm{e}^{-pT(-x) - u(T(-x) - \tau^-_{-x})}\mathbf{1}_{(T(-x)<\infty)}\middle|\mathcal{F}_{\tau^-_{-x}}\right)\right)\\ &= \mathbb{E}\left(\mathrm{e}^{-p\tau^-_{-x}}\mathbf{1}_{(\tau^-_{-x}<\infty)}\mathbb{E}\left(\mathrm{e}^{-(p+u)(T(-x)-\tau^-_{-x})}\mathbf{1}_{(T(-x)-\tau^-_{-x}<\infty)}\middle|\mathcal{F}_{\tau^-_{-x}}\right)\right)\\ &= \mathbb{E}\left(\mathrm{e}^{-p\tau^-_{-x}}\mathbf{1}_{(\tau^-_{-x}<\infty)}\mathbb{E}_{X_{\tau^-_{-x}}}\left(\mathrm{e}^{-(p+u)\tau^+_{-x}}\mathbf{1}_{(\tau^+_{-x}<\infty)}\right)\right)\\ &= \mathbb{E}\left(\mathrm{e}^{-p\tau^-_{-x}}\mathbf{1}_{(\tau^-_{-x}<\infty)}\mathrm{e}^{-\Phi(p+u)(-x-X_{\tau^-_{-x}})}\right),\end{aligned}$$

which is the required identity. Note that we have used that conditional on $\mathcal{F}_{\tau^-_{-x}}$ the time difference $T(-x)-\tau^-_{-x}$ is equal in law of first passage from $X_{\tau^-_{-x}}$ to $-x$.

(iv) Set $c=\Phi(p+u)$ and $q=p-\psi(\Phi(p+u))=-u$. Note then that since

$$\begin{aligned}\Phi_{\Phi(p+u)}(q) &= \Phi(q+\psi(\Phi(p+u)))-\Phi(p+u)\\ &= \Phi(p)-\Phi(p+u)\end{aligned}$$

we have

$$\begin{aligned}&\mathbb{E}\left(\mathrm{e}^{-pT(-x)-u(T(-x)-\tau^-_{-x})}\mathbf{1}_{(T(-x)<\infty)}\right)\\ &= \mathrm{e}^{\Phi(p+u)x}\left(Z^{(-u)}_{\Phi(p+u)}(x)-\frac{-u}{\Phi(p)-\Phi(p+u)}W^{(-u)}_{\Phi(p+u)}(x)\right).\end{aligned}$$

Hence taking limits as $u\downarrow 0$ we see that

$$\begin{aligned}\mathbb{E}\left(\mathrm{e}^{-pT(-x)}\mathbf{1}_{(T(-x)<\infty)}\right) &= \mathrm{e}^{\Phi(p)x}-\frac{1}{\Phi'(p)}\mathrm{e}^{\Phi(p)x}W_{\Phi(p)}(x)\\ &= \mathrm{e}^{\Phi(p)x}-\frac{1}{\Phi'(p)}W^{(p)}(x).\end{aligned}$$

Notice that since $\psi(\Phi(p))=p$ then by differentiating both sides with respect to p it follows that $\psi'(\Phi(p))=1/\Phi'(p)$.

Now taking limits as $x\downarrow 0$ and recalling that $W^{(p)}(0)=1/d$ if X has bounded variation with drift d and otherwise is equal to zero the stated result follows.

8.8 [9] (i) The event $\{\exists t>0: B_t=\overline{B}_t=t\}$ is equivalent to $\{\exists s>0: L^{-1}_s=H_s\}$ where (L^{-1},H) is the ascending ladder height process. However, for a Brownian motion $H_s=s$ and L^{-1}_s is a Stable-$\frac{1}{2}$ subordinator with Laplace exponent $\sqrt{2\theta}$ for $\theta\geq 0$. Let X be the difference of the latter process and a positive unit drift, then if $T(0)=\inf\{t>0:X_t=0\}$, we have

$$\mathbb{P}(\sigma<\infty)=\mathbb{P}(\exists t>0: B_t=\overline{B}_t=t)=\mathbb{P}(\exists s>0: X_s=0)=\mathbb{P}(T(0)<\infty).$$

(ii) Now note that the Laplace exponent of X is given by $\psi(\theta)=\theta-\sqrt{2\theta}$ for $\theta\geq 0$. The latter is a process of bounded variation and $\psi'(0)=-\infty$ showing that (in the usual notation) $\Phi(0)>0$. In fact $\Phi(0)$ is the largest solution of $\theta=\sqrt{2\theta}$. In other words $\Phi(0)=2$. According to Exercise 8.7 (iv) we have

$$\mathbb{P}(T(0)<\infty)=1-\psi'(\Phi(0))=1-\left(1-\frac{1}{2}\cdot\sqrt{2}\cdot 2^{-\frac{1}{2}}\right)=\frac{1}{2}.$$

8.9 (i) Let N be Poisson random measure associated with the jumps of X. Note that $N=\sum_{i=1}^n N^{(i)}$ where for $i=1,...,n$, $N^{(i)}$ is the Poisson random measure associated with the jumps of $X^{(i)}$. As $N^{(1)},...,N^{(n)}$ are independent, a little consideration of how such independent Poisson random measures are constructions reveals that they have disjoint supports with probability one. The compensation formula gives us for Borel A

[9] It is worth pointing out that the solution to this exercise can be adapted to cover the case when we replace B by any spectrally negative Lévy process in the original question.

$$
\begin{aligned}
&\mathbb{P}_x(X_{\tau_0^- -} \in \mathrm{d}y, X_{\tau_0^-} \in A, \Delta X_{\tau_0^-} = \Delta X^{(i)}_{\tau_0^-}) \\
&= \mathbb{E}_x\left(\int_{[0,\infty)}\int_{(-\infty,0)} \mathbf{1}_{(\underline{X}_{t-}>0)}\mathbf{1}_{(X_{t-}\in\mathrm{d}y)}\mathbf{1}_{(y+a\in A)}N^{(i)}(\mathrm{d}t\times\mathrm{d}a)\right) \\
&= \mathbb{E}_x\left(\int_0^\infty \mathrm{d}t\cdot\mathbf{1}_{(\underline{X}_t>0)}\mathbf{1}_{(X_t\in\mathrm{d}y)}\right)\int_{A-y}\Pi^{(i)}(\mathrm{d}a) \\
&= \int_0^\infty \mathrm{d}t\cdot\mathbb{P}_x(X_t\in\mathrm{d}y,\ \tau_0^->t)\int_A \Pi^{(i)}(\mathrm{d}z-y) \\
&= r(x,y)\int_A \Pi^{(i)}(\mathrm{d}z-y)\mathrm{d}y.
\end{aligned}
$$

Note in particular, as X is spectrally negative there is no downward creeping and 0 is irregular for $(-\infty,0]$ hence the above calculations remain valid even when x or y are equal to zero.

(ii) From Corollary 8.8 we have that

$$
r(0,y) = \mathrm{e}^{-\Phi(0)}W(0) - W(-y),
$$

where Φ is the right inverse of the Laplace exponent of X. Since $y>0$ we have $W(-y)=0$ and as X is of bounded variation then $W(0)=1/d$. Also, since X drifts to ∞ we have that $\Phi(0)=0$. The required expression now follows.

(iii) Noting that X cannot creep below the origin, we have

$$
\begin{aligned}
\mathbb{P}(\tau_0^-<\infty, \Delta X_{\tau_0^-} = \Delta X^{(i)}_{\tau_0^-}) &= \frac{1}{\mathrm{d}}\int_{(-\infty,0)}\int_0^\infty \Pi^{(i)}(\mathrm{d}z-y)\mathrm{d}y \\
&= \frac{1}{\mathrm{d}}\int_0^\infty \Pi^{(i)}(-\infty,-y)\mathrm{d}y \\
&= \frac{\mathbb{E}(X_1^{(i)})-d_1}{d} \\
&= \frac{\mu_i}{d}.
\end{aligned}
$$

8.10 (i) Starting with an integration by parts, we have

$$
\begin{aligned}
\mathbb{E}_y(\mathrm{e}^{-q\Lambda_0}) &= q\int_0^\infty \mathrm{e}^{-qt}\mathbb{P}_y(0\le\Lambda_0<t)\mathrm{d}t \\
&= q\int_0^\infty \mathrm{d}t\cdot\mathrm{e}^{-qt}\int_{[0,\infty)}\mathbb{P}_y(X_t\in\mathrm{d}x)\mathbb{P}_x(\underline{X}_\infty\ge 0) \\
&= q\int_0^\infty \theta^{(q)}(x-y)\mathbb{P}_x(\underline{X}_\infty\ge 0)\mathrm{d}x.
\end{aligned}
$$

(ii) Next recall from (8.15) that since $\lim_{t\uparrow\infty}X_t=\infty$

$$
\mathbb{P}_x(\underline{X}_\infty\ge 0) = \psi'(0+)W(x).
$$

Recall also from Corollary 8.9 that for $z\in\mathbb{R}$

$$
\theta^{(q)}(z) = \Phi'(q)\mathrm{e}^{-\Phi(q)z} - W^{(q)}(-z).
$$

We shall also need that $W^{(q)}$ is zero on $(-\infty, 0)$ and that $\int_0^\infty e^{-\beta x} W(x)dx = 1/\psi(\beta)$ for $\beta > 0$. From part (i) we thus have for $y \leq 0$

$$\begin{aligned}
\mathbb{E}_y(e^{-q\Lambda_0}) &= q\psi'(0+)\Phi'(q) \int_0^\infty e^{-\Phi(q)(x-y)} W(x)dx \\
&\quad -q\psi'(0+) \int_0^\infty W^{(q)}(y-x)W(x)dx \\
&= q\psi'(0+)\Phi'(q)e^{\Phi(q)y} \frac{1}{\psi(\Phi(q))} \\
&= \psi'(0+)\Phi'(q)e^{\Phi(q)y}.
\end{aligned}$$

Setting $y = 0$ we thus have

$$\mathbb{P}(\Lambda_0 = 0) = \lim_{q\uparrow\infty} \psi'(0+)\Phi'(q).$$

Note however that $1 = \psi(\Phi(q))' = \psi'(\Phi(q))\Phi'(q)$, that $\Phi(q) \to \infty$ as $q \uparrow \infty$ and hence the above limit is equal to $\psi'(0+)/\psi'(\infty)$. When X has paths of bounded variation with drift d, L'Hôpital's rule implies that $\lim_{\theta\uparrow\infty} \psi'(\theta) = \lim_{\theta\uparrow\infty} \psi(\theta)/\theta = d$ where the last equality follows from Exercise 2.11. If on the other hand X has paths of unbounded variation, then necessarily $\mathbb{P}(\Lambda_0 = 0) = 0$ because 0 is regular for $(-\infty, 0)$ (and hence we deduce that necessarily that for the case at hand $\psi'(\infty) = \infty$ and $\Phi'(\infty) = 0$).

(iii) Note that on $\{\Lambda_0 > 0\}$ we have $\Lambda_0 > \tau_0^-$ and hence conditioning on $\mathcal{F}_{\tau_0^-}$ and applying the Strong Markov Property

$$\begin{aligned}
\mathbb{E}_y(e^{-q\Lambda_0}\mathbf{1}_{(\Lambda_0>0)}) &= \mathbb{E}_y(e^{-q\tau_0^-} \mathbb{E}_{X_{\tau_0^-}}(e^{-q\Lambda_0})\mathbf{1}_{(\tau_0^-<\infty)}) \\
&= \psi'(0+)\Phi'(q)\mathbb{E}_y(e^{-q\tau_0^- + \Phi(q)X_{\tau_0^-}}\mathbf{1}_{(\tau_0^-<\infty)}),
\end{aligned}$$

where in the penultimate equality we have used the fact that $\mathbb{P}_y(\Lambda_0 > 0) = 1$ when $y < 0$ (by virtue of the fact that X creeps upwards) and in the final equality we have used the conclusion of part (ii).

(iv) Changing measure to $\mathbb{P}^{\Phi(q)}$ now gives

$$\begin{aligned}
\mathbb{E}_y(e^{-q\Lambda_0}\mathbf{1}_{(\Lambda_0>0)}) &= \psi'(0+)\Phi'(q)e^{\Phi(q)y}\mathbb{P}_y^{\Phi(q)}(\tau_0^- < \infty) \\
&= \psi'(0+)\Phi'(q)e^{\Phi(q)y}(1 - \psi'_{\Phi(q)}(0+)W_{\Phi(q)}(y)) \\
&= \psi'(0+)\Phi'(q)e^{\Phi(q)y} - \psi'(0+)W^{(q)}(y),
\end{aligned}$$

where in the second equality we have used Theorem 8.1 (ii) and in the final equaltiy we have used Lemma 8.4 and (8.4) and the fact that $q = \psi(\Phi(q))$ implies that $1 = \psi'(\Phi(q))\Phi'(q)$.

8.11 (i) The process Z^x will either exit from $[0, a)$ by first hitting zero or by directly passing above a before hitting zero. The process Z^x up to first hitting of zero behaves like X up to first passage below zero. These facts together with the Strong Markov Property explain the identity.

(ii) Taking the limit as x tends to zero we have

$$\mathbb{E}\left(\mathrm{e}^{-q\underline{\sigma}_a^0}\right) = \left(1 - \frac{W^{(q)}(0)}{W^{(q)}(a)} Z^{(q)}(a)\right) \mathbb{E}\left(\mathrm{e}^{-q\underline{\sigma}_a^0}\right) + \frac{W^{(q)}(0)}{W^{(q)}(a)}$$

showing that

$$\mathbb{E}\left(\mathrm{e}^{-q\underline{\sigma}_a^0}\right) = \frac{1}{Z^{(q)}(a)}.$$

Replacing this last equality into the identity in (i) we have

$$\begin{aligned}\mathbb{E}\left(\mathrm{e}^{-q\underline{\sigma}_a^x}\right) &= \left(Z^{(q)}(z) - \frac{W^{(q)}(z)}{W^{(q)}(a)} Z^{(q)}(a)\right) \frac{1}{Z^{(q)}(a)} + \frac{W^{(q)}(z)}{W^{(q)}(a)} \\ &= \frac{Z^{(q)}(z)}{Z^{(q)}(a)}.\end{aligned}$$

(iii) The path of the workload, taking account of the buffer, up to the first time the workload becomes zero is precisely that of a spectrally negative Lévy process of bounded variation reflected in its infimum run until first passage over the level c and initiated from $c - x$ units of workload. (This is best seen by drawing a sketch of the paths of the workload and rotating them by 180°). Hence, in the terminology of the previous parts of the question, the required Laplace transform is thus equal to $\mathbb{E}(\mathrm{e}^{-\underline{\sigma}_c^{c-x}}) = Z^{(q)}(x - c)/Z^{(q)}(c)$.

8.12 (i) This is a repetition of Exercise 7.7 (ii) with some simplifications. One can take advantage of the fact that $r(x, y) = r(x, 0) = \mathbb{P}(\tau_1^+ < \tau_0^-)$ which follows from spectral negativity. Recall also from Exercise 6.6 (i) that $\rho = 1/\alpha$. Reconsidering the expression for $l(x, y)$ we see that it simplifies to the given expression.

(ii) From Exercise 8.2 we know that $W(x) = \alpha x^{\alpha-1}$ and hence from Theorem 8.1 (iii), $\mathbb{P}_x(\tau_1^+ < \tau_0^-) = x^{\alpha-1}$. Plugging this into the expression derived in part (i) we have

$$\begin{aligned}&\mathbb{P}_x(-X_{\tau_0^-} \le y; \tau_0^- < \tau_1^+) \\ &= \frac{\sin\pi(\alpha-1)}{\pi}\left(\int_0^{y/x} t^{-(\alpha-1)}(t+1)^{-1}\mathrm{d}t - x^{\alpha-1}\int_0^y t^{-(\alpha-1)}(1+t)^{-1}\mathrm{d}t\right) \\ &= \frac{\sin\pi(\alpha-1)}{\pi} x^{\alpha-1}\left(\int_0^y t^{-(\alpha-1)}(x+t)^{-1}\mathrm{d}t - \int_0^y t^{-(\alpha-1)}(1+t)^{-1}\mathrm{d}t\right) \\ &= \frac{\sin\pi(\alpha-1)}{\pi} x^{\alpha-1}(1-x)\int_0^y t^{-(\alpha-1)}(x+t)^{-1}(1+t)^{-1}\mathrm{d}t.\end{aligned}$$

Chapter 9

9.1 (i) Fix $a > 0$. Note that on the event $\{T_a^+ \le \mathbf{e}_q\} = \{\overline{X}_{\mathbf{e}_q} \ge a\}$ we have that $\overline{X}_{\mathbf{e}_q} = X_{T_a^+} + S$ where stationary independent increments and the lack of memory property imply that S is independent of $\mathcal{F}_{T_a^+}$ and equal in distribution to $\overline{X}_{\mathbf{e}_q}$. We thus have

$$
\begin{aligned}
&\mathbb{E}_x\left(\mathbf{1}_{(\overline{X}_{\mathbf{e}_q}\geq a)}\left(1-\frac{\mathrm{e}^{-\overline{X}_{\mathbf{e}_q}}}{\mathbb{E}(\mathrm{e}^{-\overline{X}_{\mathbf{e}_q}})}\right)\right)\\
&=\mathbb{E}_x\left(\mathbf{1}_{(T_a^+\leq \mathbf{e}_q)}\mathbb{E}\left(1-\frac{\mathrm{e}^{-X_{T_a^+}+S}}{\mathbb{E}(\mathrm{e}^{-\overline{X}_{\mathbf{e}_q}})}\middle|\mathcal{F}_{T_a^+}\right)\right)\\
&=\mathbb{E}_x(\mathrm{e}^{-qT_a^+}(1-\mathrm{e}^{-X_{T_a^+}})\mathbf{1}_{(T_a^+<\infty)})
\end{aligned}
$$

as required.

(ii) We shall check the conditions of Lemma 9.1. It is clear from part (i) that since $x^*>0$, $v_{x^*}(x)\geq 0$ for all $x\in\mathbb{R}$. On the other hand, for $x>0$

$$
v_{x^*}(x)=(1-e^{-x})-\mathbb{E}\left(\mathbf{1}_{(\overline{X}_{\mathbf{e}_q}<x^*-x)}\left(1-\frac{\mathrm{e}^{-x-\overline{X}_{\mathbf{e}_q}}}{\mathbb{E}(\mathrm{e}^{-\overline{X}_{\mathbf{e}_q}})}\right)\right)
$$

from which one sees that the expectation on the right hand side is negative since the indicator forces $\mathrm{e}^{-x-\overline{X}_{\mathbf{e}_q}}>\mathbb{E}(e^{-\overline{X}_{\mathbf{e}_q}})$. Hence $v_{x^*}(x)\geq(1-e^{-x})$ on $x>0$. This establishes the required lower bound $v(x)\geq(1-\mathrm{e}^{-x^+})$ for all $x\in\mathbb{R}$.

Finally we need to show that $\{\mathrm{e}^{-qt}v_{x^*}(X_t):t\geq 0\}$ is a supermartingale. We again employ familiar techniques. Recall that on the event $\{\mathbf{e}_q>t\}$ we have that $\overline{X}_{\mathbf{e}_q}=(X_t+S)\vee\overline{X}_t\geq X_t+S$ where by stationary and independent increments and the lack of memory property, S is independent of $\mathcal{F}_t$ and has the same distribution as $\overline{X}_{\mathbf{e}_q}$. We now have that

$$
\begin{aligned}
v_{x^*}(x)&\geq\mathbb{E}_x\left(\mathbf{1}_{(t<\mathbf{e}_q)}\mathbf{1}_{(X_t+S\geq a)}\left(1-\frac{\mathrm{e}^{-X_t-S}}{\mathbb{E}(\mathrm{e}^{-\overline{X}_{\mathbf{e}_q}})}\right)\right)\\
&=\mathbb{E}_x\left(\mathrm{e}^{-qt}\mathbb{E}_{X_t}\left(\mathbf{1}_{(S\geq a)}\left(1-\frac{\mathrm{e}^{-S}}{\mathbb{E}(\mathrm{e}^{-\overline{X}_{\mathbf{e}_q}})}\right)\middle|\mathcal{F}_t\right)\right)\\
&=\mathbb{E}_x(\mathrm{e}^{-qt}v_{x^*}(X_t)).
\end{aligned}
$$

Using the usual arguments involving stationary and independent increments one may deduce the required supermartingale property from the above inequality.

In conclusion, Lemma 9.1 now shows that the pair (v_{x^*},T^+_{x*}) solves the given optimal stopping problem.

(iii) Note that $v_{x^*}(x^*+)=1-\mathrm{e}^{-x^*}$ and $v'_{x^*}(x^*+)=\mathrm{e}^{-x^*}$. On the other hand, from the expression for $v_{x^*}(x)$ given above we have in particular that

$$
v_{x^*}(x^*-)=(1-\mathrm{e}^{-x^*})-\mathbb{E}\left(\mathbf{1}_{(\overline{X}_{\mathbf{e}_q}=0)}\left(1-\frac{\mathrm{e}^{-x^*-\overline{X}_{\mathbf{e}_q}}}{\mathbb{E}(\mathrm{e}^{-\overline{X}_{\mathbf{e}_q}})}\right)\right)=(1-\mathrm{e}^{-x^*})
$$

and thus that there is continuity at x^*.

We also have for $x < x^*$,

$$\frac{v(x^*) - v(x)}{x^* - x} = \frac{(1 - \mathrm{e}^{-x^*}) - (1 - \mathrm{e}^{-x})}{x^* - x} + \frac{1}{x^* - x}\mathbb{E}\left(\mathbf{1}_{(\overline{X}_{\mathbf{e}_q} < x^* - x)}\left(\frac{\mathrm{e}^{-x} - \mathrm{e}^{-x - \overline{X}_{\mathbf{e}_q}}}{\mathbb{E}(\mathrm{e}^{-\overline{X}_{\mathbf{e}_q}})}\right)\right) + \frac{\mathrm{e}^{-x^*} - \mathrm{e}^{-x}}{x^* - x}\frac{1}{\mathbb{E}(\mathrm{e}^{-\overline{X}_{\mathbf{e}_q}})}\mathbb{P}(\overline{X}_{\mathbf{e}_q} < x^* - x), \quad \text{(S.18)}$$

where we have used the fact that $\mathrm{e}^{-x^*} = \mathbb{E}(\mathrm{e}^{-\overline{X}_{\mathbf{e}_q}})$. The first term on the right-hand side of (S.18) converges to e^{-x^*}. The third term converges to $-\mathbb{E}(\mathrm{e}^{-\overline{X}_{\mathbf{e}_q}})^{-1}\mathrm{e}^{-x^*}\mathbb{P}(\overline{X}_{\mathbf{e}_q} = 0) = \mathbb{P}(\overline{X}_{\mathbf{e}_q} = 0)$. For the third term, with integration by parts, we have

$$\begin{aligned}
&\lim_{x \uparrow x^*} \frac{1}{x^* - x}\mathbb{E}\left(\mathbf{1}_{(\overline{X}_{\mathbf{e}_q} < x^* - x)}\left(\frac{\mathrm{e}^{-x} - \mathrm{e}^{-x - \overline{X}_{\mathbf{e}_q}}}{\mathbb{E}(\mathrm{e}^{-\overline{X}_{\mathbf{e}_q}})}\right)\right) \\
&= \lim_{x \uparrow x^*} \frac{1}{x^* - x}\mathbb{E}\left(\mathbf{1}_{(0 < \overline{X}_{\mathbf{e}_q} < x^* - x)}\left(\frac{\mathrm{e}^{-x} - \mathrm{e}^{-x - \overline{X}_{\mathbf{e}_q}}}{\mathbb{E}(\mathrm{e}^{-\overline{X}_{\mathbf{e}_q}})}\right)\right) \\
&= \lim_{x \uparrow x^*} \frac{\mathrm{e}^{-x} - \mathrm{e}^{-x^*}}{x^* - x}\mathbb{P}(0 < \overline{X}_{\mathbf{e}_q} < x^* - x) \\
&\quad - \lim_{x \uparrow x^*} \frac{1}{x^* - x}\int_0^{x^* - x} \mathrm{e}^{-x - y}\mathbb{P}(0 < \overline{X}_{\mathbf{e}_q} < y)\mathrm{d}y \\
&= 0,
\end{aligned}$$

where in the first equality we have removed the possible atom at zero of $\overline{X}_{\mathbf{e}_q}$ as it contributes nothing to the expectation. In conclusion, returning to (S.18) we see that as $x \uparrow x^*$ we find that $v'_{x^*}(x^*-) = v'_{x^*}(x^*+) - \mathbb{P}(\overline{X}_{\mathbf{e}_q} = 0)$. Hence there is smooth fit if and only if 0 is irregular for $(0, \infty)$ in which case there is only continuous fit.

9.2 (i) Changing measure we may reformulate the given stochastic game in the form

$$v(x) = \sup_{\tau \in \mathcal{T}} \inf_{\sigma \in \mathcal{T}} \mathbb{E}^1(\mathrm{e}^{-\alpha\tau + Y_\tau^x}\mathbf{1}_{(\tau \leq \sigma)} + \mathrm{e}^{-\alpha\sigma}(\mathrm{e}^{Y_\sigma^x} + \delta)\mathbf{1}_{(\tau > \sigma)}), \quad \text{(S.19)}$$

where $\alpha = q - \psi(1)$, $Y^x = (x \vee \overline{X}) - X$ and the supremum and infimum are interchangeable.

Considering the solution to the Shepp–Shiryaev optimal stopping problem we may argue as in the other optimal stopping problems of this chapter to see that $v(x) = \sup_{\tau \in \mathcal{T}} \mathbb{E}(\mathrm{e}^{-q\tau + (\overline{X}_\tau \vee x)})$ is a convex function on account of the gain function being convex. This implies that when $1 + \delta \geq v(0+) = Z^{(q)}(x^*)$ we have that $v(x) \leq \mathrm{e}^x + \delta$. Now taking the candidate solution to the given stochastic game as the triple consisting of v, the associated optimal stopping time τ^* from the Shepp–Shiryaev optimal stopping problem and $\sigma^* = \infty$ one easily checks that all the conditions of Lemma 9.13 are all satisfied. Hence this candidate solution is in fact the solution according to Lemma 9.13.

(ii) From the definition of $Z^{(q)}$ one sees that the latter is strictly increasing with $Z^{(q)}(0) = 1$ and $Z^{(q)}(\infty) = \infty$. It follows that $Z^{(q)}(z^*) = 1+\delta$ has a unique solution and that $z^* < x^*$ as $Z^{(q)}(x^*) > 1+\delta$. According to the discussion in the proof of Theorem 9.11 f is a strictly decreasing function with $f(0) > 0$ (here we use the assumption that $d > q$), $f(\infty) = -\infty$ and there is a unique crossing point of the origin. Consequently it follows that

$$\frac{\mathrm{d}}{\mathrm{d}x}(\mathrm{e}^x Z^{(q)}(z^* - x)) = \mathrm{e}^x(Z^{(q)}(z^* - x) - qW^{(q)}(z^* - x)) = \mathrm{e}^x f(z^* - x)$$

is strictly increasing on $[0, z^*]$ and hence $v(x) := \mathrm{e}^x Z^{(q)}(z^* - x)$ is convex on $(0, z^*)$. Further, $v'(0+) > 0$ and $v'(z^*-) = \mathrm{e}^{z^*}(1 - qW^{(q)}(0)) = \mathrm{e}^{z^*}(1 - q/d) < v'(z^*+) = \mathrm{e}^{z^*}$. Since $v(x) = \mathrm{e}^x$ on (z^*, ∞) (which is clearly convex) the fact that $v'(z^*-) < v'(z^*+)$ also shows that v is convex on $(0, \infty)$.

(iii) First note that

$$\begin{aligned}
&\mathbb{E}(\exp\{-q\tau^-_{-(z^*-x)} + (x \vee \overline{X}_{\tau^-_{-(z^*-x)}})\}\mathbf{1}_{(\tau^-_{-(z^*-x)} < \tau^+_x)} \\
&\qquad + \mathrm{e}^{-q\tau^+_x}(\mathrm{e}^{(x\vee\overline{X}_{\tau^+_x})} + \delta \mathrm{e}^{X_{\tau^+_x}})\mathbf{1}_{(\tau^-_{-(z^*-x)} > \tau^+_x)}) \\
&= \mathrm{e}^x \mathbb{E}_{z^*-x}(\mathrm{e}^{-\tau^-_0}\mathbf{1}_{(\tau^-_0 < \tau^+_{z^*})}) + \mathrm{e}^x(1+\delta)\mathbb{E}_{z^*-x}(\mathrm{e}^{-q\tau^+_{z^*}}\mathbf{1}_{(\tau^+_{z^*} < \tau^-_0)}) \\
&= \mathrm{e}^x\left(Z^{(q)}(z^* - x) - W^{(q)}(z^* - x)\frac{Z^{(q)}(z^*)}{W^{(q)}(z^*)}\right) + \mathrm{e}^x(1+\delta)\frac{W^{(q)}(z^* - x)}{W^{(q)}(z^*)} \\
&= v(x),
\end{aligned}$$

where in the last equality we have used the fact that $Z^{(q)}(z^*) = 1 + \delta$. Let

$$\tau^* = \inf\{t > 0 : Y^x_t > z^*\} \text{ and } \sigma^* = \inf\{t > 0 : Y^x_t = 0\}$$

and note that $\tau^-_{-(z^*-x)} = \tau^*$ on the event that $\{\tau^* \leq \sigma^*\}$ and $\tau^+_x = \sigma^*$ on the event that $\{\tau^* > \sigma^*\}$. Changing measure as in (S.19) we thus identify the candidate triple (v, τ^*, σ^*) in terms of the reflected process Y^x once we write

$$v(x) = \mathbb{E}^1(\mathrm{e}^{-\alpha\tau^* + Y^x_{\tau^*}}\mathbf{1}_{(\tau^* \leq \sigma^*)} + \mathrm{e}^{-\alpha\sigma^*}(\mathrm{e}^{Y^x_{\sigma^*}} + \delta)\mathbf{1}_{(\tau^* > \sigma^*)}),$$

where $\alpha = q - \psi(1)$. Our objective is now to verify the conditions of Lemma 9.13.

We begin with the bounds. We are required to show that $\mathrm{e}^x \leq v(x) \leq \mathrm{e}^x + \delta$. The lower bound is trivial since $Z^{(q)}(z^* - x) \geq 1$. For the upper bound, write $v(x) = \mathrm{e}^x + qg(x)$ where $g(x) = \mathrm{e}^x \int_0^{z^*-x} W^{(q)}(y)\mathrm{d}y$. Using Lemma 8.4 and integration by parts we see that

$$\begin{aligned}
g'(x) &= \mathrm{e}^x\left(\int_0^{z^*-x} W^{(q)}(y)\mathrm{d}y - W^{(q)}(z^* - x)\right) \\
&= \mathrm{e}^x\left(\frac{1}{\Phi(q)}(W^{(q)}(z^* - x) - W_{\Phi(q)}(0))\right. \\
&\qquad \left. - \frac{1}{\Phi(q)}\int_0^{z^*-x} \mathrm{e}^{\Phi(q)y}W^{(q)\prime}(y)\mathrm{d}y - W^{(q)}(z^* - x)\right)
\end{aligned}$$

which is negative since $\Phi(q) > 1$ (which itself follows from the assumption that $q > \psi(1)$). Hence

$$v(x) \leq \mathrm{e}^x + qg(0) = \mathrm{e}^x + (Z^{(q)}(z^*) - 1) = \mathrm{e}^x + \delta$$

thus confirming the upper bound.

Next we look at when the proposed value function equals the gain functions. The distribution of $Y^x_{\tau^*}$ is concentrated on (z^*, ∞). Note that X does not creep downwards and hence Y^x cannot pass above z^* from below by hitting z^*. Further, the support of the distribution of $Y^x_{\sigma^*}$ is the single point $\{0\}$. We see that $v(x) = \mathrm{e}^x$ on the support of $Y^x_{\tau^*}$ and $v(x) = \mathrm{e}^0 + \delta = 1 + \delta$ on the support of $Y^x_{\sigma^*}$.

Now we consider the martingale requirement of Theorem 9.13. First note by the Strong Markov property that

$$\mathrm{e}^{-\alpha(t\wedge\tau^*\wedge\sigma^*)}v(X_{t\wedge\tau^*\wedge\sigma^*}) = \mathbb{E}^1(\mathrm{e}^{-\alpha\tau^* + Y^x_{\tau^*}}\mathbf{1}_{(\tau^*\leq\sigma^*)} + \mathrm{e}^{-\alpha\sigma^*}(\mathrm{e}^{Y^x_{\sigma^*}} + \delta)\mathbf{1}_{(\tau^*>\sigma^*)}|\mathcal{F}_t)$$

giving the required martingale property (note that v is continuous and hence the above process is right continuous). An argument using the Change of Variable Formula in the spirit of the calculation given in (9.28) shows that $\mathcal{L}^1 v(x) = 0$ where for any $f \in C^1(\mathbb{R})$,

$$\mathcal{L}^1 f(x) = \int_{(0,\infty)} (f(x+y) - f(x))\mathrm{e}^{-y}\Pi(\mathrm{d}y) - \mathrm{d}\frac{\mathrm{d}f}{\mathrm{d}x}(x) - \alpha f(x).$$

To show that $\{\mathrm{e}^{-\alpha(t\wedge\sigma^*)}v(X_{t\wedge\sigma^*}) : t \geq 0\}$ is a right continuous supermartingale and that $\{\mathrm{e}^{-\alpha(t\wedge\tau^*)}v(X_{t\wedge\tau^*}) : t \geq 0\}$ is a right continuous submartinagale we argue again along the lines of (9.28). We have

$$\begin{aligned}\mathrm{e}^{-\alpha t}v(Y^x_t) = v(x) &+ \int_0^t \mathrm{e}^{-\alpha s}\mathcal{L}^1 v(Y^x_t)\mathrm{d}s \\ &+ \int_0^t \mathrm{e}^{-\alpha s}(v(z^*-) - v(z^*+))\mathrm{d}L^{z^*}_t \\ &+ \int_0^t \mathrm{e}^{-\alpha s}v'(Y^x_s)\mathrm{d}(x \vee \overline{X}_s) \\ &+ M_t,\end{aligned} \tag{S.20}$$

where L^{z^*} counts the number of visits to z^* (which are almost finite in number over finite periods of time) and M is a martingale. Note that the second integral is zero since v is continuous and that $Y^x_s = 0$ if s is in the support of $\mathrm{d}(x \vee \overline{X}_s)$. Further a familiar calculation shows that for $x > z^*$ where $v(x) = \mathrm{e}^x$,

$$\mathcal{L}^1 v(x) = -q\mathrm{e}^x < 0.$$

Hence on $\{t < \tau^*\}$ we see that the first integral in (S.20) is zero whilst the third integral is non-decreasing since $v'(0+) > 0$ and so $\{\mathrm{e}^{-\alpha(t\wedge\tau^*)}v(X_{t\wedge\tau^*}) : t \geq 0\}$ is a submartingale (right continuity follows from the continuity of v and the right continuity of Y^x). On $\{t < \sigma^*\}$ the first integral in (S.20) is non-increasing and the third integral is zero showing that $\{\mathrm{e}^{-\alpha(t\wedge\sigma^*)}v(X_{t\wedge\sigma^*}) : t \geq 0\}$ is a supermartingale (right continuity follows as before).

Finally revisiting the calculations in part (ii), the presence of continuous fit and absence of smooth fit is apparent.

Chapter 10

10.1 (i) Note that as $\theta \uparrow \infty$, $\psi(\theta) \sim c\theta - \lambda$. Hence it is easy to see that

$$\int^{\infty} \frac{1}{\psi(\xi)} \mathrm{d}\xi = \infty$$

so that from Theorem 10.5 (ii) extinction occurs with probability zero.

(ii) The assumption that $\psi'(0+) > 0$ implies that $\mathbb{P}_x(\tau_0^- < \infty) = 1$ where, as usual, $\tau_0^- = \inf\{t > 0 : X_t < 0\}$ and X is the Lévy process with Laplace exponent $\log \mathbb{E}(\mathrm{e}^{-\theta X_1}) = \psi(\theta)$. This means that X has an almost surely finite number of jumps on the time interval $[0, \tau_0^-]$.

Now according to the Lamperti transform in Theorem 10.2 (i), the continuous-state branching process with branching mechanism ψ can be represented by

$$Y_t = X_{\theta_t \wedge \tau_0^-}, \ t \geq 0$$

where

$$\theta_t = \inf\left\{ s > 0 : \int_0^s \frac{\mathrm{d}u}{X_u} > t \right\}.$$

Suppose that n^* is the number of jumps that X has undergone at time τ_0^-. The jump times on $(0, \tau_0^-)$ are then denoted $T_1, T_2, ..., T_{n^*}$, with $T_0 := 0$ for convenience. Then for all $t > \int_0^{T_{n^*}} X_u^{-1} \mathrm{d}u$ it follows that under P_x for $x > 0$,

$$t = \int_0^{\theta_t} \frac{\mathrm{d}u}{X_u} = \sum_{k=1}^{n^*} \int_{T_{k-1}}^{T_k} \frac{1}{x - cu + \sum_{j=1}^{k-1} \xi_j} \mathrm{d}u + \int_{T_{n^*}}^{\theta_t} \frac{1}{x - cu + \sum_{j=1}^{n^*} \xi_j} \mathrm{d}u,$$

where $\{\xi_i : i = 1, 2, 3, ...\}$ are the independent and identically distributed sequence of jumps. For each $k = 1, 2, 3, ...$, let $S_k = \sum_{j=1}^k \xi_j$. Simple calculus then shows that

$$t = -\frac{1}{c} \log \left(\frac{x + S_{n^*} - c\theta_t}{x + S_{n^*} - cT_{n^*}} \prod_{k=1}^{n^*} \frac{x + S_{k-1} - cT_k}{x + S_{k-1} - cT_{k-1}} \right).$$

Since from the Lamperti transform and non-extinction, $Y_t = X_{\theta_t} = x + S_{n^*} - c\theta_t$, it follows that

$$Y_t = \mathrm{e}^{-ct} \Delta$$

where

$$\Delta = (x + S_{n^*} - cT_{n^*}) \prod_{k=1}^{n^*} \frac{x + S_{k-1} - cT_{k-1}}{x + S_{k-1} - cT_k}.$$

10.2 (i) Let $\mathbf{e}_q$ be an independent and exponentially distributed random variable with parameter $q > 0$ and set $g(x) = xf(x)$. We have

$$\begin{aligned} \mathbb{E}_x^{\uparrow} \left(\int_0^{\infty} \mathrm{e}^{-qt} f(X_t) \mathrm{d}t \right) &= \frac{1}{q} \mathbb{E}_x^{\uparrow}(f(X_{\mathbf{e}_q})) \\ &= \frac{1}{qx} \mathbb{E}_x(g(X_{\mathbf{e}_q}) \mathbf{1}_{(\mathbf{e}_q < \tau_0^-)}) \\ &= \frac{1}{qx} \mathbb{E}(g(x + X_{\mathbf{e}_q} - \underline{X}_{\mathbf{e}_q} + \underline{X}_{\mathbf{e}_q}) \mathbf{1}_{(-\underline{X}_{\mathbf{e}_q} < x)}), \end{aligned}$$

where $\underline{X}_{\mathbf{e}_q} = \inf_{s \leq \mathbf{e}_q} X_s$. Next recall from the Wiener–Hopf factorisation (cf. Sect. 8.1) that $X_{\mathbf{e}_q} - \underline{X}_{\mathbf{e}_q}$ and $\underline{X}_{\mathbf{e}_q}$ are independent and the former is equal in distribution to $\overline{X}_{\mathbf{e}_q}$. Further, spectral positivity also implies that $-\underline{X}_{\mathbf{e}_q}$ is exponentially distributed with parameter $\Phi(q)$ where Φ is the right inverse of ψ. It now follows that

$$\mathbb{E}_x^{\uparrow}\left(\int_0^\infty \mathrm{e}^{-qt} f(X_t)\mathrm{d}t\right)$$
$$= \frac{\Phi(q)}{qx}\int_0^\infty \mathrm{d}y \cdot \mathrm{e}^{-\Phi(q)y}\mathbf{1}_{(y<x)}\int_{[0,\infty)} \mathbb{P}(\overline{X}_{\mathbf{e}_q} \in \mathrm{d}z)\cdot g(x+z-y) \quad \text{(S.21)}$$

as required.

(ii) Changing variables in (S.21) with $u = x+z-y$ we obtain

$$\mathbb{E}_x^{\uparrow}\left(\int_0^\infty \mathrm{e}^{-qt} f(X_t)\mathrm{d}t\right)$$
$$= \frac{\Phi(q)}{qx}\mathrm{e}^{-\Phi(q)x}\int_0^\infty \mathrm{d}u \cdot \mathrm{e}^{\Phi(q)u} g(u)\int_{[(u-x)\vee 0,u)} \mathbb{P}(\overline{X}_{\mathbf{e}_q} \in \mathrm{d}z)\cdot \mathrm{e}^{-\Phi(q)z}.$$

Recall also from Chap. 8 (specifically (8.20)),

$$\mathbb{P}(\overline{X}_{\mathbf{e}_q} \in \mathrm{d}z) = \frac{q}{\Phi(q)}W^{(q)}(\mathrm{d}z) - qW^{(q)}(z)\mathrm{d}z,$$

where $W^{(q)}$ is the scale function defined in Theorem 8.1. Recall that $\lim_{q\downarrow 0} q/\Phi(q) = \lim_{q\downarrow 0}\psi(\Phi(q))/\Phi(q) = \psi'(0+)$ since $\Phi(0+) = 0$. Hence taking limits as $q \downarrow 0$ from the conclusion of the previous part, and recalling that $W^{(q)}(x) = 0$ for all $x < 0$, we see that

$$\mathbb{E}_x^{\uparrow}\left(\int_0^\infty f(X_t)\mathrm{d}t\right) = \int_0^\infty \mathrm{d}u \cdot \frac{g(u)}{x}\int_{[(u-x)\vee 0,u)} W(\mathrm{d}z)$$
$$= \int_0^\infty \mathrm{d}u \cdot \frac{g(u)}{x}(W(u) - W(u-x))$$

and the result follows once we recall that $g(u) = uf(u)$ and that f is an arbitrary continuous and compactly supported function.

(iii) From Theorem 10.10 we easily deduce that

$$\mathbb{P}_x^{\uparrow}(\tau_y^- < \tau_z^+ \wedge t) = \mathbb{E}_x\left(\mathbf{1}_{(\tau_y^- < \tau_z^+ \wedge t)}\frac{X_{\tau_y^-}}{x}\right) = \frac{y}{x}\mathbb{P}_x(\tau_y^- < \tau_z^+ \wedge t).$$

Hence taking limits as $t \uparrow \infty$ gives and using Theorem 8.1 (iii) gives us

$$\mathbb{P}_x^{\uparrow}(\tau_y^- < \tau_z^+) = \frac{y}{x}\frac{W(z-x)}{W(z-y)}. \quad \text{(S.22)}$$

Taking limits in (S.22) as $y \downarrow 0$ we see that $\mathbb{P}_x^{\uparrow}(\tau_z^+ < \infty) = 1$ for all $0 < x < z < \infty$. Now note from (8.8) that it can be deduced that $W(z)/W(z+x) \to 1$ as $z \uparrow \infty$.

Taking limits in (S.22) as $z \uparrow \infty$ and then as $x \uparrow \infty$ we thus obtain for each $y > 0$,

$$\lim_{x\uparrow\infty} \mathbb{P}_x^{\uparrow}(\tau_y^- < \infty) = \lim_{x\uparrow\infty} \frac{y}{x} = 0.$$

In conclusion, since for any $z > 0$, $\mathbb{P}_x^{\uparrow}$-almost surely

$$\liminf_{t\uparrow\infty} X_t = \liminf_{t\uparrow\infty} X_{\tau_z^+ + t} =: I_z$$

and for all $y > 0$, $\liminf_{z\uparrow\infty} I_z \geq y$, $\mathbb{P}_x^{\uparrow}$-almost surely, it follows that

$$\liminf_{t\uparrow\infty} X_t = \infty$$

$\mathbb{P}_x^{\uparrow}$-almost surely.

10.3 (i) In light of (10.22) the question is effectively asking to prove that for each $t > 0$,

$$\frac{\partial u_t}{\partial \theta}(\theta) = \frac{\psi(u_t(\theta))}{\psi(\theta)}.$$

However, this follows from (10.7) and in particular that

$$\frac{\partial}{\partial\theta}\int_\theta^{u_t(\theta)} \frac{1}{\psi(\xi)}\mathrm{d}\xi = \left\{\frac{1}{\psi(u_t(\theta))} - \frac{1}{\psi(\theta)}\right\}\frac{\partial u_t}{\partial\theta}(\theta) = 0.$$

(ii) When $\rho = 0$ we know that Y under P_x becomes extinct with probability one for each $x > 0$. Hence $u_t(\theta) \to 0$ as $t \uparrow \infty$ for each $\theta > 0$. It follows directly from part (i) that for each $x > 0$, $Y_t \to \infty$ in $P_x^{\uparrow}$-distribution and hence in $P_x^{\uparrow}$-probability. However, this implies in Lemma 10.14 (ii) that $P_x^{\uparrow}$-almost surely, $\varphi_t \to \infty$ as $t \to \infty$. Since from Exercise 10.2 (iii) we know that $\lim_{t\uparrow\infty} X_t = \infty$ under $\mathbb{P}_x^{\uparrow}$, it follows from Lemma 10.14 that $P_x^{\uparrow}(\lim_{t\uparrow\infty} Y_t = \infty) = 1$.

(iii) The first part is a straightforward manipulation. Note that the positivity of the integral follows thanks to the convexity of ψ and the fact that $\psi(0) = 0$ which together imply that $\psi(\xi) \geq \rho\xi$. Next note that

$$\int_0^{\theta x}\left(\frac{\mathrm{e}^{-\lambda} - 1 + \lambda}{\lambda^2}\right)\mathrm{d}\lambda$$

behaves like $\frac{1}{2}\theta x$ as $x \downarrow 0$ and like $\log x$ as $x \uparrow \infty$. Hence we have that

$$\int_0^\theta \frac{\psi(\xi) - \rho\xi}{\xi^2}\mathrm{d}\xi < \infty \iff \int^\infty x\log x \Pi(\mathrm{d}x) < \infty.$$

On the other hand, as $\psi(\xi) \sim \rho\xi$ as $\theta \downarrow 0$, the left-hand integral above is finite if and only

$$\int_{0+}\left(\frac{1}{\rho\xi} - \frac{1}{\psi(\xi)}\right)\mathrm{d}\xi < \infty.$$

Note that the above integral is positive also because convexity of ψ implies that $\psi(\xi) \geq \rho\xi$.

(iv) Appealing to the expression established in part (i) for $\rho > 0$, since Y becomes extinct under P_x and hence $u_t(\theta) \to 0$ as $t \uparrow \infty$, we have

$$\lim_{t \uparrow \infty} E_x^{\uparrow}(\mathrm{e}^{-\theta Y_t}) = \lim_{t \uparrow \infty} \frac{\rho u_t(\theta)}{\psi(\theta)} \mathrm{e}^{\rho t}.$$

Recalling however from (10.7) that

$$\int_{u_t(\theta)}^{\theta} \frac{1}{\psi(\xi)} \mathrm{d}\xi = t$$

we see that for t sufficiently large

$$0 \leq \int_{u_t(\theta)}^{\theta} \left(\frac{1}{\rho \xi} - \frac{1}{\psi(\xi)} \right) \mathrm{d}\xi = \frac{1}{\rho} \log \left(\frac{\theta}{u_t(\theta) \mathrm{e}^{\rho t}} \right).$$

Hence

$$\begin{aligned}
\lim_{t \uparrow \infty} \mathrm{e}_x^{\uparrow}(\mathrm{e}^{-\theta Y_t}) &= \lim_{t \uparrow \infty} \frac{\rho \theta}{\psi(\theta)} \exp \left\{ -\rho \int_{u_t(\theta)}^{\theta} \left(\frac{1}{\rho \xi} - \frac{1}{\psi(\xi)} \right) \mathrm{d}\xi \right\} \\
&= \frac{\rho \theta}{\psi(\theta)} \exp \left\{ -\rho \int_0^{\theta} \left(\frac{1}{\rho \xi} - \frac{1}{\psi(\xi)} \right) \mathrm{d}\xi \right\}
\end{aligned}$$

from which the remaining claims follow easily.

References

Alili, L. and Kyprianou, A.E. (2004) Some remarks on first passage of Lévy processes, the American put and smooth pasting. *Ann. Appl. Probab.* **15**, 2062–2080.

Andrew, P. (2005) Proof from first principles of Kesten's result for the probabilities with which a subordinator hits points. *Preprint.*

Applebaum, D. (2004) *Lévy Processes and Stochastic Calculus.* Cambridge University Press, Cambridge.

Applebaum, D., Bhat, B.V.R., Kustermans, J. and Lindsay, J.M. (2005) *Quantum Independent Increment Processes I. From Classical Probability to Quantum Stochastic Calculus.* Lecture Notes in Mathematics, Springer, Berlin Heidelberg New York.

Ash, R. and Doléans-Dade, C.A. (2000) *Probability and Measure Theory. Second Edition.* Harcourt, New York.

Asmussen, S., Avram, F. and Pistorius, M. (2004) Russian and American put options under exponential phase-type Lévy models. *Stochast. Process. Appl.* **109**, 79–111.

Asmussen, S. and Klüppelberg, C. (1996) Large deviations results for subexponential tails, with applications to insurance risk. *Stochast. Process. Appl.* **64**, 103–125.

Athreya, S. and Ney, P. (1972) *Branching Processes.* Springer, Berlin Heidelberg New York.

Avram, F., Chan, T. and Usabel, M. (2002a) On the valuation of constant barrier options under spectrally one-sided exponential Lévy models and Carr's approximation for American puts. *Stochast. Process. Appl.* **100**, 75–107.

Avram, F., Kyprianou, A.E. and Pistorius, M.R. (2002b) Exit problems for spectrally negative Lévy processes and applications to Russian, American and Canadized options. (Unabridged version of [11]). *Utrecht University Preprint.*

Avram, F., Kyprianou, A.E. and Pistorius, M.R. (2004) Exit problems for spectrally negative Lévy processes and applications to (Canadized) Russian options. *Ann. Appl. Probab.* **14**, 215–235.

Bachelier, L. (1900) Théorie de la spéculation. *Ann. Sci. École Norm. Sup.* **17**, 21–86.

Bachelier, L. (1901) Théorie mathematique du jeu. *Ann. Sci. École Norm. Sup.* **18**, 143–210.

Barndorff–Nielsen, O.E. and Shephard, N. (2005) *Continuous-time approach to financial volatility.* Cambridge Unversity Press, Cambridge.

Barndorff-Nielsen, O.E., Franz, U., Gohm, R., Kümmerer, B. and Thorbjørnsen, S. (2006) *Quantum Independent Increment Processes II Structure of Quantum Lévy Processes, Classical Probability, and Physics.* Lecture Notes in Mathematics, Springer, Berlin Heidelberg New York.

Baurdoux, E. and Kyprianou, A.E. (2005) Examples of stochastic games driven by Lévy processes. *working notes.*

Benassi, A., Cohen, S. and Istas, J. (2002) Identification and properties of real harmonizable fractional Lévy motions. *Bernoulli* **8**, 97–15.

Benassi, A., Cohen, S. and Istas, J. (2004) On roughness indexes for fractional fields. *Bernoulli* **10**, 357–373.

Bertoin, J. (1993) Splitting at the infimum and excursions in half-lines for random walks and Lévy processes. *Stochast. Process. Appl.* **47**, 17–35.

Bertoin, J. (1996a) *Lévy Processes.* Cambridge University Press, Cambridge.

Bertoin, J. (1996b) On the first exit time of a completely asymmetric stable process from a finite interval, *Bull. London Math. Soc.* **28**, 514–520.

Bertoin, J. (1997a) Regularity of the half-line for Lévy processes. *Bull. Sci. Math.* **121**, 345–354.

Bertoin, J. (1997b) Exponential decay and ergodicity of completely asymmetric Lévy processes in a finite interval, *Ann. Appl. Probab.* **7**, 156–169.

Bertoin, J. (2006) *Fragmentation processes.* Cambridge University Press, Cambridge.

Bertoin, J. and Doney, R.A. (1994a) Cramér's estimate for Lévy processes. *Stat. Probab. Lett.* **21**, 363–365.

Bertoin, J. and Doney, R.A. (1994b) On conditioning a random walk to stay non-negative. *Ann. Probab.* **22**, 2152–2167.

Bertoin, J., van Harn, K. and Steutel, F.W. (1999) Renewal theory and level passage by subordinators. *Stat. Probab. Lett.* **45**, 65–69.

Bertoin, J. and Yor, M. (2005) Exponential functionals of Lévy processes. *Probab. Surv.* **2**, 191–212.

Bingham, N.H. (1971) Limit theorems for occupation-times of Markov processes. *Zeit. Wahrsch. Verw. Gebiete* **17**, 1–22.

Bingham, N.H. (1972) Limit theorems for regenerative phenomena, recurrent events and renewal theory. *Zeit. Wahrsch. Verw. Gebiete* **21**, 20–44.

Bingham, N.H. (1975) Fluctuation theory in continuous time, *Adv. Appl. Probab.* **7**, 705–766.

Bingham, N.H. (1976) Continuous branching processes and spectral positivity. *Stochast. Process. Appl.* **4**, 217–242.

Bingham, N.H., Goldie, C.M. and Teugels, J.L. (1987) *Regular Variation.* Cambridge University Press, Cambridge.

Bingham, N.H. and Kiesel, R. (1998) *Risk-Neutral Valuation. Pricing and Hedging of Financial Derivatives.* Springer, Berlin Heidelberg New York.

Black, F. and Scholes, M. (1973) The pricing of options and corporate liabilities. *J. Polit. Econ.* **81**, 637–659.

Blumenthal, R.M. (1992) *Excursions of Markov Processes.* Birkhäuser, Basel.

Blumenthal, R.M. and Getoor, R.K. (1968) *Markov Processes and Potential Theory.* Academic, New York.

Borovkov, A.A. (1976) *Stochastic Processes in Queueing Theory.* Springer, Berlin Heidelberg New York.

Boyarchenko, S.I and Levendorskii, S.Z. (2002a) Perpetual American options under Lévy processes. *SIAM J. Control Optim.* **40**, 1663–1696.

Boyarchenko, S.I. and Levendorskii, S.Z. (2002b). *Non-Gaussian Merton- Black-Scholes theory.* World Scientific, Singapore.

Bratiychuk, N.S. and Gusak, D.V. (1991) *Boundary Problem for Processes with Independent Increments.* Naukova Dumka, Kiev (in Russian).

Bretagnolle, J. (1971) Résultats de Kesten sur les processus à accroissements indépendants. *Séminaire de Probabilités V*, 21–36. Lecture Notes in Mathematics, Springer, Berlin Heidelberg New York.

Brockwell, P.J., Chadraa, E. and Lindner, A. (2005) Continuous-time GARCH processes. *To appear in the Ann. Appl. Probab.*

Campbell, N.R. (1909) The study of discontinuous phenomena. *Proc. Camb. Philos. Soc.* **15**, 117–136.

Campbell, N.R. (1910) Discontinuities in light emission. *Proc. Camb. Philos. Soc.* **15**, 310–328.

Carr, P., Geman, H., Madan, D. and Yor, M. (2003) Stochastic volatility for Lévy processes. *Math. Finance* **13**, 345–382.

Chan, T. (2004) Some applications of Lévy processes in insurance and finance. *Finance* **25**, 71–94.

Chan, T. and Kyprianou, A. E. (2005) Smoothness of scale functions for spectarlly negative Lévy processes. *Preprint.*

Chaumont, L. (1994) Sur certains processus de Lévy conditionnés à rester positifs. *Stoch. Stoch. Rep.* **47**, 1–20.

Chaumont, L. (1996) Conditionings and path decomposition for Lévy processes. *Stochast. Process. Appl.* **64**, 39–54.

Chaumont, L. and Doney, R.A. (2005) On Lévy processes conditioned to stay positive. *Electron. J. Probab.* **10**, 948–961.

Chesney, M. and Jeanblanc, M. (2004) Pricing American currency options in an exponential Lévy model. *Appl. Math. Finance* **11**, 207–225.

Chiu, S.K. and Yin, C. (2005) Passage times for a spectrally negative Lévy process with applications to risk theory. *Bernoulli* **11**, 511–522.

Chow, Y.S., Robbins, H. and Siegmund, D. (1971) *Great Expectations: The Theory of Optimal Stopping.* Boston, Houghton Mifflin.

Christyakov, V.P. (1964) A theorem of sums of independent positive random variables and its applications to branching random processes. *Theory Probab. Appl.* **9**, 640–648.

Chung, K.L. and Fuchs, W.H.J. (1951) On the distribution of values of sums of random variables. *Mem. Am. Math. Soc.* **6**, 1–12.

Cramér, H. (1994a) *Collected Works. Vol. I.* Edited and with a preface by Anders Martin–Löf. Springer, Berlin Heidelberg New York.

Cramér, H. (1994b) *Collected Works. Vol. II.* Edited and with a preface by Anders Martin–Löf. Springer, Berlin Heidelberg New York.

Cont, R. and Tankov, P. (2004) *Financial Modeling with Jump Processes.* Chapman and Hall/CRC, Boca Raton, FL.

Darling, D.A., Liggett, T. and Taylor, H.M. (1972) Optimal stopping for partial sums. *Ann. Math. Stat.* **43**, 1363–1368.

De Finetti, B. (1929) Sulle funzioni ad incremento aleatorio. *Rend. Acc. Naz. Lincei.* **10**, 163–168.

Dellacherie, C. and Meyer. P.A. (1975–1993) *Probabilités et Potentiel.* Chaps. I–VI, 1975; Chaps. V–VIII, 1980; Chaps. IX–XI, 1983; Chaps. XII–XVI, 1987; Chaps XVII–XXIV, 1993. Hermann, Paris.

Doney, R.A. (1987) On Wiener–Hopf factorisation and the distribution of extrema for certain stable processes. *Ann. Probab.* **15**, 1352–1362.

Doney, R.A. (1991) Hitting probabilities for spectrally positive Lévy processes. *J. London Math. Soc.* **44**, 566–576.

Doney, R.A. (2005) Some excursion calculations for spectrally one-sided Lévy processes. *Séminaire de Probabilités XXXVIII.* 5–15. Lecture Notes in Mathematics, Springer, Berlin Heidelberg New York.

Doney, R.A. (2006) *Ecole d'été de Saint-Flour.* Lecture notes in Mathematics, Springer, Berlin Heidelberg New York, forthcoming.

Doney, R.A. and Kyprianou, A.E. (2005) Overshoots and undershoots of Lévy processes. *Ann. Appl. Probab.* **16**, 91–106.

Dube, P., Guillemin, F. and Mazumdar, R. (2004) Scale functions of Lévy processes and busy periods of finite capacity. *J. Appl. Probab.* **41**, 1145–1156.

Duquesne, T. (2003) Path decompositions for real Lévy processes. *Ann. I. H. Poincaré* **39**, 339–370.

Duquesne, T. (2006) Continuum random trees and branching processes with immigration. To appear in Stoch. Processes and Appl.

Duquesne, T. and Le Gall, J-F. (2002) *Random Trees, Lévy Processes and Spatial Branching Processes* Astérisque nr. 281.

Durrett, R. (2004) *Probability: Theory and Examples, 3rd Edition.* Duxbury.

Dynkin, E. (1961) Some limit theorems for sums of independent random variables with infinite mathematical expectations. In *Selected Translations Math. Stat. Prob.* **1**, 171–189. Inst. Math. Satistics Amer. Math. Soc.

Dynkin, E. (1963) The optimum choice of the instant for stopping a Markov process. *Soviet Math. Dokl.* **4**, 627–629.

Dynkin, E.B. (1969) A game-theoretic version of an optimal stopping problem. *Dokl. Akad. Nauk SSSR* **185**, 16–19.

Embrechts, P., Goldie, C. M. and Veraverbeke, N. (1979) Subexponentiality and infinite divisibility. *Zeit. Wahrsch. Verw. Gebiete.* **49**, 335–347.

Embrechts, P., Klüppelberg, C. and Mikosch, T. (1997) *Modelling Extremal Events for Insurance and Finance.* Springer, Berlin Heidelberg New York.

Emery, D.J. (1973) Exit problems for a spectrally positive process, *Adv. Appl. Probab.* **5**, 498–520.

Erickson, K.B. (1973) The strong law of large numbers when the mean is undefined. *Trans. Am. Math. Soc.* **185**, 371–381.

Erickson, K.B. and Maller, R.A. (2004) Generalised Ornstein–Uhlenbeck processes and the convergence of Lévy processes. *Séminaire de Probabilités XXXVIII*, 70–94. Lecture Notes in Mathematics, Springer, Berlin Heidelberg New York.

Etheridge, A.M. (2000) *An Introduction to Superprocesses.* American Mathematical Society, Providence, RI.

Etheridge, A.M. and Williams, D.R.E. (2003) A decomposition of the $(1+\beta)$-superprocess conditioned on survival. *Proc. R. Soc. Edinburgh* **133A**, 829–847.

Feller, W. (1971) *An Introduction to Probability Theory and its Applications. Vol II, 2nd edition.* Wiley, New York.

Fristedt, B.E. (1974) Sample functions of stochastic processes with stationary independent increments. *Adv. Probab.* **3**, 241–396. Dekker, New York.

Gapeev, P. and Kühn, C. (2005) Perpetual convertible bonds in jump–diffusion models. *Stat. Decisions* **23**, 15–31.

Gerber, H.U. and Shiu, E.S.W. (1994) Martingale approach to pricing perpetual American options. *Astin Bull.* **24**, 195–220.

Geman, H., Madan, D. and Yor, M. (2001) Asset prices are Brownian motion: only in business time. *Quantitative Analysis in Financial Markets.* pp. 103–146, World Scientific, Singapore.

Gikhman, I.I. and Skorokhod, A.V. (1975) *The Theory of Stochastic Processes II.* Springer, Berlin Heidelberg New York.

Gnedin, A. and Pitman, J. (2005) Regenerative composition structures. *Ann. Probab.* **33**, 445–479.

Good, I.J. (1953) The population frequencies of species and the estimation of populations parameters. *Biometrika* **40**, 273–260.

Greenwood, P.E. and Pitman, J.W. (1980a) Fluctuation identities for Lévy processes and splitting at the maximum. *Adv. Appl. Probab.* **12**, 839–902.

Greenwood, P.E. and Pitman, J. W. (1980b) Fluctuation identities for random walk by path decomposition at the maximum. Abstracts of the Ninth Conference on Stochastic Processes and Their Applications, Evanston, Illinois, 6–10 August 1979, *Adv. Appl. Prob.* **12**, 291–293.

Grey, D.R. (1974) Asymptotic behaviour of continuous time, continuous state-space branching processes. *J. Appl. Probab.* **11**, 669–667.

Grosswald, E. (1976) The student t-distribution of any degree of freedom is infinitely divisible. *Zeit. Wahrsch. Verw. Gebiete* **36**, 103–109.

Gusak, D.V. and Korolyuk, V.S. (1969) On the joint distribution of a process with stationary independent increments and its maximum. *Theory Probab. Appl.* **14**, 400–409.

Hadjiev, D.I. (1985) The first passage problem for generalized Ornstein–Uhlenbeck processes with nonpositive jumps. *Séminaire de probabilités, XIX* 80–90. Lecture Notes in Mathematics, Springer, Berlin Heidelberg New York.

Halgreen, C. (1979) Self-decomposability of the generalized inverse Gaussian and hyperbolic distributions. *Zeit. Wahrsch. Verw. Gebiete* **47**, 13–18.

Harrison, J. and Kreps, D. (1979) Martingales and arbitrage in multiperiod security markets. *J. Econ. Theory* **2**, 381–408.

Harrison, J. and Pliska, S. (1981) Martingales and stochastic integrals in the theory of continuous trading. *Stochastic Process. Appl.* **11**, 215–316.

Harrison, J.M. and Williams, R.J. (1987) Brownian models of open queueing networks with homogeneous customer populations. *Stoch. Stoch. Rep.* **22**, 77–115.

Heyde, C.C. and Seneta, E. (1977) I. J. Bienaymé: Statistical theory anticipated. In *Studies in the History of Mathematics and Physical Sciences.* No. 3. Springer, Berlin Heidelberg New York.

Hopf, E. (1934) *Mathematical Problems of Radiative Equilibrium.* Cambridge tracts, No. 31. Cambridge University Press, Cambridge.

Horowitz, J. (1972) Semilinear Markov processes, subordinators and renewal theory. *Zeit. Wahrsch. Verw. Gebiete* **24**, 167–193.

Hougaard, P. (1986) Survival models for heterogeneous populations derived from stable distributions. *Biometrika* **73**, 386–396.

Huzak, M., Perman, M., Šikić, H. and Vondracek, Z. (2004a) Ruin probabilities and decompositions for general perturbed risk processes. *Ann. Appl. Probab.* **14**, 1378–397.

Huzak, M., Perman, M., Šikić, H. and Vondracek, Z. (2004b) Ruin probabilities for competing claim processes. *J. Appl. Probab.* **41**, 679–90.

Ismail, M.E.H. (1977) Bessel functions and the infinite divisibility of the Student t-distribution. *Ann. Probab.* **5**, 582–585.

Ismail, M.E.H. and Kelker, D.H. (1979) Special functions, Stieltjes transforms and infinite divisibility. *SIAM J. Math. Anal.* **10**, 884–901.

Itô, K. (1942) On stochastic processes. I. (Infinitely divisible laws of probability). *Jpn. J. Math.* **18**, 261–301.

Itô, K. (1970) Poisson point processes attached to Markov processes. *Prof. 6th Berkeley Symp. Math. Stat. Probab. III*, 225–239.

Itô, K. (2004) *Stochastic processes.* Springer, Berlin Heidelberg New York.

Jacod, J. and Shiryaev, A.N. (1987) *Limit Theorems for Stochastic Processes.* Springer, Berlin Heidelberg New York.

Jirina, M. (1958) Stochastic branching processes with continuous state space. *Czech. Math. J.* **8**, 292–312.

Johnson, N.L. and Kotz, S. (1970) *Distributions in Statistics. Continuous Univariate Distributions. Vol 1.* Wiley, New York.

Jørgensen, B. (1982) *Statistical Properties of the Generalized Inverse Gaussian Distribution.* Lecture Notes in Statistics vol. 9. Springer, Berlin Heidelberg New York.

Kallsen, J. and Tankov, P. Characterization of dependence of multidimensional Lévy processes using Lévy copulas. *To appear in the J. Multivariate Analy.*

Kella, O. (1993) Parallel and tandem fluid networks with dependent Lévy inputs. *Ann. Appl. Prob.* **3**, 682–695.

Kella, O. and Whitt, W. (1992) Useful martingales for stochastic storage processes with Lévy input. *J. Appl. Probab.* **29**, 396–403.

Kella, O. and Whitt, W. (1996) Stability and structural properties of stochastic storage networks. *J. Appl. Prob.* **33**, 1169–1180.

Kennedy, D. (1976) Some martingales related to cumulative sum tests and single server queues. *Stochast. Process. Appl.* **4**, 261–269.

Kesten, H. (1969) Hitting probabilities of single points for processes with stationary independent increments. *Mem. Am. Math. Soc.* **93**.

Khintchine A. (1937) A new derivation of one formula by Levy P., Bull. Moscow State Univ., **I**, No 1, 1–5.

Kingman, J.F.C. (1964) Recurrence properties of processes with stationary independent increments. *J. Aust. Math. Soc.* **4**, 223–228.

Kingman, J.F.C. (1967) Completely random measures. *Pacific J. Math.* **21**, 59–78.

Kingman, J.F.C. (1993) *Poisson Processes.* Oxford University Press, Oxford.

Klüppelberg, C., Kyprianou, A.E. and Maller, R.A. (2004) Ruin probabilities for general Lévy insurance risk processes. *Ann. Appl. Probab.* **14**, 1766–1801.

Klüppelberg, C., and Kyprianou, A.E. (2005) On extreme ruinous behaviour of Lévy insurance risk processes. *Preprint.*

Klüppelberg, C., Lindner, A. and Maller, R. (2004) A continuous-time GARCH process driven by a Lévy process: stationarity and second order behaviour. *J. Appl. Prob.* **41**, 601–622.

Klüppelberg, C., Lindner, A. and Maller, R. (2006) Continuous-time volatility modelling: COGARCH versus Ornstein-Uhlenbeck models. In: Kabanov, Y., Lipster, R. and Stoyanov, J. (Eds.), *From Stochastic Calculus to Mathematical Finance. The Shiryaev Festschrift.* Springer, Berlin Heidelberg New York, to appear.

Kolmogorov, N.A. (1932) Sulla forma generale di un processo stocastico omogeneo (un problema di B. de Finetti). *Atti Acad. Naz. Lincei Rend.* **15**, 805–808.

Konstantopoulos, T., Last, G. and Lin, S.J. (2004) Non-product form and tail asymptotics for a class of Lévy stochastic networks. *Queueing Syst.* **46**, 409–437.

Konstantopoulos, T. and Lin, S.J. (1998). Macroscopic models for long-range dependent network traffic. *Queueing Syst.* **28**, 215–243.

Konstantopoulos, T. and Richardson, G. (2002) Conditional limit theorems for spectrally positive Lévy processes. *Adv. Appl. Probab.* **34**, 158–178.

Korolyuk, V.S. (1974) Boundary problems for a compound Poisson process. *Theory Probab. Appl.* **19**, 1–14.

Korolyuk V.S. (1975a) *Boundary Problems for Compound Poisson Processes.* Naukova Dumka, Kiev (in Russian).

Korolyuk, V.S. (1975b) On ruin problem for compound Poisson process. *Theory Probab. Appl.* **20**, 374–376.

Korolyuk, V.S. and Borovskich, Ju. V. (1981) *Analytic Problems of the Asymptotic Behaviour of Probability Distributions.* Naukova Dumka, Kiev (in Russian).

Korolyuk, V.S., Suprun, V.N. and Shurenkov, V.M. (1976) Method of potential in boundary problems for processes with independent increments and jumps of the same sign. *Theory Probab. Appl.* **21**, 243–249.

Koponen, I. (1995) Analytic approach to the problem of convergence of truncated Lévy flights towards the Gaussian stochastic process. *Phys. Rev. E* **52**, 1197–1199.

Kyprianou, A.E. and Palmowski, Z. (2005) A martingale review of some fluctuation theory for spectrally negative Lévy processes. *Séminaire de Probabilités XXXVIII.* 16–29. Lecture Notes in Mathematics, Springer, Berlin Heidelberg New York.

Kyprianou, A.E. and Surya, B. (2005) On the Novikov–Shiryaev optimal stopping problem in continuous time. *Electron. Commun. Probab.* **10**, 146–154.

Kyprianou, A.E., Schoutens, W. and Wilmott, P. (2005) *Exotic Option Pricing and Advanced Lévy Models.* Wiley, New York.

Jacob, N. (2001) *Pseudo-Differential Operators and Markov Processes. Vol. 1: Fourier Analysis and Semigroups.* Imperial College Press, London.

Jacob, N. (2002) *Pseudo-Differential Operators and Markov Processes. Vol. 2: Generators and Their Potential Theory.* Imperial College Press, London.

Jacob, N. (2005) *Pseudo-Differential Operators and Markov Processes. Vol. 3: Markov Processes and Applications.* Imperial College Press, London.

Lambert, A. (2000) Completely asymmetric Lévy processes confined in a finite interval. *Ann. Inst. H. Poincaré* **36**, 251–274.

Lambert, A. (2001) *Arbres, excursions, et processus de Lévy complètement asymétriques.* Thèse, doctorat de l'Université Pierre et Marie Curie, Paris.

Lambert, A. (2002) The genealogy of continuous-state branching processes with immigration. *Probab. Theory Relat. Fields* **122**, 42–70.

Lamberton, D. and Lapeyre, B. (1996) *Introduction to Stochastic Calculus Applied to Finance.* Chapman and Hall, Boca Raton, FL.

Lamperti, J. (1962) An invariance principle in renewal theory. *Ann. Math. Stat.* **33**, 685–696.

Lamperti, J. (1967a) Continuous-state branching processes. *Bull. Am. Math. Soc.* **73**, 382–386.

Lamperti, J. (1976b) The limit of a sequence of branching processes. *Zeit. Wahrsch. Verw. Gebiete* **7**, 271–288.

Lebedev, N.N. (1972) *Special Functions and Their Applications.* Dover Publications, Inc., New York.

Le Gall, J.-F. (1999) *Spatial Branching Processes, Random Snakes and Partial Differential Equations.* Lectures in Mathematics, ETH Zürich, Birkhäuser.

Lévy, P. (1924) Théorie des erreurs. La loi de Gauss et les lois exceptionelles. *Bull. Soc. Math. France.* **52**, 49–85.

Lévy, P. (1925) *Calcul des Probabilités.* Gauthier–Villars, Paris.

Lévy, P. (1934) Sur les intégrales dont les éléments sont des variables aléatoires indépendantes. *Ann. Scuola Norm. Pisa* **3**, 337–366; and **4**, 217–218.

Lévy, P. (1954) *Théorie de l'addition des variables aléatoires, 2nd edition.* Gaulthier–Villars, Paris.

Lindner, A. and Maller, R.A. (2005) Lévy integrals and the stationarity of generalised Ornstein–Uhlenbeck processes. *Stoch. Proc. Appl.* **115**, 1701–1722.

Lukacs, E. (1970) *Characteristic Functions.* Second edition, revised and enlarged. Hafner Publishing Co., New York.

Lundberg, F. (1903) *Approximerad framställning av sannolikhetsfunktionen. Återförsäkring av kollektivrisker.* Akad. Afhandling. Almqvist. och Wiksell, Uppsala.

Lyons, R., Pemantle, R. and Peres, Y. (1995) Conceptual proofs of $L \log L$ criteria for mean behaviour of branching processes. *Ann. Probab.* **23**, 1125–1138.

Madan, D.P. and Seneta, E. (1990) The VG for share market returns *J. Business* **63**, 511–524.

Maisonneuve, B. (1975) Exit systems. *Ann. Probab.* **3**, 399–411.

Maulik, K. and Zwart, B. (2006) Tail asymptotics for exponential functionals of Lévy processes. *Stochast. Process. Appl.* **116**, 156–177.

McKean, H. (1965) Appendix: A free boundary problem for the heat equation arising from a problem of mathematical economics. *Ind. Manag.Rev.* **6**, 32–39.

Merton, R.C. (1973) Theory of rational option pricing. *Bell J. Econ. Manage. Sci.* **4**, 141–183.

Mikhalevich, V.S. (1958) Baysian choice between two hypotheses for the mean value of a normal process. *Visnik Kiiv. Univ.* **1**, 101–104.

Mikosch, T., Resnick, S.I., Rootzén, H. and Stegeman, A. (2002). Is network traffic approximated by stable Lévy motion or fractional Brownian motion? *Ann. Appl. Probab.* **12**, 23–68.

Mordecki, E. (2002) Optimal stopping and perpetual options for Lévy processes. *Finance Stoch.* **6**, 473–493.

Mordecki, E. (2005) Wiener–Hopf factorization for Lévy processes having negative jumps with rational transforms. *Submitted.*

Nelson, R.B. (2006) *An Introduction to Copulas.* Springer, Berlin Heidelberg New York.

Nguyen-Ngoc, L. and Yor, M. (2005) Some martingales associated to reflected Lévy processes. *Séminaire de Probabilités XXXVII.* 42–69. Lecture Notes in Mathematics, Springer, Berlin Heidelberg New York.

Novikov, A.A. (2004) Martingales and first-exit times for the Ornstein–Uhlenbeck process with jumps. *Theory Probab. Appl.* **48**, 288–303.

Novikov, A.A. and Shiryaev, A.N. (2004) On an effective solution to the optimal stopping problem for random walks. *Theory Probab. Appl.* **49**, 344–354.

Øksendal, B. (2003) *Stochastic Differential Equations. An Introduction with Applications.* Springer, Berlin Heidelberg New York.

Øksendal, B. and Sulem, A. (2004) *Applied Stochastic Control of Jump Diffusions.* Springer, Berlin Heidelberg New York.

Patie, P. (2004) *On some first passage time problems motivated by financial applications.* Ph.D. Thesis, ETH Lecture Series Zürich.

Patie, P. (2005) On a martingale associated to generalized Ornstein–Uhlenbeck processes and an application to finance. *Stochast. Process. Appl.* **115**, 593–607.

Payley, R. and Wiener, N. (1934) Fourier transforms in the complex domain. *Am. Math. Soc. Colloq. Pub.* **19**.

Percheskii, E.A. and Rogozin, B.A. (1969) On the joint distribution of random variables associated with fluctuations of a process with independent increments. *Theory Probab. Appl.* **14**, 410–423.

Peskir, G. and Shiryaev, A.N. (2000), Sequential testing problems for Poisson processes. *Ann. Statist.* **28**, 837–859.

Peskir, G. and Shiryaev, A.N. (2002) Solving the Poisson disorder problem. *Advances in Finance and Stochastics.* pp. 295–312, Springer, Berlin Heidelberg New York.

Peskir, G. and Shiryaev, A. N. (2006) *Optimal Stopping and Free-Boundary Value Problems.* Lecture Notes in Mathematics, ETH Zürich, Birkhäuser.

Pistorius, M.R. (2004) On exit and ergodicity of the completely asymmetric Lévy process reflected at its infimum. *J. Theor. Probab.* **17**, 183–220.

Pistorius, M.R. (2006) On maxima and ladder processes for a dense class of Lévy processes. *J. Appl. Probab.* **43** no. 1, 208-220.

Port, S.C. and Stone, C.J. (1971) Infinitely divisible processes and their potential theory I, II. Annales de l'Institut Fourier. **21**, 157–275 and **21**, 179–265.

Prabhu, N.U. (1998) *Stochastic Storage Processes. Queues, Insurance Risk, Dams and Data Communication. 2nd Edition.* Springer, Berlin Heidelberg New York.

Protter, P. (2004) *Stochastic Integration and Differential Equations. 2nd Edition.* Springer, Berlin Heidelberg New York.

Rogers, L.C.G. (1990) The two-sided exit problem for spectrally positive Lévy processes, *Adv. Appl. Probab.* **22**, 486–487.

Rogozin, B.A. (1968) The local behavior of processes with independent increments. (Russian) *Theory Probab. Appl.* **13**, 507–512.

Rogozin, B.A. (1972) The distribution of the first hit for stable and asymptotically stable random walks on an interval. *Theory Probab. Appl.* **17**, 332–338.

Rubinovitch, M. (1971) Ladder phenomena in stochastic processes with stationary independent increments. *Zeit. Wahrsch. Verw. Gebiete* **20**, 58–74.

Samorodnitsky, G. and Taqqu, M.S. (1994) *Stable Non-Gaussian Random Processes.* Chapman and Hall/CRC, Boca Raton, FL.

Samuleson, P. (1965) Rational theory of warrant pricing. *Ind. Manage. Rev.* **6**, 13–32.

Sato, K. (1999) *Lévy Processes and Infinitely Divisible Distributions.* Cambridge University Press, Cambridge.

Schmidli, H. (2001) Distribution of the first ladder height of a stationary risk process perturbed by α-stable Lévy motion. *Insurance Math. Econom.* **28**, 13–20.

Schoutens, W. (2003) *Lévy Processes in Finance. Pricing Finance Derivatives.* Wiley, New York.

Schoutens, W. and Teugels, J.L. (1998) Lévy processes, polynomials and martingales. *Commun. Stat.–Stochast. Models.* **14**, 335–349.

Shepp, L. and Shiryaev, A.N. (1993) The Russian option: reduced regret, *Ann. Appl. Prob.* **3**, 603–631.

Shepp, L. and Shiryaev, A.N. (1994) A new look at the pricing of the Russian option, *Theory Probab. Appl.* **39**, 103–120.

Shiryaev, A.N. (1978) *Optimal Stopping Rules.* Springer, Berlin Heidelberg New York.

Shtatland, E.S. (1965) On local properties of processes with independent incerements. *Theory Probab. Appl.* **10**, 317–322.

Silverstein, M.L. (1968) A new approach to local time. *J. Math. Mech.* **17**, 1023–1054.

Snell, J.L. (1952) Applications of martingale system theorems. *Trans. Am. Math. Soc.* **73**, 293–312.

Spitzer, E. (1956) A combinatorical lemma and its application to probability theory. *Trans. Am. Math. Soc.* **82**, 323–339.

Spitzer, E. (1957) The Wiener–Hopf equation whose kernel is a probability density. *Duke Math. J.* **24**, 327–343.

Spitzer, E. (1964) *Principles of Random Walk.* Van Nostrand, New York.

Steutel, F.W. (1970) *Preservation of Infinite Divisibility under Mixing and Related Topics.* Math. Centre Tracts, No 33, Math. Centrum, Amsterdam.

Steutel, F.W. (1973) Some recent results in infinite divisibility. *Stochast. Process. Appl.* **1**, 125–143.

Suprun, V.N. (1976) Problem of ruin and resolvent of terminating processes with independent increments, *Ukrainian Math. J.* **28**, 39–45.

Takács, L. (1966) *Combinatorial Methods in the Theory of Stochastic Processes.* Wiley, New York.

Tankov, P. (2003) Dependence structure positive multidimensional Lévy processes. `http://www.math.jussieu.fr/~tankov/`

Thorin, O. (1977a) On the infinite divisibility of the Pareto distribution. *Scand. Actuarial J.* 31–40.

Thorin, O. (1977b) On the infinite divisibility of the lognormal distribution. *Scand. Actuarial J.* 121–148.

Tweedie, M.C.K. (1984) An index which distinguishes between some important exponential families. *Statistics: Applications and New Directions (Calcutta, 1981), 579–604, Indian Statist. Inst., Calcutta.*

Vigon, V. (2002) Votre Lévy rampe-t-il ? *J. London Math. Soc.* **65**, 243–256.

Watson, H.W. and Galton, F. (1874) On the probability of the extinction of families. *J. Anthropol. Inst. Great Britain Ireland* **4**, 138–144.

Winter, B.B. (1989) Joint simulation of backward and forward recurrence times in a renewal process. *J. Appl. Probab.* **26**, 404–407.

Zolotarev, V.M. (1964) The first passage time of a level and the behaviour at infinity for a class of processes with independent increments, *Theory Prob. Appl.* **9**, 653–661.

Zolotarev, V.M. (1986) *One Dimensional Stable Distributions.* American Mathematical Society Providence, RI.

Index

Universitext

Aguilar, M.; Gitler, S.; Prieto, C.: Algebraic Topology from a Homotopical Viewpoint

Aksoy, A.; Khamsi, M. A.: Methods in Fixed Point Theory

Alevras, D.; Padberg M. W.: Linear Optimization and Extensions

Andersson, M.: Topics in Complex Analysis

Aoki, M.: State Space Modeling of Time Series

Arnold, V. I.: Lectures on Partial Differential Equations

Arnold, V. I.: Ordinary Differential Equations

Audin, M.: Geometry

Aupetit, B.: A Primer on Spectral Theory

Bachem, A.; Kern, W.: Linear Programming Duality

Bachmann, G.; Narici, L.; Beckenstein, E.: Fourier and Wavelet Analysis

Badescu, L.: Algebraic Surfaces

Balakrishnan, R.; Ranganathan, K.: A Textbook of Graph Theory

Balser, W.: Formal Power Series and Linear Systems of Meromorphic Ordinary Differential Equations

Bapat, R.B.: Linear Algebra and Linear Models

Benedetti, R.; Petronio, C.: Lectures on Hyperbolic Geometry

Benth, F. E.: Option Theory with Stochastic Analysis

Berberian, S. K.: Fundamentals of Real Analysis

Berger, M.: Geometry I, and II

Bliedtner, J.; Hansen, W.: Potential Theory

Blowey, J. F.; Coleman, J. P.; Craig, A. W. (Eds.): Theory and Numerics of Differential Equations

Blowey, J.; Craig, A. (Eds.): Frontiers in Numerical Analysis. Durham 2004

Blyth, T. S.: Lattices and Ordered Algebraic Structures

Börger, E.; Grädel, E.; Gurevich, Y.: The Classical Decision Problem

Böttcher, A; Silbermann, B.: Introduction to Large Truncated Toeplitz Matrices

Boltyanski, V.; Martini, H.; Soltan, P. S.: Excursions into Combinatorial Geometry

Boltyanskii, V. G.; Efremovich, V. A.: Intuitive Combinatorial Topology

Bonnans, J. F.; Gilbert, J. C.; Lemaréchal, C.; Sagastizábal, C. A.: Numerical Optimization

Booss, B.; Bleecker, D. D.: Topology and Analysis

Borkar, V. S.: Probability Theory

Brunt B. van: The Calculus of Variations

Bühlmann, H.; Gisler, A.: A Course in Credibility Theory and Its Applications

Carleson, L.; Gamelin, T. W.: Complex Dynamics

Cecil, T. E.: Lie Sphere Geometry: With Applications of Submanifolds

Chae, S. B.: Lebesgue Integration

Chandrasekharan, K.: Classical Fourier Transform

Charlap, L. S.: Bieberbach Groups and Flat Manifolds

Chern, S.: Complex Manifolds without Potential Theory

Chorin, A. J.; Marsden, J. E.: Mathematical Introduction to Fluid Mechanics

Cohn, H.: A Classical Invitation to Algebraic Numbers and Class Fields

Curtis, M. L.: Abstract Linear Algebra

Curtis, M. L.: Matrix Groups

Cyganowski, S.; Kloeden, P.; Ombach, J.: From Elementary Probability to Stochastic Differential Equations with MAPLE

Dalen, D. van: Logic and Structure

Da Prato, G.: An Introduction to Infinite-Dimensional Analysis

Das, A.: The Special Theory of Relativity: A Mathematical Exposition

Debarre, O.: Higher-Dimensional Algebraic Geometry

Deitmar, A.: A First Course in Harmonic Analysis

Demazure, M.: Bifurcations and Catastrophes

Devlin, K. J.: Fundamentals of Contemporary Set Theory

DiBenedetto, E.: Degenerate Parabolic Equations

Diener, F.; Diener, M.(Eds.): Nonstandard Analysis in Practice

Dimca, A.: Sheaves in Topology

Dimca, A.: Singularities and Topology of Hypersurfaces

DoCarmo, M. P.: Differential Forms and Applications

Duistermaat, J. J.; Kolk, J. A. C.: Lie Groups

Dumortier.: Qualitative Theory of Planar Differential Systems

Edwards, R. E.: A Formal Background to Higher Mathematics Ia, and Ib

Edwards, R. E.: A Formal Background to Higher Mathematics IIa, and IIb

Emery, M.: Stochastic Calculus in Manifolds

Emmanouil, I.: Idempotent Matrices over Complex Group Algebras

Endler, O.: Valuation Theory

Erez, B.: Galois Modules in Arithmetic

Everest, G.; Ward, T.: Heights of Polynomials and Entropy in Algebraic Dynamics

Farenick, D. R.: Algebras of Linear Transformations

Foulds, L. R.: Graph Theory Applications

Franke, J.; Härdle, W.; Hafner, C. M.: Statistics of Financial Markets: An Introduction

Frauenthal, J. C.: Mathematical Modeling in Epidemiology

Freitag, E.; Busam, R.: Complex Analysis

Friedman, R.: Algebraic Surfaces and Holomorphic Vector Bundles

Fuks, D. B.; Rokhlin, V. A.: Beginner's Course in Topology

Fuhrmann, P. A.: A Polynomial Approach to Linear Algebra

Gallot, S.; Hulin, D.; Lafontaine, J.: Riemannian Geometry

Gardiner, C. F.: A First Course in Group Theory

Gårding, L.; Tambour, T.: Algebra for Computer Science

Godbillon, C.: Dynamical Systems on Surfaces

Godement, R.: Analysis I, and II

Goldblatt, R.: Orthogonality and Spacetime Geometry

Gouvêa, F. Q.: p-Adic Numbers

Gross, M. et al.: Calabi-Yau Manifolds and Related Geometries

Gustafson, K. E.; Rao, D. K. M.: Numerical Range. The Field of Values of Linear Operators and Matrices

Gustafson, S. J.; Sigal, I. M.: Mathematical Concepts of Quantum Mechanics

Hahn, A. J.: Quadratic Algebras, Clifford Algebras, and Arithmetic Witt Groups

Hájek, P.; Havránek, T.: Mechanizing Hypothesis Formation

Heinonen, J.: Lectures on Analysis on Metric Spaces

Hlawka, E.; Schoißengeier, J.; Taschner, R.: Geometric and Analytic Number Theory

Holmgren, R. A.: A First Course in Discrete Dynamical Systems

Howe, R., Tan, E. Ch.: Non-Abelian Harmonic Analysis

Howes, N. R.: Modern Analysis and Topology

Hsieh, P.-F.; Sibuya, Y. (Eds.): Basic Theory of Ordinary Differential Equations

Humi, M., Miller, W.: Second Course in Ordinary Differential Equations for Scientists and Engineers

Hurwitz, A.; Kritikos, N.: Lectures on Number Theory

Huybrechts, D.: Complex Geometry: An Introduction

Isaev, A.: Introduction to Mathematical Methods in Bioinformatics

Istas, J.: Mathematical Modeling for the Life Sciences

Iversen, B.: Cohomology of Sheaves

Jacod, J.; Protter, P.: Probability Essentials

Jennings, G. A.: Modern Geometry with Applications

Jones, A.; Morris, S. A.; Pearson, K. R.: Abstract Algebra and Famous Inpossibilities

Jost, J.: Compact Riemann Surfaces

Jost, J.: Dynamical Systems. Examples of Complex Behaviour

Jost, J.: Postmodern Analysis

Jost, J.: Riemannian Geometry and Geometric Analysis

Kac, V.; Cheung, P.: Quantum Calculus

Kannan, R.; Krueger, C. K.: Advanced Analysis on the Real Line

Kelly, P.; Matthews, G.: The Non-Euclidean Hyperbolic Plane

Kempf, G.: Complex Abelian Varieties and Theta Functions

Kitchens, B. P.: Symbolic Dynamics

Kloeden, P.; Ombach, J.; Cyganowski, S.: From Elementary Probability to Stochastic Differential Equations with MAPLE

Kloeden, P. E.; Platen; E.; Schurz, H.: Numerical Solution of SDE Through Computer Experiments

Kostrikin, A. I.: Introduction to Algebra

Krasnoselskii, M. A.; Pokrovskii, A. V.: Systems with Hysteresis

Kurzweil, H.; Stellmacher, B.: The Theory of Finite Groups. An Introduction

Kyprianou, A. : Introductory Lectures on Fluctuations of Lévy Processes with Applications

Lang, S.: Introduction to Differentiable Manifolds

Luecking, D. H., Rubel, L. A.: Complex Analysis. A Functional Analysis Approach

Ma, Zhi-Ming; Roeckner, M.: Introduction to the Theory of (non-symmetric) Dirichlet Forms

Mac Lane, S.; Moerdijk, I.: Sheaves in Geometry and Logic

Marcus, D. A.: Number Fields

Martinez, A.: An Introduction to Semiclassical and Microlocal Analysis

Matoušek, J.: Using the Borsuk-Ulam Theorem

Matsuki, K.: Introduction to the Mori Program

Mazzola, G.; Milmeister G.; Weissman J.: Comprehensive Mathematics for Computer Scientists 1

Mazzola, G.; Milmeister G.; Weissman J.: Comprehensive Mathematics for Computer Scientists 2

Mc Carthy, P. J.: Introduction to Arithmetical Functions

McCrimmon, K.: A Taste of Jordan Algebras

Meyer, R. M.: Essential Mathematics for Applied Field

Meyer-Nieberg, P.: Banach Lattices

Mikosch, T.: Non-Life Insurance Mathematics

Mines, R.; Richman, F.; Ruitenburg, W.: A Course in Constructive Algebra

Moise, E. E.: Introductory Problem Courses in Analysis and Topology

Montesinos-Amilibia, J. M.: Classical Tessellations and Three Manifolds

Morris, P.: Introduction to Game Theory

Nikulin, V. V.; Shafarevich, I. R.: Geometries and Groups

Oden, J. J.; Reddy, J. N.: Variational Methods in Theoretical Mechanics

Øksendal, B.: Stochastic Differential Equations

Øksendal, B.; Sulem, A.: Applied Stochastic Control of Jump Diffusions

Poizat, B.: A Course in Model Theory

Polster, B.: A Geometrical Picture Book

Porter, J. R.; Woods, R. G.: Extensions and Absolutes of Hausdorff Spaces

Radjavi, H.; Rosenthal, P.: Simultaneous Triangularization

Ramsay, A.; Richtmeyer, R. D.: Introduction to Hyperbolic Geometry

Rautenberg, W.; A Concise Introduction to Mathematical Logic

Rees, E. G.: Notes on Geometry

Reisel, R. B.: Elementary Theory of Metric Spaces

Rey, W. J. J.: Introduction to Robust and Quasi-Robust Statistical Methods

Ribenboim, P.: Classical Theory of Algebraic Numbers

Rickart, C. E.: Natural Function Algebras

Roger G.: Analysis II

Rotman, J. J.: Galois Theory

Rubel, L. A.: Entire and Meromorphic Functions

Ruiz-Tolosa, J. R.; Castillo E.: From Vectors to Tensors

Runde, V.: A Taste of Topology

Rybakowski, K. P.: The Homotopy Index and Partial Differential Equations

Sagan, H.: Space-Filling Curves

Samelson, H.: Notes on Lie Algebras

Schiff, J. L.: Normal Families

Sengupta, J. K.: Optimal Decisions under Uncertainty

Séroul, R.: Programming for Mathematicians

Seydel, R.: Tools for Computational Finance

Shafarevich, I. R.: Discourses on Algebra

Shapiro, J. H.: Composition Operators and Classical Function Theory

Simonnet, M.: Measures and Probabilities

Smith, K. E.; Kahanpää, L.; Kekäläinen, P.; Traves, W.: An Invitation to Algebraic Geometry

Smith, K. T.: Power Series from a Computational Point of View

Smoryński, C.: Logical Number Theory I. An Introduction

Stichtenoth, H.: Algebraic Function Fields and Codes

Stillwell, J.: Geometry of Surfaces

Stroock, D. W.: An Introduction to the Theory of Large Deviations

Sunder, V. S.: An Invitation to von Neumann Algebras

Tamme, G.: Introduction to Étale Cohomology

Tondeur, P.: Foliations on Riemannian Manifolds

Toth, G.: Finite Möbius Groups, Minimal Immersions of Spheres, and Moduli

Verhulst, F.: Nonlinear Differential Equations and Dynamical Systems

Weintraub, S.H.: Galois Theory

Wong, M. W.: Weyl Transforms

Xambó-Descamps, S.: Block Error-Correcting Codes

Zaanen, A.C.: Continuity, Integration and Fourier Theory

Zhang, F.: Matrix Theory

Zong, C.: Sphere Packings

Zong, C.: Strange Phenomena in Convex and Discrete Geometry

Zorich, V. A.: Mathematical Analysis I

Zorich, V. A.: Mathematical Analysis II